T0289189

HANDBOOK OF LUMINESCENCE DATING

HANDBOOK OF
LUMINESCENCE DATING

edited by

MARK D. BATEMAN

Whittles Publishing

Published by
Whittles Publishing,
Dunbeath,
Caithness KW6 6EG,
Scotland, UK

www.whittlespublishing.com

© 2019 text and illustrations M.D. Bateman
and authors (unless otherwise specified)

ISBN 978-184995-395-5

*The publisher and authors have used their best efforts in preparing this book, but assume no
responsibility for any injury and/or damage to persons or property from the use or implementation of
any methods, instructions, ideas or materials contained within this book. All operations should be
undertaken in accordance with existing legislation, recognized codes and standards and trade
practice. Whilst the information and advice in this book is believed to be true and accurate at the time
of going to press, the authors and publisher accept no legal responsibility or
liability for errors or omissions that may have been made.*

Printed by Melita Press, Malta

CONTENTS

AUTHORS

Simon Armitage is a Reader in Quaternary Science at Royal Holloway University of London, UK. He received his undergraduate degree from Oxford University (1998), where he was inspired to pursue a career in luminescence dating by the late Stephen Stokes. His PhD (2003) on Mozambican coastal dunes was supervised by Ann Wintle and Geoff Duller at Aberystwyth University, after which he returned to Oxford to work on luminescence dating of deep-sea cores. Since 2006 he has directed the Royal Holloway Luminescence Laboratory. His current research interests include climate change in drylands, environmental influences on ancient human dispersal and African Middle Stone Age archaeology. Since 2017 he has been a Principal Investigator in the Centre for Early Sapiens Behaviour (SapienCE) at the University of Bergen, Norway

Ian Bailiff is a Professor at Durham University (UK). His main interests lie in the development and application of luminescence techniques for the dating of archaeological artefacts and deposits and in radiation dosimetry. He is an interdisciplinary scientist with over 25 years involvement in experimental science and publishes in both fields of research, directing a research laboratory whose activities include the dating of sediments, lithics and heated materials, the use of surrogate materials for dosimetry following radiological emergencies and basic luminescence investigation. He is Co- Editor-in-Chief of the journal Radiation Measurements.

Mark Bateman is professor and deputy head of department in the Department of Geography at the University of Sheffield, Sheffield, UK. Since 1995, he has set up and directed the Luminescence Dating laboratory at Sheffield. His research ranges from work in arctic Canada in the north to Chile and South Africa in the south. He has published extensively (over 160 journal articles and book chapters) on luminescence dating from dunes (coastal and deserts), periglacial structures, mega floods and tsunami, former glaciations as well as chronologies for archaeological sites. He also has a special interest in the affects of bioturbation on site preservation and luminescence dates. He has won awards for best papers and the Sorby Medal prize for his work on the Glacial history of Yorkshire. A former International Association of Sedimentologists Bureau member, he currently sits on the editorial boards of three journals and hosted the UK National Luminescence conference in 2007 and 2018.

Laine Clark-Balzan obtained a B.A. in Physics from Princeton University, and a D.Phil. from the University of Oxford in Archaeological Science. She specialized in methodological developments in luminescence dating, and during her D.Phil., she published the first successful ultra-low-light imaging study of ultraviolet luminescence emissions from quarts grains. She also focused on improving the accuracy and precision of luminescence ages via applied Bayesian modelling. During her first post-doctoral position at the University of Oxford, she worked with the Oxford-based Palaeodeserts team and dated many sites in Saudi Arabia, harnessing both Bayesian modelling and multiple-mineral dating in order to successfully date sites near the upper age range of luminescence dating. Following this work, she continued to develop imaging analysis software and methods during her Marie Skłodowska-Curie Individual Fellowship, based at the University of Freiburg with Professor Frank Preusser. Today she continues to innovate in image processing research, applying machine learning to commercial applications for unmanned aerial vehicles.

Alastair Cunningham is a post-doctoral researcher at Aarhus University, and the Technical University of Denmark. He read Geography (Plymouth) and Quaternary Science (London) in the UK, before receiving his PhD from Delft University of Technology, the Netherlands, in 2011. His PhD research focused on the luminescence dating of coastal sediment in the Netherlands. Subsequently, his research has focused on technique development in luminescence dating, with particular regard to dose rate measurements and statistical data analysis. His current work also includes the application of luminescence methods in coastal science and archaeology, and in the application of rock surface dating.

Regina DeWitt is Associate Professor in the Department of Physics at East Carolina University. She has been involved in luminescence related research for more than 15 years. In 2009 she received a grant from NASA to develop a breadboard OSL dating instrument for deposits on Mars. She also developed a confocal scanning instrument for dating of rock surfaces and dose mapping. Her current research interests include OSL dating of cobbles and sediment from Antarctica, microscopic dose rate variations in sediments, impact of ion-irradiation on luminescence characteristics, and characterizing new materials for OSL applications. Since 2014 she has been editor of Ancient TL.

Kathryn Fitzsimmons currently leads the Research Group for Terrestrial Palaeoclimates at the Max Planck Institute for Chemistry in Mainz, Germany. She completed her PhD in Quaternary geology (desert palaeoenvironments; 2007) at the Australian National University, and an Habilitation in Physical Geography (prehistoric human-environmental interactions on desert margins; 2016) at the University of Leipzig, Germany. She has managed three luminescence dating laboratories in two countries (Australia and Germany), and was awarded the 2014 German National Albert Maucher Prize for Geosciences. She has published more than 80 articles and book chapters, and presented 85 conference talks and invited seminars, largely on the subject of luminescence dating in aeolian environments. Her current research interests include quantitative palaeoclimate reconstruction on the desert margins of Central Asia and Australia.

Markus Fuchs is professor at the department of geography, Justus-Liebig-University Giessen, Germany. As a geomorphologist and Quaternary Scientist, his research is focused on earth surface processes, paleoenvironments of the Quaternary and geoarchaeology. For his PhD he did his research on luminescence dating of sediments at the Max Planck Institute for Nuclear Physics in Heidelberg, Germany, receiving his PhD in 2001 at the Ruprecht-Karls-University Heidelberg. Since then, his methodological focus is on luminescence dating techniques. He is actually chair of the German Working Group for Geomorphology and vice president of the German Quaternary Association (DEUQUA).

Georgina King is an Assistant Professor in the Institute of Earth Surface Dynamics at the University of Lausanne, Switzerland. After studying for a BA in Geography from the

University of Oxford, she completed an MSc in Quaternary Science at Royal Holloway University London and continued to a PhD in Earth Sciences at the University of St Andrews which she completed in 2012. She has held post-doctoral research positions in Scotland, Wales, Switzerland and Germany. Her research is focussed on the development of novel luminescence methods to constrain landscape evolution, and she has most recently been working on the development of luminescence rock surface exposure dating and luminescence thermochronometry.

Benjamin Lehmann is a PhD student in Institute of Earth Surface Dynamics at the University of Lausanne (Switzerland). After graduating with an MSc in Solid Earth geosciences (Joseph Fourier University, Grenoble, France), Benjamin worked at the Glacioclim Observatory of La Paz in Bolivia (www.great-ice.ird.fr) where he was employed as an engineer, managing glaciological, hydrological and meteorological field observations of glaciated drainage basins (Zongo and Charquini Sur). Benjamin started his PhD research in 2014 in the field of Earth surface dynamics, with a specific study area of the Mer de Glace glacier located in the Mont Blanc massif (European Alps). His PhD research has focussed on reconstructing variations in ice extent and constraining deglaciated bedrock surface erosion rates over the last interglacial. He has achieved this through combining optically stimulated luminescence (OSL) surface exposure dating and terrestrial cosmogenic nuclide dating (TCN).

Shannon Mahan is a research geologist with the United States Geological Survey (USGS) in Denver, Colorado (USA). She has been involved in luminescence related research for more than 25 years. Her work in the USGS has led to more than 150 publications, maps, reports, and collaborative projects across many federal agencies. She is the Director of the USGS Luminescence Geochronology lab and her work ranges from basic inquiries of geologic mapping and the timing of sediment deposition to fault history and trench sampling, dosimetry associated with the measurement of elemental concentrations using gamma spectrometry, ICP-MS, XRF, and neutron activation, paleontological timelines, and archeological investigations.

Ed Rhodes graduated in Geology from the University of Oxford (University College), followed by a DPhil from the same institution studying the optical dating of quartz from sediments in 1990. After post-doctoral positions at Oxford and Cambridge, he was appointed Lecturer in Geography at Royal Holloway, University of London in 1992. Following five years at the Research Laboratory for Archaeology and the History of Art, Oxford, from 1998, he was a fellow at the Australian National University in 2003, then at Manchester Metropolitan University, before taking up a position of Full Professor, University of California, Los Angeles in 2009. After a joint position at the University of Manchester 2013-14, he was appointed as Professor at Sheffield, starting in July 2014. He also holds the position of Adjunct Professor of Geology at UCLA.

Thomas Stevens is a Senior Lecturer at the Department of Earth Sciences, Uppsala University, Sweden. He received his DPhil at the Oxford University (Jesus College), UK, in 2007, and since then has been Lecturer or Senior Lecturer at Kingston University London, Royal Holloway University of London, and now Uppsala. His current research interests include loess provenance, loess luminescence dating, dust and climate interactions, and long-term landscape evolution. He has been Editor in Chief of the International Association of Sedimentologists Special Publications, and a Bureau member of the association. He has authored >70 articles in scientific journals and has given multiple invited keynote talks at arrange of international conferences.

Toru Tamura is a Senior Researcher at the Geological Survey of Japan, National Institute of Advanced Industrial Science and Technology. He received his PhD in 2004 at Kyoto University, Japan, on sedimentology of Holocene coastal barriers in eastern Japan based on radiocarbon chronology. He was then appointed as a Researcher at Geological Survey of Japan to work for coastal sedimentology in Japan and Southeast Asia. From 2009 to 2011, he stayed in the University of Sheffield, UK, and started luminescence dating of coastal sediments under the supervision of Mark Bateman. Since 2013, he has directed a luminescence dating laboratory he built at Geological Survey of Japan to date sediments in various settings in Japan and overseas.

Pierre Valla is a research scientist at the Institute of Earth Sciences (ISTerre - Université Grenoble Alpes / CNRS, France). His research combines geomorphology, earth surface geochronology, and numerical modelling to understand the interplay between tectonics and climate in mountain erosion and landscape development. During his PhD work (Univ. Grenoble, 2011) and successive post-doctoral fellowships in Switzerland, he used a multidisciplinary approach to assess the impact of glaciations on the evolution of the Earth's surface, participating in the development of low-temperature thermochronometry tools and rock-surface exposure methods. His current research is still focused on Quaternary glaciations and their impact on mountain relief evolution, while continuing methodological development of luminescence tools in geochronology.

Richard T Walker is professor at the Department of Earth Sciences, University of Oxford, UK. He has almost twenty years of experience in the study of active faults and of large destructive earthquakes across the interior of Asia (modern, historical, and prehistoric). His studies combine satellite image analysis, fieldwork, and dating techniques to uncover the locations of active faults, determine their average rates of movement, and to unravel the history of earthquakes upon them. He is an investigator in the UK COMET Institute, which is one of the leading worldwide centres of excellence in the study of earthquakes, volcanoes, and tectonics, and in several multi-institute consortia in earthquake science and hazard.

Jakob Wallinga is chair of the Soil Geography and Landscape group of Wageningen University, and director of the Netherlands Centre for Luminescence dating. He (co-) authored over 100 articles in scientific journals, mostly on luminescence dating methods and applications. In his research he focusses on reconstructing the dynamics of soils and landscapes, and exploring how the insights obtained can be used in sustainable soil, water and landscape management. He has a keen interest in nature based solutions, as these may provide ways to service human needs in a sustainable way by making effective use of natural processes.

PREFACE

Luminescence dating is now widely applied by Quaternary geologists and archaeologists to obtain ages for events as diverse as past earthquakes, desertification and cave occupation sites. Using quartz or feldspar minerals found in almost ubiquitous sand and finer sediments, luminescence can provide ages from over 500,000 years ago to modern. However, the technique has been described by non-practitioners as 'the dark art'. Such a label may be due to the fact that as samples are light sensitive much of the laboratory work is undertaken under controlled lighting in 'dark rooms'. It may be also alluding to the fact that the underlying principles of the technique lie in solid state physics and the field of semi-conductors and dosimetry which can be at times impenetrable to understand. Technical luminescence books, papers and reviews abound in the academic literature and dating text books often include a chapter on luminescence dating. At present none try to present in detail the principles, practicalities and potential applications of luminescence dating in a manner which is accessible to anyone who might want to have something dated or is learning to use the technique first hand. This is what this book aims to do.

This new approach has been written by some of the foremost experts in luminescence dating. The first chapter gives the background to the technique in simple terms so that the range of potential applications, limits and issues can be understood. Chapter 2 focuses on what, where and how to sample, in order to optimise the successful application of luminescence dating. It also provides guidance on how luminescence ages can be interpreted and published. Chapter 3 follows with an outline of how luminescence can best be used within chronological frameworks. Chapters 4 to 11 cover the application of luminescence dating to glacial, tectonic, fluvial, aeolian, coastal and marine environments as well as use in dating rock exposure and in archaeological sites. Each setting has different challenges and limitations for luminescence dating and the chapters outline these and provide practical advice on how issues might be avoided in sampling or mitigated by requesting different laboratory measurement approaches or analysis. All are accompanied by fully illustrated case studies. The final chapter, as luminescence dating is still developing, provides information on up and coming more experimental approaches which may help expand the range of chronological problems that luminescence dating can be routinely applied to in the near future.

I would like to note my thanks and debt to a range of people who have led me to the point where I could help produce this book. Firstly I wish to acknowledge the ongoing support and patience of my parents, wife Marion and children Rebecca and Elizabeth. Secondly I will forever by indebted to Professors Helen Rendell and Peter Townsend for their training and guidance as I learnt luminescence dating, without which I could not have had the confidence to set up the University of Sheffield Luminescence Laboratory back in 1995. Thirdly, thanks must go to the multitude of academic and professional colleagues who have allowed me to be part of their projects and given me the opportunity to apply luminescence dating to so many different parts of the world and in such a diverse range of settings. It's been fun and I have learnt a lot. Finally I would like to thank those who have authored and reviewed the chapters of this book, for sharing their knowledge, for their time commitment and for their patience.

Mark Bateman

1 PRINCIPLES AND HISTORY OF LUMINESCENCE DATING

SHANNON MAHAN[1] AND REGINA DEWITT[2]

[1] United States Geological Survey, Denver Federal Center, Bldg 15, MS 974, Box 25046, Denver, Colorado 80225-5046, U.S.A. Email: smahan@usgs.gov
[2] Department of Physics, East Carolina University, Howell Sciences C-209, 1000 E. 5th Street, Greenville, North Carolina 27858, U.S.A. Email: dewittr@ecu.edu

ABSTRACT: Luminescence dating has revolutionised archaeological and geological sciences. It is the only dating technique that spans historical applications of the last hundred years and then broadens its range to include sediment deposition into the Middle Pleistocene (a stage of geologic time from 780,000 to 125,000 years ago). It is also the only dating technique with principles that can be used in art authenticity, medical dosimetry, radiation therapy and tracking, wildfire mapping, ceramic dating, the thermal evolution of geologic units, or space explorations. This chapter reviews the basic principles of luminescence dating. It defines commonly used terms and answers the following questions: What is the physical basis for luminescence dating? How did the method evolve? Why is sample analyses so labour-intensive? What do the terms 'equivalent dose' or 'palaeodose' and 'dose rate' mean? What steps are involved in the luminescence measurements and instrumentation? How can I use the technique and where can I learn more about the method?

KEYWORDS: age models, equivalent dose (D_E), (environmental) dose rate (D_R), dosimetry, infrared-stimulated luminescence (IRSL), optically stimulated luminescence (OSL), thermoluminescence (TL).

1.1 INTRODUCTION

Dating occupies a central place in answering the scientific question of 'when'. Understanding the magnitude and duration over which processes shaping the landscape we live in is important in almost any scientific field. The last decade has seen a remarkable increase in development and accuracy of a range of different dating techniques, such as *in situ* cosmogenic radionuclides, radiocarbon, the family of uranium-series (U-series), short-lived isotopic chronometers such as lead-210 (^{210}Pb), caesium-137 (^{137}Cs), as well as luminescence dating. These techniques provide measures of the rates of geomorphic processes acting over a wide range of time and spatial scales. In particular, these dating techniques, often coupled with stable isotope proxy-based palaeo-climate and palaeoenvironmental reconstructions, are emerging as powerful diagnostic and quantitative tools revolutionising our understanding and knowledge of past environmental changes in the Quaternary (the last 2.57 Ma of Earth's history). Furthermore, these tools also play important roles in the emerging globally linked geological and archaeological disciplines, including those focusing on climate change, landscape evolution, human development, and paleontological progressions.

The techniques included in the term luminescence dating are well established for determining the timing of the last sunlight exposure or heating event of minerals, sediment, or archaeological artefacts. Luminescence dating covers the techniques thermoluminescence (TL), optically stimulated luminescence (OSL) and infrared-stimulated luminescence (IRSL). They have been applied to an ever-expanding range of samples including Quaternary sediments such as soil, rocks, and rock surfaces as well as archaeological samples including mortar, carved blocks, ceramics and other fired materials. Because the method relies on measurement of a property found in the very commonly occurring minerals of quartz and potassium feldspar, samples for dating are usually readily obtained (for further details on sampling see Chapter 2). Thus, for ages from essentially modern up to 300,000 years, luminescence dating is often the dating technique of choice or necessity.

Luminescence has become widely applied as a method complementary to (extending beyond the chronological range of isotopic methods) and with (overlaps the time range of) other dating techniques, such as cosmogenic nuclides (^{3}He, ^{10}Be ^{26}Al, ^{36}Cl, etc), for four significant reasons:

1. its application is ideal for most settings because it **does not require the presence of organic material** for dating, leading to a far wider range of potential sample material than what is available with radiocarbon dating.

2. luminescence dating is **typically able to date a wide range of targets** older than the maximum age of radiocarbon (~40,000 years, occasionally older; Bronk Ramsey *et al.* 2012) and within range of cosmogenic nuclides (hundreds to millions of years (Ma); Darville 2013) and U-series (~5,000 years to 500,000 years; Dickin 2005). This is because, in favourable environments, OSL and IRSL can be reliably obtained from 25 years to 300,000 years (Buylaert *et al.* 2012; Rittenour 2008) with notable older exceptions (Rhodes *et al.* 2006).

3. the measurement of the luminescence signal is accomplished by means of relatively **inexpensive and widely available machines** (Lapp *et al.* 2012; Richter *et al.* 2015; Sanderson and Murphy 2010), although the maintenance and running of the equipment is not without its unique challenges (Yukihara *et al.* 2014).

4. the coupling of chronological methods provides useful information for **understanding sediments that may have undergone complex geological histories** with the zeroing of grains for luminescence occurring at variable times. This often gives insight into reworking issues affecting some categories of components (e.g. sand grains, charcoal, or man-made artefacts re-located by bioturbation or soil-forming processes) or contamination issues.

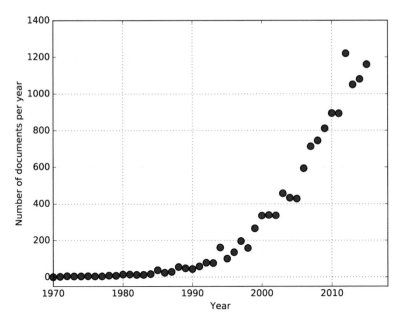

Figure 1.1
Number of publications per year reporting luminescence dating results since 1970.

As luminescence has developed and become established as a chronological technique, more and more people are applying the related scientific principles to solve research questions. This can be judged by the number of publications reporting luminescence results, which has increased substantially since the early days of TL dating (the root of the luminescence dating techniques), to the more recent development of single-aliquot and single-grain methods in OSL dating (Fig. 1.1).

In order to fully appreciate the present state and future of luminescence (i.e. where we are going), it is beneficial to understand the development and history of the technique (i.e. where we have been). This chapter provides users of luminescence dating with a succinct summary of the principles, methodology, and theoretical and experimental limitations before considering the practical application of the technique in a variety of situations, ranging from all forms of terrestrial environments to fluvial, lacustrine, marine, and the numerous depositional environments in between. This book will explain the techniques, inform the reader on what can and what cannot be currently achieved, elucidate the most appropriate way to apply the process to current scientific projects, and clarify how to obtain the most useful information from collected samples.

1.2 THE PHYSICAL BASIS OF LUMINESCENCE DATING

Environmental background radiation from the decay of unstable isotopes ionises atoms in minerals. The word isotope is from the Greek *isos*, meaning 'same,' and *topos*, signifying 'place'. Isotopes are defined as two or more atoms that have the same atomic number (the same number of protons and electrons) but a different number of neutrons and thus slightly different physical properties. The nucleus of a stable isotope will only be stable for certain combinations of protons and neutrons; the number of neutrons is not just random. Most chemical elements have several stable isotopes. Radioactive isotopes, or radionuclides, are any of several species of the same chemical element with different masses that have nuclei that are unstable and dissipate excess energy by spontaneously emitting radiation in the form of particles and rays.

Minerals receive a continuous low level of radiation in nature from either radioactive elements (such as uranium) in soil or sediment or from space in the form of cosmic radiation, and this fact is the basis for luminescence dating. To obtain a luminescence age, it is necessary to understand two things:

1. the background environmental radiation, which is ionising or 'dosing' the grains in the sediment

2. how sediments and archaeological artefacts can store up this radiation dose and, under the appropriate conditions, release it as measurable light.

Environmental background radiation is all around us and has been since the Earth formed; humans have literally evolved in a stew of low background radiation. Natural terrestrial radiation arises from the low level emissions of uranium (U), thorium (Th) and their radioactive daughters, potassium (K), and, to some extent, rubidium (Rb). In their decay processes, these elements emit alpha (α), beta (β), and gamma (γ) radiation. The other principal source of radiation in the near surface environment is cosmic rays. Cosmic rays are primarily protons or alpha particles as well as other, less common, heavier nuclei originating from cosmic and solar sources. When cosmic particles encounter Earth's upper atmosphere and collide with a nucleus, this collision produces a chain reaction of other, lighter particles and contributes to surficial radiation of sediment (150–500 g/cm²; Burow 2018; Castelvecchi 2017; Ferrari and Szuszkiewicz 2009; Prescott and Hutton 1994; Crookes and Rastin 1972).

All of these forms of ionising radiation initiate the luminescence phenomenon within any solid insulator or semi-conductor; metals do not exhibit luminescence properties (McKeever 1985). The basic principle is illustrated in Figure 1.2. The radiation transfers its energy to the electrons in the material but only to those outermost electrons orbiting the nuclei. This transfer of energy lifts electrons from the ground state (or valence energy band) across the band gap (or forbidden energy zone that does not correspond to any allowed combinations of atomic orbitals) to the conduction energy band where the electrons are physically mobile within the crystal lattice. A vast proportion of these excited electrons will return to the valence band (since this state is ultimately unstable), recombine, and release unmeasured, transient energy as heat or light; however, a small proportion of the electrons will encounter lattice defects and become localised at the defects or trapped. Over geologic timescales, the trapped charge population steadily accumulates. It is this slowly increasing charge population that provides the time clock for luminescence dating. Of course, this trapped charge population will also be released as heat or light, but if this happens under controlled laboratory conditions, it can be measured.

Some traps are so close to the conduction band that ambient temperature is sufficient to release the trapped charges. These traps are generally empty in a buried mineral; they are 'thermally unstable' and cannot be used for the dating process. In other traps, charges will accumulate until they are released by exposure to heat or light. Under these conditions,

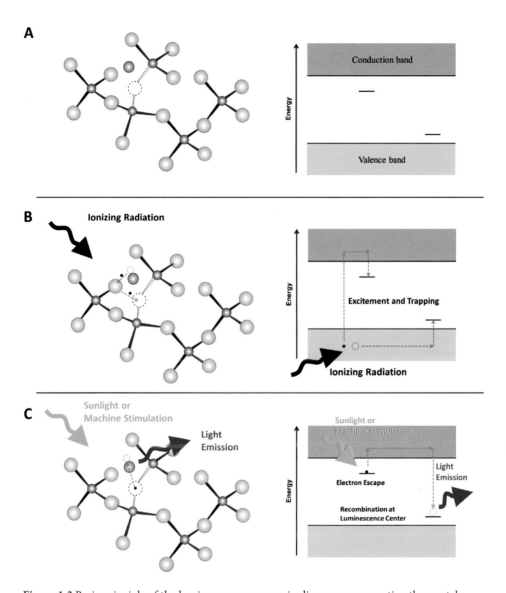

Figure 1.2 Basic principle of the luminescence process, in diagrams representing the crystal structure (left) and the energy band structure (right). a) The energy diagram of a crystal is characterised by the lower valence band, separated from the upper conduction band by a band gap. Crystal defects that act as electron donors, in this case the 'red' interstitial atom, have energy levels that are just above the valence band. Defects that attract electrons, in this case the missing atom, have energy levels just below the conduction band. b) Ionising radiation transfers energy to electrons (black dots). Some excited electrons get attracted to the site of the missing atom and get 'trapped.' Electrons from the donor atoms become excited as well and leave a hole (lack of an electron where one could exist; hollow circle). c) Upon stimulation with light or heat, electrons are released from their traps. They recombine with the holes and luminescence is emitted.

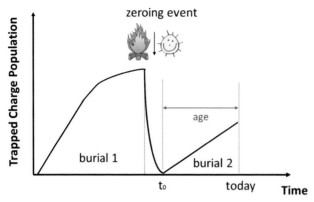

Figure 1.3 Evolution of the trapped charge population during periods of burial, resetting and another burial until a luminescence sample is taken. This cartoon illustrates the transport and depositional cycles that take place in sediment transport processes. t_0 is the beginning of the burial period, i.e. the time to be determined.

they become mobile again and when they recombine, light or luminescence, is emitted. The luminescence intensity is directly correlated with the number of trapped charges and thus the burial time. The wavelength of the emitted luminescence depends on the energy difference between electrons and the electron holes. Since a single mineral can have a variety of impurities, it can emit luminescence at multiple wavelengths because the type of impurity determines the colour of the luminescence. For instance, typical impurities in quartz (in decreasing percentage; titanium, aluminium, iron, manganese, magnesium, and calcium; Gaines *et al.* 1997) generate ultraviolet emission while those in feldspars (in decreasing percentage; barium, titanium, ferric iron, lithium, rubidium, etc.; Gaines *et al.* 1997) emit violet, yellow-orange, and sometimes deep red (Berger 2008).

Figure 1.3 illustrates the evolution of the trapped charge population over time. Upon burial, the charge population increases linearly at a rate based on the geochemical or mineralogical properties of the sediment. The finite number of traps available to store electrons means those traps will fill with electrons at some point in time and this leads to a saturation effect with continuing radiation exposure. This results in an upper practical limit of luminescence dating. The time from deposition of the sediment to luminescence saturation depends on the number and type of defects in the crystal lattice and the rate at which ionising radiation is released from the environment. During sediment transport and exposure to sunlight, or heating of an artefact, trapped charges are released and recombine. When all charges have recombined, the sample has been fully 'reset,' 'zeroed' or 'bleached.' In Figure 1.3, the scale of this time period is variable. While it takes many thousands of years (ka) for a sample to reach saturation, zeroing from a saturated level takes seconds to a few minutes of exposure to sunlight. Trapped charges accumulate again upon renewed burial (denoted in the figure by t_0) until a sample is obtained for dating.

Because luminescence is used to evaluate the trapped charge population that accumulated over time, a luminescence age reflects the time elapsed since the last zeroing event, i.e. the last sunlight exposure or heating event. To obtain an age, two separate steps are necessary. It is the first step that gives the method its name, luminescence dating, since it involves the emission of light. The measurement in the laboratory mimics the natural process except that it is performed on human timescales. The samples are either heated or exposed to light under controlled conditions. In both cases, trapped charges are released and recombine. The resulting luminescence is termed TL in the case of heating and IRSL or OSL in the case of light exposure (Fig. 1.4).

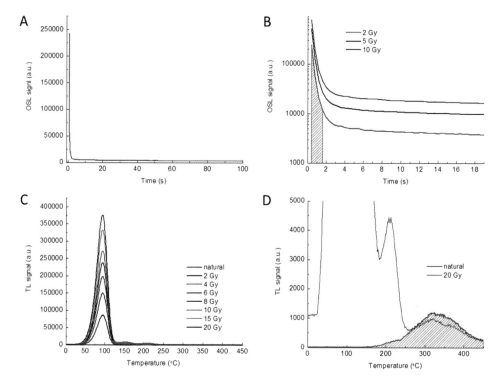

Figure 1.4 Examples of TL and OSL signals. A) OSL signal of a quartz sample after irradiation with 2 Gy. The signal is bleached (reset to zero) after a few seconds of light exposure. B) OSL signals of the same sample after irradiation plotted on a logarithmic scale with the doses indicated in the legend. The shaded area indicates the part of the signal that might be used as the OSL signal for dating since it very quickly reaches a baseline after a few seconds. The OSL intensity increases with exposure to radiation, either natural or laboratory. C) TL curves for the same quartz sample after irradiation with the doses indicated in the legend. The magnitude of the TL signal also increases with exposure to radiation. D) Comparison of the signal after irradiation in nature and after laboratory irradiation with 20 Gy. The shaded area indicates the part of the signal that might be used as the TL signal for dating. Thermally unstable traps, i.e. traps that empty at low temperatures, are empty for the natural sample, while they are filled during laboratory irradiation. Only the thermally stable part of the signal can be used for dating.

In TL, the temperature is increased over time and emission of photons (i.e. the signal) is recorded versus stimulation temperature. In OSL or IRSL, the light source is switched on, the sample is often kept at an elevated temperature to provide thermal assistance, and the signal is recorded over time. An optical filter is used to separate the luminescence from the stimulation light. The intensity of the resulting luminescence signal is directly correlated with the radiation received, or, in other terms, the absorbed radiation dose (Figs. 1.4b and 1.4c). Signal intensity can be determined in different ways: peak area or peak height in TL, initial signal or area under a selected part of the curve for OSL (compare shaded areas in Figs. 1.4b and 1.4d). In practice, luminescence dating uses artificial laboratory irradiation to mimic natural radioactive processes to measure the response of a sample to radiation exposure. These dose–response measurements make it possible to calculate the amount of stored charge because the measured natural luminescence signal is compared to that induced in subsamples or aliquots by known radiation doses from calibrated laboratory sources, thus scaling the natural signal and yielding an equivalent dose (D_E) in gray (Gy, where 1 Gy = absorbed dose of 1 joule/kg)) (Berger 2008). The measurement of the D_E is described in detail in Section 1.5.

The D_E depends on the rate at which ionising radiation is released from the environment, i.e. the background radiation dose rate (D_R), and the time since last exposure to heat or light. The D_R, in turn, depends on the concentrations of K, U, and Th in the surrounding sediments as well as exposure to cosmic rays. Procedures for obtaining the D_R are outlined in Section 1.6. Luminescence ages, whether they are derived from TL, IRSL, or OSL, are calculated by dividing the D_E by the environmental dose rate D_R (in Gy/ka):

$$\text{Age (ka)} = \frac{D_E\,(\text{Gy})}{D_R\,(\text{Gy ka}^{-1})}$$

Most laboratory instrumentation is geared towards the precise and accurate measurement of the D_E. This measurement calls for careful preparation of the sample in a dark-room environment and most luminescence readers are housed in some sort of darkened space as samples cannot be exposed to anything other than very controlled (wavelength restricted) lighting.

1.3 HISTORICAL DEVELOPMENTS AND EARLY WORKERS

In order to fully appreciate the current state of luminescence dating, it is beneficial to understand the physical phenomenon that creates luminescence, how it was discovered, the accurate ways to measure the dose in the mineral of choice (D_E), and the rate at which energy is absorbed from the radiation flux (D_R). A lengthy review of the history of luminescence dating can be found in McKeever (1985) and a more recent and concise version by Zöller and Wagner (2015) with online versions in the Encyclopaedia Britannica (Gundermann, K-D. 2000).

The earliest historical record of luminescence is by Robert Boyle (Fig. 1.5), an Irish-Anglo writer, philosopher, and chemist. Concerning a large brown diamond, he reported on 28 October 1663 to the Royal Society:

I also brought it to some kind of Glimmering Light, by taking into bed with me, and holding it to a good while upon a warm part of my Naked Body

(cited after Aitken 1985; Boyle 1664).

What is not known is the ultimate fate of this diamond, how Boyle obtained such a large diamond, or indeed why this sample emitted light at such low temperatures. A few years later in Berlin, Johann Sigismund Elsholtz (1676) described a similar property for the more common and mundane mineral of fluorspar. There are also much earlier reports on luminescent gems and stones that glow in the

Figure 1.5 Sir Robert Boyle (1627–1691) by Johann Kerseboom, 1689[1].

1 Robert Boyle. Chemical Heritage Foundation. N.p., 21 Feb. 2017. Web. 10 July 2017.

dark after being exposed to light, probably due to a kind of photoluminescence (Harvey 1957).

The first historical usage of the word 'Lumineszenz' was in 1888 by the German physicist Eilhardt Wiedemann as he described *'all those phenomena of light which are not solely conditioned by the rise in temperature,'* i.e. the condition of incandescence or 'hot light' in contrast to luminescence or 'cold light' (cited after Harvey 1957; Zöller and Wagner 2015). Weidman differentiated various kinds of luminescence according to the method of stimulation, such as photo-luminescence, thermoluminescence, tribo-luminescence and chemi-luminescence. When irradiating various minerals with cathode rays, Wiedemann and his colleague Gerhart Schmidt, (Wiedemann and Schmidt 1895) observed a light emission after gentle heating, and thus first described the phenomenon of TL, which was thermally stimulated radioluminescence induced by artificial ionising radiation (Zöller and Wagner 2015). Trowbridge and Burbank (1898) removed the natural TL from fluorite by heating it and then re-excited it by exposing the mineral to X-rays (McKeever 1985).

In 1945, Randall and Wilkins formalised the theory of TL by assuming that once released from the trap, the electron will undergo recombination instead of re-trapping (Randall and Wilkins 1945), which subsequently energised the study of TL. Once the relationship between TL intensity and absorbed radiation dose was established, it was only a short time until Farrington Daniels and co-workers at the University of Wisconsin proposed to use TL for geological and archaeological age determination (Daniels *et al.* 1953). Archaeological dating on ceramics and burned bricks was successfully tested by Norbert Grögler and colleagues at the University of Bern (Grögler *et al.* 1958, 1960), followed by Kennedy and Knopff (1960) at the University of California. During the 1960s, luminescence dating evolved into a powerful tool for heated archaeological materials due to the efforts by Martin Aitken and his team at the University of Oxford (Aitken *et al.* 1964; Aitken 1968, 1985). The two essential procedures of pottery TL-dating were developed in the late 1960s: the quartz inclusion technique (Fleming 1966) and the fine-grain technique (Zimmerman 1967). An important byproduct of these methodological developments was the TL authenticity-testing of ceramic objects (Fleming *et al.* 1970). In addition to baked clay, TL dating has also been applied to heated stones, in particular to flint artefacts (Göksu *et al.* 1974).

The observation that the natural TL intensities from sediments increase with geological age, galvanised researchers from the former Soviet Union to propose TL as a tool for the chronology of loess sediments. Pioneer research was done by Shelkoplyas and Morozov at Kyiv in 1965 (Dreimanis *et al.* 1978). The lack of translatable and on-line publications by the Kyiv group makes it difficult to fully understand sample preparation and measurement procedures, however. Wintle and Huntley (1979, 1980) first demonstrated that sunlight exposure bleaches the TL signal of sediments, boosting the development of TL dating for geological applications. When daylight bleaching was identified as the main resetting mechanism, further work concentrated on TL dating of aeolian sediments (those dominantly transported by wind, such as loess or dunes).

The procedures for measuring the TL signal went through many iterations, such as the 'partial bleach' or R-γ method (R-ß if a beta source is used) and the regeneration technique to correct for this non-zeroed component (Wintle and Huntley 1980), before settling on additive and regenerative 'total bleach' methods for aeolian sediments. The resetting of the TL signal at deposition (or creation in the case of pottery or fired objects) is called 'totally bleached' if there is no measurable signal above background. If the resetting mechanism is inadequate and residual TL is still present after creation or deposition this is known as 'partial bleaching'.

The most common method in the 1980s and 1990s was the multiple aliquot additive dose (MAAD). It was the method with the simplest procedures and was evaluated by Huntley and others in their 1985 trials. This method requires that the luminescence stemming from natural accumulation during burial is measured on several subsamples (typically five), termed aliquots, of a sample (these are often referred to as naturals and commonly denoted by *N*). Several other groups of aliquots are each given progressively increasing doses of laboratory radiation (*N*+ dose) and the luminescence response is measured. All the aliquots are then preheated and measured, usually within 24 hours. The luminescence intensity is then plotted as a function of laboratory radiation dose. A function is fitted to these data, and extrapolated to where it intercepts the dose axis, the intercept point being defined as the D_E (Aitken 1985; Preusser *et al.* 2008; Fig. 1.6a).

Because the aliquots in the set are not identical (due to different numbers of grains and the different luminescent properties of individual grains), a process of normalisation was employed: either a short shine of OSL before preheat and laboratory dose or a small dose post-primary measurement. Each aliquot was bleached by exposure in the laboratory to a light source (i.e. solar simulator, light-emitting diode (LED), or laser) to completely erase any OSL signal, heated to a temperature above 160 °C for 10 seconds, then exposed to a radioactive source of known concentration and this signal is measured or the aliquots are given a small dose after the natural measurement. These measurements are then used for an average or normalised value against which individual aliquots are compared. In all cases, the laboratory doses are uniformly applied and it is assumed the difference in OSL responses (i.e. curve height) are related to grain differences in the aliquots. See Lian (2007) for a detailed discussion on the limitations and advantages of the method.

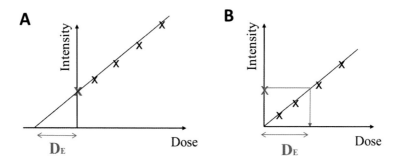

Figure 1.6 Dose response curves. A) Additive method: Symbols indicate signals resulting from irradiation in nature (blue) and after different laboratory irradiations are added to the natural dose (black). The dose response is extrapolated to obtain the D_E. B) With regenerative methods the natural signal (blue) is interpolated on the dose response curve to obtain the D_E. In this case, laboratory irradiations (or dose) are carried out after the aliquot has been fully reset.

The excitement of the early 1980s was slowly moderated (beginning in 1985) when a pattern of significant TL age underestimates from loess older than 30,000–100,000 years ago was noticed in the published data (Zöller and Wagner 2015). The poorly understood phenomenon of 'anomalous fading' from feldspars (see Section 1.5.2 for details) was often thought to be responsible for this, but some authors maintained that reliable ages up to 300,000 years or even 800,000 years could be obtained in spite of anomalous fading (e.g. Berger *et al.* 1992; Singhvi *et al.* 1989). To some extent, non-standardised laboratory procedures contributed to the uncertainty, but the large majority of researchers simply could not duplicate such old ages due to a variety of factors: source geology, limited time and resources for experimentation, or poorly understood sediment transport and post-depositional processes. Zöller and colleagues (1988, 1991) were able to extend the reliable TL age range from loess to a minimum of about 100,000 years. In addition to anomalous fading, the 'effective mean lifetime' of an electron trap (Wagner 1998), and 'dose-dependent sensitivity change' (Wintle 1985) were also suggested, to explain observed age underestimates from older loess.

The common occurrence of TL age overestimates from sediments is due to incomplete optical resetting (partial bleaching or zeroing) during transport (Godfrey-Smith *et al.* 1988; Preusser *et al.* 2008). This meant that the TL dating technique has survived to this day mostly for heated archaeological artefacts (e.g. pottery, burned flint; see Chapter 10 for further details) or for specialised applications such as sediments and rocks heated by volcanic eruptions (e.g. burnt flint, Göksu *et al.* 1974; microdosimetry problems of flint tools, Richter 2016; or examining wildfire temperatures, Rengers *et al.* 2017).

Because TL dating on sediment is so problematic, research instead began to focus on development of OSL dating. The recognition that not only heating, but also exposure to sunlight, resets the luminescence signal (both TL and OSL; Wintle and Huntley 1979), meant that luminescence dating became applicable to the dating of light-exposed sediments. The observation that instead of heating quartz, one could expose it to intense green illumination and stimulate luminescence (Huntley *et al.* 1985) meant practical utilisation of the phenomenon of OSL for geochronology. Soon afterwards, the luminescence stimulation by infrared illumination (800–900 nm) of feldspar (IRSL) was discovered by Galena Hütt (Hütt *et al.* 1988). In the late 1990s, TL dating was largely replaced by IRSL on fine-grained (4–11 microns) polymineral silt (Forman *et al.* 2000; Personius and Mahan 2000; Porat *et al.* 1997) with a gradual evolution to OSL on quartz sand-sized grains using either green or blue wavelengths or feldspar grains using infrared wavelengths.

1.4 SAMPLE PREPARATION

To prevent light exposure of samples and unintentional resetting of the luminescence signal within the grains, sample preparation was first performed under dim red or orange lights, similar to those used in photographic laboratories, since these were found to have the least effect on luminescence properties; (Spooner 1992; Lamothe 1995; Spooner and Prescott 1986; Huntley and Baril 2002). With the advent of cheap and plentiful LEDs, optimum peak wavelength, giving the best clarity for non-dark adapted vision for the least trapped charge loss within the wavelength range, is 590–630 nm (Sohbati *et al.* 2017). Modern laboratory lighting is recommended to be of peak emission at 594 nm since this was the most efficient for luminescence purposes, because the relative perception of brightness by the eye at this wavelength (orange) is ~76%, twice the perception at 621 nm but with half the power density (Berger and Kratt 2008; Sobhati *et al.* 2017). Modern-day LED amber opticals are also found to be suitable (Berger and Kratt 2008; Ed Rhodes personal communication 2017).

The goal of the preparation procedure is to isolate material suitable for the luminescence measurement (Wintle 1997) which comprises, in the majority of cases, quartz and feldspar minerals. Only selected grain sizes are used for OSL dating: coarse grain and fine grain (see also Section 1.6.1). The term coarse grain usually encompasses grain sizes from 63 μm to 250 μm, depending on the texture of the sample; fine grains are usually 4–11 μm in size. Grains in the desired size range are obtained by sieving and/or Stokes settling. To remove carbonates and organics, the samples are treated with hydrochloric acid (HCl) and hydrogen peroxide (H_2O_2) either before or after sieving; there are merits to both approaches.

Different mineral phases of the coarse grain fraction can be obtained by density separation. Quartz is extracted in one of two ways: as either a two-step process in heavy liquid by isolating minerals between 2.62 and 2.75 g/cm³ (the specific gravity of quartz is 2.62–2.65 g/cm³) or by removing the magnetic and para-magnetic minerals of heavier density on a magnetic separator and then quartz and feldspar are isolated by submersion in a 2.58–2.60 g/cm³ heavy liquid with potassium feldspar floating and quartz sinking. The final preparation step is that quartz is etched with hydrofluoric acid (HF) to remove the outer ~20 μm, i.e. the shell affected by alpha radiation. Etching takes place with highly concentrated HF for a period of about an hour. Plagioclase feldspars (calcium or sodium feldspars; specific gravity of 2.62–2.76 g/cm³) that sometimes remain in the quartz fraction are generally preferentially dissolved along cleavage planes during the etching process and require re-sieving to remove the newly reduced grains. Potassium feldspars have densities between 2.53 and 2.58 g/cm³ and are isolated in a similar procedure. The small grain sizes of fine grains do not allow for mineral separation and are usually measured in a polymineral mixture. However, quartz can be isolated by etching samples with hexafluorosilicic acid (H_2SiF_6) for 1–2 weeks, during which time feldspars dissolve. A wash with warm (50 °C) dilute HCl is applied after etching to remove any precipitated fluorides. Further details can be found in Preusser *et al.* (2008).

After preparation, samples are mounted on sample carriers, often stainless steel or aluminum disks, 10 mm in diameter, for analysis. Coarse grains are fixed using silicone oil. The quantity of grains is controlled by dictating the surface of the disk that is covered with silicone oil. Fine grains are pipetted from suspension in acetone or water onto the disks, forming a thin uniform layer that covers the entire surface of the disk and then the liquid is evaporated to leave behind the altered grains.

1.5 MEASUREMENT OF THE EQUIVALENT DOSE (D_E) AND OPTICALLY STIMULATED LUMINESCENCE

Most of the time and effort in obtaining a luminescence age is related to measuring the D_E. In simple terms, the D_E is measured by recording the sample's emitted luminescence intensity versus increasing doses of laboratory administered ionising radiation, commonly using a calibrated beta source (Lian 2007; Fig. 1.6). The term D_E was introduced because the laboratory radiation used to construct a sample's dose response consists only of beta particles (from a calibrated $^{90}Sr/$ ^{90}Y beta source; rarely from gamma rays or X-rays), whereas in the natural environment the radiation absorbed by a sample comes from a mixture of alpha and beta particles and gamma and cosmic rays (Huntley 2001). Equivalent dose is variously represented by the notation D_E, ED, D_{eq}, or D_e, and is sometimes referred to incorrectly as palaeodose (Lian 2007).

By 1985, the term 'OSL dating' was firmly established in the scientific literature and continues to dominate the titles and common use of any luminescence technique apart from TL. OSL in the narrower sense is limited to stimulation with visible light (i.e. ~385–700 nm; mostly violet, blue and green). Stimulation with photons in the infrared range leads to emission of IRSL; thus the initial argument that the technique should be called 'photon stimulated luminescence' (Aitken 1998; Berger *et al.* 2004). The favoured stimulation method depends on the mineral under investigation and the age range of the sample.

1.5.1 Instrumentation

A rudimentary OSL 'reader' consists of two essential elements: a light source to stimulate the OSL dosimeter (lamps, lasers, or light emitting diodes (LEDs)) and a light detector to measure the emitted luminescence (photomultiplier tube (PMT) or charge coupled devices (CCD); Yukihara and McKeever 2011, their Fig. 2.23). Additional common elements in an OSL reader are a heater to control the sample temperature, a radiation source (beta or X-ray) for laboratory irradiation of the sample, and a sample changer to allow measurement of multiple samples (turntable, robotic arm). Current instrumentation and 'off the shelf but customer modified' analytical software allows for hundreds of D_E values to be efficiently determined from one gram (or less, down to 0.2 g) of prepared sediment (Lian 2007).

In OSL, light is used to stimulate light emission from a sample. This approach requires techniques that allow separation between luminescence and stimulation light. The light source is chosen to have the optimal wavelength for the mineral that is being measured. The sample will emit OSL at different wavelengths. Detection filters are placed between the PMT and the sample to block stimulation light and to isolate specific OSL emission bands. Filter combinations are chosen to optimise the signal-to-noise ratio. Popular filter combinations are listed by Liritzis *et al.* (2013, their Table 2.2). The nature of the OSL technique allows monitoring or detecting the OSL at wavelengths *shorter* than the wavelength of stimulation light. Thus, feldspar will be stimulated with infrared wavelengths but detected with blue wavelengths while quartz is stimulated with blue wavelengths but detected in the ultraviolet wavelength.

There are two main commercial sources for OSL readers. Risø National Labs[2] (Denmark) have developed and sold OSL readers for nearly thirty years including their recent model Risø TL/OSL-DA-20. The second source is Freiberg Instruments[3] (Germany) which offer several different models and sizes of their lexsyg reader. Daybreak Nuclear is a US-based company that produced automated irradiators, alpha counters, and TL/OSL readers in the early 1980s through to the early 2000s but is no longer available for commercial applications. All of these companies sell instrumentation meant for the laboratory since

---

2 http://www.nutech.dtu.dk/english/products-and-services/radiation-instruments/tl_osl_reader; June 30, 2017
3 http://www.lexsyg.com; June 30, 2017

they contain delicate electronics, need a continuous power supply, and are outfitted with a radioactive source.

1.5.2 IRSL, feldspars, and polymineral fine-grains

The potassium feldspars of orthoclase, microcline, and sanidine, and several of the plagio-clase feldspars are sensitive to infrared light (IR, 800–900 nm) and generate IRSL (Hütt *et al.* 1988; Buylaert *et al.* 2013). Although quartz grains can emit luminescence signals when they are stimulated by infrared light (Bøtter-Jensen *et al.* 2003; Li and Li 2011), potassium feldspar vastly dominates the weak quartz signal if the minerals are mixed.

The higher saturation dose of feldspar (measured using IRSL) when compared to the lower quartz saturation level meant that an upper dating limit of 200,000–300,000 years or even more was possible. However, a limiting factor was found in the phenomenon of 'anomalous fading', a loss of signal over time that cannot be explained by thermal influence (first described by Wintle (1973) and later Spooner (1992, 1994)). Procedures to correct for anomalous fading were proposed by Huntley and Lamothe (2001) and Auclair *et al.* (2003) but they are expected to yield reliable results only in the linear part of the dose–response curve, i.e. samples not exceeding 20,000 to 50,000 years, which greatly limits the correction (Huntley and Lian 2006). This correction method for anomalous fading was given the term 'g-value'. The method measures the percentage of IRSL signal decay after storage of varying periods following the same laboratory irradiation. The decay is then converted to a g-value in terms of percentage loss of signal per decade. This allows for a correction for the age underestimation induced by IRSL anomalous fading (Aitken 1985; Huntley and Lamothe 2001).

IRSL stimulation is generally carried out at a temperature slightly above room temperature (e.g. 50 °C or 60 °C) to provide a stable thermal environment. In recent years, an extended procedure for measurement of the IRSL signal in well-bleached sediments has been adopted. In this 'post-IR IRSL' (pIR-IRSL) procedure, IR stimulation is first carried out in the conventional way near room temperature, and is then repeated at elevated temperatures (e.g. 225 °C or 290 °C; Buylaert *et al.* 2009; Thiel *et al.* 2011; Buylaert *et al.* 2012; Thomsen *et al.* 2008). This pIR-IRSL signal was

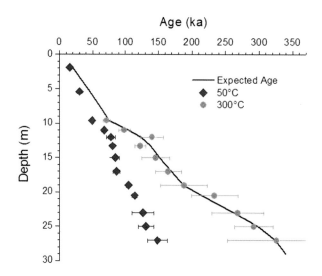

Figure 1.7 Comparison of IRSL and post-IR IRSL ages for a Chinese Loess section (modified Fig. 5 from Li and Li, 2012). The IR ages (50°C, blue) suffer from fading and underestimate the independently verified ages considerably, while post-IR IRSL ages measured at 300°C (orange) show good agreement.

found to have significantly reduced fading, allowing reasonable age estimates of a few hundred thousand years (e.g. Li and Li 2012; Li *et al.* 2013, Guérin *et al.* 2015; Yi *et al.* 2016; compare also Fig. 1.7). It was also found useful in regions where quartz proved problematic due to low sensitivity because the sediment eroded from bedrock, deposited nearby, and the grains contained inclusions of feldspar, such as in southern California (Lawson *et al.* 2012).

1.5.3 Bleaching rates of minerals

Published studies have shown that the OSL from quartz and the IR signal from feldspar bleach at different rates (e.g. Godfrey-Smith *et al.* 1988; Aitken 1998, their Table 6.1; Thomsen *et al.* 2008). Godfrey-Smith showed that the quartz OSL signal was reduced to 1% of the signal after bleaching for 90 seconds and Aitken showed OSL from quartz reached 10% of its original value within 2 seconds while potassium feldspar reached this value after 60–90 seconds. More recently, Colarossi *et al.* (2015) showed that South African quartz was reduced to 5% of the original luminescence after 10 seconds of bleaching using a Honle™ SOL-2 simulator. It took roughly 24 hours of sunlight for feldspar (at IR 50 °C) to reach the same level of bleach that quartz reached in 10 seconds (Colarossi *et al.* 2015). The situation is even worse for the pIR-IRSL signals (Buylaert *et al.* 2013; Colarossi *et al.* 2015). After 10 seconds of exposure, the pIR-IRSL signal lost only 2% of the signal. After 1,000 seconds of exposure, the pIR-IRSL signal still had 90% of its signal, losing only 10% (Murray *et al.* 2012; for a synopsis of current bleaching studies on IRSL and pIR-IRSL read Smedley *et al.* 2015).

As Murray *et al.* (2012) acknowledge, all of the data were collected using the full solar spectrum, so it is not clear how big the differences in bleaching rates would be in locations where the shape and intensity of the solar spectrum change. This would be the case for sediments transported underwater, in a watery slurry, in debris flows, or in a body of aeolian dust where the shape of the spectrum would change and become less intense (Berger 1990). Since the stimulation cross-sections of quartz and feldspar depend differently on wavelength (Spooner 1994a, b; Bøtter-Jensen *et al.* 1994) even the qualitative effects of time-variant simultaneous changes in spectrum and intensity are difficult to predict (Murray *et al.* 2012). When one considers the variability of feldspar grains in a sample, there must also be consideration for the different rates of bleaching in the grains (Smedley *et al.* 2015). Within a geomorphic context, the short transport distances and/or exposures to a relatively non-attenuated spectrum (such as those that occur during sheet erosion or colluvial transport on a fault scarp) would probably dominate over the much longer exposures to an attenuated spectrum (e.g. after entering a river).

Studies on the rate of luminescence bleaching during sediment transport in attenuated and non-attenuated mediums are a current topic of intense research (Gray and Mahan 2015; Gray *et al.* 2017) but surprisingly little has been published on the effect of elevation, cloud cover, shade, water depth, grain variation, and turbidity on light attenuation to sediment. A simplified table highlighting the advantages and disadvantages of quartz and/or feldspar for OSL dating is therefore provided (Table 1.1; Lian 2007).

Table 1.1 Advantages and disadvantages of quartz and feldspar for luminescence dating (modified Fig. 7 from Lian, 2007)

QUARTZ		FELDSPAR	
Advantages	**Disadvantages**	**Advantages**	**Disadvantages**
Highly resistant to weathering	Relatively low luminescence intensity; some quartz samples do not emit measurable luminescence	Luminescence saturates at higher radiation doses than does that from quartz	Weathers more readily than quartz from the environment
Luminescence signal bleaches more rapidly in sunlight than that from feldspar	Luminescence saturates at lower radiation doses compared to that emitted from feldspar	Luminescence intensity may be orders of magnitude higher than that emitted from quartz	Bleaches more slowly than quartz
Does not suffer from anomalous fading	Thermal transfer can be higher in quartz than in feldspar	IRSL can be stimulated preferentially in quartz/feldspar mixtures	Suffers from anomalous fading; rate of fading must be measured and modification acommodated

1.5.4 Quartz and the SAR procedure

Since 2000, the preferred method for age determination of sediments is OSL on quartz sand especially for young sediments (Zilberman *et al.* 2000; Arnold *et al.* 2007). OSL (i.e. blue and green stimulated luminescence) is associated with easy to bleach traps in the crystalline lattice and thus avoids the problems of the hard to bleach traps of TL/IRSL. Blue-green light (~ 470 nm) is typically used for quartz OSL stimulation with measurement of the resultant luminescence in the ultraviolet wavelength (300–380 nm).

The previously mentioned MAAD techniques (Section 1.3) have the disadvantage that the aliquots have to be normalised to allow comparison. The method also requires extrapolation of the dose response curve, which is always associated with larger uncertainties. Duller (1991) was the first to propose practical methods where all of the measurements necessary to determine D_E could be made using a single aliquot. Duller (1995) went on to develop a variety of single-aliquot methods for feldspars and Murray *et al.* (1997) did the same for quartz. This led to the refinement of the single-aliquot regenerative-dose (SAR) technique. This technique uses a single aliquot to obtain a D_E value. First the 'natural signal' of the aliquot, i.e. the signal resulting from irradiation in nature, is measured. This step is followed by repeated cycles of irradiation with increasing doses and measurement, to obtain a regenerative dose response for the aliquot (Fig. 1.6b).

One major drawback to the OSL technique is that during the OSL measurement there is no way of selecting only thermally stable electron traps like there is during TL. While thermally unstable traps are generally empty when a sample is collected, they are filled during laboratory irradiation (Fig. 1.4d). To overcome this problem, the sample aliquots are heated (preheated) prior to each repeated cycle of measurements. The objective of the preheat is to empty the thermally unstable traps, with little to no impact on thermally stable traps.

The D_E is then determined by interpolating the natural signal onto the dose response. While the method does not require normalisation of different aliquots, it was found that sensitivity changes with each measurement cycle (the luminescence signal) even if the same dose is given. The SAR procedure monitors a sample's OSL response to a small radiation dose (the test dose) that is administered at the end of each regenerative cycle (Roberts *et al.* 1998, Murray and Wintle 2000 2003; Wintle and Murray 2006). The critical development in Murray and Wintle (2000) was in introducing a protocol which made it possible to combine all of these measurements using repeatedly generated OSL signals (Duller 2015).

The SAR procedure includes various systematic tests, which are generally applied to a sample set to test its suitability for the procedure. These standard tests, as outlined by Duller (2008), are:

1. the preheat plateau test, which determines the most suitable pre-heat temperature to avoid measurement of thermally unstable traps (Murray and Wintle 2000; Fig. 1.8A);

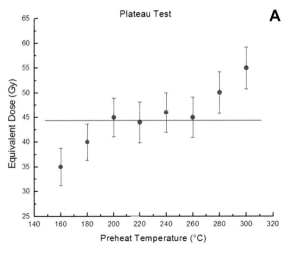

A

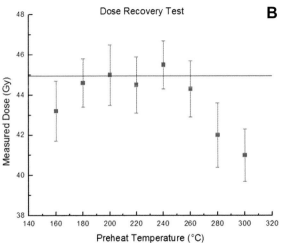

B

Figure 1.8 A) Example of a plateau test. The D_E is measured using different preheat temperatures. Several aliquots are used for each temperature and the average dose is calculated. Error bars indicate the standard error. In this example, the measured dose is constant for temperatures from 200 °C to 260 °C; this is the preheat plateau, as indicated by the solid line. B) Example of a dose recovery test. Aliquots are bleached and irradiated with a known dose (here 45 Gy, indicated by the solid line). The dose is measured using different preheat temperatures. For temperatures from 180 °C to 260 °C the measured dose agrees with the given dose within error limits and therefore any preheat within those limits is acceptable.

2. dose recovery test, which measures whether a known laboratory given dose can be determined accurately with the chosen OSL protocol; (Wintle and Murray 2006; Fig. 1.8B).

Every aliquot or grain of a sample should pass additional quality assurance tests in order to be deemed reliable for age calculation (Murray and Wintle 2003; Wintle and Murray 2006):

1. The recycling ratio test for repeatability of the measurement: One of the regenerative doses is administered for a second time and the signal is measured. The ratio of the first and second signal should not deviate more than 10% from unity.

2. The recuperation test: The luminescence signal is measured without prior irradiation. If a signal is observed, it is an unwanted result of the heating steps in the measurement procedure. This recuperation signal should be less than 5% of the natural signal.

3. IR depletion test: One of the regenerative doses is administered for a second time and the signal is measured, but after prior stimulation with infrared. The ratio of first and second signal is determined. A ratio smaller than 0.9 indicates significant signal depletion by IR and thus contamination with feldspar.

4. The natural signal must be below 20% of the saturated level otherwise errors become unacceptably large.

If the sample fails these tests, it precludes the determination of an age or indicates that a special statistical treatment is necessary to obtain a reliable age (Rhodes 2011). The most common errors that exclude aliquots from inclusion in datasets are due to either an unacceptably high recycling ratio or D_E values at or close to saturation. The SAR procedure enables multiple D_E measurements for a single sample by measuring a large number of aliquots. It is currently the method favoured by most optical dating practitioners because the method provides a routine procedure for a large number of luminescence measurements (Rhodes 2011). Statistical methods, as outlined in Section 5.6, are applied to obtain a final D_E value for the sample and to discriminate between well and poorly bleached subsamples. This opened up the ability to obtain meaningful dating results from sediments or geomorphologic events not datable by luminescence before, such as periglacial slope deposits (Hülle et al. 2009; see Chapter 6 this volume) or tsunami sediments (Brill et al. 2012; see Chapter 8 this volume). The simultaneous development of single-aliquot (Murray and Wintle 2000; Wallinga et al. 2000) and single-grain dating capabilities (Section 1.5.5), along with commercialised and robust instrumentation, have greatly expanded archaeological and geological applications of OSL dating in the last several decades and led to a rapid increase in the precision, accuracy, and applicability of luminescence techniques (Murray and Roberts 1997; Erfurt and Krbetschek 2003; Duller 2008; Wagner et al. 2010; Preusser et al. 2014).

The rather low saturation dose of quartz OSL (~ 150 Gy) limits the upper dating range to <150,000 years, depending on the natural radioactivity of the sediment. To extend the dating limit of quartz OSL by an order of magnitude, a 'thermal transfer-OSL' (TT-OSL)

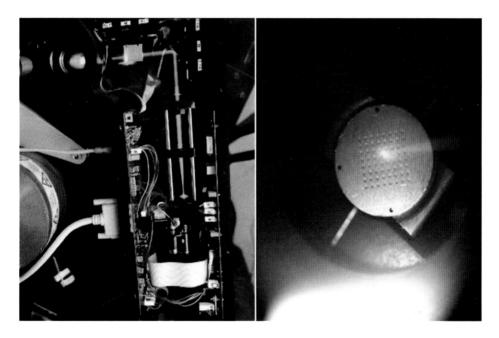

Figure 1.9 Focused laser used to stimulate single grains of sand (left) and sample holder with laser spot (right). Photographs courtesy of DTU Nutech (Riso).

procedure using recuperated OSL was suggested (Wang *et al.* 2006, 2007; Porat *et al.* 2009) but results from testing the procedure still appear ambiguous (e.g. Zander and Hilgers 2013). These innovations show promise to extend the age range of luminescence over an order of magnitude if properly mitigated (Duller 2012; Brown and Forman 2012; Duller 2015; see chapter 12).

1.5.5 Single-grain options

Aliquot sizes are either quoted by the diameter of the circular area covered with grains (e.g. 3 mm) or by their estimated number of grains (tens to few hundreds; Heer *et al.* 2012). The OSL signal measured from an aliquot is the sum of the signals emitted by all grains, therefore the most pragmatic approach recommends measuring about 50 single-aliquot values (Rodnight 2008; Galbraith and Roberts 2012) so that an averaging effect can be robustly applied. More grains on an aliquot leads to more signal to measure and a better signal to background noise ratio unless the grains have received different doses or have been bleached to substantially different degrees. To circumvent this single grain has been introduced.

Single-grain options for both quartz OSL (Murray and Roberts 1997) and for feldspar IRSL or pIR-IRSL (Reimann *et al.* 2012) allow for statistical isolation of signals from only the best-bleached grains and more robust age determinations in otherwise difficult to date deposits such as alluvium, colluvium, and younger fluvial sediments. Despite this significant development in OSL methodology, and the introduction of purpose-built measurement equipment (Duller *et al.* 1999; Bøtter-Jensen *et al.* 2000; Fig. 1.9), there has been limited use of single-grain OSL for routine age determination. The main reason for this is probably the significantly greater measurement and data analysis times involved, and technical expertise in dealing with the equipment and data reduction demands when making single-grain determinations (Rhodes 2007). Another factor is the presence in most sediment of a large proportion of grains that contribute very little luminescence signal, and therefore contribute very little useful age information yet extend the measurement time greatly. In most single-grain approaches, it is not possible to avoid the measurement of these relatively insensitive grains (Rhodes 2007).

1.5.6 Graphical representations and age models

Single-aliquot and single-grain procedures result in a large number of measured D_E values for a sample. Averaging multiple values generally reduces the uncertainty for the final D_E value that is used for age calculation. The most difficult portion of data analysis is to interpret the spread in these D_E values which is the result of the sample bleaching history, differences in local dosimetry during deposition, and variations in the intrinsic sensitivity of the grain(s) to laboratory measurements that make up an aliquot (Duller 2008). Even samples that consist of grains that initially had all received extended sunlight exposure can show a spread in D_E values that is beyond that attributed to analytical uncertainties alone; this spread is termed overdispersion (OD).

A lower OD percentage indicates high internal consistency in D_E values within 2 σ errors. The reported range of OD values from single-grain studies on sedimentary samples which are thought to have been well bleached prior to deposition is 9–22% (Arnold and Roberts 2009; Jacobs *et al.* 2008). OD values >20% may indicate mixing of grains of various ages (e.g. through post-depositional disturbance (Bateman *et al.* 2003) or partial solar resetting of grains (Galbraith *et al.* 1999)). For single-aliquot analyses, OD values between 10–30% are reported for well-bleached samples. Single-aliquot datasets should display lower OD values for a given degree of mixing or partial bleaching than single-grain datasets, due to the averaging of OSL signals from a large number of grains in the single-aliquot measurement.

Once a sufficiently large number of reliable aliquots or grains has been measured (the number can range from ~20 to ~100, depending on the spread in D_E values), the D_E datasets are displayed in a graph (e.g. radial plots, probability plots, and/or histograms; Fig. 1.10). By these means, the distribution of measurements is easily observed and allows the evaluation of the likelihood of partial bleaching or post-depositional mixing (e.g. Olley *et al.* 1999; Bateman *et al.* 2003; Duller 2008). Figure 1.11 illustrates dose distributions for several scenarios in which grains have been mixed after deposition or were incompletely bleached prior to deposition (Bateman *et al.* 2003).

Galbraith and Roberts (2012) present a powerful argument for the use of radial plots as the method of choice in evaluating D_E values. Radial plots were introduced by Rex Galbraith (1988a, 1988b, 1994) and are a form of an *xy* scatter plot of the standardised estimates against the reciprocals of the standard errors (see Fig. 1.10), usually with a 'radial' axis added to show the actual estimates. These plots are complicated to properly construct and understand, but comparisons between estimates are visually simpler and statistically robust. This graphical method uses both angular and radial extent to convey measurements with differing precisions in an unbiased statistical manner. Thus, several to many measured quantities can be examined for heterogeneity, even if they have different standard errors (Galbraith and Roberts 2012).

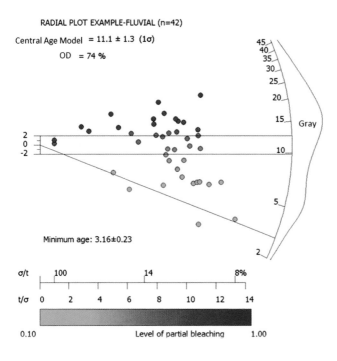

RADIAL PLOT EXAMPLE-FLUVIAL (n=42)

Central Age Model = 11.1 ± 1.3 (1σ)

OD = 74 %

Minimum age: 3.16±0.23

Gray

σ/t   100   14   8%

t/σ 0   2   4   6   8   10   12   14

0.10   Level of partial bleaching   1.00

Figure 1.10 Galbraith plot or radial plot showing each D_E value with a unit standard deviation on the y-axis (the 'standardised estimate') and points estimated with higher precision generate shorter confidence intervals on the curved (radial) D_E scale (i.e. values to the right of the graph have more precise errors than values to the left). In this example, D_E is on a log scale and relative standard errors are indicated on the x–axis, σ/τ = relative standard error and τ/σ = precision. Colours of points correspond to increasing D_E values (green circles are lower values).

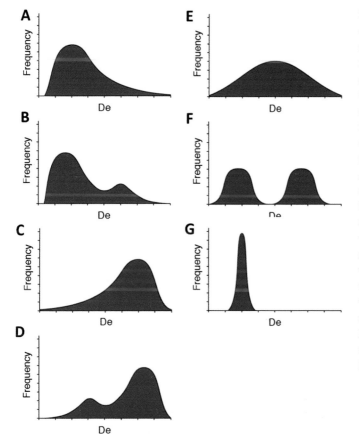

Figure 1.11 Histograms of dose distributions for a sample into which older material has been mixed giving rise to a high D_E tail or skewed bimodal distribution (A, B); a sample into which younger and exhumed material has been mixed giving rise to a low D_E tail or skewed bimodal distribution (C, D); a sample in which mixing has caused near homogeneity giving a wide range of D_E values with a low frequency at any single value or 2+ definable peaks (E, F); and an undisturbed well-bleached sample with a small D_E distribution and high reproducibility (G) (modified from Figure 3, Bateman *et al.* 2003).

According to Galbraith and Roberts (2012), the key points of the radial plot are:

1. **Horizontal or *x*-axis** – indicates precision; the D_E values are divided by their standard errors (i.e. reciprocals) and evaluated against a chosen reference value on the *x*-axis. A scale of standard errors can be added to this axis so that these can also be easily shown and read. A log scale for this axis is common because there is usually a wide range of D_E values. The relative standard error is indicated by σ/τ and precision is indicated by τ/σ (Fig. 1.10).

2. **Vertical or *y*-axis** – represents a scale of 'standardised estimate' centred at the reference value (Z_0) with each value having a unit standard error. If the *y*-axis extends from −2 to +2, then its total length is equivalent to a 'two σ' or 95% percent confidence level error bar applicable to each and every point. The two σ can be seen by (mentally) placing this bar on a point and extending a line from the origin through each end of the bar to intersect the D_E scale (Fig. 1.10). If the data points are consistent with a common true value, then about 95% of them should fall in a band that extends ±2 units vertically about the line drawn from the origin of this axis to the chosen value on the radial or *z*-axis.

3. **Arc, radial, or *z*-axis** – The D_E value for any particular point is given by the slope of the line from the origin (0, 0) through that point and can be read off by extending this line to a scale of slopes (the radial axis, which is usually drawn as an arc of a circle). This scale can be centred on some value of interest, such as the average D_E in units of Gy, as shown in Figure 1.10.

4. In a radial plot, the estimates are conveniently sorted, so that the **most precise estimates fall to the right** and the least precise fall to the left (Fig. 1.10). This is because points estimated with higher precision generate shorter confidence intervals on the curved (radial) D_E.

Once a given D_E dataset for a sample has gone through quality assurance tests (see Section 1.5.4) statistical analysis is required in order to obtain a final value for the D_E, either through averaging, weighted mean, or 'age models' as defined by Galbraith *et al.* 1999. The term 'age models' is a bit misleading since the majority of models are used to generate a reliable and reproducible D_E value and not the age, which is calculated later using the modelled D_E values divided by the D_R. These models were primarily modified to provide the D_E and D_R values of single grains, but the research for modelling grain specific D_R is still to be worked out (Roberts and Jacobs 2015). The routinely employed models are: the common age model (COM), the central age model (CAM), the minimum age model (MAM), and the finite mixture model (FMM). Of these, only the FMM should be exclusively used for single-grain sets of D_E because it allows identification and estimation of the parameters of interest for each grain measurement in the data, not just an overall mixed population as would be found in single aliquots. Examples of applied age models are shown in later chapters of this book.

1.6 THE DOSE RATE (D_R)

The name OSL dating refers to the luminescence process which is the basis for the D_E measurement, but neglects to mention that every OSL age has a second component of equal importance: the dose rate (D_R). In simplest terms, the D_R is calculated from the activities of the radioactive elements that occur naturally in the terrestrial environment. The leap from knowing the elemental concentrations of sediment to calculating the effective D_R is often the most overlooked and demanding component of the dating procedure. Understanding this process is as important as obtaining the D_E. The radiation environment in the subsurface has multiple components and can be quite complicated to ascertain and measure. Some of the parameters that must be quantified with regard to the D_R are that naturally occurring radioactive nuclides emit radiation at varying rates and energies, radioactive elements have different efficiencies in producing luminescence, water absorbs part of the emitted energy, and the radiation environment almost certainly changes over time.

1.6.1 Sources and types of radiation and their influence on grain size selection

Radiation in the subsurface stems from naturally occurring radioactive nuclides and, to a lesser degree, from cosmic rays (Fig. 1.12). Several elements have naturally occurring radioactive isotopes. ^{40}K is the radioactive isotope of potassium (K), and occurs as 0.012% of natural K. This isotope decays to ^{40}Ar with the emission of gamma rays with energy 1.46 MeV. Since ^{40}K occurs as a fixed proportion of K in the natural environment, these gamma rays can be used to estimate the total amount of K present if one knows that the half-life of ^{40}K is 1.3×10^9 years.

Uranium (U) occurs naturally as the radioisotopes ^{238}U and ^{235}U, which give rise to decay series that terminate in the stable isotopes ^{206}Pb and ^{207}Pb. The half-lives of ^{238}U and ^{235}U are 4.46×10^9 and 7.13×10^8 years, respectively. Thorium (Th) occurs naturally as the radioisotope ^{232}Th which initiates a decay series that terminates in the stable isotope

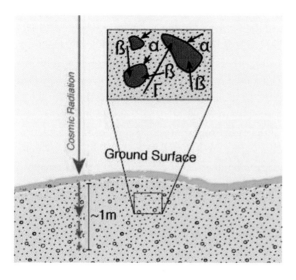

Figure 1.12 Sources of radiation in the subsurface. Alpha radiation (α) traverses very short distances (no more than 20 μm) before losing energy, beta radiation (β) is less localised and can penetrate modest thicknesses of material (~1–3 mm), while gamma radiation (γ) consists of high energy photons that can travel 30–40 cm in sediment. Cosmic ray components vary depending on the density of the sediment, the location and elevation of the site, and the sample depth from the modern surface; the cosmic dose decreases with depth in sediment but increases with elevation.

^{208}Pb. The half-life of ^{232}Th is 1.39×10^{10} years. Neither ^{238}U nor ^{232}Th emits gamma rays, and gamma ray emissions from their radioactive daughter products are used to estimate their concentrations. Average crustal abundances of these elements quoted in the literature vary widely for K; but are in the range 2–2.5% (Bunker and Bush 1967), 2.8% (Taylor and McLennan 1995), and 3.5% (Wedepohl 1995). Crustal abundances are 2–3 parts per million (ppm) U and 8–12 ppm Th (Aitken 1985, 1998; Bunker and Bush 1966, 1967; Taylor and McLennan 1995; Wedepohl 1995). In general, the ratio of U to Th will be 1:2 to 1:5 if the elements are in equilibrium; ^{40}K emits beta and gamma radiation but no alpha while U and Th emit all three types of radiation (Bunker and Bush 1966, 1967).

The nuclides emit radiation in their decays. Radiation energy is usually measured in electron volts (eV). This is the energy acquired by any charged particle carrying unit electronic charge when it falls through the potential difference of one volt (details can be read in the current Handbook of Chemistry and Physics[4]). Gammas consist of high energy photons in the range of a few keV to MeV; therefore they are of much higher energy than X-rays (Erdi-Krausz *et al.* 2003). They can penetrate ~30–40 cm into sediment. This means every sediment grain is exposed to gamma radiation from nuclides within this range, which should be taken into consideration during sampling in the field. Attention to the stratigraphic units, sampling in multiple directions and for all grain sizes, and meticulous photography and field notes will ensure best results.

Beta radiation, which consists of high energy electrons, is more localised and can penetrate modest thicknesses of material (~ 1–3 mm), depending on their energy. They are attenuated while passing through a grain so that more energy is deposited on the entry-side of the grain than on the exit-side. Since grains are exposed to beta radiation from all directions, the net effect is that the centre of a grain receives less dose from beta radiation than the outer shell. For all practical purposes, it is assumed that beta and gamma radiation produce the same amount of luminescence per energy deposited (Erdi-Krausz *et al.* 2003).

Alpha particles consist of two protons and two neutrons and have energies in the range of several MeV. This radiation can only penetrate to a very small depth into a material (~ 20 μm) before it loses all its energy. This means alpha radiation deposits a large amount of energy in a comparatively small volume. Many of the charges created in the ionisation process recombine before getting trapped. Alpha radiation is, therefore, less efficient in producing luminescence than beta or gamma radiation. The effect is allowed for by determining the *a-value* for a sample and then correcting the alpha-dose rate accordingly (Aitken 1985).

The varying luminescence properties of minerals require that they be separated from a bulk sample, but only smaller grains sizes are used for OSL dating so that mineral inclusions are rare and dosimetry of the desired mineral can be as simple and uncomplicated as possible (see Section 1.4). The etching process of coarse grains removes the outer shell affected by alpha radiation. As a result, alpha radiation is not considered for the calculation of the dose rate of etched grains. Silt-sized grains are not etched and all effects of alpha radiation have to be considered. On the other hand, one can assume that beta attenuation is negligible for these small grains; as mentioned earlier, beta radiation can penetrate through the grain centre.

---

4 https://www.crcpress.com/CRC-Handbook-of-Chemistry-and-Physics-97th-Edition/Haynes/p/book/9781498754286; July 7, 2017

1.6.2 Measurement of elemental concentrations and direct measurement of the D_R

The D_R is the calculation of the energy absorbed by a sample per unit mass and time. As explained in Section 1.6.1, all radioactive nuclides emit a well-defined amount of energy in their decay. The half-lives of the nuclides are known with great accuracy (Rumble 2017). Thus the D_R of a sample can be calculated if the nuclide concentration in the surrounding environment is known. The most common method of measuring elemental concentrations of the sediment is to count the number and type of particles emitted when placed on a detector of known efficiency. Some techniques of D_R assessment, such as beta counting, measure the total number of beta rays emitted by a sample in a fixed time (Thomsen 2015). By comparing the measured number with that from a sample of known D_R, the unknown D_R can be inferred. The method assumes that relative concentrations of different nuclides (K-U-Th, sometimes Rb) in sample and reference material are comparable. Other methods overcome the associated uncertainties by directly inferring the nuclide concentrations from the measured data.

Gamma ray energies are characteristic for each emitting nuclide. *Gamma spectrometers* count the number of gamma rays in defined energy intervals. The counts for the sample to be dated are compared, for each energy interval separately, with a reference material of well-known nuclide concentrations. When compared with a standard reference, the nuclide concentrations in the unknown sample can be determined (Knoll 2010). Laboratory measurements are generally carried out with a high-resolution germanium (Ge) gamma spectrometer (see Fig. 1.13b) that is cooled with liquid nitrogen. These instruments provide precise energy resolution, which in turn allows measurement of a large number of different nuclides. This is particularly advantageous if disequilibria in the U decay chain need to be assessed (see Section 1.6.5). Samples are usually dried before the measurement because water will preferentially absorb radiation. Sample sizes depend on the detector geometry and can range from several g to 1 kg. Due to the large range of gamma radiation (30–40 cm), if the amount of material is small (<20 g) the amount has to be representative of the entire environment affecting the sample to be dated. Sometimes more than one sample is measured to determine an effective nuclide concentration.

Figure 1.13 Photos of portable gamma specs with NaI detectors for field use and the laboratory gamma spectrometers with Ge detectors for higher precision counting. a) Portable gamma spectrometer. b) High-resolution germanium (Ge) gamma spectrometer. Note that these detectors require cooling with liquid nitrogen and often lead shielding if counting low level sediment samples.

An alternative for non-uniform sampling sites is a *portable gamma spectrometer* (Fig. 1.13a). These gamma spectrometers range from a few to tens of cm in diameter and can be buried directly at the sampling site. They provide *in situ* measurements of the radiation within approximately one hour. However, their sodium iodide (NaI) or lanthanum bromide (LaBr) crystals have considerably lower energy resolution than Ge. As a result, many nuclides contribute to a single energy interval. Only few selected nuclides can be detected without overlap, so that portable gamma spectrometers can provide an effective concentration of U, Th and K, but no information about the individual nuclides in the decay chains (Duller 2015). The measured concentrations are influenced by the moisture on the day of the measurement.

Luminescence can be used to measure the environmental D_R as well. Quartz and feldspar are natural dosimeters that absorb radiation from the environment and OSL is used to measure the absorbed dose. *Artificially produced luminescence dosimeters* work on the same principle. They can be buried in the subsurface and build up a sufficient luminescence signal in just a few days in the case of carbon-doped aluminum oxide (Al_2O_3:C; Akselrod *et al.* 1998; Kalchgruber and Wagner 2006) or months (e.g. Calcium sulfate doped with dysprosium ($CaSO_4$:Dy); Aitken 1985). TL or OSL are then used to measure the absorbed dose. Since the duration of burial is known, the dose can be divided by time to obtain the D_R. The advantages of luminescence dosimeters is that they directly measure the D_R and do not require indirect calculations via nuclide concentrations. They are particularly useful in non-uniform environments because multiple dosimeters can be buried at a single sampling site, allowing separate measurement of beta and gamma dose rates (Kalchgruber and Wagner 2006). Disadvantages are that they measure the effective D_R and the result is influenced by the water content, they do not allow measurements of the alpha dose rate, and require a return to the site to retrieve the buried dosimeters.

Other laboratory methods directly measure the isotope concentrations in a sample. *Inductively Coupled Plasma Mass Spectrometry (ICP-MS,* Sylvester 2015) is an analytical technique used for elemental determinations. An ICP-MS combines a high-temperature ICP source with a mass spectrometer. The ICP source converts the atoms of the elements in the sample to ions. These ions are then separated and detected by the mass spectrometer. The sample is typically introduced into the ICP plasma as an aerosol, either by aspirating a liquid, dissolving a solid sample into a nebuliser, or using a laser to directly convert solid samples into an aerosol. As the ions enter the mass spectrometer, they are separated by their mass-to-charge ratio (Sylvester 2015). The majority of luminescence dating labs use ICP-MS to measure the elemental concentrations of the bulk sediment associated with an OSL sample because the process is commercially available, cheap, easy, accurate, and quick.

Another method that offers high resolution, but also allows only small sample volumes is *neutron activation analysis* (NAA; DeBey *et al.* 2012[5]; Hancock 2015). NAA is a technique that relies on the measurement of gamma rays emitted from a sample that was irradiated by neutrons, providing a measurement of the parents of the radioactive series. The rate at which gamma rays are emitted from an element in a sample is directly proportional to the concentration of that element. The major advantages of NAA are: it is a multi-element

5 https://geology.cr.usgs.gov/facilities/gstr/neutron_analysis.html; July 7, 2017

technique capable of simultaneously determining up to about 70 elements in many materials; it is non-destructive, and therefore does not suffer from the errors associated with yield determinations; it has very high sensitivities for most of the elements (detection limits range from 0.03 ng to 4 μg), it is highly precise and accurate (overall errors of 2–5% relative standard deviation can be achieved for many elements), and samples as small as a few micrograms can be analysed. When using either NAA or ICP-MS, samples must be thoroughly homogenised so that a representative portion ends up in the smaller sized sample for analyses.

1.6.3 Moisture content

In calculating the D_R, it is assumed that all energy emitted in the sediment is also absorbed by the sediment. Sediments are assemblages of individual grains that were deposited by water, wind, ice, or gravity, but also include pore spaces between sediment grains, so the sediments are not solid (Fetter 1988). Water present in the burial environment can fill these pore spaces and absorb part of the emitted radiation, 'diluting' the amount of radiation that the sediment grains receive. The D_R received by the sample decreases as the water content increases, and vice versa. Thus the D_R has to be corrected for the moisture of the sample. A useful rule of thumb is that a 1% increase in the water content leads to approximately a 1% increase in the luminescence age (Duller 2015).

To calculate D_R, an average moisture content for the burial period is required. To account for unknown seasonal groundwater or water table fluctuations, wide error margins are used for measured water contents. The error margins include careful consideration of special conditions during field work and the geologic setting of the sample such as relative zones of wetness, widely mixed grain sizes, or past depositional climatic regimes that may no longer be present. Considerations are aided by the fact that in many cases soil porosity places limits on the possible water content. In an arid environment, field water contents are likely to reflect water content over Holocene time. Typical porosity ranges for sediments are: well-sorted sand and gravel 25–50%, mixed sand and gravel 20–35%, glacial till 10–20%, silt 35–50%, and clay 33–60% (Fetter 1988). Table 1.2 illustrates how water content affects the D_R of a typical sample.

Table 1.2 Influence of water content on the D_R of a sample. The dose rate corresponds to an average sample with 1% K, 3 ppm U and 10 ppm Th (Aitken 1985). The grain size was assumed to be 4–11 μm (fine grain). Only external D_R values are listed. An a-value of 0.1 was used for calculation of the alpha D_R. The water content reflects the weight of water divided by the dry weight of the sample.

WATER CONTENT	GAMMA D_R (GY/KA)	BETA D_R (GY/KA)	ALPHA D_R (GY/KA)	TOTAL D_R (GY/KA)
1%	1.05	1.49	1.23	3.78
5%	1.00	1.42	1.16	3.58
10%	0.95	1.34	1.08	3.37
20%	0.86	1.21	0.96	3.02
50%	0.67	0.93	0.71	2.31

Methods to assess moisture content include measurement of sediment saturation, measurement of current moisture by weighing and drying, or estimating the porosity of the sample and assuming some amount of moisture. All of these approaches face the same difficulty; no matter how accurate the measurement or calculation, water contents vary on both seasonal and geological time scales. A sample might be collected on a rainy day in an area that is usually very arid. A lake might have dried out over geologic time. There may be a fluctuating local water table. Field gamma spectrometry and dosimetry are directly affected by the same issues. Seasonal fluctuations can mean that gamma measurements taken in the winter may differ from summer readings. Also, where excavation has been carried out much in advance of the dosimetry, readings may be affected by drying out of the sediments (sometimes called 'case hardening'). For such reasons, the measured 'field' dose rates may not accurately reflect average conditions for the duration of burial.

1.6.4 Conversion factors in calculation of dosimetry

As mentioned above (Section 1.6.2), all radioactive nuclides have well-known half-lives and emit a well-defined amount of energy. Thus, elemental concentrations can be directly converted to D_R values by using conversion factors (Guérin *et al.* 2011). D_R values are calculated individually for alpha, beta and gamma radiation. All three values have to be corrected for the moisture content, taking into account the fact that water absorbs different types of radiation at varying rates. In addition to elemental concentrations and the moisture content, D_R equations also consider the issues discussed above. The alpha component of the D_R is only considered if the samples have not been etched, i.e. for fine grains. The *a*-value is applied to correct for the smaller efficiency of alpha radiation in luminescence production. Attenuated while passing through the grains, beta radiation is accounted for by applying grain-size-dependent correction factors (Mejdahl 1979). K-feldspars contain the radioactive nuclide ^{40}K and thus the beta D_R from this internal radiation has to be

Table 1.3 Variation of the cosmic D_R. For burial depth dependence, elevation was assumed to be at sea level, at 40°N. For elevation dependence, burial depth was assumed to be 1 m, at 40°N. For latitude dependence, elevation was assumed to be at sea level, burial depth was 1 m. Longitude was in all cases 0°, sediment density was 2.0 g /cm³.

BURIAL DEPTH DEPENDENCE		ELEVATION DEPENDENCE		LATITUDE DEPENDENCE	
Burial depth (m)	Cosmic D_R (Gy/ka)	Elevation (m)	Cosmic D_R (Gy/ka)	Latitude	Cosmic D_R (Gy/ka)
0.1	0.205	0	0.192	80°N	0.192
0.2	0.204	100	0.196	60°N	0.192
0.5	0.200	200	0.200	40°N	0.192
1.0	0.192	500	0.212	20°N	0.188
2.0	0.180	1000	0.233	0°	0.154
4.0	0.157	2000	0.285	40°S	0.196

Table 1.4 Example of D_R calculation for a fine grain and a quartz coarse grain sample. All D_R values are quoted in Gy/ka.

Sample grain size	U (ppm)	Th (ppm)	K (%)	Water content (%)	Cosmic D_R (Gy/ka)	a-value
Fine grain	3	10	1	10	0.200	0.1
Coarse grain	0.5	2	0.4	10	0.200	NA

Sample grain size	Alpha D_R (Gy/ka)	Beta D_R (Gy/ka)	Gamma D_R (Gy/ka)	Total D_R (Gy/ka)
Fine grain	1.081	1.339	0.951	3.572
Coarse grain	NA	0.383	0.227	0.810

considered as well. Gamma radiation has a large range and one should take into account the overall general stratigraphy at the site, in particular non-uniformities within the gamma range (see Chapter 2 for further details). Due to its large range, the precision with which the gamma D_R can be measured is likely to dominate the error limits (Section 1.7).

The contribution to the overall D_R by cosmic rays, i.e. the cosmic-ray D_R, is estimated for the sample as a function of burial depth, altitude and geomagnetic latitude and longitude (Prescott and Hutton 1994, 1988; Table 1.3). The Earth is constantly bombarded with high-energy particles, principally protons, coming from outer space; the source of the particles remains something of a mystery. However, when cosmic rays hit atoms on the upper atmosphere, showers of secondary particles are produced (mostly muons and neutrinos by the time they reach ground level), which rain down on the surface of the Earth all the time (Griffiths 2008). The intensity is larger at the poles than the equator due to modulation by the Earth's magnetic field (Gosse and Phillips 2001). Cosmic ray components vary depending on the density of the sediment, the depth from the modern surface, and the elevation of the sample site (Fig. 1.12). Table 1.3 illustrates the variation of the cosmic D_R with burial depth, elevation and latitude. The cosmic ray component generally falls into the range of about 5%–10% of the total D_R unless the sample is very near the surface, at high elevation, and or in sediment of low natural radioactivity.

Table 1.4 shows examples for calculation of the D_R for a sediment sample. The fine-grained sample (4–11 μm) is quartz mixed with feldspar with an average nuclide concentration, taken at 1 m burial depth and 200 m elevation. The coarse-grain sample is a beach sand (90–125 μm quartz grains) with very low nuclide concentration, collected at 0 m (sea level) and 0.5 m burial depth. The same water content of 10% was used for both samples. Nuclide concentrations are obtained with gamma spectrometry. Alpha, beta and gamma D_R are calculated from the nuclide contents and the water content, using conversion factors discussed above. Coarse grains are etched and the outer rind affected by alpha radiation has been removed, while the alpha D_R has to be considered for fine grains. Both samples are quartz, which is assumed to be free of internal radioactivity, such that the internal D_R can be neglected. For the fine-grain sample, the cosmic D_R contributes only 5% to the overall D_R. For the coarse grains of low radioactivity, the contribution is 25%.

1.6.5 Radioactive disequilibria and dosimetry variations over the burial period

To calculate the D_R, average water content values (Section 1.6.3) and nuclide concentrations over the burial time are used. Nuclide concentrations might change over time, which is generally summarised under the term *radioactive disequilibrium*. Radioactive disequilibrium occurs when one or more decay products in a decay series are completely or partially removed or added to the system. Disequilibrium is not a serious problem if geochemically active environments are avoided (i.e. organic-rich sediments, samples with evidence of cementation since deposition, environments at the interface between saline and fresh water, and deep marine conditions; see Duller 2015). Th rarely occurs out of equilibrium in nature, and there are no disequilibrium problems with K. However, in the U decay series, disequilibrium is more common than not, and can occur at several positions in the ^{238}U decay process: ^{238}U can be selectively leached relative to ^{234}U; ^{234}U can be selectively leached relative to ^{238}U; ^{230}Th and ^{226}Ra can be selectively removed from the decay chain; ^{222}Rn (radon gas) is mobile and can escape from soils and rocks into the atmosphere (Thomsen 2015).

Depending on the half-lives of the radioisotopes involved, it may take days, weeks or even years for equilibrium to be restored; in extreme cases, it can take thousands to a million years. Disequilibrium in the U decay series is the most serious source of error in calculating the D_R. High resolution Ge gamma spectrometry can give insight into the disequilibrium for some nuclides, e.g. loss of ^{222}Rn can be detected by comparing concentrations of ^{226}Ra and ^{214}Bi and ^{214}Pb. A ratio of ^{230}Th and ^{226}Ra different from 1 can indicate U loss or uptake. Other nuclides such as ^{238}U and ^{234}U are not directly accessible with the method. Estimates of U concentration are therefore usually reported as 'equivalent uranium' (eU) since these estimates are based on the assumption of equilibrium conditions. Marine sediments often suffer from radioactive disequilibria and will be discussed in Chapter 8.

Varied approaches can be used to take radioactive disequilibria into account, but in each case a possible scenario for the origin of the disequilibrium is evaluated (e.g. Guibert *et al.* 2009). When the disequilibrium has existed for the entire burial period of a sediment, the D_R can be calculated with the individual measured nuclide concentrations, i.e. instead of using an average U value, the D_R contribution is determined for each daughter nuclide individually. In other cases, minimum and maximum values for the average U concentration are used to widen the error limits for the nuclide concentration. Software packages have been developed to allow modelling of time-varying concentrations and can

Table 1.5 Example of a sample with U-series disequilibrium. (A) Nuclide concentrations for Th, K and different nuclides in the U decay series obtained with high resolution gamma spectrometry. (B) calculated beta and gamma D_R based on different assumptions about the U concentration.

A

Nuclide	Concentration
Th (ppm)	0.963 ± 0.060
K (%)	0.082 ± 0.004
U-series (ppm):	
^{234}Th	1.05 ± 0.11
^{214}Bi and ^{214}Pb	0.283 ± 0.02
^{210}Pb	0.50 ± 0.19
Average U	0.34 ± 0.025

B

	Beta D_R (Gy/ka)	Gamma D_R (Gy/ka)
1) average U	0.133 ± 0.005	0.105 ± 0.004
2) individual U-nuclides	0.164 ± 0.008	0.102 ± 0.004
3) max U (^{234}Th)	0.214 ± 0.014	0.184 ± 0.012
4) min U (^{214}Bi and ^{214}Pb)	0.125 ± 0.005	0.098 ± 0.004

be easily modified for individual use (e.g. Durcan *et al.* 2015; Guérin *et al.* 2012; Grün 2009; Kulig 2005) These packages provide a web-based open access research tool that is standardised, constantly updated, provides transparent calculations so that users can trace methods used to calculate dose rate values, improve inter-laboratory comparisons, and diminishes the potential for miscalculation (Durcan *et al.* 2015).

Table 1.5 illustrates the influence of U-series disequilibrium on the calculated D_R of a sample for which the concentrations of the various nuclides were obtained with high resolution gamma spectrometry. For a sample in equilibrium, all nuclides of the U decay chain have the same activities. The example (Table 1.5A) indicates an elevated content of U/Th (as determined by ^{234}Th) compared to the daughter nuclides ^{214}Bi and ^{214}Pb; ^{210}Pb is elevated as well. Possible scenarios include intake of U, loss of daughter nuclides, and intake of ^{210}Pb. Table 1.5B lists beta and gamma D_R that were calculated with the given values for Th and K and different U values: (1) average U concentration calculated from all daughter nuclides, neglecting the disequilibrium; (2) D_R contributions from individual U daughters as measured; (3) maximum U concentration as determined by ^{234}Th; (4) minimum U concentration as determined by ^{214}Bi and ^{214}Pb. The quoted errors are 1σ. To obtain D_R values for age calculation, the scenarios discussed above (i.e. whether the disequilibrium has existed from the initial burial or whether it is a result of leaching over time) have to be considered and the error bars widened accordingly.

1.7 PRECISION AND ACCURACY OF OSL

Uncertainties for luminescence ages are normally quoted at one standard deviation; that is, at the 68% confidence interval (1σ). Increasingly, a large number of samples or meta-datasets are the norm in multi-chronometer studies or where the scientist wishes to employ Bayesian-type modelling (see Chapter 3). It is often most useful to provide luminescence ages at the 95% confidence level (2σ) because this will be expected in the programming or required to properly evaluate the range of dates from all the geochronology. The values for OSL ages are often better than 10% of the age (near 5% in many cases; Duller 2007; Murray and Olley 2002) and for TL dating near 15–20% of the age. Age uncertainties are normally based directly on the error of the D_E dataset as well as the error of the D_R, although the quoted uncertainties should include both random and systematic errors. Random errors follow a statistical distribution and influence the precision of the age, i.e. the scatter of the results if the sample was to be analysed multiple times under the same conditions and using the same assumptions. Systematic errors influence the accuracy of the age, i.e. how close is the measured age to the 'true' age of the sample.

Uncertainties for the D_E include statistical errors of measured signals, dose errors for individual aliquots or grains, scatter in dose distributions of all measured aliquots or grains, and systematic errors in calibration of the beta or gamma source. In the past, any uncertainties arising from systematic errors have tended to be swamped by those arising from random errors. SAR procedures have reduced D_E uncertainties to less than 5%, so that increasing attention is given to the uncertainties arising from systematic effects, including the suitability of the sediment (or material) for use with the SAR procedure (Duller 2007).

D_R uncertainties have small contributions from random errors, such as the counting statistics in gamma spectrometry. Instead, the D_R is dominated by systematic errors including water content of the sediment over time, the varying cosmic dose component, radioactive disequilibria, concentration errors in reference materials, and errors in the factors for conversion from nuclide concentration to dose rate. The uncertainty of the D_R often dominates the overall error and thus the accuracy of the age. It is therefore essential that as much dosimetry measurement information as possible is included in any report or publication dealing with luminescence ages.

In any comparison of luminescence ages with other independent dating results, it is important that uncertainties on the luminescence ages include contributions from all known components (of course the same applies to the independent ages, but such problems are outside the scope of this book). It should be clear that it is difficult to obtain a luminescence age with an overall or combined standard uncertainty of much less than 5% at one σ and 10% at two σ (Duller 2007).

FURTHER READING

A concise summary of the luminescence method, history, and instrumentation can be found in a series of articles in the *Springer Encyclopedia of Scientific Dating Methods* (Rink and Thompson 2015). Reviews of luminescence dating and its application in geology and archaeology as well as methodological advances have been published by Lian (2007), Duller (2008), Preusser et al. (2008), Wintle (2008), Rhodes (2011), and Duller (2015). Numerous scientific papers informing the reader about latest progress in luminescence dating can be found in *Ancient TL*, *Archaeometry*, *Quaternary Geochronology*, and *Radiation Measurements*. Proceedings of the triennial international conference on Luminescence and ESR Dating (LED) are published in special volumes of *Quaternary Geochronology* and *Radiation Measurements*.

REFERENCES

Aitken, M.J., Tite, M.S., Reid, J. 1964. Thermoluminescent dating of ancient ceramics. *Nature* 202, 1032–1033.

Aitken, M.J. 1968. Low-level environmental radiation measurements using natural calcium fluoride, Proceedings of the 2nd International Conference on Luminescent Dosimetry, Gatlinburg; edited by J.A. Auxier, K. Becker, and E.M. Robinson, CONF-680920: U.S. National Bureau Standards, Washington, DC, 281–290.

Aitken, M.J. 1985. *Thermoluminescence Dating.* Academic Press, Oxford, 359 pp.

Aitken, M.J. 1998. An Introduction to optical dating. The dating of quaternary sediments by the use of photon-stimulated luminescence. Oxford, New York, Tokyo: Oxford University Press.

Akselrod, M.S., Lucas A.C., Polf J.C., McKeever S.W.S. 1998. Optically stimulated luminescence of Al2O3. *Radiation Measurements* 29, 391–399.

Arnold, L., Bailey, R., Tucker, G. 2007. Statistical treatment of fluvial dose distributions from southern Colorado arroyo deposits. *Quaternary Geochronology* 2, 162–167.

Arnold, L.J. and Roberts, R.G. 2009. Stochastic modelling of multi-grain equivalent dose (D_E) distributions: implications for OSL dating of sediment mixtures. *Quaternary Geochronology* 4, 204–230.

Auclair, M., Lamothe, M., Huot, S. 2003. Measurement of anomalous fading for feldspar IRSL using SAR. *Radiation Measurements* 37, 487–492.

Bateman, M.D., Frederick, C.D., Jaiswal, M.K., Singhvi, A.K. 2003. Investigations into the potential effects of pedoturbation on luminescence dating. *Quaternary Science Reviews* 22, 1169–1176.

DeBey, T.M., Roy, B.R., Brady, S.R. 2012. The U.S. Geological Survey's TRIGA® reactor. *U.S. Geological Survey Fact Sheet 2012–3093.*

Berger, G.W. 1990. Effectiveness of natural zeroing of the thermoluminescence in sediments. *Journal of Geophysical Research* 95, 12375–12397.

Berger, G.W. 2008. Dating techniques, luminescence. In Gornitz, V (ed). *Encyclopedia of Paleoclimatology and Ancient Environments Springer*, Springer Science and Business Media.

Berger, G.W., Pillans, B.J., Palmer, A.S. 1992. Dating loess up to 800 ka by thermoluminescence. *Geology* 20, 403–406.

Berger, G.W., Henderson, K.T., Banerjee, D., Nials, F.L. 2004. Photonic dating of prehistoric irrigation canals at Phoenix, Arizona, U.S.A. *Geoarcheology* 19, 1–19.

Berger, G.W. and Kratt, C. 2008. LED laboratory lighting. *Ancient TL* 26 (1), 11–14.

Bøtter-Jensen, L. 2000. Development of Optically Stimulated Luminescence Techniques using Natural Mineral and Ceramics, and their Application to Retrospective Dosimetry. Riso-R-1211(EN), DSc Thesis, Riso National Laboratory.

Bøtter-Jensen, L., Poolton, N.R.J., Willumsen, F., Christiansen, H. 1994. A compact design for monochromatic OSL measurements in the wavelength range 380–1020nm. *Radiation Measurements* 23, 519–522.

Bøtter-Jensen, L., Bulur, E., Duller, G.A.T., Murray, A.S. 2000. Advances in luminescence instrument systems. *Radiation Measurements* 32, 523–528.

Bøtter-Jensen, L., Andersen, C.E., Duller, G.A.T., Murray, A.S. 2003. Developments in radiation, stimulation and observation facilities in luminescence measurements. *Radiation Measurements* 37, 535–541.

Bøtter-Jensen, L., McKeever, S.W.S., Wintle, A.G. 2003. *Optically Stimulated Luminescence Dosimetry.* Elsevier, Amsterdam.

Boyle, R. 1664. *Experiments and Considerations upon colours with observations on a Diamond that shines in the dark.* Henry Herringham, London.

Brill, D., Klasen, N., Jankaew, K., Brückner, H., Kelletat, D., Scheffers, A., Scheffers, S. 2012. Local inundation distances and regional tsunami recurrence in the Indian Ocean inferred from luminescence dating of sandy deposits in Thailand. *Natural Hazards and Earth System Science* 12, 2177–2192.

Bronk Ramsey, C., Staff, R.A., Bryant, C.L., Brock, F., Kitagawa, H., van der Plicht, J., Schlolaut, G., Marshall, M.H., Brauer, A., Lamb, H.F., Payne, R.L., Tarasov, P.E., Haraguchi, T., Gotanda, K., Yonenobu, H., Yokoyama, Y., Tada, R., Nakagawa, T., 2012. A complete terrestrial radiocarbon record for 11.2 to 52.8 kyr B. P. *Science* 338 (6105), pp. 370–374. DOI: 10.1126/science.1226660

Brown, N.D. and Forman, S.L. 2012. Evaluating a SAR TT-OSL protocol for dating fine-grained quartz within Late Pleistocene loess deposits in the Missouri and Mississippi river valleys, United States. *Quaternary Geochronology* 12, 87–97.

Buylaert, J.P., Murray, A.S., Thomsen, K.J., Jain, M. 2009. Testing the potential of an elevated temperature IRSL signal from K-feldspar. *Radiation Measurements* 44, 560–565.

Buylaert, J. P., Jain, M., Murray, A.S., Thomsen, K. J., Thiel, C., Sohbati, R. 2012. A robust feldspar luminescence dating method for Middle and Late Pleistocene sediments. *Boreas* 41, 435–451.

Buylaert, J.P., Murray, A.S., Gebhardt, A.C., Sohbati, R., Ohlendorg, C., Thiel, C., Wastegard, S., Zolitschka, B. 2013. Luminescence dating of the PASAFO core 5022-1D from Laguna Potrok Aike (Argentina) using IRSL signals from feldspar. *Quaternary Science Reviews* 7, 70–80.

Bunker, C.M. and Bush, C.A. 1966. Uranium, thorium, and radium analyses by gamma-ray spectrometry (0.184–0.352 million electron volts) in Geol. Survey Research, 1966. *U.S. Geological Survey Professional Paper* 550-B, p. B176–B181.

Bunker, C.M. and Bush, C.A. 1967. A comparison of potassium analyses by gamma-ray spectrometry and other techniques, in Geol. Survey Research 1967. *U.S. Geol. Survey Professional Paper* 575-B, pp. B164–B169.

Burow, C. 2018. Calc_CosmicDoseRate(): calculate the cosmic dose rate. Function version 0.5.2. In: Kreutzer, S., Burow, C., Dietze, M., Fuchs, M.C., Schmidt, C., Fischer, M., Friedrich, J. *Luminescence: Comprehensive Luminescence Dating Data Analysis*. R package version 0.8.2. https://CRAN.R-project.org/package=Luminescence

Colarossi, D., Duller, G.A.T., Roberts, H.M., Tooth, S., Lyons, R. 2015. Comparison of paired quartz OSL and feldspar post-IR IRSL dose distributions in poorly bleached fluvial sediments from South Africa. *Quaternary Geochronology* 30 (Part B), 233–238.

Crookes, J.N., Rastin, B.C. 1972. An investigation of the absolute intensity of muons at sea-level. *Nuclear Physics* B 39, 493– 508.

Castelvecchi D. 2017. High-energy cosmic rays come from outside our Galaxy. *Nature* Sep 21; 549(7673), 440– 441. doi: 10.1038/nature.2017.22655.

Daniels, F., Boyd, C.A., Saunders, D.F. 1953. Thermoluminescence as a research tool. *Science* 117, 343–349.

Darville, C. M. 2013. Cosmogenic nuclide analysis. *British Society for Geomorphology*, Geomorphic Techniques Chapter 4, Sec 2.10, 1–25.

Dreimanis, A., Hütt, G., Raukas, A., Whippey, P.W. 1978. Dating methods of Pleistocene deposits: Thermoluminescence. *Geoscience Canad*a 5, 55–60.

Dickin, A.P. 2005. U-series dating. *Radiogenic Isotope Geology*, Cambridge University Press, 324–352.

Duller, G.A.T. 1991. Equivalent dose determination using single aliquots. *Nuclear Tracks and Radiation Measurements* 18, 371– 378.

Duller, G.A.T. 1995. Luminescence dating using single aliquots: methods and applications. *Radiation Measurements* 24, 217– 226.

Duller, G.A.T. 2007. Assessing the error on equivalent dose estimates derived from single aliquot regenerative dose measurements. *Ancient TL* 25, 15–24.

Duller, G.A.T. 2008. Luminescence Dating: Guidelines on Using Luminescence Dating in Archaeology. English Heritage, Swindon.

Duller, G.A.T. 2008. Single-grain optical dating of Quaternary sediments: why aliquot size matters in luminescence dating. *Boreas* 37, 589–612.

Duller, G.A.T. 2015. Luminescence dating. In Rink, J.W., Thompson J.W. (eds.), *Encyclopedia of Scientific Dating Methods*, 390–404.

Duller, G., Bøtter-Jensen, L., Murray, A., Truscott, A. 1999. Single grain laser luminescence (SGLL) measurements using a novel automated reader. *Nuclear Instruments and Methods in Physics Research Section B: Beam Interactions with Materials and Atoms* 155, 506–514.

Duller, G.A.T., Wintle and A.G. 2012. A review of the thermally transferred optically stimulated luminescence signal from quartz for dating sediments. *Quaternary Geochronology* 7, 6–20.

Durcan, J.A., King, G.E., Duller, G.A.T. 2015. DRAC: Dose Rate and Age Calculator for trapped charge dating. *Quaternary Geochronology* 28, 54–61.

Elsholtz, J.S., 1676. De Phosphoris Quatuor Observations.

Erdi-Krausz, G., Matolin, M., Minty, B., Nicolet, J.P., Reford, W.S., Schetselaar, E.M. 2003. Guidelines for radioelement mapping using gamma ray spectrometry data: also as open access e-book. (IAEA-TECDOC; Vol. 1363). Vienna: International Atomic Energy Agency (IAEA).

Erfurt, G. and Krbetschek, M.R. 2003. IRSAR – A single-aliquot regenerative-dose dating protocol applied to the infrared radiofluorescence (IR-RF) of coarse-grain K-feldspar. *Ancient TL 21*, 35.

Ferrari, F. and Szuszkiewicz, E. 2009. Cosmic rays: a review for astrobiologists. *Astrobiology* 9(4), 413–436. doi: 10.1089/ast.2007.0205.

Fetter, C.W. 1988. *Applied Hydrogeology* (2nd edition). Merrell Publishing company.

Fleming, S.J. 1966. Study of thermoluminescence of crystalline extracts from pottery. *Archaeometry* 9, 170–173.

Fleming, S.J., Moss, H.M., Joseph, A. 1970. Thermoluminescence authenticity testing of some six dynasties figures. *Archaeometry* 12, 57–68.

Forman, S.L., Pierson, J., Lepper, K. 2000. Luminescence Geochronology. In Noller, J.S., Sowers, J.M., Lettis, W.R. (eds) *Quaternary Geochronology: Methods and Applications.* American Geophysical Union, Washington, DC.

Gaines, R.V., Skinner, H.C.W., Foord, E.E., Mason, B., Rosenzweig, A., King, V.T., Dowty, E. 1997. *Dana's New Mineralogy*, Eighth edition, New York, John Wiley and Sons.

Galbraith, R.F. (1988a). Graphical display of estimates having differing standard errors. *Technometrics* 30 (3), 271–281.

Galbraith, R.F. (1988b). A note on graphical presentation of estimated odds ratios from several clinical trials. *Statistics in Medicine* 7 (8), 889–894.

Galbraith, R.F. 1994. Some applications of radial plots. *Journal of the American Statistical Association* 89 (428), 1232–1242.

Galbraith, R.F., Roberts, R.G., Laslett, G.M., Yoshida, H., Olley, J.M. 1999. Optical dating of single and multiple grains of quartz from Jinmium rock shelter, northern Australia, Part I: Experimental design and statistical models. *Archaeometry* 41, 339–364.

Galbraith, R.F. and Roberts, R.G. 2012. Statistical aspects of equivalent dose and error calculation and display in OSL dating: an overview and some recommendations. *Quaternary Geochronology* 11, 1–27.

Griffiths, D. 2008. Introduction to Elementary Particles, 2nd edition, Wiley-VCH, 454 pages.

Godfrey-Smith, D.I., Huntley, D.J., Chen, W.H. 1988. Optical dating studies of quartz and feldspar sediment extracts. *Quaternary Science Reviews* 7, 373–380.

Göksu, H.Y., Fremlin, J.H., Irwin, H.T., Fryxell, R. 1974. Age determination of burned flint by a thermoluminescence method. *Science* 183, 651–654.

Gosse, J.C. and Phillips, F.M. 2001. Terrestrial *in situ* cosmogenic nuclides: theory and application. *Quaternary Science Reviews* 20, 1475–1560.

Gray, H.J. and Mahan, S.A. 2015. Variables and potential models for the bleaching of luminescence signals in fluvial environments. Quaternary International 362, 42–49. http://dx.doi.org/10.1016/j.quaint.2014.11.007

Gray, H.J., Tucker, G.E., Mahan, S.A., McGuire, C., Rhodes, E.J. 2017. On extracting sediment transport information from measurements of luminescence in river sediment. *Journal of Geophysical Research Earth Surface* 122, 654–677.

Grögler, N., Houtermans. F.G., Stauffer, H. 1958. Radiation damage as a research tool for geology and prehistory. Supplemento agli Atti del Congresso Scientifico, Sezione Nucleare, 5a Rassegna Internazionale Elettronica e Nucleare, Roma, 275–285.

Grögler, N., Houtermans, F.G., Stauffer, H. 1960. Über die Datierung von Keramik und Ziegel durch Thermolumineszenz. *Helvetica Physica Acta. 33*, 595–596.

Grün, R. 2009. The 'AGE' program for the calculation of luminescence age estimates. *Ancient TL* 27, 45–46.

Guérin, G., Mercier, N., Adamiec, G. 2011. Dose rate conversion factors: Update. *Ancient TL* 29, 5–8.

Guérin, G., Mercier, N., Nathan, R., Adamiec, G., Lefrais, Y. 2012. On the use of the infinite matrix assumption and associated concepts: a critical review. *Radiation Measurements* 47, 778–785.

Guérin, G., Frouin, M., Talamo, S., Aldeias, V., Bruxelles, L., Chiotti, L., Dibble, H.L., Goldberg, P., Hublin, J.-J., Jain, M., Lahaye, C., Madelaine, S., Maureille, B., McPherron, S.J.P., Mercier, N., Murray, A.S., Sandgathe, D., Steele, T.E., Thomsen, K.J., Turq, A. 2015. A multi-method luminescence dating of the Palaeolithic sequence of La Ferrassie based on new excavations adjacent to the La Ferrassie 1 and 2 skeletons. *Journal of Archaeological Science* 58, 147–166.

Guibert, P., Lahaye, C., Bechtel, F. 2009. The importance of U-series disequilibrium of sediments in luminescence dating: a case study at the Roc de Marsal cave (Dordogne, France). *Radiation Measurements* 44, 223–231.

Gundermann, K.-D. 2000. Luminescence in the *Encyclopaedia Britannica*, on-line version. Accessed April 2, 2018. https://www.britannica.com/science/luminescence/Luminescence-physics

Hancock, R. 2015. Neutron activation analysis. In Rink, J.W., Thompson, J.W. (eds.) *Encyclopedia of Scientific Dating Methods*, 607–608.

Harvey, E.N. 1957. *A History of Luminescence from the Earliest Times Until 1900*. The American Philosophical Society. Philadelphia, PA.

Heer, A.J., Adamiec, G., Moska, P. 2012. How many grains are there on a single aliquot? *Ancient TL* 30, 9–16.

Hülle, D., Hilgers, A., Kühn, P., Radtke, U. 2009. The potential of optically stimulated luminescence for dating periglacial slope deposits: A case study from the Taunus area, Germany. *Geomorphology* 109, 66–78.

Huntley, D.J. 2001. Some notes on language. *Ancient TL* 19, 27–28.

Huntley, D.J., Godfrey-Smith, D.I., Thewalt, M.L.W. 1985. Optical dating of sediments. *Nature* 313, 105–107.

Huntley, D.J. and Lamothe, M. 2001. Ubiquity of anomalous fading in K-feldspars and the measurement and correction for it in optical dating. *Canadian Journal of Earth Science* 38, 1093–1106.

Huntley, D.J. and Baril, M.R. 2002. Yet another note on laboratory lighting. *Ancient TL* 20, 39–40.

Huntley, D. and Lian, O.B. 2006. Some observations on tunneling of trapped electrons in feldspars and their implications for optical dating. *Quaternary Science Reviews* 25, 2503–2512.

Hütt, G., Jaek, I., Tchonka, J. 1988. Optical dating: K-feldspars optical response stimulation spectra. *Quaternary Science Reviews* 7, 381–385.

Jacobs, Z., Wintle, A.G., Duller, G.A.T., Roberts, R.G., Wadley, L. 2008. New ages for the post-Howiesons Poort, late and final Middle Stone Age at Sibudu, South Africa. *Journal of Archaeological Science* 35, 1790–1807.

Kalchgruber, R., Wagner, G.A. 2006. Separate assessment of natural beta and gamma dose rates with TL from α-Al2O3:C single-crystal chips. *Radiation Measurements* 41, 154–162.

Kennedy, G.C. and Knopff, L. 1960. Dating by thermoluminescence. *Archaeology* 13, 147–148.

Knoll, G.F. 2010. *Radiation detection and measurement* (4th edition). John Wiley and Sons, Hoboken.

Kulig, G. 2005. Erstellung einer Auswertesoftware zur Altersbestimmung mittels Lumineszenzverfahren unter spezieller Berücksichtigung des Einflusses radioaktiver Ungleichgewichte in der 238U-Zerfallsreihe [Creation of a software for luminescence dating with special attention to the influence of radioactive disequilibria in the 238U decay chain] (Technische Bergakademie Freiberg, unpublished BSc thesis).

Lamothe, M. 1995. Using 600–650 nm light for IRSL sample preparation. *Ancient TL* 13, 1–4.

Lapp, T., Jain, M., Thomsen, K.J., Murray, A.S., Buylaert, J.P. 2012. New luminescence measurement facilities in retrospective dosimetry. *Radiation Measurements* 47, 803–808.

Lawson, M.J., Roder, B.J., Stang, D.M., Rhodes, E.J. 2012. OSL and IRSL characteristics of quartz and feldspar from southern California, USA. *Radiation Measurements* 47, 830–836.

Li, B., Li, S.-H. 2011. Luminescence dating of K-feldspar from sediments: a protocol without anomalous fading correction. *Quaternary Geochronology* 6, 468–479.

Li, B., Li S.-H. 2012. Luminescence dating of Chinese loess beyond 130 ka using the non-fading signal from K-feldspar. *Quaternary Geochronology* 10, 24–31.

Li, B., Jacobs, Z., Roberts R.G., Li, S.-H. 2013. Extending the age limit of luminescence dating using the dose-dependent sensitivity of MET-pIRIR signals from K-feldspar. *Quaternary Geochronology* 17, 55–67.

Lian, O.B. 2007. Luminescence dating: optically-stimulated luminescence. In Elias, S.A (ed.) *Encyclopedia of Quaternary Science*. Elsevier, Amsterdam, 1491–1505.

Liritzis, I., Singhvi, A.K., Feathers, J.K., Wagner, G.A., Kadereit, A., Zacharias, N., Li, S.H. 2013. *Luminescence Dating in Archaeology, Anthropology, and Geoarchaeology: An Overview*. Springer, London, 1–70.

McKeever, S.W.S. 1985. *Thermoluminescence of Solids*. Cambridge University Press, Cambridge.

Mejdahl, V. 1979. Thermoluminescence dating: beta-dose attenuation in quartz grains. *Archaeometry* 21, 61–73.

Murray, A.S. and Roberts, R.G. 1997. Determining the burial time of single grains of quartz using optically stimulated luminescence. *Earth Planet. Sci. Lett.* 152, 163–180.

Murray, A.S., Roberts, R.G., Wintle, A.G. 1997. Equivalent dose measurements using a single aliquot of quartz. *Radiation Measurements* 27, 171–184.

Murray, A.S. and Olley J.M. 2002. Precision and accuracy in the optically stimulated luminescence dating of sedimentary quartz: a status review. *Geochronometria* 21, 1–16.

Murray, A.S. and Wintle, A.G. 2000. Luminescence dating of quartz using an improved regenerative-dose protocol. *Radiation Measurements* 32, 57–73.

Murray, A.S. and Wintle, A.G. 2003. The single aliquot regenerative dose protocol: Potential for improvements in reliability. *Radiation Measurements* 37, 377–381.

Murray, A., Thomsen, K., Masuda, N., Buylaert, J., Jain, M. 2012. Identifying well-bleached quartz using the different bleaching rates of quartz and feldspar luminescence signals. *Radiation Measurements* 47, 688–695.

Olley, J.M., Caitcheon, G.G., Roberts, R.G. 1999. The origin of dose distribution in fluvial sediments and the prospect of dating single grains from fluvial deposits using optically stimulated luminescence. *Radiation Measurements* 30, 207–217.

Personius, S.F., Mahan, S.A. 2000. Paleoearthquake Recurrence on the East Paradise Fault Zone, Metropolitan Albuquerque, New Mexico. *Bulletin of the Seismological Society of America* 90, 357–369.

Porat, N., Amit, R., Zilberman, E., Enzel, Y. 1997. Luminescence dating of fault-related alluvial fan sediments in the southern Arava Valley, Israel. *Quaternary Science Reviews* 16, 397–402.

Porat, N., Duller, G.A., Roberts, H., Wintle, A. 2009. A simplified SAR protocol for TT-OSL. *Radiation Measurements* 44, 538–542.

Prescott J.R. and Hutton, J.T. 1994. Cosmic ray contributions to dose rates for luminescence and ESR dating: large depths and long-term time variations. *Radiation Measurements* 23, 497–500.

Prescott, J.R. and Hutton, J.T. 1988. Cosmic ray and gamma ray dosimetry for TL and ESR. *Nuclear Tracks and Radiation Measurements* 14, 223–227.

Preusser, F., Degering, D., Fuchs, M., Hilgers, A., Karereit, A., Klasen, N., Kbrbetschek, M., Richter, D., Spencer, J.Q.G. 2008. Luminescence dating :basics, methods and applications. *Eiszeitalter und Gegenwart* 57, 95–149.

Preusser, F., Muru, M., Rosentau, A. 2014. Comparing different post-IR IRSL approaches for the dating of Holocene coastal foredunes from Ruhnu Island, Estonia. *Geochronometria 41*, 342–351.

Randall, J.T., Wilkins, M.H.F. 1945. Phosphorescence and electron traps. *Proceedings of the Royal Society of London A* 184, 366–407.

Reimann, T., Thomsen, K.J., Jain, M., Murray, A.S., Frechen, M. 2012. Single-grain dating of young sediments using the pIRIR signal from feldspar. *Quaternary Geochronology* 11, 28–41.

Rengers, F., Pagonis, V., Mahan, S. 2017. Can Thermoluminescence be used to determine soil heating from a wildfire? *Radiation Measurements*, 107, 119–127.

Rhodes, E.J. 2007. Quartz single grain OSL sensitivity distributions: implications for multiple grain single aliquot dating. *Geochronometria* 26, 19–29.

Rhodes, E.J. 2011. Optically stimulated luminescence dating of sediments over the past 200,000 years. *Annual Review of Earth and Planetary Sciences* 39, 461–488.

Rhodes, E.J., Singarayer, J.S., Raynal, J-P., Westaway, K.E., Sbihi-Alaoui, F.Z. 2006. New age estimates for the Palaeolithic assemblages and Pleistocene succession of Casablanca, Morocco. *Quaternary Science Reviews* 25, 2569–2585.

Richter, D. 2016. Chronostratigraphy. In Gilbert, A. S. (ed.), *Encyclopedia of Geoarchaeology (Encyclopedia of Earth Sciences Series)*, Springer, Netherlands, 139–141.

Richter, D., Richter, A., Dornich, K. 2015. Lexsyg smart: a luminescence detection system for dosimetry, material research and dating application. *Geochronometria* 42, 202–209.

Rink, W.J. and Thompson, J.W. 2015. Encyclopedia of scientific dating methods. Springer, Dordecht, p978.

Rittenour, T.M., 2008. Luminescence dating of fluvial deposits: applications to geomorphic, palaeoseismic and archaeological research. *Boreas* 37, 613–635.

Roberts, R., Yoshida, H., Galbraith, R., Laslett, G., Jones, R., Smith, M. 1998. Single-aliquot and single-grain optical dating confirm thermoluminescence age estimates at Malakunanja II rock shelter in northern Australia. *Ancient TL* 16, 19–24.

Roberts, R.G. and Jacobs, Z. 2015. Luminescence dating, single-grain dose distribution. In Rink, J.W., Thompson J.W. (eds.), *Encyclopedia of Scientific Dating Methods,* 435–440.

Rodnight, H. 2008. How many equivalent dose values are needed to obtain a reproducible distribution? *Ancient TL* 26, 3–9.

Rumble, J.R. 2017. *Handbook of Chemistry and Physics* (98th edition). CRC Press

Sanderson, D.C.W, Murphy, S. 2010. Using simple portable OSL measurements and laboratory characterisation to help understand complex and heterogeneous sediment sequences for luminescence dating. *Quaternary Geochronology* 5, 299–305.

Singhvi, A.K., Bronger, A., Sauer, W., Pant, R.K. 1989. Thermoluminescence dating of loess–paleosol sequences in the Carpathian basin (East-Central Europe): a suggestion for a revised chronology. *Chemical Geology: Isotope Geoscience Section* 73, 307–317.

Smedley, R.K., Duller, G.A.T., Roberts, H.M. 2015. Bleaching of the post-IR IRSL signal from individual grains of K-feldspar: implications for single-grain dating. *Radiation Measurements* 79, 33–42.

Sohbati, R., Murray, A., Lindvold, L., Buylaert, J-P., Jain, M. 2017. Optimization of laboratory illumination in optical dating. *Quaternary Geochronology* 39, 105–111.

Spooner, N. 1992. Optical dating: preliminary results on the anomalous fading of luminescence from feldspars. *Quaternary Science Reviews* 11, 139–145.

Spooner, N. 1994a. On the optical dating signal from quartz. *Radiation Measurements* 23, 593–600.

Spooner, N. 1994b. The anomalous fading of infrared-stimulated luminescence from feldspars. *Radiation Measurements* 23, 625–632.

Spooner, N.A. and Prescott, J.R. 1986. A caution on laboratory illumination. *Ancient TL* 4, 46–48.

Sylvester, P. J. 2015. Laser ablation inductively coupled mass spectrometer (LA ICP-MS). In Rink, J.W., Thompson J.W. (eds.), *Encyclopedia of Scientific Dating Methods*, 1–2.

Taylor, S.R. and McLennan, S. M. 1995. The geochemical evolution of the continental crust. *Reviews of Geophysics* 33 (2), 241–265.

Thiel, C., Buylaert, J., Murray, A., Terhorst, B., Hofer, I., Tsukamoto, S., and Frechen, M. 2011. Luminescence dating of the Stratzing loess profile (Austria): testing the potential of an elevated temperature post-IR IRSL protocol. *Quaternary International* 234, 23–31.

Thomsen, K.J. 2015. Luminescence dating, instrumentation. In Rink, W., Thompson, J. (eds.), Earth Sciences Series. *Encyclopedia of Scientific Dating Methods*. Springer-Verlag Berlin Heidelberg.

Thomsen, K.J., Murray, A.S., Jain, M., Bøtter-Jensen, L. 2008. Laboratory fading rates of various luminescence signals from feldspar-rich sediment extracts. *Radiation Measurements* 43, 1474–1486.

Trowbridge, J. and Burbank, J.E. 1898. Phosphorescence produced by electrification. *Am J Sci. Series* 4 (5), 55–56.

Wagner, G.A. 1998. *Age Determination of Young Rocks and Artifacts.* Springer, New York.

Wagner, G.A., Krbetschek, M., Degering, D., Bahain, J-J., Shao, Q., Falguères, C., Voinchet, P., Dolo, J-M., Garcia, T., Rightmire, G.P. 2010. Radiometric dating of the type-site for *Homo heidelbergensis* at Mauer, Germany. *Proceedings of the National Academy of Sciences 107,* 19726–19730.

Wallinga, J., Murray, A., Duller, G. 2000. Underestimation of equivalent dose in single-aliquot optical dating of feldspars caused by preheating. *Radiation Measurements* 32, 691–695.

Wang, X. L., Lu, Y. C., Wintle, A.G. 2006. Recuperated OSL dating of fine-grained quartz in Chinese loess. *Quaternary Geochronology* 1, 89–100.

Wang, X.L., Wintle, A.G., Lu, Y.C. 2007. Testing a single-aliquot protocol for recuperated OSL dating. *Radiation Measurements* 42, 380–391.

Wedepohl, K.H. 1995. The composition of the continental crust. *Geochimica et Cosmochimica Acta 59,* 1217–1232.

Wiedemann, E., Schmidt, G.C. 1895. Ueber Luminescenz. Annalen der Physik und Chemie, *Berlin 54,* 604–625.

Wintle, A.G. 1973. Anomalous fading of thermoluminescence in mineral samples. *Nature* 245, 143–144.

Wintle, A.G. 1985. Stability of TL signal in fine grains from loess. *Nuclear Tracks* 10, 725–730.

Wintle, A.G. 1997. Luminescence dating: Laboratory procedures and protocols. *Radiation Measurements* 27, 769–817.

Wintle, A. G. 2008. Fifty years of luminescence dating. *Archaeometry* 50, 276–312.

Wintle, A.G. and Huntley, D.J. 1979. Thermoluminescence dating of a deep-sea core. *Nature* 279, 710–712.

Wintle, A.G. and Huntley, D.J. 1980. Thermoluminescence dating of ocean sediments. *Canadian Journal of Earth Sciences* 17, 348–360.

Wintle, A.G. and Murray, A.S. 2006. A review of quartz optically stimulated luminescence characteristics and their relevance in single-aliquot regeneration dating protocols. *Radiation Measurements* 41, 369–391.

Yi, S., Buylaert, J.-P., Murray, A.S., Lu, H., Thiel, C., Zeng, L. 2016. A detailed post-IR IRSL dating study of the Niuyangzigou loess site in northeastern China. *Boreas* 45, 644–657.

Yukihara, E.G. and McKeever, S.W.S. 2011. *Optically Stimulated Luminescence: Fundamentals and Applications.* Wiley, Sussex.

Yukihara, E.G. McKeever, S.W.S., Akselrod, M.S. 2014. State of art: optically stimulated dosimetry: frontiers of future research. *Radiation Measurements* 71, 15–24.

Zander, A., Hilgers, A. 2013. Potential and limits of OSL, TT-OSL, IRSL and pIRIR$_{290}$ dating methods applied on a Middle Pleistocene sediment record of Lake El'gygytgyn, Russia. *Climate of the Past* 9, 719–733.

Zilberman, E., Amit, R., Heimann, A., Porat, N. 2000. Changes in Holocene Paleoseismic activity in the Hula pull-apart basin, Dead Sea Rift, northern Israel. *Tectonophysics* 321, 237–252.

Zimmerman, D.W. 1967. Thermoluminescence from fine grains from ancient pottery. *Archaeometry* 10, 26–28.

Zöller, L., Stremme, H. E., Wagner, G. A. 1988. Thermolumineszenz – Datierung an Löß-Paläoboden-Sequenzen von Nieder-, Mittel- und Oberrhein. *Chemical Geology, Isotope Geoscience* 73, 39–62.

Zöller, L., Conard, N. J., Hahn, J. 1991. Thermoluminescence dating of middle Palaeolithic open air sites in the middle Rhine valley/Germany. *Naturwissenschaften* 78, 408–410.

Zöller, L. and Wagner, G.A. 2015. Luminescence Dating, History. In Rink, W., Thompson, J. (eds.), Earth Sciences Series. *Encyclopedia of Scientific Dating Methods.* Springer-Verlag Berlin Heidelberg.

2 FROM SAMPLING TO REPORTING

MARK D. BATEMAN

Geography Department, University of Sheffield, Winter St., Sheffield S10 2TN. Email: m.d.bateman@sheffield.ac.uk

ABSTRACT: So how do you collect a sample for luminescence dating without exposing it to sunlight? What is the best material to collect and how much? What other information and materials are required before a luminescence sample can be sent to a laboratory for analysis? Once results are received, how should they be reported? As explained in this chapter, getting sampling correct can have a huge impact on the resultant accuracy and precision of luminescence ages. Options on sampling from collecting samples in tubes through to coring are looked at. Guidance is set out on where within a site it is best to sample for luminescence to avoid unnecessary complexities and what other field measurements are needed. Information is given on sending samples for luminescence measurement and, finally, what key information is needed to understand the reliability of the ages and for reporting.

KEYWORDS: coring, site selection, post-depositional disturbance, dosimetry

2.1 INTRODUCTION

Sampling for luminescence has potentially a huge impact on the accuracy and precision of a luminescence age. Large age underestimates could occur where unintentional light exposure during sampling happened. Worse still, poor sampling could lead to no results where inappropriate materials or insufficient material are collected. In such instances it may be obvious that something has gone wrong and so, whilst wasting time and money, data can be excluded from interpretations. Harder to deal with are samples collected from poorly considered sites or poorly chosen stratigraphic positions within sites from which it is possible to produce luminescence ages but from which the luminescence data is equivocal as some of the underpinning assumptions upon which the ages are based have not been met. In such cases, poor choices made when sampling may lead to additional (and potentially avoidable) complexities in sample measurement and data analysis and ultimately may reduce the certainty to which events can be dated.

The starting point before sampling has to be knowing what question luminescence dating will help answer. This might sound obvious, but if sediment is collected, luminescence dating undertaken and an age produced, knowing what this number means is crucial. With the exception of rock exposure dating (see Chapter 11), luminescence dating provides the

Figure 2.1 Knowing the question. Sampling the base of the large dune will date when this large feature stopped moving, perhaps giving an indication of changing regional aridity levels. Sampling the sigmoidal dunes on the slopes or the vegetated coppice dunes in the foreground may date more recent minor dune activity.

age of resetting. For sediments, the age is a burial age, and for archaeological materials when they were heated or fired. Knowing what event is being dated and how this might relate to sediment burial or firing is important. It also might not be obvious. To give a hypothetical example in the context of dunes, increased climatic aridity may have allowed them to form and move. As they move, sediment within the dune will be overturned through time and exposed to sunlight. Every time this occurs the luminescence time clock will be reset. This situation could continue until the dunes stabilise and the sediment within the dunes are finally buried and preserved. In this example, whilst the event of interest might be when aridity caused desertification, luminescence dating will only provide the date when the dune stopped moving, which may relate to the end of desertification or latter minor reworking. For short-lived events, their starting and stabilising will be little different and may be within the reported uncertainties of luminescence ages, but in some instances there could be quite a difference in the timing between initiation and burial.

As can be seen from above, getting sampling right is really important. This chapter looks at where, what and how to sample in order to get back the best possible luminescence dating results. In doing so, it also looks at other samples and data that need to be collected at the point of sampling. Advice is given on how to get samples safely to luminescence laboratories and what to send. Finally, when luminescence ages are received, this chapter covers how to interpret the reliability of the results and the key information necessary to publish them.

2.2 SAMPLING

Sampling for luminescence knowing that the slightest exposure to light causes problems can sound daunting. Whilst it has been known for researchers to have waited for a moonless night to go out to sample under blackout covers using only specialist lighting, in most instances this is not necessary. In fact, it could be argued such a semi-blind sampling approach is more likely to lead to inappropriate samples or researchers having accidents or getting lost. It is far better to see what is being sampled to check it is of an appropriate material size and mineral composition (as per Chapter 1). Sampling can be viewed in four

Figure 2.2 Different strategies for accessing material to sample for luminescence dating. Top left: machine-dug exposure through a sandy mantle in Texas, USA with three luminescence samples shown. Top middle: a natural gully in the loess plateau, China from which extensive luminescence sampling has taken place. Top right: machine-dug exposure through flood plain sediment, Florida USA with luminescence sampling taking place. Bottom left: an exposure in a sea cliff, South Africa with a luminescence sample being taken bottom left. Bottom right: a hand-dug excavation through a shell midden, South Africa with a luminescence sample being taken from the underlying sand.

stages: accessing material, choosing where to sample within this, physically taking the luminescence sample and finally collecting other measurements and materials required to support the luminescence sample.

2.2.1 Sampling sites

The first stage of sampling has to be getting access to the material to be dated, be it quartz/feldspar, fine sand or silt. By far the best for this is using vertical exposures through sediments/sites. Naturally occurring cliffs, river banks or gully walls have been used to collect luminescence samples, as have artificial cuts and those from quarrying activities (Fig. 2.2). There are two key advantages of utilising natural exposures. The first is the minimal labour required to make such exposures suitable for sample collection. This maybe of appreciable important with very large sites, e.g. the long luminescence dating records from the Chinese loess plateau (e.g. Li *et al*. 2016; Lu *et al*. 2004; see Chapter 5). All that is required often is removal of any vegetation and scrapping back by a few centimetres (starting at the top and moving down) the entire sediment exposure which is to be sampled. This avoids accidentally sampling unwanted organic material and also material which has been disturbed and/or exposed to sunlight via transmission through sediments or via cracking. The second significant advantage is that in having a full site/sediment sequence exposed it is much easier to see the boundaries of units and how they relate to each other. It is also possible through using standard field techniques to quickly characterise each unit and make an initial interpretation of their level of preservation and depositional origin. Both have a bearing on where within an exposure should be sampled (see Section 2.2).

Where natural exposures are not available or only cover part of what needs to be dated, sampling can also be undertaken by creating vertical exposures through site excavation or hand-dug pits (Fig. 2.2). Whilst it is often tempting to limit the size of such excavations to minimise time, cost or environmental impact, it is worth bearing in mind that sampling will still benefit from the understanding of the full range of units and their relationships to each other. This may mean that the limited excavation opened up for luminescence sampling is supplemented with more wide-ranging core or test pit data at the time of sampling or from previous site work. In archaeological contexts, a good understanding of units may also allow for the more destructive luminescence sampling to be moved laterally to less sensitive areas whilst maintaining the clear association of what is being dated to the archaeological artefacts. As with all excavations, appropriate stepping down and shoring should be applied to avoid luminescence sampling dislodging sediments and causing collapses. With excavated exposures near to the natural water table it is also worth ensuring that excavations provide sufficient additional depth below where is to be sampled to create a natural sump for water, or that pumping is undertaken. This is particularly important where *in situ* dosimetry measurements using electrically driven gamma spectrometers are being taken (see Section 2.4).

If sampling of exposures is not an option or the water table too high, luminescence sampling can be carried out by coring. This allows for sampling of sediment at the bottom of lakes or oceans as well as for sampling depths beyond the limit of safe excavation. With coring, the challenge for sampling is to avoid light contamination and also avoid sampling sediment translocated up or down during coring. Coring can be undertaken with adapted

Figure 2.3 Different coring strategies for luminescence dating. Top left: the Dormer Engineering sand drill with hydraulic head, ideal for drilling through slightly damp sand and silt but not for coarser or clay-sized sediment. Top right: the luminescence sampling head of the Dormer Engineering sand drill which collects sand in a ~80 mm diameter × ~250 mm long metal tube from specified depths; in this case, sampling dune systems in Vietnam. Bottom left: the British Geological Survey's Dando Terrier percussion rig sampling for luminescence in East Anglia, UK. Bottom right: vibracoring into aluminium tubing lake marginal sediments in South Africa.

commercially available coring equipment such as the Dormer Engineering sand drill (http://dormersoilsamplers.com/) which has a luminescence sampling head (Fig. 2.3; Munyikwa *et al.* 2011). This system allows for the drilling out of a borehole to the desired sampling depth before a light tight metal sampling tube is driven in at the base of the borehole to retrieve the sediment for luminescence dating. Extensive use of this has been made luminescence sampling in Southern Africa (e.g. Burrough *et al.* 2009; Thomas *et al.* 2009) where samples have been retrieved from up to ~15 m (e.g. Telfer and Thomas 2007). With such systems, care should be taken to case the uppermost part of the borehole as it is easy when pulling the corer out or putting it in the borehole to dislodge sediment, potentially causing contamination issues. Drilling of boreholes for sampling is not possible in free-running sediment (where

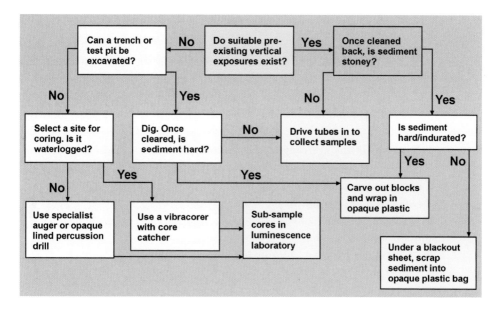

Figure 2.4 Decision-making tree to guide luminescence sampling strategy for sediment samples. Starting with the question in the orange box should lead to one of four sampling strategies.

water may be need to be pumped in) or sediment beneath the water table. Alternatively, samples can be collected by driving in metal tubing using a vibracorer (Fig. 2.3; Rittenour *et al.* 2003, Carr *et al.* 2006; Mallinson *et al.* 2011). Mechanical percussion coring has also been used for luminescence sampling using, for instance, a Dando Terrier rig (Fig. 2.3). Such systems can reach down 10–15 m but struggle beneath the water table with core retrieval and liquefaction of sediments. Luminescence samples have been retrieved from far larger drilling equipment with sampled retrieved from over 140 m below the surface in Oman by Pruesser *et al.* (2002). For systems where core casings are required, black opaque ones should be used to avoid light contamination. Beneath the water table, core catchers will be required leading to the possibility of sediment disruption on the outer edges of cored sediments. This should be factored in when subsampling for luminescence. Where coring systems retrieve opaque cores it becomes challenging to assess whether a site contains material suitable for luminescence dating, where in the core this material occurs and whether there is sufficient for a luminescence sample. Whilst doubling the work, paired coring is often useful with one set of cores either split and logged on site or collected in transparent liners for splitting and logging in the laboratory. Whatever coring system is adopted, core diameters of 80 mm or greater should be used and the laboratory undertaking the dating should be informed of the method adopted. This allows for appropriate removal and discarding of cross-contaminated and/or disturbed sediment adhering to the outer surface of the cores which retain sufficient material for dating purposes.

As shown in Figure 2.4, the sampling strategy for luminescence dating of sediment samples is very much dependent on availability of exposures, stoniness of the material, and how hard it is. In the archaeological context, added complexities may well exist with restrictions on what can be excavated and where and how destructive sampling can be.

FROM SAMPLING TO REPORTING

45

2.2.2 Where and how many to sample

Where to sample within a site clearly depends on the exact nature of the sediments, their thicknesses and their relationship to the event in question that requires dating (Fig. 2.5). The following are guidelines to ensure the appropriate samples are taken without adding unnecessary complexity to measurements and potentially diminishing the accuracy and precision of the resultant luminescence ages.

1. **Stick to within the age range of the technique.** Sampling usually will be constrained to sediment/artefacts thought to be within the age range of luminescence dating (see Chapter 1 for details). Clearly this is often not known, and the exact upper limit varies depending on the level of background radiation and or the exact luminescence measurement approach taken (see Chapter 1). Samples from near the upper limits of the technique may be worth trying to date but this must be on the basis of an increased risk of an age not being returned or an age with high uncertainties being returned.

2. **Only consider appropriate materials for the technique.** Sediment samples must contain and the appropriate minerals (feldspar/quartz) within the correct size range. This is either fine silt (4–11 μm in diameter) or more commonly fine sand (90–250 μm in diameter). Artefacts must have been fired or heated and need to be >10 mm thick and ideally at least 30 mm across. For standard luminescence dating it is best to avoid

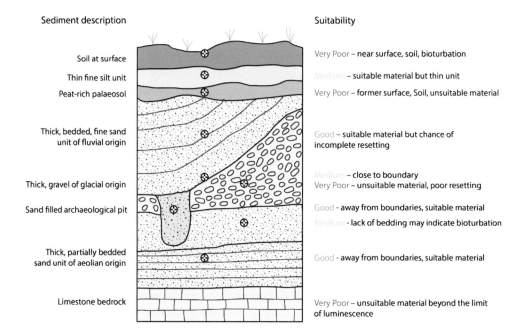

Figure 2.5 Example evaluation of luminescence sample suitability based on their position and a range of sediment types (see Chapter 1 for further details).

carbonates, organics and iron rich sediments, e.g. laterites, as these greatly extend the processing time in the laboratory. Sediment dominated by sizes or minerals other than these can still be sampled but may require larger volume samples to enable sufficient material in the appropriate size range or minerals to be extracted for measurement.

3. **Sample sediment which were reset prior to burial.** Sampling should consider which of the sediment or artefacts are most likely to have had their luminescence signal reset prior to burial. If a site has a range of archaeological artefacts, which have been fired or heat-treated to reset the luminescence in the artefacts? Fired pottery should be fine, material caught in fire may be ok, but common building materials such as brick mostly retain a signal not reset during their manufacture (Feathers *et al.* 2008). Commonly, archaeological sediments also are not well reset due not only to disturbance but also due to construction methods, e.g. buckets or baskets (e.g. Frederick and Bateman 1998). If a site contains a range of sedimentary units, based on sediment and bedding, which will come from a depositional setting more likely to have reset the grains before burial? For some sites this is straightforward, e.g. those with thick aeolian units within which sampling could take place in a range of localities. For other sites it may be less clear if sediments formed into bar top surfaces will have been reset or not. (For specific details on how different depositional settings affect luminescence dating, see Chapters 4–10)

4. **Sample away from sediment boundaries and thin sediment layers.** Sampling should consider how homogeneous and constant the background radioactivity levels are likely to have been as this is a fundamental assumption of luminescence dating. This might sound difficult to achieve when on site sampling. However, if you consider that the background radioactivity is largely a function of geology providing the radioactive minerals, distance from radioactive emitters and also sediment size, then it is reasonable to suppose that different sedimentary units are likely to have different background radioactivities. Ideally, all samples would be taken from the middle of thick (greater than 50 cm), homogeneous sediment units with a background dose rate reading taken *in situ*. Alternatively, it is just as good to sample from stacked similar sediments which have accumulated through time (e.g. Leighton *et al.* 2013). If this is the case, then radioactive variability between units is not problematic (see sample 3 in Figs. 2.6A or 2.6B). However, often the chronometric question posed is when did events start or finish. In such cases there is a tendency to want to sample close to unit boundaries. As gamma radiation is attenuated only slowly over distance, it can provide a dose 20 cm from its source (Aitken 1985). As such, by sampling at a boundary the sediment collected will have received gamma dose not only from the unit sampled but also from adjacent ones (see samples 2 and 4 in Figs. 2.6A or 2.6B).

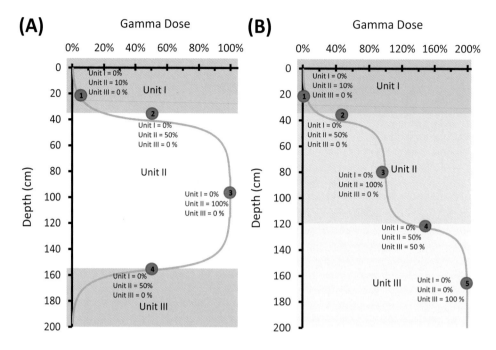

Figure 2.6 The blue line shows the effects on the gamma field moving across from one sediment unit to another. A) In this scenario Units I and III are radioactively inert. Unit II has an average environmental dose. B) In this scenario Unit I is inert and Unit III has twice the dose of Unit II. Based on data from Aitken (1985) with the blue line assuming a weighted average based on 20% dose coming from potassium, 50% from thorium and 30% from uranium.

One way round this is to avoid sampling at boundaries, as this avoids the above complexities in the background dose rate. Given, in most cases sedimentation rates are relatively high unless the samples for dating are very young, moving the sampling away from the boundaries will not change the age significantly when the uncertainties associated with luminescence dating are factored in. A second alternative is to use a portable luminescence reader to rapidly assess relative ages down profile and where temporal hiatuses are (e.g. Bateman *et al.* 2014). A third alternative, if sampling at boundaries is critical or the unit in question is thin, is to measure the background gamma dose rate *in situ* and base the beta dose rate only on the sediment sampled (beta dose attenuates to zero within ~2 mm; (Aitken 1985). In doing this, even if the adjacent unit has a different radioactivity this will have been taken into consideration for the purposes of luminesce age calculation.

5. **Sample away from the present-day and former land surfaces**. Distance from the present and former land-surfaces affects the background dose rate. This applies to sediment that is at or was formerly within ~20 cm of the surface, which received an enhanced cosmic component (both hard and soft muons) for which it is difficult to accurately calculate a dose rate (Prescott and Hutton 1994). Unless the sediment is in a very low dose rate environment and at high altitude this will probably only be a small inaccuracy

within the annual dose rate. Former land surfaces are more problematic in that the sediment near them will only have been receiving background gamma radioactivity from the unit it is in for part of its burial time and then from this and the overlying unit for the rest of its burial time. Basing the gamma dose rate just on the sampled unit or from the sampled unit and that above it will both be wrong. Unless the time and duration of the palaeosurface is known, it is difficult to correct for the change in dose rate and so if left uncorrected the resultant luminescence ages may be highly inaccurate.

A number of other issues should also be considered before sampling.

Post-depositional disturbance. A lack of bedding structures in near-surface or formerly near-surface unconsolidated sediments may be an indication that post-depositional disturbance by, for example, insects (bioturbation) or freeze–thaw (cryoturbation) has taken place (e.g. Bateman *et al.* 2003; 2007). Processes post-depositionally moving sediment up and down as well as potentially exhuming sediment will impact on luminescence dates by incorporating older or younger sediment into a unit. Such sediment should be avoided if possible. Where they can not be, luminescence measurement at the single-grain level should be requested from the laboratory undertaking the dating.

Soils. Soils, both presently active and palaeosols, can also be considered problematic. Not only will bioturbation moving grains around have occurred in them whilst the soils were active, but soil chemistry can cause mobility (either leaching or concentration) of the more mobile radioactive elements. As explained in Chapter 1, changes to background radiation levels through time are also problematic for the derivation of accurate luminescence ages. As a result, the luminescence age may reflect parent material burial, an average residence time of grains between exhumations or a combination of both.

Cementation. If, after sediment deposition, cementation has occurred with the addition of carbonates iron, or silicates (e.g. calcretes, silcretes, ferricretes or calcites), these require further consideration before sampling. The reason for this is that in infilling pore spaces cement changes the background radioactivity received to grains. Cement will increase attenuation of radiation passing through it (compared to air in the pore spaces) and if background radioactivity levels are measured with it included, this will lead to a different dose rate than that when the uncemented sediment was deposited. Depending on whether the cement is radioactively inert or not, this dose rate will be an over- or underestimate. If cementation occurred rapidly on deposition, as argued for the Wilderness aeolianites of South Africa (Bateman *et al.* 2011), or cementation is recent then its effect on the dose rate can be corrected for either by basing the dose rate on measurements with or without the cement respectively. Where it is uncertain when within the sediment's burial history the cement formed or whether the cement formed in multiple phases, such samples are best avoided if at all possible.

Sediment subject to significant water movement. The percolation of water through sediments can lead to the selective removal of soluble uranium whilst leaving more immobile thorium. In marsh deposits, cyclical wetting–drying can also cause mobility of radium. Both can cause secular disequilibrium in the decay chains altering through time the background dose rate. Samples with a high proportion of peat or organic content should be avoided as it has been shown that mobile uranium can be concentrated within them (e.g. Frechen *et al.* 2007). Likewise, sediment directly adjacent to peats and within the range of gamma radiation emitted from them should also be avoided. Where there is evidence for significant movement of groundwater this may make establishing the palaeo-moisture value used in luminescence age calculation more difficult.

Tectonically active areas. If working in tectonically active, volcanic, hydrothermal or microcrystalline source areas, research has shown quartz in these areas to have poor luminescence properties which lead to age underestimation (e.g. Pruesser *et al.* 2006; Steffan *et al.* 2009). This is compounded in volcanic areas with volcanic glass which cannot be easily separated from quartz (it has the same density) but it not crystalline so does not behave well in terms of luminescence (Fattahi and Stokes 2003). However, often in these areas feldspars appear as good dosimeters for luminescence dating (Lawson *et al.* 2012; Chapter 9). Thus, whilst sampling should not be precluded from such areas, consultation with a luminescence specialist should take place to ensure the correct minerals are used for dating purposes.

How many samples to collect will be partly a function of time, finance and whether other independent chronological control is available (e.g. radiocarbon). It needs also to reflect both the complexity of the site, precision to which the event in question needs to be dated, and suitability of samples for luminescence dating. A single sample for luminescence is rarely considered good practice. As the likelihood of issues arising from partial resetting, getting close to the upper age range, or changes in background dose rate through time increases, so increasing the number of samples taken should be considered. Why? Taking multiple samples from the same sediment unit/archaeological event horizon should produce the same luminescence ages, and if it doesn't it allows isolation of problematic samples. Taking multiple samples down a profile allows the Law of Superposition to come into play. Thus samples from lower in the profile should either be within errors of those above (in a rapidly accumulating environment) or older. Age reversals can indicate issues arising from violation of the luminescence technique's underpinning principles, e.g. dose rate disequilibirum, poor bleaching on deposition, post-depositional disturbance, where further luminescence analysis is needed.

The above are guidelines only and are there to be ignored if the sample is critical enough, providing key information about the samples context, composition and potential problems are passed on to those undertaking the luminescence dating. In many cases, sample preparation, measurement and analysis applied to the sample can be modified to mitigate the effects of insufficient appropriate grain size, minerals or partial resetting.

2.2.3 Collecting the luminescence sample

Once a site has been accessed and cleaned back, perhaps the simplest approach for luminescence sampling of sediments is to drive, either by pushing or hammering, an opaque tube into the sediments until it is completely filled (Fig. 2.7, lower left panel). This tube can then be carefully extracted and the ends sealed with light-tight caps or tape (Fig. 2.7, middle left and top right panels). To be extra secure, all tubes should be placed in a thick opaque black plastic bag, sealed and kept out of direct sunlight or sources of radiation (e.g. portable XRF, airport X-ray scanners). Complete filling of the tube is important as clearly sediment at the tube ends gets light exposure with this method of sampling. If the tube is part filled, movement of sediment during transport back to the luminescence laboratory could mix this light-exposed material in, causing samples to give erroneously young ages. If on extraction the tube is not completely filled, it can be packed with plastic before sealing. It is worth noting that most plastic bin liners are too thin to be considered opaque and should not be used to wrap samples.

The material that the tube is composed of partly depends on the hardness of the sediment. For unconsolidated soft sediments polyvinyl chloride (PVC) tubing is fine providing it is of a dark colour (brown, grey or black) and sufficiently thick walled to prevent light penetration. Sealing a small torch into a tube and standing with it in a fully blacked out room is a quick way to see if the PVC is thick enough. These can easily be obtained from hardware stores as domestic external down piping. Whilst light and cheap to make, PVC tubing will distort, causing problems if driven into sediment which is too hard. Metal tubes are good for consolidated sediments and if thick-walled enough, a cutting edge can be ground onto the end to be driven in. Such tubing may be obtainable from car exhaust/Muffler stores. However, they are heavy and therefore may cost more to freight back to the luminescence laboratory.

Irrespective of the tube composition the diameter and length of the sampling tube needs to be tailored to the sediment size. As a guide, samples should be around 500 g of the appropriate-sized material. Typically, tubes are a minimum of 50 mm in diameter and 120 mm in length, but larger tubes may be needed where there is limited fine silt or fine sand in the material to be sampled. All tubes should be clearly labelled in large, clear lettering and noting that red ink is invisible under the lighting conditions used in luminescence laboratories (Fig. 2.7 middle right panel).

Where sediment is too hard or contains stones, it may not be possible to use tubes for sampling. In such cases, one option is to carefully scrape sediment which has not been exposed to sunlight directly into a thick opaque black plastic bag. This operation should be done under a large black tarpaulin to exclude sunlight. The person doing the sampling should instead use a dim red light (e.g. red head LED torch or rear bicycle light). LEDs should have emissions in the 590–630 nm range but if this is not possible to determine, or only white LEDs are available, filtering can use Lee 106 red plastic filter gel which is widely available. All samples collected in this way should be double-bagged and again well labelled. Such an approach is also useful in archaeological contexts, where sediment directly associated with a structure requires sampling, such as under a stone structures (e.g. Feathers 2012). Alternatively, it may be possible to carve out blocks of sediment using a cold-chisel or angle grinder (Fig. 2.7 top left and middle panels). Blocks should be at least 1000 cm^3 (10 cm × 10 cm × 10 cm). As the outer surfaces of the block will have been exposed to sunlight during sampling, care should be taken to ensure the block remains intact and without cracking. This can be achieved by first wrapping in thick black opaque plastic and foil before taping up. If the block is composed of moist and loosely consolidated material, wrapping it in tissue paper and placing it somewhere safe to allow it to dry out before wrapping it in opaque plastic for transportation can increase the cohesiveness of the block.

Where coring has taken place, splitting of cores should only be undertaken once the cores have been delivered to the luminescence laboratory. If paired coring or on-site logging has been done, the laboratory should be provided with clear details of sampling positions so that short core lengths can be sliced out to form a luminescence sample. As per above, the exact core length to be cut out will depend on the core diameter and the sediment contained within it, but as a guide 100 mm would not be unreasonable. Sediments at core ends or at the core lining may have been contaminated from other horizons or been exposed

Figure 2.7 Different sampling techniques. Top left: using a cold chisel and hammer to carve out a block of indurated sediment. Top middle, a luminescence block 10 × 15 × 20 cm. Top right: a 50 mm diameter opaque PVC tube driven into soft sediment. Upper middle left: where large volumes of material are needed, multiple tubes may be used. Upper middle right: where sediment may be heterogeneous or sample layers are thin, smaller diameter tubes may be used. In this example they are 1 cm diameter and 12 cm long. Lower middle left: for harder sediments, metal tubing can be driven in to collect a sample. Lower middle right: using the Dormer Engineering coring system to core down for a sample. Lower left and right: inserting a PVC tube into the Dormer OSL sample head to collect a sample from depth.

to light and should be avoided. Where the exact position of luminescence samples is not known, then cores will have to be split lengthwise so that the most appropriate sediments pertinent to the event to be dated can be identified. As core splitting may disturb sediment, subsampling from cores for luminescence should avoid not only the outer edges but also clean back the split surface.

In archaeological contexts, the sampling of brick can be taken using a 50 mm diameter diamond-tipped core drill, ensuring the sample extends at least 100 mm into the material (Bailiff 2007; see Chapter 10). Exposure to light of pottery sherds or heated material during excavation is not a problem, providing they are opaque. As a result, these can be collected by hand and placed in opaque black plastic bags to protect them.

2.2.4 Other material required

For each luminescence sample, a number of additional information, measurements and samples are required.

1. **Information.** To enable accurate calculation of the cosmic dose received by each sample, the site's longitude, latitude and altitude are required, as well as the sample's depth below the surface. As Figure 2.8 shows, especially for near surface samples, accurately measuring the burial depth to the nearest 10 cm is important. Where recent erosion or deposition has occurred, e.g. soil removal prior to site excavation or dumping of quarry spoil, depth measurements should be reported as the long-term depths ignoring these changes. If there is site information which indicates that burial depth has changed significantly in the past this information should also be noted. If the site is partially obscured by mountains, e.g. a rock shelter, then the site orientation and percentage of the horizon obscured should be given as this also will affect cosmogenic dose rates.

In addition, details of the sediment type, bedding and stratigraphy should also be logged and a photo record made of what was sampled from where. Particular attention

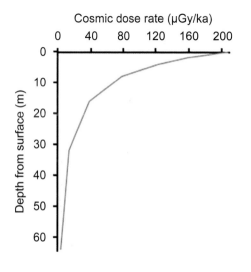

Figure 2.8 Cosmogenic dose declines with depth below surface due to shielding of overlying sediment. Data based on the algorithm of Prescott and Hutton (1994). As the biggest changes are for near-surface sediments, accurately determining the burial depth to the nearest 10 cm is important.

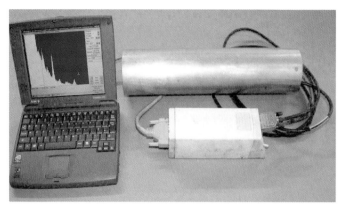

Figure 2.9 Example of a portable gamma spectrometer. Left: an EG and G Micronomad system with a typical gamma spectra for a natural sediment sample. Right: the system in use in the field with the probe inserted fully into the sediment in order to collect the full gamma field

should be made to any evidence which might indicate that present-day moisture contents of the samples may not represent long-term levels, e.g. gleying, mottling and evidence of site drainage.

2. **Measurements**. Background radioactivity measurements should be undertaken wherever there is the likelihood of a complex radiation field around a sample. This may be due to proximity of the sample to boundaries or the heterogeneity of the context it was taken from, e.g. pottery shard in a silt matrix. Two approaches to this measurement are taken. One is to bury small (10 mm) dosimeters made of artificial phosphors (one sensitive to radiation such as calcium fluoride or aluminium oxide) at each sampled point. Examples of these are produced by Landauer (http://www.landauer.co.uk/). These should be buried at least 30 cm deep so as to get the full gamma radiation field. After between a few months to a year, these capsules need to be recovered and then the background dose rate can be measured from them. Whilst causing fairly low levels of site destruction, which may be a big consideration at archaeological sites, such an approach does slow down getting the luminescence dating results and requires two site visits. The more common approach is the use of a portable gamma spectrometer (e.g. EG and G micronomad system: Fig. 2.9). These use scintillation of a sodium iodide (NaI) crystal to detect different energy gamma rays which can then, with calibration, be attributed to concentrations of K, U and Th and a dose rate determined. All that is required for measurement is that the sampled location is augered out so the probe can be inserted to a depth of 30 cm. Measurement times vary depending on levels of background radiation, but are typically between 20 and 60 minutes in length per sample.

3. **Additional samples.** Where sediment samples have not been taken from thick homogeneous sedimentary units then an additional sample for each luminescence sample collected is required upon which to base gamma dose rates. This should be taken from the same point that was sampled and be collected in sufficient quantities that it is a representative bulk sample (typically 200–500 g in weight). Once collected, it should be bagged and labelled so it is clear which luminescence sample it relates to. If sampling from core material, then it is unlikely such quantities of sediment will be available and it is possible to base beta dose rates on much smaller quantities of material by using inductively coupled plasma mass spectrometry techniques (ICP-MS). For archaeological artefacts, collection of a bulk sample from the sediments they were found in is always required. Where samples are from thin units where gamma radiation from adjacent sediments may contributed to the dose rate received by a sample (Fig. 2.6) and no *in situ* gamma dose measurements were made, then further sampling of adjacent sediments is required. These should be similar in size to the beta dose rate sample but clearly labelled showing their stratigraphic relationship to the luminescence sample, i.e. unit above, unit below. All units within 30 cm above and below the sample should be subsampled and bagged separately.

The exact field equipment needed for luminescence sampling will vary greatly depending on how the site is being accessed, and the hardness, size and stoniness of the sediments, but Figure 2.10 provides a general check list of equipment which is often taken for sampling.

2.3 SENDING A SAMPLE

Before sending samples for luminescence dating, contact should be made with the appropriate laboratory to ensure they have capacity to take them and to arrange what to send, when and costs. Before packaging samples make sure the samples have unique and clear labels (not OSL1, OSL2...) which will be able to be read under the darkroom lighting conditions of the luminescence laboratory. Use permanent black or dark coloured pens (not red). Including an itemised list of what is contained within the package is good practice. All luminescence samples should be double-bagged in thick black plastic and labelled to indicate the contents should not be exposed to light as they contain light-sensitive materials. Other samples can be left in transparent bags for visual inspection.

In terms of paper work to accompany the samples, to allow appropriate sample preparation and measurement approaches, it is useful to include details of sample sediment size (silt, sand etc.), whether it contains carbonates or organics, and notes, photos and drawing indicating depositional interpretations and relationship of samples to one another. If the samples are on the upper or lower limits of the luminescence dating technique then this should also be communicated to the luminescence laboratory making the measurements. Often all the above information is requested via a sample submission form provided by the laboratory (Fig. 2.11).

Figure 2.10 Luminescence sampling equipment checklist.

Luminescence Sampling Equipment Check List

- Trowel - for scraping back sediments
- Tape measure for sample depth and unit thickness
- GPS/Map to establish site co-ordinates
- Sampling tubes - opaque, metal or plastic with end caps
- Portable gamma spectrometer/dosimeter
- Duct tape for sealing tubes
- Hammer for driving tubes into sediment
- Metal or wooden block to protect tube end when hammering
- Sealable clear plastic bags for dosimetry/moisture samples
- Black, thick plastic bag to wrap tubes in
- Permanent marker pen to label samples

Figure 2.11 Example sample submission sheet detailing the range of information required in addition to the samples.

Luminescence Submission Sheet

The University Of Sheffield.

Sheffield Luminescence Laboratory

Contact Details

Name:		Address:
Title		
Tel:		
E-mail:		

Site Details

Site Name:	
Longitude: (Degrees, Minutes)		Latitude: (Degrees, Minutes)	
Total Number of OSL samples in batch:		Altitude (m):	

Sample Details (fill out this part for each sample)

Field Code	
Sediment Size		Sediment Contamination	
Sampling style		Details:
Is there any reason why the present-day sample moisture should not be used?		none
Details:		
Have stratigraphic site details, photos and the relationship of this sample to any other OSL samples been provided?		Yes

Notes:

* Depth is from present-day surface (to within 10 cm) unless recent made ground or erosion in
Longitude and Latitidue should not be degrees with decimal places but Degrees and Minutes

Print and send with samples
One form per sample required.

In terms of sending samples, overland is better where possible, as it avoids airport security X-raying and additional cosmogenic dose from travelling at high altitude if air-based carriage is used. However, for all but the very young samples, the additional dose received during flying will be negligible. Given the time and money often invested in collecting samples, use of a reputable freight courier is advised. When shipping samples across national borders, if the replacement costs of resampling are declared this may incur import tax on the receiving end. Given that often the value of the sediment is negligible, clearing customs is faster if no commercial value is declared. If sending samples internationally, make sure all export/import regulations are complied with and appropriate licences are obtained. If in doubt, co-ordinate with the luminescence laboratory they are being sent to. Avoid referring to samples using the terms 'soil' or 'artefact' unless this is truly the case as these terms have clear customs implications in many countries. For most sediment samples with no archaeological or organic material, labelling them as 'inert geological material to be tested to destruction' can be helpful.

As samples cannot be opened without exposing them to light and thereby destroying them, a cover note carefully explaining what is being shipped, the research they are addressing and the people and institutions involved is a good idea. To forestall problems with customs checks, a polite request not to open without contacting either the sender or the receiving luminescence laboratory can do no harm as smugglers rarely put their name and contact details on their consignments! Finally, get the receiving laboratory to acknowledge safe receipt of the samples so that any lost samples can immediately be chased up.

2.4 REPORTING

When all the measurements and calculations are complete, the luminescence laboratory should provide the results and the all-important ages. Many laboratories will provide full details of the methods employed, the data measured and evaluate the reliability of this data based on replicate measurements. This is often detailed, containing not only information useful for publication but other contextual information which may or may not be required for supplementary information in a publication. Care should be taken to read all the provided information in order to understand whether there have been problems with the luminescence samples and what assumptions have been made, i.e. is the statistical model applied to the palaeodose replicates reasonable? This should give a better understand of the reliability of the ages. Depending on the original sampling strategy, seeing whether samples from the same unit have conformable ages or that ages increase with site depth or site stratigraphy are simple first steps in evaluating luminescence datasets. Information should also be provided in the report for each sample indicating upon how many replicates the age is based, whether the data was normally distributed around a mean and therefore considered well reset prior to burial or not. If the latter has been found, the report should detail how the effects of incomplete resetting have been mitigated either through different measurements and/or different approaches in analysing the palaeodose data. Finally, the report should flag up any issues arising from the background radioactivity data which may have affected the ages. Now is the time to raise any questions of understanding or correct any assumptions.

Table 2.1 Example table for reporting luminescence ages. Data extracts taken from Bateman *et al.* (2011).

Site/Lab code	Depth from surface (m)	Water content (%)	K (%)	U (ppm)	Th (ppm)	Cosmic dose rate (Gy ka)	Total dose rate (µGy a⁻¹)	D_E (Gy)	N^1	Age (ka)
Sedgefield Ridge 1										
Shfd04277	2.2	4.2	0.12	2.66	3.26	0.155±0.008	1.057±0.050	2.46±0.04	33	2.35±0.13
Shfd04278	4.2	2.8	0.12	2.74	3.02	0.121±0.006	1.038±0.051	3.14±0.08	23	3.02±0.17
Shfd04279	5.2	4.8	0.12	2.41	5.44	0.106±0.005	1.153±0.056	3.36±0.08	23	2.91±0.16
Shfd04280	6.1	3.4	0.10	2.74	3.49	0.096±0.005	0.993±0.048	2.52±0.07	31	2.54±0.14
Shfd04281	7.1	2.8	0.15	1.62	3.19	0.086±0.005	0.830±0.039	2.19±0.08	24	2.64±0.16
Shfd04282	8.1	2.8	0.13	2.70	4.92	0.077±0.004	1.127±0.056	3.05±0.09	31	2.96±0.16
Shfd04283	9.2	2.7	0.14	2.74	5.21	0.069±0.003	1.151±0.058	3.43±0.07	25	3.13±0.17
Castle Rock, Brenton-on-the-Sea										
Shfd04275	2.8	4.0	0.11	2.57	2.58	0.142±0.007	1.023±0.044	7.02±0.13	16	6.9±0.4
Kirsten Tulleken Quarry										
Shfd04271	8.1	2.4	0.10	1.97	1.80	0.076±0.004	0.766±0.040	103±3.7	15	133±9
Shfd04259	9.15	3.5	0.12	1.22	1.45	0.069±0.004	0.571±0.027	81.6±1.8	13	142±7
Shfd04258	13.05	5.8	0.12	1.36	1.67	0.048±0.002	0.577±0.029	88.5±3.1	12	149±9
Shfd04257	16.15	4.4	0.14	0.95	1.68	0.037±0.002	0.507±0.024	68.1±1.3	14	129±7

[1] N is the number of replicate aliquots measured for a given age or number of grains if measured at the single-grain level.

In terms of what to publish, the amount of luminescence detail will obviously reflect publication type, be it site assessment report, PhD or journal. It should also reflect any complexities encountered when measuring and calculating the ages. The following should be seen as a minimum guide and many luminescence laboratories can assist in writing for publication if asked. Any methods pertaining to the luminescence should state the laboratory where the samples were analysed and the mineral used (usually quartz or feldspar). In terms of measurement, the method of luminescence stimulation (e.g. OSL/IRSL/TL) and the aliquot size (single grain, small aliquot or standard 9.6 mm diameter aliquots) should be included. For most measurements, the palaeodose (D_E) will have been measured using the single-aliquot regeneration (SAR) protocol (Murray and Wintle 2003) so this, or the alternative method used, should be briefly outlined. With SAR, the preheat used should have been optimised for each site and this should be stated, as should the results from any dose-recovery experiments which prove the method is working for the samples in question. In terms of dose rate measurement, the method used to measure this should be stated (e.g. field gamma spectroscopy, ICP analysis, thick source beta counting) as should the palaeo-moisture values used for attenuation. In terms of D_E analysis, many replicates for each sample will have been undertaken so the method used to derive a single value for age calculation purposes should be stated (e.g. CAM, see Chapter 1 for details). Where the D_E replicates are well clustered and normally distributed it should be the case that ages are based on a central or mean value. For samples where partial beaching or post-depositional disturbance has been detected, this should be reflected in the sample D_E data scatter and/or it having D_E multiple components. In such cases, ages should have been based on minimum models or single components within the dataset. For such samples, a justification of the approach adopted should be given as well. For example, the D_E data was skewed and partial bleaching suspected so ages are based on the minimum age model extracted D_E thought to contain the most fully rest sediment. The data associated with the above and the ages should all be included within the main body of the report/paper as a table. An example of such a sample is set out in Table 2.1. The data should have been provided by the luminescence laboratory to populate this table. It is worth noting at this stage that whilst there is no standardisation, the precision most laboratories quote luminescence ages to will reflect the D_E uncertainty. If this uncertainty is large then ages and their uncertainties will be rounded to reflect this. For example, the Kirsten Tullekan Quarry samples in Table 2.1 are old and reported as integers as their D_E uncertainty exceeds one gray, whereas those from Sedgefield Ridge 1 are young, with a low D_E uncertainty and therefore are reported with uncertainties to two decimal places.

Some reports and journals may require inclusion of plots of sample replicate D_E data, either as examples or for all samples. Providing the measurement of the samples was straightforward, this is perhaps better appearing in an appendix or supplementary information. The form of these plots will probably be dependent on what was supplied by the luminescence laboratory but may take the form of radial or combined probability plots (see Chapter 1).

In terms of reporting the actual ages, wherever these are used in the text they should appear both with their 1 standard deviation uncertainties and their labcode. Luminescence ages should not normally be quoted as BP as this is a radiocarbon convention with the datum set at 1950. There is no equivalent datum for luminescence ages and so all ages are usually calculated from the year of measurement. Given most luminescence ages are

100s–1000s of years old, this is not a major issue but where sample ages are less than 100 years the year of measurement should be stated in the outline of methods as this may make a difference to interpretations in the future. If ages are to be reported as BP then they should be readjusted using the appropriate time difference between luminescence measurement and 1950. Interpretations using luminescence ages should reflect that the ages are burial ages.

2.5 SUMMARY

Like most scientific research, establishing a luminescence-based chronology for past events requires investment of time and money. This investment, along with the results, can be greatly improved with careful framing of the dating question to be investigated and locating an appropriate place to sample. Even then, effective collection of luminescence samples ensuring the collection of appropriate material and size of sample as well as optimising the location of the sample is required to avoid unnecessary complications. All this has to be carried out in such a way as to avoid exposing samples to sunlight. By following the guidance laid out in this chapter, many problems can be avoided such as sampling poorly reset or post-depositional disturbed material, exposing samples during collection, mixing/breaking of samples during transportation and failure to measure or sample complex background radiation fields. Most times, involvement of a luminescence specialist prior to sampling can also be helpful.

REFERENCES

Aitken, M.J. 1985. *Thermoluminescence Dating*. London: Academic Press.

Bailiff,I.K. 2007. Methodological developments in the luminescence dating of brick from English late-medieval and post-medieval buildings. *Archaeometry* 49, 827–851.

Bateman, M.D., Frederick, C.D., Jaiswal, M.K., Singhvi, A.K. 2003. Investigations into the potential effects of pedoturbation on luminescence dating. *Quaternary Science Reviews* 22, 1169–1176.

Bateman, M.D., Boulter, C.H., Carr, A.S., Frederick, C.D., Peter, D., Wilder, M. 2007. Detecting post-depositional sediment disturbance in sandy deposits using optical luminescence. *Quaternary Geochronology* 2, 57–64.

Bateman, M.D., Carr, A.S., Dunajko, A.C., Holmes, P.J., Roberts, D.L., McLaren, S.J., Bryant, R.G., Marker, M.E., Murray-Wallace, C.V. 2011. The evolution of coastal barrier systems: a case study of the Middle-Late Pleistocene Wilderness barriers, South Africa. *Quaternary Science Reviews* 30, 63–81.

Bateman, M.D., Stein, S., Ashurst, R.A., Selby, K. 2015. Instant Luminescence Chronologies? High resolution luminescence profiles using a portable luminescence reader. *Quaternary Geochronology* 30, 141–146.

Burrough, S.L., Thomas, D.S.G., Bailey, R.M. 2009. A mega-lake in the Kalahari: A late Pleistocene record of the Palaeolake Makgadikgadi system. *Quaternary Science Reviews* 28, 1392–1411.

Carr, A.S., Thomas, D.S.G., Bateman, M.D., Meadows, M.E., Chase, B. 2006. Late Quaternary palaeoenvironments of the winter-rainfall zone of southern Africa: palynological and sedimentological evidence from the Agulhas Plain. *Palaeogeography, Palaeoclimatology, Palaeoecology* 239, 147–165.

Fattahi, M., Stokes, S. 2003. Dating volcanic and related sediments by luminescence methods: a review. *Earth-Science Reviews* 62, 229–264.

Feathers, J.K 2012. Luminescence dating of anthropogenic rock structures in the Northern Rockies and adjacent high plains, North America: A Progress Report. *Quaternary Geochronology* 10, 399–405.

Feathers, J.K., Johnson, J., Kembel, S.R. 2008. Luminescence dating of monumental stone architecture at Chavín de Huántar, Perú. *Journal of Archaeological Method and Theory* 15, 266–296.

Frechen, M., Sierralta, M., Oezen, D., Urban, B. 2007. Uranium-series dating of peat from central and Northern Europe Developments. *Developments in Quaternary Science* 7, 93–117.

Frederick, C.D., Bateman, M.D. 1998. The potential applications of optical dating to the sandy uplands of east Texas and northwest Louisiana. *Journal of North-east Texas Archaeology* 11, 133–147.

Lawson, M.J., Roder, B.L., Stang, D.M., Rhodes, E.J. 2012. OSL and IRSL Characteristics of quartz and feldspar from Southern California, USA. *Radiation Measurements* 47, 830–836.

Leighton, C.L., Bailey, R.M., Thomas, D.S.G. 2013. The utility of desert sand dunes as Quaternary chronostratigraphic archives: evidence from the northeast Rub' al Khali. *Quaternary Science Reviews* 78, 303–318.

Li, Y., Song, Y., Zhongping, L. 2016. Rapid and cyclic dust accumulation during MIS 2 in Central Asia inferred from loess OSL dating and grain-size analysis. *Scientific Reports* 6, No. 32365.

Lu, H.Y., Wang, X.Y., Ma, H.Z., Tan, H., Vandenberghe, J., Miao, X., Li, Z., Sun, Y., An, Z., Cao. G. 2004. The Plateau Monsoon variation during the past 130 kyr revealed by loess deposit at northeast Qinghai-Tibet (China). *Global and Planetary Change* 41, 207–214.

Mallinson, D.J., Smith, C.W, Mahan, S., Culver, S.J., McDowell, K. 2011. Barrier Island response to Late Holocene climate events, North Carolina, USA. *Quaternary Research* 76, 46–57.

Munyikwa, K., Telfer, M.W., Baker, I., Knight, C. 2011. Core drilling of Quaternary sediments for luminescence dating using the Dormer Drillmite. *Ancient TL* 29, 15–24.

Murray, A.S., Wintle, A.G., 2003. The single aliquot regenerative dose protocol: potential for improvements in reliability. *Radiation Measurements* 37, 377–381.

Prescott, J.R., Hutton, J.T. 1994. Cosmic ray contributions to dose rates for luminescence and ESR dating: large depths and long-term variations. *Radiation Measurements* 23, 497–500.

Preusser, F., Radies, D. Matter, A., 2002. A 160,000-year record of dune development and atmospheric circulation in southern Arabia. *Science* 296, 2018–2020.

Preusser, F., Ramseyer, K., Schlüchter, C. 2006. Characterization of low OSL intensity quartz from the New Zealand Alps. *Radiation Measurements* 41, 871–877.

Rittenour, T.M, Ronald J., Goble, R.J., Blum, M.D. 2003. An optical age chronology of late Pleistocene fluvial deposits in the northern lower Mississippi valley. *Quaternary Science Reviews* 22, 1105–1110.

Steffan, D., Preusser, F., Schlunegger, F. 2009. OSL quartz age underestimation due to unstable signal components. *Quaternary Geochronology* 4, 353–362.

Telfer, M.W. and Thomas, D.S.G. 2007. Late Quaternary linear dune accumulation and chronostratigraphy of the southwestern Kalahari: implications for aeolian palaeoclimatic reconstructions and predictions of future dynamics. *Quaternary Science Reviews* 26, 2617–2630.

Thomas, D.S.G., Bailey, R., Shaw, P.A., Durcan, J.A., Singarayer, J.S. 2009. Late Quaternary highstands at Lake Chilwa, Malawi: frequency, timing and possible forcing mechanisms in the last 44 ka. *Quaternary Science Reviews* 28, 526–539.

3 INCORPORATING LUMINESCENCE AGES INTO CHRONOMETRIC FRAMEWORKS

LAINE CLARK-BALZAN

Institute of Earth and Environmental Sciences, University of Freiburg, Alberstr. 23-B, Freiburg im Breisgau, 79104, Germany. Email: l.clarkbalzan@gmail.com

ABSTRACT: New statistical methods such as applied Bayesian inference, the proliferation of chronometric information, and the development of user-friendly age modelling software have led to a significant rise in the number of chronological frameworks published over the past 20 years. Though originally confined primarily to techniques such as radiocarbon dating, age models increasingly incorporate luminescence ages, and both site-specific primary data and regional meta-analyses can benefit from such approaches. This chapter discusses the benefits of Bayesian statistical inference for creating chronological frameworks, issues arising from the use of luminescence ages in particular, and practical steps including framework design, legacy data quality assessment, and model construction.

KEYWORDS: Bayesian modelling, data quality, hypothesis testing

3.1 INTRODUCTION

Computational advances and the development of numerical simulations have enabled increasingly sophisticated analyses of data in a wide variety of fields. Many applications in the sciences and social sciences have benefitted from these developments, particularly the implementation of Bayesian inference. Bayesian methods give explicit instructions for adapting hypotheses in light of new evidence via a simple mathematical relationship (Bayes' Theorem). Bayes' Theorem was originally presented to the Royal Society due to its relevance to the burgeoning field of experimental philosophy (Bayes 1763), but the mathematics were too complex for exact calculation except in a small subset of cases. Interest in Bayesian inference has grown substantially post-World War II, with the concomitant developments of computing and numerical calculation methods such as Monte Carlo simulation (see Fienberg (2006) for more details).

Within such fields as archaeology, palaeoecology, and palaeoenvironmental reconstruction, Bayesian inference has become particularly influential in chronological analysis. Adopted first due to its application to radiocarbon calibration (Steel 2001), it has since become more generally applied to the creation of chronometric frameworks and quantification of uncertainty for inferred events (Bronk Ramsey 2009a). Typically, Bayesian inference will be used to combine prior knowledge concerning the relative order of events, often stratigraphic units within an archaeological site or sample depths within a sediment core, with new information provided by chronometric techniques in order to provide an updated estimate for the chronology of some event. This type of procedure provides multiple benefits (Millard 2008; Parnell *et al.* 2008; Rhodes *et al.* 2003), which include:

1. Forcing assumptions about likely chronology to be made explicit
2. Incorporation of age data from multiple chronometric techniques
3. Increasing precision in age estimates where a sufficient number have been assessed (so that their uncertainties overlap)
4. Quantifying uncertainty for indirectly dated events, which may be bracketed by or occur within a unit with multiple dated samples.

Primarily luminescence age-based frameworks are not yet as advanced as those that depend solely or primarily upon radiocarbon data (Higham *et al.* 2014), due to differences in the precision and derivation of the ages. Nevertheless, interest is growing in the use of Bayesian statistical frameworks that incorporate luminescence ages, due to the long time range covered by the technique, ubiquity of appropriate samples (mineral grains), and direct dating of sediment deposition events. Luminescence age data has already been incorporated into a number of informative chronometric frameworks at both local and regional scales. At a site-specific and local level, Bayesian methods may increase the precision of luminescence ages in order to allow stronger comparisons between climatically driven ecological changes and human behaviours such as prey choice (Discamps *et al.* 2011; Guibert *et al.* 2008) or marine resource use (Veth *et al.* 2017). They can be used to hypothesise links between human landscape use and fluvial evolution (Brown 2008), or improve historical chronologies with current legal ramifications, such as the history of water rights usage by native peoples in the western United States (Huckleberry *et al.* 2016). On the other end of the scale, data mining and meta-analyses – large and systematic 'secondary' analyses which compare data gathered and analysed by multiple 'primary' sources (Glass 1976) – are increasingly utilised to create chronometric frameworks and test hypotheses across regional or continental scales. Incorporating such data may be difficult, as it requires explicit and critical analysis of the quality of all relevant dating studies, but this approach is more likely to illuminate underlying trends that could otherwise be masked by local variation (Bailey and Thomas 2014). Incorporation of luminescence-based chronologies for maar lake cores containing pollen proxies into a continental database and model for pollen dating records, for example, is expected to significantly enhance systematic climate response comparisons across Central and South America (Flantua *et al.* 2016).

This chapter will discuss the process of building chronological frameworks on several scales. The use of 'dates as data' via 'stacking' methods, i.e. assessment of peaks in histograms of dates or summed probability density functions, is not addressed. Instead, this chapter focuses upon chronological models utilising Bayesian inference to rigorously assess the most likely dating of a particular event of interest. Sections 3.2 and 3.2.1 cover important background issues, including data transmission and loss through geological time, and difficulties arising from the incorporation of luminescence ages into chronometric frameworks. Section 3.2.2 comprises a gentle introduction to Bayes' Theorem and Bayesian inference, as well as a discussion of some of the impediments to treating luminescence ages in a robust statistical framework. Given the number of user-friendly Bayesian programs now available, a full understanding of the mathematical underpinnings of Bayesian statistics is not necessary to build robust models. However, it is useful to understand the philosophical basis of the methods and particularly the pitfalls which can affect the use of luminescence ages in these. Section 3.3 covers model design and approaches to compilation and analysis of previously published data, and Section 3.4 discusses the practical process of building a model, including constraints, input of ages, outlier analysis, and querying the model. Finally, in Section 3.5, case studies are considered that highlight the potential for high precision, high resolution dating at a single site (Riwi Cave, Australia), and meta-analyses of sometimes poorly constrained but high value published data (African and Near Eastern hominin fossil chronology) along with project design to maximise the information derived from new, primary studies (BRITICE-CHRONO and the British–Irish Ice Sheet chronology).

3.2 CHRONOLOGICAL FRAMEWORKS

Chronometric data is gathered and analysed so that inferences can be made about the timing and (potentially) spatial distribution of some past event. This process is examined here as it relates to luminescence dating of sediment, but the general principles of information recording and loss are applicable to other sorts of dating as well. For instance, one could imagine a similar set of stages that describe archaeological inferences based on thermoluminescence dating of pottery from museum collections (Zink and Porto 2005). Bailey and Thomas (2014) offer a useful discussion of this framework as it pertains to dryland dune deposition, and Parnell *et al.* (2008) describe pollen-based environmental inferences in core records.

To begin, an event of interest occurs at a particular point in space during a certain span of time. Such an event may be a transition between stable states, e.g. Middle/Upper Palaeolithic archaeological transition, or a 'short-lived excursion from the longer-term mean state of some component(s) of the system,' e.g. a tephra layer (Parnell *et al.* 2008, p. 1873). Often this event will occur in multiple locations, either synchronously or with some delay between each occurrence. At some or all of these locations, evidence for the event will be recorded, whether directly via sediment deposition (such as a tephra layer or flood deposit) or indirectly, via co-deposition of proxies/fossils/lithics and sediment. This is the first stage in which 'noise' can affect the event 'signal': there may be random or systematic variation in each location's response time to the underlying event. Evidence for the event must then survive over geological timescales in order to be measured, but taphonomic processes may result in direct alteration/decay processes that result in loss of the event marker (e.g. pollen decay) or the disturbance of the sedimentological record which contains it (if applicable). This latter process can include erosive processes and mixture of sediment layers via bio-, pedo-, or cryoturbation (see Chapter 2). Over time, the combination of such processes may yield very complex depositional records, such as those found in densely and repeatedly utilised archaeological cave sites (Hunt *et al.* 2015).

At the time of sampling, site records can be considered fixed, and information must then be extracted from these samples via scientific analyses.

Measurement and inference involves two pathways: first, detection and measurement of evidence for the event, and second, analysis of site chronology. Both pathways are significantly affected by study design, particularly sampling strategy and choice of measurement techniques. From the first set of data, a measurable quantity must be defined to indicate an observation of the underlying event of interest, and from the second set, we must infer an age for this occurrence. Uncertainty in both event and chronological observations are inherent in the measurement process and therefore unavoidable. Finally, many such records may be studied and compiled in order to increase the sample set and provide better estimators for the underlying event parameters. Specific combinations of these events may be examined to infer synchroneity, palaeoenvironmental or sedimentological forcing factors, the spread and adoption of archaeological technocomplexes, or a number of other questions about the past.

Depending upon the context, luminescence ages may offer the best option for dating the event of interest. The process of sampling, signal measurement, and final age calculation, however, introduces some method-specific difficulties in the inference stage, which are discussed next.

3.2.1 Challenges in the use of luminescence dating for chronological frameworks

3.2.1.1 Precision and percentile error

Two concerns are commonly cited regarding the inclusion of luminescence ages in chronological frameworks. Perhaps the most common is the relatively low precision of the ages, particularly in contexts where radiocarbon dates or varve counting may be applicable (Jones *et al.* 2015; Újvári *et al.* 2014). Uncertainty tends to vary between 5 and 10% for ages determined via most luminescence dating protocols and applications. Hesse (2016) reported modal coefficients of variation of approximately 8% for pools of OSL (n = 342) and TL ages (n =312) measured for Australian dune samples, while similar mean precisions of 9% are reported by Thomas and Burrough (2016) for OSL dune chronologies from southern Africa. Arnold *et al.* (2015) also report similar average relative errors of 8–9% from a compilation of several hundred dates determined via modern post-IR-IR feldspar dating protocols. Uncertainties can be much greater, as well, with high relative standard error tails in compiled values of greater than 50% (Hesse 2016; Thomas and Burrough 2016).

Consideration of the errors involved in dose rate determination led Guérin *et al.* (2013) to suggest that the minimum achievable precision is likely to be near 5%. Murray and Olley (2002) came to a similar conclusion based on a systematic literature review. They suggest all samples will share 3% uncertainty, deriving from elemental concentration conversion factors, 3% from calibration of equipment for concentration determinations, and 2% from beta dose attenuation factors, as well as 2% for beta source calibration (related to D_E determination). Likely significant systematic sources of uncertainty such as water content and determination of the cosmic ray dose rate will increase this minimum relative precision for certain samples. It is unlikely that the precision of luminescence ages will increase much beyond this range. Murray and Olley (2002) note that, while improvements

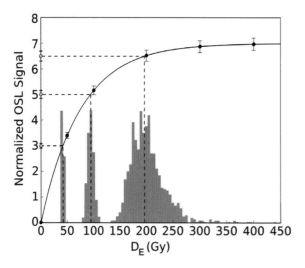

Figure 3.1 Monte Carlo-derived (1000 cycles) luminescence error distributions for three simulated D_Es from a typical quartz single-aliquot regeneration OSL growth curve; Normalised OSL signals equalling 3, 5, and 6.5 have been used, with uncertainties between 2 and 4%. Error distributions are vertically scaled to aid comparison. It is apparent that higher D_Es are associated with increasingly asymmetric uncertainties and decreasing precision. This is due to the properties of interpolation onto an exponential curve.

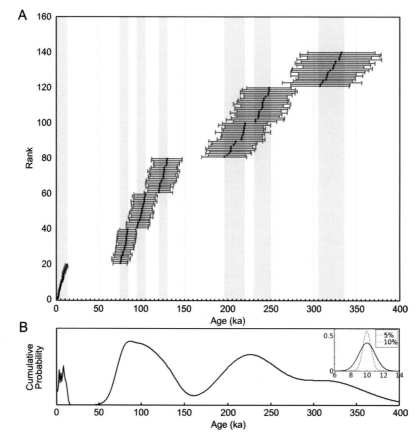

Figure 3.2 The effect of luminescence age uncertainty on event estimation: A) ten ages have been randomly selected from uniform distributions representing seven time periods (grey-shaded), with age uncertainties between 5% and 10%. B) The summed probability density function (PDF) for these ages is displayed in (B), with the inset showing two example PDFs for ages of 10 ± 0.5 ka and 10 ± 1 ka (5% and 10% uncertainty, respectively). It is clear that chronological events will become more difficult to visually distinguish as ages increase.

in D_E determination such as the single-aliquot regeneration protocol have provided D_E values with significantly improved precision, unavoidable uncertainties in standard dose rate calculations will prohibit the increase in luminescence precision beyond a certain point.

Additionally, certain properties of luminescence signal measurement and propagation of uncertainty from equivalent dose (D_E) and dose rate (D_R) quantities result in the correlation of age and uncertainty magnitude. This is described more fully by Galbraith and Roberts (2012). However, it can briefly be noted that in addition to the percentile uncertainty described above, precision often decreases as D_E increases due to the mathematics of interpolation onto an exponential curve (Murray and Funder 2003). This effect is particularly significant near signal saturation levels, as is apparent in Figure 3.1 (Duller *et al.* 2000; Duller 2012). Such effects can be limited by including ages only with D_E lower than a specific threshold (e.g. the suggestion of $2D_0$ by Wintle and Murray 2006). However, this will systematically affect the precision of estimates made via chronometric frameworks that cover age ranges of several orders of magnitude (Fig. 3.2).

3.2.1.2 Sampling bias

Biased sample coverage may occur in both temporal and spatial dimensions, for a variety of reasons. In general, targeted sampling is desired in the interests of research time and economic efficiency (see, for instance, the design of sampling strategies by BRITICE, Section 3.5.3). For meta-analyses and data-mining purposes, however, it is important to recognise such bias, and accurate inferences may depend on correcting for these. Luminescence signal properties guarantee a maximum obtainable D_E due to saturation (variable between grains/samples) and therefore age (depending upon D_R). Other important drivers of biased sampling include nonsystematic sampling and loss of datable deposits due to erosive events.

Nonsystematic sampling may be caused by a variety of factors including transportation and political concerns, research interests, or difficulty in reaching potentially more deeply buried, older deposits. The effects of such bias on age databases have been highlighted most notably by several participants in the INQUA Dunes Atlas project, which collated available chronometric data for inland and continental dune accumulation (Lancaster *et al.* 2016). Geographic bias is ubiquitous, due often to research interests of projects functioning in each area (Bristow and Armitage 2016; Thomas and Burrough 2016; Tripaldi and Zárate 2016); it may be extreme, as for luminescence chronologies from the Arabian Peninsula, for which 67% of ages are derived from UAE dunes (Duller 2016). Temporal bias of the resulting ages is similarly apparent (Li and Yang 2016; Thomas and Burrough 2016). Though some of this is due to taphonomic processes (Lancaster *et al.* 2016), Hesse (2016) showed that the upper and basal portions of Australian dunes are systematically under-sampled by normalising sampling depth by dune thickness. Within a given region, multiple dated records are necessary to average out random fluctuations in accumulation and preservation for environmental hypothesis testing. Any studies attempting to use such luminescence ages for analysis of dune accumulation through time must recognise and correct for these biases.

Taphonomic processes such as erosive events provide their own bias to the age determination process, as older sediments are less likely to be preserved for sampling. Considering the arid environments under discussion, Bailey and Thomas (2014) created a one-dimensional model of sediment surface height to investigate the effect of forcing factors including windspeed, precipitation, and sediment sources upon dune accumulation. They find a preservation bias for sediments with a relationship described approximately as $1/\sqrt{age}$. Exponential loss of sediments is also suggested by the frequency distribution of sample ages collated by the INQUA Dunes Atlas (Lancaster *et al.* 2016), with older sample ages occurring much less frequently in the database, although this is also likely to result in part due to the common undersampling of dune bases (Hesse 2016). Bias due to such relationships can be modelled. However, correlations between erosive events and particularly strong environmental drivers may complicate this process (Munyikwa 2005; Leighton *et al.* 2014).

3.2.2 Comparison of legacy luminescence ages

Even in situations where luminescence dating provides the most direct age for the event in question, it may be overlooked due to the complexity of comparing and assessing the quality of 'legacy' luminescence ages. In general, legacy data refers to data that has been collected, measured, or calculated based on methods and assumptions that have since been superseded. Due to the pace of development in luminescence dating and the assumptions which must be used to calculate D_R, any chronological framework incorporating previously published ages will need to develop guidelines for handling this data. Approaches may include clear and consistent quality control criteria combined with outlier assessment and recalculation of data (Small *et al.* 2017).

D_E measurement protocols, data handling, and supplemental quality assurance experiments for both quartz and feldspar have advanced significantly during the past several decades (see Chapter 1 for more information). The development of optically stimulated luminescence (OSL) and infrared stimulated luminescence (IRSL) (Huntley *et al.* 1985; Huntley *et al.* 1991) and single-aliquot and single-grain measurement techniques (Duller *et al.* 1999; Murray and Wintle 2000; Murray and Wintle 2003) have enabled systematic studies of processes affecting luminescence D_E distributions including partial bleaching, sediment mixing, and microdosimetric variation (Nathan *et al.* 2003; Singarayer *et al.* 2005; Mayya *et al.* 2006; Bateman *et al.* 2007; Cunningham *et al.* 2012). Concomitantly, statistical models ('age models') have been created to account for such processes in order to yield more accurate ages (Galbraith and Green 1990; Galbraith *et al.* 1999; Roberts *et al.* 2000; Guérin *et al.* 2017a). Quality assurance standards for D_E data (c.f. Clarke *et al.* 1999 and Gliganic *et al.* 2012) and growing understanding of luminescence properties – for example, signal components in quartz OSL (Steffen *et al.* 2009) and anomalous fading of feldspars (Huntley and Lamothe 2001; Auclair *et al.* 2003, Thomsen *et al.* 2008) – have also led to numerous changes in the calculation of luminescence ages over the years. Therefore, legacy ages for which D_Es have been calculated via older techniques, may not be considered as reliable as modern ages, and may be excluded from a framework entirely or incorporated as less secure data. The final approach will depend the environment under consideration. For example, a dune sample age based upon OSL D_Es measured from large multigrain aliquots is more likely to be reliable than the age of a glaciofluvial sample calculated from similar large aliquots; this is due to the probability of partial bleaching in the latter case (see Chapter 6).

D_R calculations have altered less significantly over the years. However, various quantities necessary for D_R calculation have undergone multiple revisions, including factors for conversion of radioisotope concentrations to alpha, beta, and gamma dose rates, grain size attenuation factors, and alpha-efficiency factors (see discussion and references in Durcan *et al.* 2015). Different attenuation and conversion factors will shift central ages by several percent. Compilation and comparison of legacy data is also complicated by the individual choices made during calculation of dose rates, of which the most important is average burial water content. This will be discussed further below, but it has been suggested that legacy ages are recalculated before modelling in order to reconcile such differences (Millard 2006b; Small *et al.* 2017). Otherwise, systematic differences in ages due to laboratory practice may be conflated with the underlying processes of interest.

It should be apparent that the proliferation of calculation techniques and protocols makes assessing relative data quality complex and time-intensive. Given the numerous environmental and archaeological hypotheses that can only be tested with large datasets, however, legacy data is a very valuable resource. Moreover, as will be seen in Section 3.3, progress in modelling chronological data has yielded several techniques for incorporating legacy ages in a robust manner.

3.2.2.1 Systematic vs. random error and independent vs. dependent quantities

Each measured or estimated parameter involved in the calculation of a luminescence age has an associated uncertainty; these uncertainties can be classified as random or systematic, and this classification has important ramifications for statistical treatment of luminescence ages.

Random uncertainty is the nonreproducible fluctuation in a measurement due to variations in measurement condition and, at a more fundamental level, to quantisation of the measured value (Bevington and Robinson 2003). For example, the magnitude of the luminescence signal is intrinsically variable around some average value due to the random nature of the atomic transitions. Random uncertainty in measurements of this signal combines this signal noise with measurement noise from the detector's electronics, as well as minute variations in the distance of the PMT and aliquot, differences in optical stimulation power, fluctuations in aliquot reflectivity, etc., all of which may combine to yield several percent uncertainty in replicate D_E measurements (Thomsen *et al.* 2005; Truscott *et al.* 2000). Importantly, combining multiple measurements affected only by random uncertainty will approach the true underlying value as the number of measurements increases. All quantities involved in calculating a luminescence age are affected by random error. In luminescence measurements, counting errors from measurement of luminescence signals, elemental concentrations, gamma spectrometer measurements, beta counting values, etc., can be propagated through the age calculations to give a precision-only error (Wood *et al.* 2016).

By contrast, systematic uncertainty involves an offset between the true underlying value and the measured quantity, which will not disappear as measurements are aggregated (Bevington and Robinson 2003). Rhodes *et al.* (2003) subdivide this class further, into 'unshared' (USS) and 'shared' systematic error. The former is essentially the sample-specific systematic error, while the latter is shared by all samples in the group under consideration. Systematic error may contribute to any value used in the calculation of a luminescence age. However, the values most commonly mentioned in dating papers are luminescence reader calibration (*c.* 2–3%) and average sediment water content values, which have a larger effect (Millard 2006b; Murray and Olley 2002). Other dose rate values are sometimes also suggested to suffer from a systematic offset, including cosmic dose rates, particularly in cases of unusual geometry, such as caves (Jacobs *et al.* 2015), or when overall dose rates are very low, such as the nearly pure quartz sands of the Middle Kalahari (Burrough *et al.* 2009). In general, dose rate calculations deal with this by assigning symmetric error values large enough to cover the expected spread in systematic error, but this is often not discussed explicitly. Systematic uncertainties in D_E determination tend to be assessed separately and 'corrected' as far as possible via the choice of an appropriate age model. Nevertheless, large legacy databases may include ages calculated from D_Es for which a systematic offset is suspected. For instance, older fluvial or glacial ages may be more likely to suffer from systematic offsets due to partial bleaching if they are based upon D_E measurements from multiple aliquot or large SAR aliquot data (Rodnight *et al.* 2006; Stokes *et al.* 2015). Similarly, feldspar IRSL ages may systematically underestimate true burial age for sediments if ages have not been appropriately corrected for anomalous fading, or quartz OSL ages may systematically underestimate the true age if too close to saturation (Duller 2016; Rosenberg *et al.* 2011). Overall, Murray and Olley (2002) find that ages are often not published with full analyses of random and systematic uncertainties, and they note that this absence is particularly significant when such ages are compared with independent chronologies.

3.2.2.2 The advantages of Bayesian inference

Bayes' Theorem was first proposed via a 1763 reading to the Royal Society titled: 'An Essay towards solving a problem in the doctrine of chances.' Presented after its discovery in the papers of the deceased Reverend Thomas Bayes, its kernel is a probabilistic theorem that mathematically updates a previous hypothesis given a new set of measurements. In other words, it is simply a rule that describes exactly how to calculate the probability of a hypothesis in light of new evidence. The fundamental relationship is expressed as:

$$P(\theta|y) = \frac{P(\theta)P(y|\theta)}{P(y)}$$

where $P(-)$ denotes the probability of the quantity inside the brackets, θ is a parameter about which we seek information, observations are denoted by y, and $P(\theta|y)$ is the conditional probability of given . Given that the denominator is the total probability of event and therefore a constant, this equation can be simplified and restated:

posterior α *likelihood · prior,*

the posterior probability ('posterior') is proportional to the likelihood function times the prior probability ('prior'). If probabilities are treated as distributions (or densities) rather than discrete values, the power of this relationship but also the complexity of the mathematics increase substantially.

A simple example shows how we can calculate a posterior probability and update it with new information. Let us suppose that we are excavating an archaeological site and have found a pottery sherd of a particular fabric that is decorated with a distinctive zig-zag pattern. Assuming we have no other information about the production date of this pottery sherd, we would like to know how likely it is to date from a certain time period (T_i) given its decoration and fabric. We know that all pottery made of this same type of fabric was created within five time periods (T_1–T_5), and that its popularity changed over time. Therefore, we expect that 5%, 10%, 35%, 30%, and 20% of the pottery made from this fabric was created during time periods T_1, T_2, T_3, T_4, and T_5, respectively. This value is our prior probability. We would like to further narrow down the age of the pottery by applying a stylistic chronology based on the zig-zag design. Recently, we have discovered that the popularity of the zig-zag pattern changed over time, so that it is found on 40% of the pottery made during time T_1, 20% of the pottery from time T_2, and 10% from time T_3; as our functional chronological data, these values become our likelihood function (frequencies are indicated graphically in Fig. 3.3A). As seen in Figure 3.3C, we can apply Bayes' Theorem by inserting the appropriate probability values into the equation, where the denominator is simply the full probability of finding a pottery sherd of this description dating from any time period. Therefore, in Case A(Fig. 3.3C), we calculate the posterior probability that this sherd belongs to the group of ceramics from time period i (C_i) given that it has a zig-zag decoration (Zz). Applying Bayes' Theorem to each of the five possible time periods in turn gives us a posterior probability of 26.67% that the sherd dates from time periods T_1 or T_2, and a 46.67% probability that it is from T_3. The sherd was then sampled for luminescence dating, and we have now received an age that can be used to improve our chronology (Case B: Fig. 3.3D): this new observation suggests that the sherd has a 60% chance of dating to time period T_3, and 20% each for time periods T_2 and T_4. Therefore we update our previous hypothesis by again applying Bayes' Theorem. The posterior probabilities from Case A are now the prior, and the luminescence age probabilities serve as the likelihood. The new posterior probability can be seen in Figure 3.3D. By combining all sources of information, the dating of the sherd is more precise than if the stylistic and chronometric data are each considered alone.

Bayes' Theorem forms the basis of a powerful statistical technique known as Bayesian inference, which can be described as a multi-step, iterative process (Gelman *et al.* 2004):

1. Creation of the model, which comprises a joint probability distribution for all relevant observed and unobserved quantities,

2. Updating the prior probabilities given the observed likelihoods to give posterior probability distributions,

3. Evaluation of the model outcomes via fit and conclusion, and possible repetition of all steps.

Bayesian inference, then, understands the gathered data to be fixed, which in turn update the beliefs about the system as described by probabilistic parameters. This statistical technique has significant advantages in the context of constructing chronological frameworks. Given the formalism discussed above, we can now interpret the parameters of interest in a Bayesian framework as follows (Millard 2006b):

The benefits of applying such a statistical technique to luminescence ages and chronometric frameworks are manifold. Chronometric ages used to create a framework do not exist in a vacuum. As in the example above, it is often the case that dating can become more robust and precise by combining chronometric data with accompanying information such as stratigraphic order, independent chronological data, stylistic variation, or other expert information. This may improve the accuracy of the conclusions that are based upon the framework. By using Bayesian inference and combining these sources of data in a prescribed manner, hypotheses can be compared in a quantitative manner between laboratories and researchers. This makes an often-implicit process of quality assessment explicit and repeatable. This is particularly useful for chronometric frameworks, as the

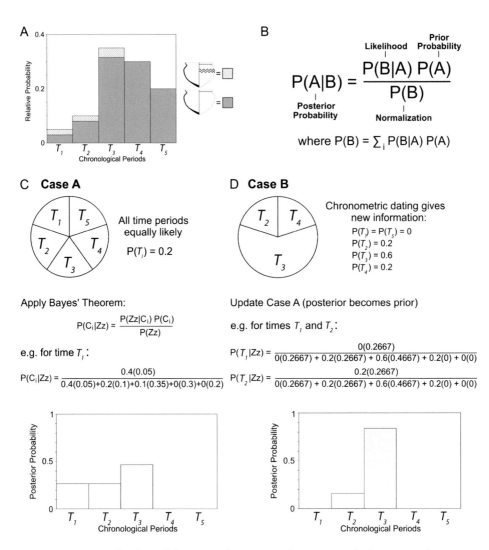

Figure 3.3 An example of Bayes' theorem and Bayesian inference, as applied to pottery dating. Please see text for details.

complex web of associations between dated samples can be simplified and described numerically in a reproducible manner.

3.3 CHRONOLOGICAL FRAMEWORK CREATION: PRELIMINARY STEPS

If the creation and use of a chronometric framework falls within the scope of a larger project, then consideration of model design and the analysis of legacy data should take place at the inception of a project, and continual communication through all stages is key (Brauer *et al.* 2014). Ideally, any new luminescence samples that are collected can be specifically targeted: sample locations should improve dated constraints for the event(s) of interest and minimise the effect of systematic errors (see Combès and Philippe (2017) and Section 3.5.3 for further discussion). This latter approach includes collection of closely associated samples to be dated via independent chronometric techniques (Clark-Balzan *et al.* 2012), the use of semi-independent luminescence techniques involving multiple minerals or luminescence signals (Arnold *et al.* 2015; Guérin *et al.* 2017b), or if possible, collection of multiple samples from outcrops of a definitively identified unit (Alexanderson *et al.* 2014). Frameworks based solely on newly collected, primary data will be principally concerned with model specification, and less so with database creation and quality control of legacy data. Nevertheless, it is useful for such studies to ensure that they meet minimum publication requirements for potential inclusion in later meta-analyses.

3.3.1 Model specification

Complex hypothetical scenarios can be expressed as clear relationships between events of interest, available chronometric data, and temporal constraints. In some cases, modelled posterior probabilities for chronometric data or groups of ages will be the desired outcome (Douka *et al.* 2014), while other models may use these probabilities as the basis for more sophisticated queries (Chiverrell *et al.* 2009; Parnell *et al.* 2008). Initial model specification includes identification of the events of interest, compilation of legacy and new data, consideration of quality control criteria both during database building and as an iterative process during modelling, and an assessment of coding feasibility and model sensitivity testing. We will discuss these subjects in turn.

Event definition, that is, the stable state transition or short-term excursion for which we wish to provide a date, is straightforward in many cases. Most often, the event is a palaeontological, archaeological, or palaeoenvironmental occurrence that can be localised to a specific stratigraphic position, possibly with some uncertainty. In the case of an event that relies on the interpretation of palaeoenvironmental proxies, for instance, pollen frequency, one can define a localised inflection point via mathematical interpretations of the data (Parnell *et al.* 2008). Luminescence ages may directly or indirectly date this event of interest. OSL, TL, or IRSL ages measured from mineral grains in sediment only directly date clastic sediment burial, e.g. palaeoflood events, dune or loess accumulation, deposition of glacial landforms, or direct crystallisation of minerals within subaerial volcanic rock. Though it is common to assume the coeval deposition of lithics or fossils contained within such sediment has occurred essentially simultaneously with last use/occupation or an animal's death, it is important to remember that this may be incorrect due to redeposition or mixing (Bueno *et al.* 2013). Careful consideration of exactly what is being dated by specific chronometric techniques is necessary, particularly in cases when independent dating methods yield seemingly conflicting results.

3.3.2 Compiling a database

Model design directly influences the choice of parameters recorded in a project database, including any associated chronometric data and the level of detail used for recording luminescence ages. Database design is a complex field, and the creation of a useful, robust, informative database to serve many different specialisms is a difficult problem with a significant body of literature. Bronk Ramsey *et al.* (2014) outline the creation of such a database for the INTIMATE project, which attempts to integrate proxy palaeoenvironmental information with chronostratigraphic alignment and associated uncertainty across ice, marine, and terrestrial cores for the past 60,000 years. Simpler approaches can be equally as useful for more focused projects. However, it is vital that all necessary information is recorded in order to appropriately treat any systematic errors and harmonise dose rate calculations. Thorough data compilations will require significant professional time. For projects involving multiple data entry specialists, it is vital that clear guidance and definitions are communicated to all collaborators (Lancaster *et al.* 2016). While the development of automated data scrapers may help in the initial compilation of data, older papers may not have been digitised in the correct format and important data is often interspersed through the text instead of within tables.

Large datasets highlight publication variability in details provided concerning sample collection, treatment, and dating methodologies. In luminescence dating, a common absence is publication of the reporting datum for the luminescence ages presented, the units of which are typically 'years ago' or 'thousands of years ago' (a, ka) (Brauer *et al.* 2014; Halfen *et al.* 2016). Specific sample details that may be used to judge age quality are also poorly represented in many papers, as methodological development of quality control tests and sample reporting standards have evolved over time. For example, Thomas and Burrough's (2016) assessment of >30 papers with dune dates from southern Africa yielded only 34% of samples measured via the SAR protocol for which dose recovery tests were reported, and even fewer samples for which overdispersion values were reported. They note that even basic information such as the number of aliquots measured is only reported 45% of the time. In light of this, it may be useful at first to include all data that meets a certain, very limited standard of publication, with enough basic information such that specialists using the information can devise their own set of quality control assessments for data ranking (Small *et al.* 2017). This is the approach taken by the INQUA Dunes Atlas database: data inclusion required only minimum published details (sample location coordinates, deposit type, and dating method), but significant further data was recorded when available.

3.3.3 Quality assessment

Data quality is of paramount importance when creating a chronological framework that includes legacy data, as ages which are not accurate or do not securely relate to the events under consideration will bias the model. The problem must be approached from several perspectives, including methodological assessment (one aspect of this may be the use of different techniques) and the security of an age's association with the event of interest and any other age data (i.e. constraints), or 'chronometric hygiene' and 'stratigraphic hygiene' (Millard 2008; Spriggs 1989). It is vital that the published source of the chronometric data provides enough information that the database compiler can assess the chronometric and stratigraphic security of the data, otherwise it is prudent to remove the data from the model. It will be desirable in many cases, however, to assign multiple levels of security to age information, rather than using a binary inclusion/exclusion principle. These rankings can then either be used to create several distinct models, with posterior probabilities compared afterwards, or used in more sophisticated Bayesian 'outlier analyses' during the modelling process (Blaauw and Christen 2011; Bronk Ramsey 2009b). This will be discussed further in Sections 3.4.3 and 3.5.3. More information concerning quality assurance criteria can also be found in discussions of archaeological (Higham *et al.* 2014), palaeoenvironmental (Brauer *et al.* 2014), and palaeontological contexts (Rodríguez-Rey *et al.* 2015).

Overall, there is general agreement about the issues that can affect the quality of luminescence data in various geological settings. However, explicit guidelines for ranking the security of legacy data depend on the specific application. Methodological assessment requires evaluation not only of the original study's measurement protocols and age calculation methods, but also of their appropriateness for the sample set under consideration. It is crucial that the first assessment of data quality is kept as independent as possible from evaluation of the 'fit' of the age with any hypothesised chronometric patterns (Small *et al.* 2017). We cannot offer a full set of quality control criteria for all cases, but some important questions to consider include:

Is there sufficient evidence available that the methodology is appropriate for the particular mineral sample under consideration? This has already been mentioned in Section 3.2.2, and it is discussed fully in Chapter 1 and the references therein. Briefly, therefore, one may wish to consider some significant issues including the following. OSL ages on unheated sediment are typically considered to be more suitable than TL ages due to the problem of partial bleaching, and single-aliquot or single-grain techniques more reliable than multiple aliquots due to heterogeneity of crystal behaviour. For feldspar ages, has anomalous fading been considered and measured? Early feldspar ages may assume no anomalous fading, rather than measuring experimentally and attempting to correct for it. Are measured D_Es near saturation? Have data been treated rigorously and in the same manner? That is, are age models applied to D_E data or are 'outliers' rejected with no explicit criteria? Have age models been used appropriately? For example, a particular age model known as the finite mixture model (Galbraith and Green 1990) has sometimes been used to calculate D_E components from small aliquot measurements even though simulations show that this leads to 'phantom' distributions (Arnold and Roberts 2009). Finally, highest quality data will include supporting experiments, such as dose recoveries, and data analysis techniques (all protocols used, aliquot size, numbers of aliquots, etc.) that are also published in detail.

Are there geological factors that might cause age inaccuracy, and have these been accounted for? Ages may be inaccurate due either to the D_E estimate or the dose rate calculation, and the degree to which geological factors will adversely affect either or both values depends upon the sample type and its environmental setting. The applications chapters throughout this book should be consulted for detailed assessments of the most common risk factors. However, some major issues are mentioned here. For settings in which partial bleaching is considered likely, have appropriate measurement protocols and data handling procedures been applied (e.g. Chapters 6 and 7)? Have the samples remained intact since burial, or is it possible post-depositional mixing has affected the ages (e.g. Chapters 1, 6 and 7)? Dose rate reliability should also be assessed. Some problems that may occur include decay chain disequilibrium, particularly for waterlain sediments and open-system behaviour/radioisotope migration. Gamma dose rate heterogeneity may also be a problem if on-site gamma spectrometer measurements have not been made (Chapter 2). If dating feldspars, the assumed potassium content may bias the age if not measured. Finally, it is possible for alpha and beta microdosimetric variation in sediment to cause D_E distributions that mimic partial bleaching or bioturbation (see Chapters 4, 6 and 7). Due to the complexity of this topic, it is a good idea to identify the factors that are most likely to affect collected ages, either in discussion with a luminescence dating practitioner or via a literature survey. At a minimum, publications with legacy data should be read carefully to note any difficulties highlighted by the authors.

Is there any supporting chronometric data at the site? Dating samples that form a local stratigraphic sequence or have been collected from the same geographic unit across a wider region can highlight issues that are not necessarily apparent in individual sample data. Evidence for age inaccuracy may include reversals in the age–depth relationship and age outliers, that is, single data points separated by some defined amount from the main body of data. In general, one can imagine an order of reliability for a site as follows (from least to most reliable): single luminescence age, stratigraphic sequence of luminescence ages (one mineral type and protocol), sequence of luminescence ages (multiple mineral types or protocols: semi-independent data), both luminescence ages and independent chronometric data. It must be noted here that this criterion is not completely independent of assessment of the chronological likelihood of the ages, and in general, there is often some circularity between luminescence age calculation and 'expert knowledge' relating to site chronology. That is, an assessment of D_E distribution as evidence for partial bleaching rather than microdosimetry or bioturbation is likely to, in part, depend on that sample's relationship to other associated ages. This criterion overlaps significantly with outlier analysis, and will be discussed further below.

3.4 BUILDING A MODEL

There are several free Bayesian modelling packages online that can be used to model lumines-cence ages, either simply, as calendar dates with some associated uncertainty, or more flexibly, with scope for treating systematic errors and dependent quantities.[1] For users with coding experience, highly flexible development packages include JAGS (Just Another Gibbs Sampler) (Plummer 2003; see example model in Combès *et al.* 2015), Stan (Stan Development Team 2015), and the more specialized, primarily luminescence age modelling package, BayLum for R (Christophe *et al.*, 2017; Philippe *et al.*, 2019). More user-friendly programs that can include specification of luminescence ages include WinBugs (Lunn *et al.* 2000, 2012; see model in Millard 2008), BCal (Buck *et al.* 1999) and OxCal (Bronk Ramsey 2009a; see models in Chiv-errell *et al.* 2013; Douka *et al.* 2014). These have both a graphical user interface designed to simplify the input of ages and constraints, as well as the possibility of coding for more complex input. Some have been specifically designed for age–depth modelling, e.g. BPeat (Blaauw and Christen 2005, see model in Ampel *et al.* 2008; Wohlfarth *et al.* 2008), Bacon (Blaauw and Christen 2011, 2013), and BChron (Haslett and Parnell 2008, see model in Livsey *et al.* 2016).

In this section, specific examples have been modelled in OxCal, in order to highlight a few basic methods for linking chronological and stratigraphic information. Figures 3.4–3.6 show three different models created for a hypothetical cave site that is being excavated and dated, presented in order of simplest to most complex. In each case, the code used to create the model is given in (A), a schematic of the model assumptions is shown in (B), and the likelihoods and posterior probabilities are displayed in (C). For reference, Table 3.1 lists the syntax for a number of commonly used commands in OxCal's Chronological Query Language (Bronk Ramsey 2009a). A non-coding user interface is also available for OxCal. However, some very light coding allows the creation of more complex models (e.g. with parametrisation, see Section 3.4.2).

---

1 Programs/packages mentioned here are available online from the following sites: Baylum for R (https://cran.r-project.org/package=Baylum), JAGS (http://mcmc-jags.sourceforge.net/), Stan (http://mc-stan.org/), WinBugs (https://www.mrc-bsu.cam.ac.uk/software/bugs/), BCal (http://bcal.sheffield.ac.uk/), OxCal (https://c14.arch.ox.ac.uk/oxcal.html), BPeat and Bacon (http://chrono.qub.ac.uk/blaauw/wiggles/), BChron (https://cran.r-project.org/web/packages/Bchron/index.html).

3.4.1 Groupings, constraints and priors

Groups, constraints, and priors form the overall structure of the model, by relating the dated samples to each other and to the event of interest. 'Groups,' as the name suggests, are collections of ages that can be treated in the same manner. Constraints have been mentioned above in Section 3.1, and priors (Section 3.2.2) are essentially the mathematical expressions of these constraints. For further details about the mathematics underlying the information presented in this section, please see Bronk Ramsey (2008 and 2009a).

In OxCal, there are four key commands for defining temporal relationships between dated samples and events of interest (Table 3.1):

1. Phase
2. Sequence
3. Boundary
4. *Terminus post quem* (TPQ)/*terminus ante quem* (TAQ)

A phase is a method of creating a group of associated ages; no chronological order is imposed on the dated samples or events within a phase. Sequences impose a chronological order on individual samples or groups of samples. Both phases and sequences must be bracketed by boundaries, which impose mathematical 'cut-offs' such that ages are not modelled as unrealistically far apart – completely independent dated samples are less likely to be closely grouped in time than associated ones. Phases and sequences can also be modified by the introduction of specific priors, which reflect beliefs about the frequency of dated samples through time. Certain priors have been encoded in specific 'Boundary' commands. A standard 'Boundary' assigns a uniform prior to a phase or sequence, such that samples are considered to have been randomly selected from a uniform probability distribution between a certain start and end date. However, other distributions are possible. Finally, the *terminus post quem* and *terminus ante quem* are used to insert chronometric data, which define maximum and minimum ages, respectively, in the modelled sequence (see for example Hobo *et al.* 2014). Cross-correlations of boundaries or inferred events can also be inserted to link various sets of constraints (Fig. 3.6, Section 3.4.4). Examples of the use of each of these commands is given in Table 3.1.

The commands defined in Table 3.1 are simple relationships that can be nested to build complex models, in a manner akin to creating the well-known Harris matrix for an archaeological site (Millard 2006b). Most often, the principle of superposition is used to define stratigraphy priors: that is, an overlying layer must be younger than the layer underneath. The flexibility offered by combining sequences and phases is indispensable in complex stratigraphies, where few assumptions can be made about the continuity and rate of accumulation processes (Macken *et al.* 2013). They can also be used to create nested constraints exist where necessary, such as in re-excavated and resampled archaeological sites (Clark-Balzan *et al.* 2012). Phases represent a useful alternative to the use of a weighted mean as a means of combining age data from a given sedimentary unit, which (perhaps incorrectly) suggests essentially instantaneous deposition (Millard 2008). Though phases and sequences generally reflect stratigraphic constraints, a number of projects have also based models on 'pseudo-stratigraphic' relationships (Chiverrell *et al.* 2009), in which an order of events can be inferred from some other type of evidence. This type of approach has

Table 3.1 Selected commands from OxCal's Chronological Query Language, adapted from Bronk Ramsey (2009a) and OxCal v4.3 Manual. [Name] indicates an assigned name for the parameter, which can be used for cross-references, and [Expression] stands in for the assigned probability density function (PDF) or value.

COMMAND	USAGE
Groupings, constraints, priors	
Sequence([Name]){ … };	Orders events and groups
Phase([Name]){ … };	Groups unordered events
Boundary([Name], [Expression]);	Used with Sequences and Depositional Sequences; defines a group of events (uniform prior if combined with another Boundary)
Tau_Boundary([Name], [Expression]);	As Boundary, but pair with Boundary to define exponential event frequency
Sigma_Boundary([Name], [Expression]);	As Boundary, but pair with Sigma_Boundary to define normally distributed event frequency or with Boundary for truncated normal distribution
After([Name], [Expression]){ … };	*Terminus post quem* ('limit after which', TPQ); constrains event PDF to occur after all grouped elements
Before([Name], [Expression]){ … };	*Terminus ante quem* ('limit before which', TAQ); PDF for event occurring before all grouped elements
Depositional models	
P_Sequence([Name], k0, [Interpolation], … [log10(k/k0) expression]){ … };	Depth model with poisson distributed deposition
U_Sequence([Name], [Interpolation]){ … };	Depth model with uniform deposition
Age input	
C_Date([Name], Cal Date, Uncertainty);	Defines calendar age with normally distributed uncertainty
Age([Name], Expression);	Converts an expression/PDF into an age
Prior(Name, [Filename]);	Defines PDF from file or from elsewhere in model
Probability distributions and operations	
N(Name, mu, sigma, [Resolution]);	Normal distribution (PDF)
U([Name], From, To, [Resolution]);	Uniform distribution (PDF)
Combine([Name]){ … };	Combines PDFs from independent parameters
Difference(Name, Parameter1, … Parameter2, [Expression]);	Finds difference between parameter PDFs
Probability([Name], At, Distribution);	Returns probability, can be used with sequences or phases

Table 3.1 (Cont.)

COMMAND	USAGE
Outlier analysis	
Outlier_Model([Name], Distribution, … [Magnitude], [Type (t/r/s)]);	Outlier model definition (see text and references)
Outlier([Name], [Probability]);	Assigns prior probability of being an outlier to a likelihood

been used to assess depositional periods for geographically ordered beach ridges in Phra Thong (Thailand; Brill *et al.* 2015) and ice sheet advance and retreat (see Section 3.5.3), but other types of information such as stylistic criteria in archaeological settings may also be used. This latter type of evidence must be carefully assessed for validity, as more theoretical chronologies may be less secure.

Figures 3.4–3.6 demonstrate scenarios in which various constraints may be most appropriate due to the varying levels of knowledge available. In Figure 3.4, the relationships between dated samples are simple. Samples have been collected from a single stratigraphic sequence within a cave, which built up slowly over time. Therefore, there should be a direct link between stratigraphic position and age order (based on stratigraphic superposition), which allows all ages to be input directly into a sequence. By contrast, Figure 3.5 supposes that we have been able to distinguish three sedimentary units in the cave (Units A–C), with Unit A the youngest and Unit C the oldest. In this case, samples have been collected throughout the cave, so each sample cannot be related directly to each other sample via direct superposition. Instead, we can only identify which unit each sample was collected from. A sequence of phases, as shown in Figure 3.5, is a good approach in this scenario. Each phase is equivalent to a sedimentary unit, reflecting our expert knowledge, and these have been placed in a stratigraphic sequence. Finally, Figure 3.6 shows how these two disparate models might be linked by new information; this will be discussed further below.

A second type of model commonly used is a 'depositional model,' in which sample depths (rather than simply stratigraphic order) are assumed to yield some chronological information (Blaauw and Christen 2005; Bronk Ramsey 2008). These types of models have been implemented in multiple Bayesian modelling programs (see BACON, BPEAT, and OxCal), and they are most often used to improve precision and derive accumulation models for cored lake records (Ampel *et al.* 2008; Blockley *et al.* 2008; Wohlfarth *et al.* 2008). Some studies have also used them to provide accumulation rates for other environments in which deposition is considered to be well-understood, such as loess exposures (Li *et al.* 2015). It should be noted, however, that environments with high magnitude, pulsed accumulation, such as dunes, are probably better suited by modelling with either only simple sequences or phases (Leighton *et al.* 2013). Bronk Ramsey (2008) notes that changes in mode of deposition via sedimentological information and likely deposition rate due to underlying processes must be considered when such a model is being created. Examples of the CQL code used in OxCal to create such a model are given in Table 3.1.

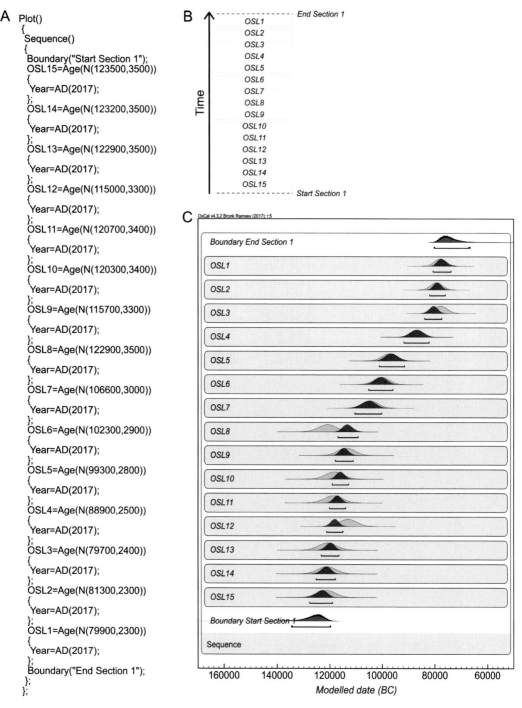

Figure 3.4 A simple Bayesian model created for a simulated archaeological section in a complex cave site (Section 3.1). Fifteen samples (OSL1-OSL15) were collected in a directly associated stratigraphic sequence, and these have been dated with OSL. They are modelled in OxCal as a simple bounded sequence: A) CQL code is given, B) a model schematic. In the schematic, single ages are shown as grey boxes with no border, and boundaries are dotted lines. C) shows modelled posterior probabilities (dark grey) for ages and boundaries and original age information (light grey likelihood)

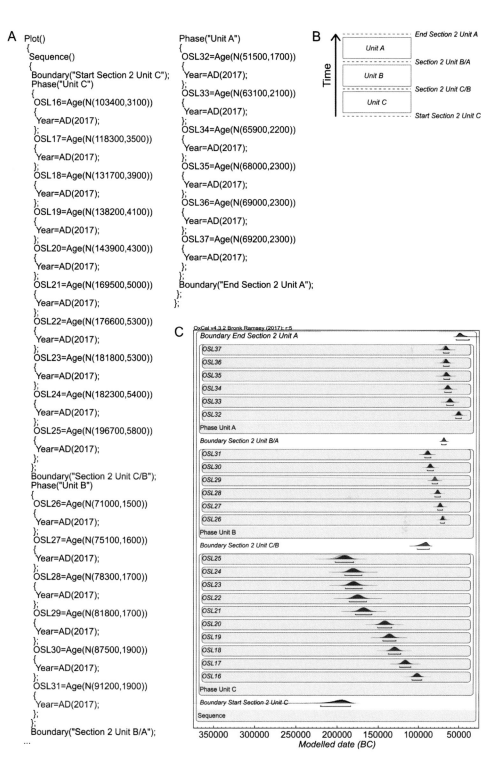

Figure 3.5 Simulated luminescence age samples from the same cave site as in Figure 3.4. Rather than in a simple sequence, these samples were collected from identifiable sedimentary units spread throughout the excavations. That is, they can be identified as belonging to Units A, B, or C, but there is no secure stratigraphic order within these unit assignments. Therefore, these ages have been modelled as a sequence of bounded phases (unordered groups of dates), A) with the CQL shown, B) a model schematic (key as Figure 3.4, with phases as outlined boxes), C) model results.

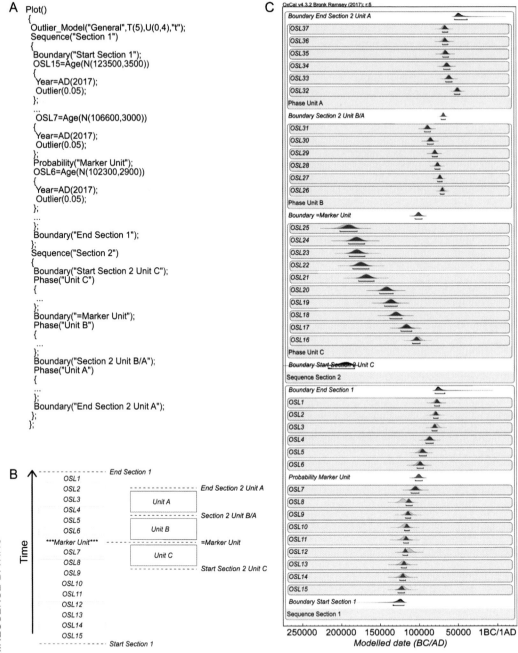

Figure 3.6 Simulated luminescence ages from Figures 3.4 and 3.5 have been combined in a more complex model. In this scenario, a clearly identifiable tephra unit is known to occur in the sediment in Section 1 between samples OSL6 and OSL7, and this layer is also found between Units B and C throughout the cave. Therefore, the simple sequence of ages (Fig. 3.4) and ordered groups (Fig. 3.5) have been linked via a probability query (to extract a PDF for the tephra layer) and cross-reference (A, B). An outlier model has also been run. However, all sample ages were already acceptable. C) Model results.

It is important to note that the same scenario might be modelled in various fashions, depending on the information required. For example, the ages in Figure 3.4 could be incorporated into a larger model by simply grouping them into a phase. By using this approach, we choose not to build in the information about their relationships to each other within the model. If we do not need to extract posterior probabilities for particular ages or events within that sequence, though, this is a perfectly valid alternative scenario.

3.4.2 Age input

Two methods have been used to input ages into user-friendly Bayesian modelling programs: 'calendar age' definition and parametrisation/hierarchical modelling (Table 3.1, Fig. 3.4). The first of these, in which a calendar age and associated error (normally distributed) are assigned, is the most commonly used, as it is simple and does not require full publication of D_E and dose rate calculations. This is demonstrated in the code given in Figures 3.4–3.6. Systematic error and dependent quantities are not treated rigorously, however, and in general, modelled age uncertainties will be underestimated (Millard 2008). If recalculation of dose rates is desired in order to harmonise legacy data, this must be also be done separately. Two methods have been published that can address some of the shortcomings of this technique. Rhodes *et al.* (2003) suggested a 'pragmatic' approach to calendar age input to reduce the effects of systematic errors. This involves stripping out shared systematic error from luminescence age estimates, so that the uncertainty used in the model comprises only random error and unshared systematic (USS) error added in quadrature (see Section 3.2.2.1 for definitions). A suitable value for the unshared systematic error can be obtained by running several models with various USS magnitudes, and observing the effect on the agreement index (*A*-value), see below. Shared systematic errors are recombined in quadrature after the model has been run, in order to obtain final age estimates. Due to its simplicity, this method of modelling via precision-only uncertainties has been used by multiple studies (Barton *et al.* 2009; Feathers *et al.* 2006; Jacobs *et al.* 2008), albeit sometimes without the incorporation of the USS term. It is important to note that many legacy studies do not necessarily differentiate and publish systematic and random error quantities. Therefore, in practice, this pragmatic approach may still be difficult. Moreover, the shared systematic error must be included in the Bayesian model if the ages are incorporated into a model with independent chronological control (Cunningham and Wallinga 2012).

The second method involves defining a likelihood probability distribution other than the standard Gaussian (see also discussion in Zeeden *et al.*, 2018). Cunningham and Wallinga (2012) proposed a method to robustly estimate the uncertainty associated with the use of the minimum age model (MAM) in a Bayesian framework. Their procedure is based on bootstrap resampling of measured D_Es, calculation of the MAM with a randomly assigned over-dispersion, incorporation of an unshared systematic error term and smoothing, to generate a final likelihood distribution. Hobo *et al.* (2014) have used this procedure to model OSL ages for cores collected from the River Waal, a Rhine distributary, and estimate sedimentation rates and budgets. Cunningham and Wallinga's bootstrap-based likelihood or another related procedure could also be used to create an age likelihood that reflects issues such as asymmetric uncertainty or be generalised to the use of other age models (see for example Christophe *et al.*'s (2016) Bayesian Gaussian mixture model for poorly bleached sediment).

Alternatively, parameterisation involves the definition of independent parameters that can be shared appropriately within a hierarchical model to calculate ages within the Bayesian framework (Millard 2006a, 2006b). Disadvantages include coding complexity, model running time, and the number of legacy publications that may not yield enough data to develop a fully parametrised model. Use of parameters, however, allows much more rigorous treatment of shared systematic error and dependent quantities. Parametrisation also allows the choice of more appropriate probability distributions for particular parameters. Millard (2008) has suggested the use of a log-normal distribution for both water content, as it cannot provide unrealistic negative water contents, and elemental concentrations in certain cases, such as those measured from sediment samples from Skhul, Israel and used to recalculate flint TL ages. Parametrisation has not been commonly used in past publications as it was not included as an option in more user-friendly modelling programs. However, OxCal now offers parametrisation capabilities (Bronk Ramsey 2017). Explicit choice of the parameters used will depend on site characteristics (e.g. expected shared water contents) and quantity measurement/age calculation choices as these control the dependencies of interest. If possible, parametrisation is desirable due to the more rigorous treatment of uncertainties. Millard (2006b) compared a full hierarchical model produced for ESR age data from Border Cave with an OxCal 'calendar age' model that treats all values as independent, and notes that mean ages for phase boundaries are within 1,000 years of each other. However the OxCal calendar age model probably yields unrealistically higher precision boundary estimates.

Another issue worthy of consideration is the treatment of replicate ages, whether these are measured from closely associated, replicate samples collected from the same stratigraphic unit and expected to share the same bleaching, mixing, and dose rate conditions, or closely related age estimates, such as different scale (single and multiple grain) or different signal (TL and OSL) D_Es measured from grains extracted from the same sample. If one age can be judged more secure, for stratigraphic or methodological reasons or due to the amount of detailed publication information, this age may simply replace the less secure age in the model. For instance, if we imagine a fluvial sequence measured with multiple-grain aliquots and re-measured via single grains, it is possible that the single-grain age would be chosen in order lessen the issue of partial bleaching. Alternatively, if single-grain data suggests that no mixing or partial bleaching is likely to affect the sediment, then multiple-grain ages for the same sample may be chosen as more representative of the sediment and a more appropriate match for the bulk dose rate calculations, as with data from Blombos Cave (Millard 2008). Similarly, a sequence measured with quartz OSL near the signal saturation limit might be eschewed in favour of pIR-IRSL feldspar ages, which would be less likely to underestimate the true age. If there is no *a priori* reason to reject one set of ages or samples, a weighted mean of the quantities might be used, such as in the case of indistinguishable D_Es measured via TL and OSL on the same sample (Millard 2006b).

OxCal provides a third option, the use of the command 'combine,' which is used to apply several distributions to the same parameter. It has been used to combine ages obtained for samples collected side-by-side in the same unit within a cave setting (Barton *et al.* 2009) or from the same identified stratigraphic unit in a sand dune (Leighton *et al.* 2013). 'Combine' has also been used to associate single and multiple-grain ages measured from the same sample (Burrough and Thomas 2013; Clark-Balzan *et al.* 2012), in cases where both D_E sets are considered equally valid and are congruent with each other. Fu *et al.* (2017) also use 'Combine' for single-grain OSL and TT-OSL ages measured for a sample in their sequence. Luminescence age modelling would benefit from further investigation into statistically rigorous input strategies for such semi-dependent values.

3.4.3 Outlier analysis

Outliers are data points that are distinctly separate from a group of other data, according to some defined criterion. In luminescence dating, outliers are commonly defined by reference to replicate measurements upon subsamples, e.g. aliquot or grain D_Es. Ages may also be defined as outliers if, for example, they are reversed with respect to an observed age–depth model (after uncertainty is taken into account at a one σ or two σ level), or they do not cluster with a group of ages expected to date the same essentially instantaneous event. Identification of outliers should be accompanied by the criteria for defining these, such as a quantitative homogeneity test (Galbraith 2003; Ward and Wilson 1978).

Identification and treatment of outlying D_Es is reasonably complex, as both single-grain and multiple-grain measurements for a given sample nearly always scatter more than can be explained by the uncertainties on the measurements (Galbraith and Roberts 2012). This excess scatter, known as the overdispersion, is an important quantity often used to identify 'problematic' samples in conjunction with site analyses. Depending on the inferred source of the overdispersion, whether sediment mixing, partial bleaching, or microdosimetric variation, it is likely that the publication author will have attempted to take this into account by calculating the final sample D_E using an appropriate age model. It is useful, however, for a modeller to be able to judge the extent of the scatter in the data in order to assess data quality.

Treatment of outlying ages is usually handled differently. The homogeneity test mentioned above may be used upon groups of ages as well as D_Es. Jacobs et al. (2016) applied Galbraith's homogeneity test to ages obtained for each stratigraphic unit, before combining self-consistent ages with precision errors only via weighted means to get a best guess for the depositional age of each unit (though see discussion above about the use of phases versus weighted means). In addition to such tests, OxCal's agreement indices can be used to judge the extent to which an age may be an outlier (Bronk Ramsey 2009a). The individual agreement index is a measurement of the similarity of the likelihood and posterior probabilities for each age. Following the modelling protocol of Rhodes et al. (2003), unshared systematic error values can be enlarged iteratively until agreement indices for all ages rise above some nominal value, often 60%. The authors specifically note that there is often no a priori reason to reject this data, so this may be a better way of incorporating such outliers in cases where a single age would require the USS to rise to a significantly higher value. However, it may be more useful to exclude this age from the model. Feathers et al. (2006), for example, apply iterative culling of the dataset using such agreement indices until a given measure of internal consistency has been reached. Bronk Ramsey (2009a), though, notes that one in twenty sample ages is statistically likely to have an individual agreement index <60%. He therefore advocates inspection of the overall model agreement index, and suggests that if this is >60% outlier rejection is not necessary.

A newer option is the ability to statistically treat the possibility that a given age or date is an outlier within the model (Table 3.1, Fig. 3.6). A rigorous Bayesian approach for analysis of outliers with respect to radiocarbon dating was first outlined by Christen (1994), and related approaches are now offered by packages including BPeat (Blaauw et al. 2007), BChron (Haslett and Parnell 2008) and OxCal (Bronk Ramsey 2009b), some of which can be applied to luminescence data. In OxCal, an outlier model is first created with three key parameters (Table 3.1): the outlier distribution (typically student T or normal), outlier magnitude (can cover several orders of magnitude), and outlier type. Type 't', which assumes that the association between the calculated sample age and the true age is offset for some reason, provides the best means of incorporating outlying luminescence data. The 'outlier' command can then be used to assign a prior probability that a given age is an outlier; this probability can be the same value for all samples when there is no *a priori* information (e.g. 0.05), or individual sample values can be defined if data is included with differing levels of chronological security (see Section 3.5.3). After the model is run, a percentage probability that each age is an outlier is produced. Multiple studies have used this approach, with varying levels of analysis of the resulting identified outlier probabilities (Chiverrell *et al.* 2013; Fu *et al.* 2017; Leighton *et al.* 2013; Mischke *et al.* 2017).

3.4.4 Querying, assessing, and reporting

There are several ways in which modelled parameters can be used for hypothesis testing. Modelled boundaries have sometimes been used as the event of interest (Davies *et al.* 2016), but OxCal offers the ability to extract multiple sources of information beyond modelled date probabilities, including posterior probabilities for inferred events (Table 3.1, Fig. 3.6), event orders, and further operations on modelled probability distributions (see Section 3.5.1). These operations are known as 'queries'. In Figure 3.6, this capability has been used to define a posterior PDF for a tephra marker unit whose age is tightly constrained by luminescence dating samples collected from Section 1. This same tephra layer forms the boundary between Units C and B, which therefore provides new chronological control for our phases from Section 3.2. This is built into the model by cross-referencing the boundary between these unit phases with the PDF we extracted for the tephra from Section 3.1. Simply put, we have constrained the modelled chronologies of these two parameters to be identical and thus linked our two sections. In published models, queries have been used to suggest durations for particular archaeological and palaeoenvironmental events (Douka *et al.* 2014), or to calculate event frequency, such as in Greenbaum *et al.*'s (2006) study of flood frequency in the Negev desert. Agreement indices can also provide useful information if examined carefully. Rhodes *et al.* (2003) point out that models can be used to estimate USS magnitude, as the USS can be enlarged iteratively until agreement indices are considered acceptable. Shared systematic error can also be estimated in a similar manner if an independent chronology is available.

It is also desirable to test model sensitivity to the inclusion of various dated samples and the applied constraints (Bronk Ramsey 2009a). Explicit and systematic sensitivity testing is not common in the literature. However, authors often test several models prior to finalising the published data. Input ages and constraints may be altered, and even the results of various modelling programs may be compared (Kempf *et al.* 2017; Millard 2008). In some cases, it may become apparent that model chronology relies heavily upon one or two dated samples (Millard 2008). In the best case, this may result in collection of new data, but in any case, it is crucial information for any chronology-based hypothesis. It is also important to recognise that a particular physical scenario may be expressed as several models, depending on the types of constraints chosen. For example, the stratigraphic information for archaeological layers in cave sites can often be modelled in two main ways. First, identified sedimentary units can be used to relate events and chronometric data. These may be complex, and the most secure relationships could be influenced by the excavation history of the site. Second, the sequence of archaeological technocomplexes at the site might instead be used to define the grouping and ordering of the ages. Both models may be useful in different ways (Aubry *et al.* 2014), and it is common for these two types of models to be compared (Douka *et al.* 2014; Millard 2006a).

Finally, models must be reported fully in the literature. Software used (with version), sample ages and age input strategy, constraints, outlier treatment, sensitivity testing, and scientific rationales for all of the above must be reported fully (Bayliss 2015). Ideally, model code should be shared in the paper or supplementary information (e.g. supplementary information provided by Higham *et al.* 2014), otherwise, a full description is required (Clark-Balzan *et al.* 2012; Douka *et al.* 2014). The importance of including all input data and explicit rationales for assumptions used can be seen in cases where there is disagreement – see, for instance, disagreement concerning rejection of certain dates in chronological models created by Kennett *et al.* (2015a) for the controversial Younger Dryas Boundary impact theory (Boslough *et al.* 2015; Holliday 2015; Kennett *et al.* 2015b).

3.5 CASE STUDIES

Several case studies have been chosen to illuminate particular points concerning the creation and use of chronological frameworks. The first, that of Riwi Cave (Australia), demonstrates the potential benefits of Bayesian modelling at a single site, both to increase precision of chronometric data and to query the existence of systematic error. The second (hominin fossil chronology) and third (reassessment of the British–Irish ice sheet chronology) provide information about projects based entirely or in a large part on legacy data, with interesting instances of age recalculation, outlier analysis, and the use of pseudo-stratigraphic constraints.

3.5.1 The settling of Australia: Riwi Cave

Determining the date of the first human occupation of Australia is a complex research topic, with which luminescence dating has a venerable though slightly motley history (Fullagar *et al.* 1996; Galbraith *et al.* 1999; Olley *et al.* 2006; Roberts *et al.* 1990, 1999). Recent reviews have suggested that the earliest dated occupation layers are certainly older than 40,000 years ago (O'Connell and Allen 2004, 2015), and potentially earlier than 50,000 years ago (Clarkson *et al.* 2015), with interesting implications for the global spread of modern Homo sapiens (Groucutt *et al.* 2015), and – within Australia – human adaptation to challenging environments and interactions with megafauna (O'Connell and Allen 2015; Hughes *et al.* 2017). Precise dating has been complicated by a lack of sites with early Pleistocene archaeological remains, the closeness of the earliest chronology to the functional boundary of radiocarbon dating methods (*c.* 50–55 ka), and the lack of precision of other dating methods, such as luminescence (Balme 2000; O'Connell and Allen 2015). Several recent studies have therefore used Bayesian techniques or age models to assess the chronometric data (Clarkson *et al.* 2015; Hamm *et al.* 2016; Veth *et al.* 2017). Wood *et al.* (2016) have used a particularly interesting approach for their analysis of Riwi Cave, western Australia, which demonstrates the power of allying Bayesian methods and high-resolution dating studies.

Riwi Cave ('camping place') is a large limestone cave situated in the base of a cliff (Lawford Range), in an area traditionally controlled by the Gooniyandi people (Fig. 3.7). In 1999, a

1 × 1 m test pit was excavated here, revealing a >1 m sequence of fine quartz sediment and hearths with cultural materials including lithics, shells, ochre, and bone (Balme 2000). Six charcoal pieces collected during this excavation were radiocarbon dated to the Holocene and between 30,000 and >40,000 years ago, highlighting the value of this cave for a high resolution study of Pleistocene archaeology. Re-excavation occurred in 2013. The original test pit (Square 1) and three further 1 m × 1 m test pits (Squares 3, 4, 5) were excavated to bedrock (>1 m for the interior test pits, and c. 65 cm at the mouth of the cave). Like the original excavation, the 2013 re-excavation excavated in quadrants and units of arbitrary depth (2 cm for cultural levels, 3–5 cm for underlying sterile sediments), while singular features such as pits and hearths were excavated as units. Excavated material was dry sieved with 5 mm and 1.5 mm mesh. Twelve stratigraphic units ('SU's) were identified: from top to bottom, SU1–2 were grey/ashy, and SU3–12 were brown, fine sands, predominantly aeolian.

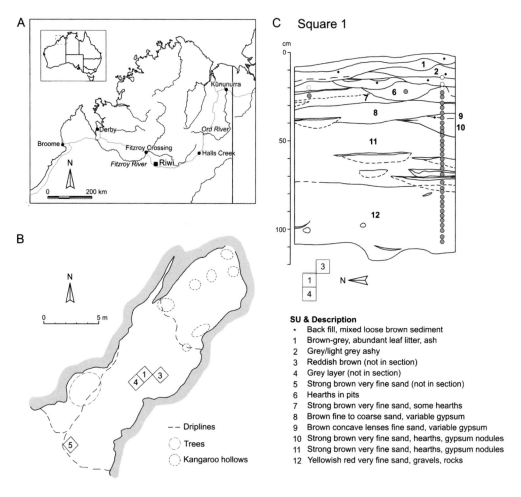

Figure 3.7 A) Location of Riwi Cave, B) cave plan, C) stratigraphic section showing the positions of dated OSL samples (circles) and several radiocarbon samples (black diamonds). OSL samples are grey-filled if they were used in the site model, white with solid outlines if they are dated to the Holocene, and white with dotted outlines if they are considered mixed. This figure has been adapted from Figures 1, 2, and 3 in Wood *et al.* (2016).

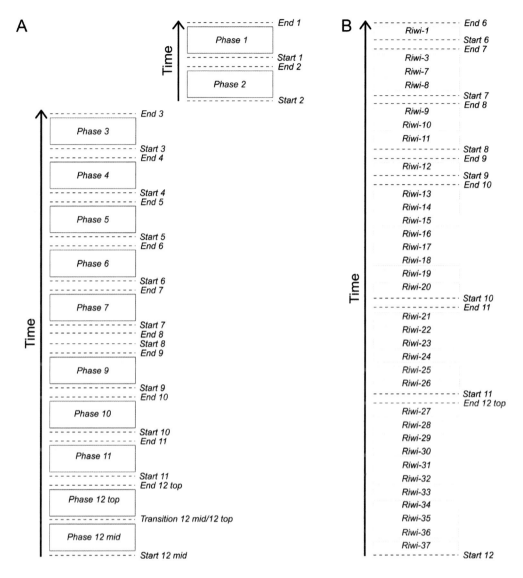

Figure 3.8 Schematics for A) the radiocarbon dating model and B) the OSL dating model, from Riwi Cave. Boundaries, phases, and individual ages are as explained in Figures 3.4–3.6.

A high-resolution dating study was initiated during this second season. Radiocarbon dates were obtained for 33 charcoal fragments, some of which were collected directly from the section wall (26 samples), and some from *in situ* excavation (2 samples) or from sieved material (5 samples). Stringent pretreatment protocols were applied to older samples to remove modern contamination, weaker and stronger pretreatment regimes were compared in order to investigate systematic offsets, and replicate samples were prepared and measured. 37 high-resolution luminescence samples were also collected, via smaller than usual collection tubes (2 cm diameter, 10 cm long); 34 of these were collected from a single stratigraphic column approximately 1 m high, while three samples were collected from nearby units (Fig. 3.7c). Single quartz grains were prepared and measured via well-reported methods, including preheat treatments tested via dose recoveries and elimination of grains via explicit rejection criteria. Decay curves and growth curves were also shown for representative grains, and radial plots were presented for examination of sample D_E distributions.

Data compilation for this site was relatively simple due to the lack of legacy data and the restricted nature of the chronological framework. The six legacy radiocarbon dates were subjected to similar pretreatments as the newer dates, with no methodological concerns noted by the authors, and depths and excavation units given in the original publication could be identified with the newly defined SUs. Four radiocarbon dates were excluded from the database prior to modelling: one 'very obvious outlier' (Wood *et al.* 2016: pp. 15–25) from a heavily bioturbated unit (included in preliminary models), a non-finite age, and two samples from the interface of two stratigraphic units. The treatment of the luminescence data, all from the current study, is slightly unusual. Two mixed samples, Riwi-6 and Riwi-2, were identified via D_E distributions as likely containing two grain populations (fit via the finite mixture model); this is supported by the stratigraphy, as both samples were collected from SU7, immediately underneath an unconformity and overlain by SU2. Other samples were termed 'scattered', and based on the suggestion that grains might have been post-depositionally reworked, an outlier screening process based on median absolute deviations was applied to the D_E data. D_Es were converted into natural logs, normalised median absolute deviations (nMAD) were calculated, and those with values >1.5 (i.e. above the value of 1.48 which characterises a normal distribution) were removed from the data. After this process, between 3% and 23% of D_Es had been removed from each sample prior to age calculation, and new D_Es ranged from 3% less than the original D_E to 19% greater. This process is not standard in luminescence dating studies, and its effects should be further studied, particularly as the interactions between D_E distribution, scatter and microdosimetry are slowly being unraveled, and this process may increase precision at the risk of decreasing accuracy. Nevertheless, the age models constructed are interesting and worthy of consideration.

Wood *et al.* (2016) presented two Bayesian age models incorporating the chronometric data, one each for the radiocarbon dates and the OSL ages, based on the stratigraphic sequence (SUs) identified in the cave. Model construction in OxCal reflected the different information available from the sampling strategy used for each dating method. As shown in Figure 3.8, the radiocarbon model grouped dates into phases to correspond with each dated stratigraphic unit; therefore, the assumption is that the order of the dates cannot be defined any further within each unit. This reflects the excavation strategy of arbitrary spits and allows for movement of charcoals within each archaeological unit. Samples for luminescence dating, by contrast, were collected primarily in a single column through the stratigraphy in square 1, with three samples collected from clearly overlying/incut layers further west in a contiguous stratigraphic section of square 1. Radiocarbon dates were input via R_Date and replicate-dated samples were combined with R_Combine. Luminescence ages were input with C_Date prior to the 1950 datum, with one-σ uncertainties comprising only random uncertainties. An outlier model (general *t*-type) was applied to both models, with prior probabilities of 5% assigned to each date. In both models, two boundaries were used to bracket the dates associated with each stratigraphic unit, providing a start and end estimate for the stratigraphic units that form the events of interest. One exception to this occurred in the radiocarbon date model, where unit 12 was broken into two phases

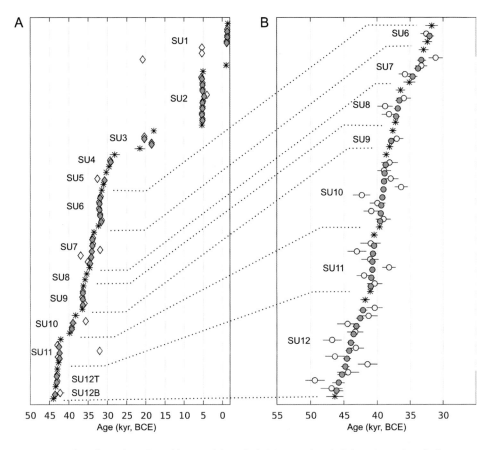

Figure 3.9 A) Radiocarbon dates (diamonds), and B) OSL ages (circles) from Riwi Cave before (light grey) and after (dark grey) modelling. Ages are shown in stratigraphic order; radiocarbon order within phases is notional only. Boundary ages are indicated via stars.

(mid and top) to provide a tighter estimate for the earliest hearth unit in the cave. As discussed in Section 3.4.1, grouping and ordering of dates in the final model is based upon the prior stratigraphic information available via excavation, and it is optimised to provide chronological information for the archaeological events of interest.

Archeologically, the probable age for use of the earliest hearth feature was identified with the mid/top SU12 boundary (46.4–44.6 cal kBP), directly bracketed by three radiocarbon dated samples: one legacy date from mid-SU12, and two new dates from the top of SU12, one of which was collected from the feature itself (Fig. 3.9). It was suggested that 920–5210 years (68.2% uncertainty range) of aeolian accumulation are present prior to this first secure archaeological finding (isolated lithics appear lower, but are not considered secure finds). Hiatuses in sedimentation were also suggested at *c.* 30–21 cal BP, 21–7 cal BP, and 7–1 cal BP. However, a hearth feature dated to *c.* 21 cal BP has been found which suggests that occupation occurred at least occasionally during these periods.

OxCal was then used to calculate the differences between the modelled SU boundary ages derived from the radiocarbon and the OSL models (Fig. 3.10). These were entered as normal distributions, with means and standard deviations determined from the two independent age models; the authors note that in their opinion this approximation was reasonable. The Difference command was used to calculate the difference in the probability distributions of the radiocarbon and OSL boundaries before and after a 3.5% average systematic error was added in quadrature for the OSL boundary ages. If 0 was included within the 68.2% or 95.4% highest probability range of the difference distribution, then the modelled boundaries were indistinguishable from each other at the specified level of uncertainty. All difference distributions included 0 within their 95.4% probability range when the OSL systematic error term was included, and only two did not include 0 when a more restrictive range (68.2%) was considered.

This approach allowed the authors to note several findings of interest. First, the differences between the radiocarbon and OSL ages are not systematically offset throughout the whole of the sequence, though examination of Figure 3.10 indicates that there is some pattern in the difference probabilities with depth. Examination of the posterior outlier probabilities allowed six radiocarbon dates to be identified as outliers with a probability >80%; these were all detrital charcoals rather than ones collected from contained archaeological features. The authors suggested this correlation provides evidence for minor charcoal movement through sedimentary sequences, a process that has been suggested to occur but has been rarely tested for. Five OSL ages were identified as outliers with between 10 and 18% probability, even after the D_E winnowing described above, but these were both too old and too young and occurred randomly throughout the sequence. By building and comparing two models, the authors investigated the accuracy of two sequences of high-resolution, independent chronometric data. This approach also solved the issue of 'weighting' of models by high precision dates. That is, if a single age model incorporates ages with dramatically different precisions (e.g. luminescence ages and radiocarbon dates), the final model tends to be determined by the higher precision data. In such a case, it is difficult to recognise the existence of a small systematic offset between ages provided by two chronometric techniques. It should be noted that it is possible to implement this

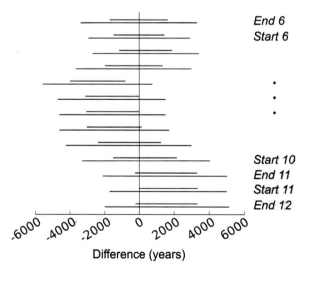

Figure 3.10 Difference between modelled boundary ages calculated via radiocarbon dates and OSL ages from Riwi Cave (see Fig. 3.9), shown at one σ and two σ uncertainties.

comparative dual-model procedure within OxCal by creating the two models with named boundaries, as well as a separate section that calls the boundary probabilities and calculates the difference between them. This approach allows the Difference command to directly access the calculated boundary posteriors rather than requiring approximation via normal distributions. Nevertheless, the published approach is novel and interesting, and it provides an example of the more sophisticated analyses available via model querying (see Section 3.4.4).

3.5.2 Hominin fossil chronology

Hominin fossils provide the most direct data available for testing hypotheses concerning evolutionary trajectories, behavioural adaptations, and palaeoenvironmental drivers (Blome *et al.* 2012; Kingston and Harrison 2007; Nespoulet *et al.* 2008). Well-dated fossils, however, are relatively rare, which is problematic when one wishes to assess relationships between sparse remains across continental-scale regions (McBrearty and Brooks 2000). Associated dates, too, may rely on early chronometric data and insecure stratigraphic correlations, which are not necessarily apparent in current literature. This section will introduce Millard's (2008) wide-ranging, well-documented critical study of the available dating evidence for hominin fossils from Arica and the Near East believed to date from between 500 and 50 ka. Though newly published fossils and new dating evidence are available in this rapidly progressing field (Dirks *et al.* 2017; Grine *et al.* 2007; Richter *et al.* 2017), these will not be covered here due to limitations of space.

For this research, compilation of data involved repeated winnowing of dating evidence according to explicitly reported criteria, as discussed in Section 3.3. First, Millard eliminated papers with no chronometric dating available for the site itself or from an associated site, including those where chronological data was limited to general faunal or archaeological associations. Second, data from the remaining sites was subjected to a critical review on the basis of both 'chronological hygiene' and 'stratigraphic hygiene', parameters for each of which were discussed explicitly. Both luminescence ages and electron spin resonance ages were analysed with respect to the security of external dose rate assessment, particularly if measurements were made on lithics or fossil remains from museum collections. Due to partial bleaching issues, Millard proposed that luminescence ages from unheated sediment should preferably use OSL rather than TL signals, and measurements on single grains or several aliquots should have been analysed by study authors in order to show that no partial bleaching has occurred. If feldspars were measured, fading experiments must also have been reported. In some cases, legacy data were considered completely unreliable based on the current state of knowledge. Millard suggested that it was safest to avoid uranium-series ages on mollusc shells, for instance, due to a lack of knowledge about uptake history. Other data, however, could still be included if the interpretation was slightly altered. For instance, finite radiocarbon dates obtained in the early 1960s could be interpreted as minimum ages given what is now known about the effects of modern contamination on radiocarbon dates near the limit of the method (Wood 2015). Similarly, stratigraphic constraints were judged based on the security with which the position and relationship of the dating samples and hominin fossils were known. Millard noted that constraints are most secure if they relate directly to samples at the site of interest, slightly less secure but still strong if distinct markers such as tephra layers are used across a region, and finally, least secure if continental-scale constraints such as biostratigraphy are used. The strength of faunal correlations was discussed on a site-by-site basis in a qualitative manner, though Millard noted that quantitative correlation strategies have been developed in some cases and would be a useful approach for the African case. Finally, if primary publications included too little information about the chronometric methods or stratigraphic associations so that the quality of the dating could not be assessed, the data was excluded from further analysis.

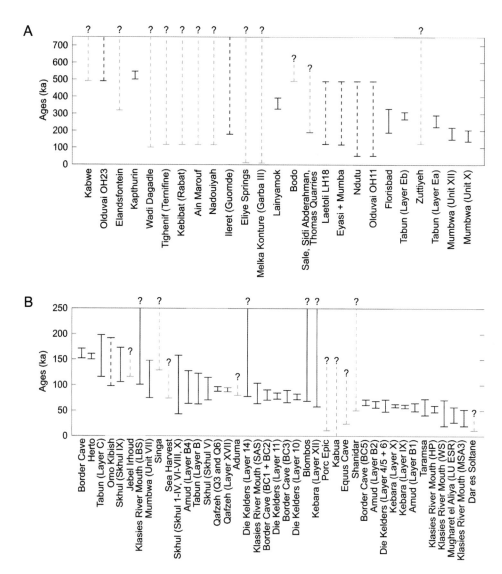

Figure 3.11 Hominin fossil ages from Millard's (2008) data reassessment in approximate chronological order (adapted from Figs. 3.3 and 3.4). Dotted, light-grey lines indicate unmodelled ages (due to lack of secure chronological information), dotted black lines indicate ages with partial chronologies, and solid black lines are fully modelled ages. Question marks indicate uncertainty about dating constraints.

After rejection of insufficiently secure data, Millard followed one of two approaches for modelling of each site. The preferred approach involved two steps. First, chronometric evidence was inspected for data outliers via a statistical test for data spread (Ward and Wilson 1978). In this paper, Millard applied this step only to argon-argon single crystal laser fusion data. However, this approach could also be used with luminescence age data. Following this, data was entered into a hierarchical Bayesian model coded in WinBUGS (v. 1.3 or 1.4) which allowed both a simplified parametrised description of the chronological likelihoods for TL/OSL, U-series, ESR and the input of stratigraphic constraints (Millard 2006a, 2006b). Unfortunately, it was not uncommon for relevant publications to omit some of the information necessary to perform recalculations and rigorous analysis. Millard noted that for six of 34 sites, luminescence ages could not be recalculated due to a lack of detail in the relevant publications. In these cases, and when errors for dates at a site were considered to be independent, OxCal (v. 3.10) was used to create the model (ages input as calendar ages) with the caution that uncertainties would be underestimated if there were some dependency between the ages.

From 66 distinct fossil groups (as identified by Millard), Millard modelled posterior probabilities for only 38 (Fig. 3.11). The majority of the rest did not contain enough chronometric data after quality control, or the chronometric analyses were only partially published. Recalculation of trapped charge ages was undertaken most often to harmonise water content and cosmic dose rate content within associated groups of luminescence ages or between closely associated samples collected for luminescence and ESR age determination, as for Mugharet el Aliya (Morocco), Die Kelders (South Africa), and Klasies River Mouth (South Africa). This procedure also allowed flexibility for more unusual cases, such as the mathematical combination of both TL and OSL D_Es measured for quartz grains extracted from sediment samples from Die Kelders (Feathers and Bush 2000) and re-assessment of dose rates from ESR samples taken from excavation archives for Skuhl (Israel) and Tabun (Israel).

Millard's stringent reanalysis of the available chronometric data for Pleistocene hominin fossils is to date one of the most ambitious publications of a Bayesian chronometric framework involving luminescence dating. First and foremost, this reanalysis provided a more rigorous means by which to discuss the ages of important hominin fossils (Fig. 3.11). In complex scenarios, wherein chronologies are inferred via a web of stratigraphic associations linking a few dated sites, it can be difficult to combine such constraints robustly to provide a 'best dating estimate'. Bayesian inference provides a quantitative method of balancing these associations, and, importantly, assessing uncertainties. This study also highlighted gaps in the literature, both in the primary publications and due to loss of information through long secondary citation 'chains,' and it pinpointed sites where further chronometric samples should be collected if possible. As can be seen in Figure 3.11, significant new dating projects are necessary to improve constraints on most of the older (>200 ka) fossils. Finally, it is worth stressing that though Bayesian analysis often increases precision via age modelling of chronometric likelihoods, this is not the only possible outcome. In some cases, a rigorous analysis of uncertainty will yield lower precisions than those generally cited in literature. For instance, modelling a number of ages obtained from closely associated contexts as a phase rather than averaging them together is likely to increase the age uncertainty of an associated fossil (dated event). However, in many cases this will provide a more accurate assessment. Millard noted that Vermeersch *et al.* (1998) performed weighted averaging of OSL ages from samples of aeolian deposits within and overlying Late Middle Palaeolithic chert extraction pits to obtain a best guess of 55.5 ± 3.7 ka BP, while modelling these ages as a phase provides a 95% probability ranging from 72.7/72.9 to 40.7 ka (depending on whether a uniform or uniform span prior is used). Though less precise, the modelled age range more accurately represents the knowledge available, and it would therefore be of greater use in a chronological framework. While no two chronological frameworks require precisely the same approaches, the methods and outcomes of this publication exemplify the possibilities of legacy-based analysis.

3.5.3 British–Irish ice sheet chronology (BRITICE-CHRONO)

The activity of palaeo-ice sheets can be used to inform ice sheet models and provide palaeoenvironmental information. Constraints on palaeo-ice sheet behaviour have improved due to advances in remote sensing, the availability of bathymetric data, and chronological techniques. Ice sheet models may incorporate hundreds or thousands of dates, and in a recent review, Stokes *et al.* (2015) noted that one of the biggest current challenges is that of uncertainty quantification for chronological constraints on ice sheet margins. The BRITICE-CHRONO project is a large consortium effort which has adopted an integrated approach to legacy data collection, analysis, and the targeted collection of new data in order to chronologically constrain the last advance and retreat of the British–Irish Ice Sheet. As part of this project, explicit and robust quality control criteria (Small *et al.* 2017) have been developed for application to a database of more than 1000 ages compiled by Hughes *et al.* (2011, 2016), including 106 luminescence ages. Bayesian models will be created from this quality-assessed database for hypothesis testing, and targeted dating studies have been implemented to fill in 'gaps' in knowledge that have been revealed by this process (e.g. re-dating of evidence for the Irish Sea Ice Stream (ISIS) from the Isles of Scilly by Smedley *et al.* 2017). To examine the effect of legacy data quality control, we will discuss two closely related Bayesian models created to constrain the extent and retreat of the Irish Sea Ice Stream. The first of these was published by Chiverrell *et al.* (2013). Small *et al.* (2017) published an updated version of this model using the same constraints and same overall construction, but applied the quality control criteria and 'traffic-light' system employed by the BRITICE project. This second model acts as a sampler of the expected BRITICE output. Each model is discussed in turn.

Chiverrell *et al.* (2013) identified 26 sites along the coast of Ireland, England, Wales, and the Isles of Scilly where radiocarbon dates, luminescence ages, and cosmogenic nuclide ages directly date or constrain the age of identifiable glacial deposits (Fig. 3.12). In this case, a site could be identified either as a single geographic location or as a set of neighbouring landforms expected to share same glacial history (e.g. moraines traceable over several kilometres) (Clark *et al.* 2012; Hughes *et al.* 2011). Each applicable dating technique was discussed with the implication of the use of quality control exclusion criteria, but such criteria were not explicitly stated, nor was it clear which or how many ages were excluded from consideration. For example, incomplete bleaching was acknowledged as an issue affecting luminescence ages upon glacigenic sediments, therefore the authors noted that all OSL ages used were calculated from D_Es measured via the SAR protocol either upon single grains or small aliquots (<5 grains). Studies published between 1986 and 2012 yielded 56 ages, 13 of which were luminescence ages for glaciofluvial sands.

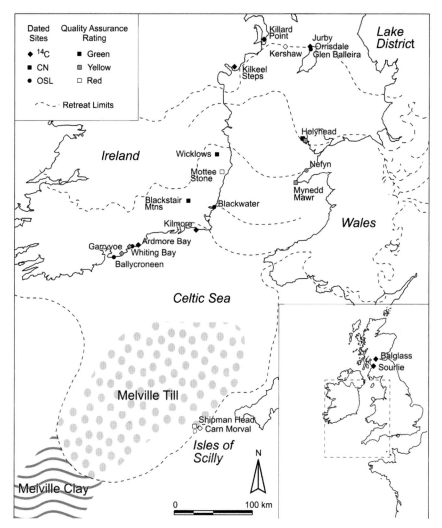

Figure 3.12 Map of sites with dated samples pertaining to the advance and retreat of the Irish Sea Ice Stream (adapted from Fig. 1 from Chiverrell *et al.* 2013). Data quality as re-assessed by Small *et al.* (2017) is indicated.

Chiverrell *et al.*'s relatively complex model was constructed via constraints combining geomorphological analyses, geographic relationships ('pseudo-stratigraphy'), and an understanding of the specific events dated by each chronometric technique (Fig. 3.13). The backbone of the model comprised a sequence of phases that grouped dates which were most closely associated with glacial presence during advance or retreat, including cosmogenic nuclide ages on quartzites eroded by glacial activity and redeposited boulders (Bowen *et al.* 2002; McCarroll *et al.* 2010), and OSL ages on glaciofluvial sands (Ó Cofaigh *et al.* 2012; Scourse *et al.* 2006; Thrasher 2009; Thrasher *et al.* 2009). These phases (and sometimes individual dates) were ordered via geographic location, geomorphic, and stratigraphic evidence into advance, maximum, and retreat stages. Events of interest were implicitly associated with boundaries separating these phases; these boundaries corresponded to inferred glacial presence at a particular geographic location. Other radiocarbon and cosmogenic nuclide dates or groups of dates were included as maximum or minimum

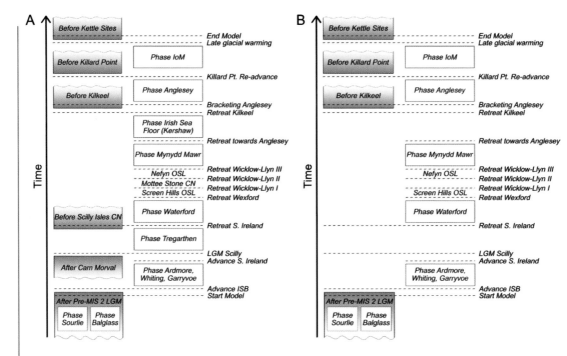

Figure 3.13 A) Model schematics for original Bayesian model from Chiverrell *et al.* (2013), B) reassessed model from Small *et al.* (2017) for advance and retreat of the Irish Sea Ice Stream. Phases, boundaries, and individual dates are displayed as in Figures 3.4–3.6; 'Before' and 'After' constraints are indicated by gradient-filled blocks. In (B), 'red' ages are not shown, 'yellow' ages are indicated by half-grey, half-white fill, and 'green' ages are indicated with grey fill.

constraints, via the 'After' and 'Before' commands, respectively. Organics underlying glacigenic sediments near the ISIS ice source region (Sourlie, Ayrshire (Bos *et al.* 2004) and Balglass Burn (Brown *et al.* 2007)) provided an initial *terminus post quem* (TPQ) for ice sheet advance, and a radiocarbon date on organics underlying glacial deposits at Carn Morval (Scilly) provided a TPQ for the maximal ISIS extent. Minimum ages (TAQs) were provided by radiocarbon dates on foraminifera in marine muds, suggesting an absence of ice, from Killard Point and the Kilkeel Steps (McCabe *et al.* 2007); geography determined their placement in the model, with the Kilkeel steps being more southerly and therefore providing an earlier TAQ for the retreat. Dated plant macros collected from kettle holes (Isle of Man) (Roberts *et al.* 2006) provided a final TAQ and the latest constraint in the model. Chronometric data was excluded from this first model on methodological grounds, e.g. bulk radiocarbon dates from marine carbonates and cosmogenic nuclide data where <4 samples from the same context were measured, and or due to suspected outlier status, e.g. those which yielded low individual agreement indices in previous models (not published). The final model was run with outlier detection; some ages with low agreement indices were still included but assigned an elevated prior probability of being an outlier. No information was given as to how the prior was assigned, or why these particular ages were not excluded.

Small *et al.* (2017) reanalysed the above data via a three-tier, 'semi-quantitative' quality assessment traffic light system (see Table 3.2). With this approach, they identified 30 of the original ages as high quality (green), seven as amber (moderately secure), and seven as red

Table 3.2. Quality control criteria used to screen legacy data for constraining British–Irish Ice Sheet chronology; reproduced and adapted from Table 3 in Small *et al.* (2017).

Technique	Quality	Criterion
All	Prerequisite	-Sufficient data for recalculation or recalibration
Radiocarbon	Green	-Multiple, consistent macro/microfossil samples -Reservoir addressed -Clear stratigraphic relationship to event of interest
	Yellow	-Single macro/microfossil sample -Strat. consistent bulk sample dates -Reservoir addressed -Good stratigraphic relationship to event of interest
	Red	-Single macrofossil/microfossil -Single bulk sample -No internally consistent ages -Poor stratigraphic security
Cosmogenic nuclide dating	Green	-Acceptable reduced chi-square -Ages directly related to event of interest
	Yellow	-Only 2 internally consistent ages from a site ->2 samples not directly related to event of interest
	Red	-Single sample -No internally consistent ages
Luminescence dating	Green	-Sensitivity normalised protocol (e.g. SAR) used -Partial bleaching addressed using small aliquot/single grain measurements -Supported by other chronological data (luminescence or independent) from same site -Good stratigraphic relationship to event of interest
	Yellow	-Partial bleaching possible and not assessed -Supported by other chronological data (luminescence or independent) from same site -Good stratigraphic relationship to event of interest
	Red	-Preliminary ages or experimental protocol used -Feldspar ages with no fading experiments/correction -Single sample, no supporting chronology from same site -Insufficient depositional context -Insufficient methodological details -Poor stratigraphic relationship to event of interest

(not usable). The authors re-ran Chiverrell *et al.*'s model with an outlier model, but this time assigned all ages outlier prior probabilities of 0.05 (green), 0.2 (amber), and 1.0 (red). As Chiverrell *et al.* had already down-weighted outlying ages in the 2013 model, modelled boundary distributions from both models were similar. Small *et al.* (2017), though, noted that the data quality ranking strategy resulted in higher boundary precisions, with a

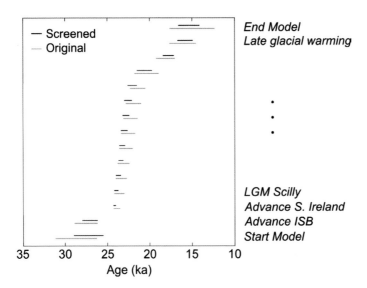

Figure 3.14 Modelled boundaries from original (Chiverrell *et al.* 2013) and quality screened (Small *et al.* 2017) Bayesian models of the advance and retreat of the Irish Sea Ice Stream.

consistent increase in precision of at least 500 years (Fig. 3.14). This demonstrates the utility of such an approach, in which outliers were weighted purely on methodological security, stratigraphic/geomorphic security, and the existence of other supporting data from a site, whilst remaining neutral with respect to agreement or disagreement of the chronometric data with the prevailing hypothesis. Such an approach is more likely to eliminate dates/ages that are 'conformable' with the chronometric framework, but rely on specious data, and thereby may adversely influence the accuracy and precision of the final age model.

Based on the earlier 2013 model, Chiverrell *et al.* calculated that the Irish Sea Ice sheet expanded from its source area after 34.0 ka, and reached a maximum extent near the Isles of Scilly between 24.3 and 23.1 ka. The authors used the modelled boundary ages and a distance model in order to calculate geographically variable expanse and retreat rates, and hypothesised links between retreat rates and topographic control, and eustatic sea-level rise linked with Heinrich Event 2. Improving the precision of the available models by improving quality control assessments and implementing rigorous outlier handling should significantly improve the validity of such hypotheses. Given the outlier analyses now available in several modelling programs (Section 3.4.3), the approach implemented in this study is a useful benchmark for projects incorporating significant legacy data.

3.6 CONCLUSIONS

There is increasing acknowledgement of the utility of Bayesian inference for constructing chronological frameworks which include luminescence ages. These approaches may lead to further use of luminescence age data, even in fields for which such ages are traditionally considered problematic, such as fluvial development and landscape evolution (Rixhon *et al.* 2016). Further developments in user-friendly software and data-mining techniques will enable larger-scale meta-analyses, but it is apparent that Bayesian modelling techniques can also be useful at a much smaller scale. Similarly, there is clearly room for development of more sophisticated approaches to treating luminescence ages in particular. Age input techniques that utilise bootstrapping methods (Cunningham and Wallinga 2012), parametrisation (Millard 2006b), or other methods such as covariance matrices (Combès and Philippe 2017) should allow more robust assessment of shared uncertainties and assumptions in both D_E and D_R calculations. Research into more rigorous methods of combining linked data, such as single grain and multiple-grain D_Es, and development of more sophisticated modelling methods, such as the development of new informative priors and model averaging methods (Bronk Ramsey and Lee 2013), are also desirable.

Finally, it worth re-emphasising that, just as chronometric and stratigraphic data must be published with sufficient detail for critical reappraisals, so too must the quality control criteria, constraints, and model design of any published chronological framework. One of the great strengths of Bayesian inference lies in its reproducibility when input data are provided and assumptions are made explicit. It will always be possible to model a scenario in various ways, however, and it is crucial that the distinction between original data and interpretation is made clear.

REFERENCES

Alexanderson, H., Backman, J., Cronin, T.M., Funder, S., Ingólfsson, Ó., Jakobsson, M., Landvik, J.Y., Löwemark, L., Mangerud, J., März, C., Möller, P., O'Regan, M., Spielhagen, R.F. 2014. An Arctic perspective on dating Mid-Late Pleistocene environmental history. *Quaternary Science Reviews* 92, 9–31.

Ampel, L., Wohlfarth, B., Risberg, J., Veres, D. 2008. Paleolimnological response to millennial and centennial scale climate variability during MIS 3 and 2 as suggested by the diatom record in Les Echets, France. *Quaternary Science Reviews* 27, 1493–1504.

Arnold, L.J., Demuro, M., Parés, J.M., Pérez-González, A., Arsuaga, J.L., Bermúdez de Castro, J.M., Carbonell, E. 2015. Evaluating the suitability of extended-range luminescence dating techniques over early and Middle Pleistocene timescales: published datasets and case studies from Atapuerca, Spain. *Quaternary International* 389, 167–190.

Arnold, L.J., Roberts, R.G. 2009. Stochastic modelling of multi-grain equivalent dose (D_e) distributions: implications for OSL dating of sediment mixtures. *Quaternary Geochronology* 4, 204–230.

Aubry, T., Dimuccio, L.A., Buylaert, J.-P., Liard, M., Murray, A.S., Thomsen, K.J., Walter, B. 2014. Middle-to-Upper Palaeolithic site formation processes at the Bordes-Fitte rockshelter (Central France). *Journal of Archaeological Science* 52, 436–457.

Auclair, M., Lamothe, M., Huot, S. 2003. Measurement of anomalous fading for feldspar IRSL using SAR. *Radiation Measurements* 37, 487–492.

Bailey, R.M., Thomas, D.S.G. 2014. A quantitative approach to understanding dated dune stratigraphies. *Earth Surface Processes and Landforms* 39, 614–631.

Balme, J. 2000. Excavations revealing 40,000 years of occupation at Mimbi Caves, south central Kimberley, Western Australia. *Australian Archaeology* 51, 1–5.

Barton, R.N.E., Bouzouggar, A., Collcutt, S.N., Schwenninger, J.-L., Clark-Balzan, L. 2009. OSL dating of the Aterian levels at Dar es-Soltan I (Rabat, Morocco) and implications for the dispersal of modern Homo sapiens. *Quaternary Science Reviews* 28, 1914–1931.

Bateman, M.D., Boulter, C.H., Carr, A.S., Frederick, C.D., Peter, D., Wilder, M. 2007. Preserving the palaeoenvironmental record in Drylands: bioturbation and its significance for luminescence-derived chronologies. *Sedimentary Geology* 195, 5–19.

Bayes, T. 1763. An essay toward solving a problem in the doctrine of chances. Communicated by R. Price. *Philosophical Transactions of the Royal Society of London* 53, 370–418.

Bayliss, A. 2015. Quality in Bayesian chronological models in archaeology. *World Archaeology* 47, 677–700.

Bevington, P.R., Robinson, D.K. 2003. *Data Reduction and Error Analysis for the Physical Sciences.* 3rd edition, McGraw-Hill Higher Education, New York.

Blaauw, M., Bakker, R., Christen, J.A., Hall, V.A., van der Plicht, J. 2007. A Bayesian framework for age modelling of radiocarbon-dated peat deposits: case studies from the Netherlands. *Radiocarbon* 49, 357–367.

Blaauw, M., Christen, J.A. 2013. *Bacon manual – v2.2.* Version 2.2.

Blaauw, M., Christen, J.A. 2011. Flexible paleoclimate age–depth models using an autoregressive gamma process. *Bayesian Analysis* 6, 457–474.

Blaauw, M., Christen, J.A. 2005. Radiocarbon peat chronologies and environmental change. *Journal of the Royal Statistical Society, Series C (Applied Statistics)* 54, 805–816.

Blockley, S.P.E., Bronk Ramsey, C., Lane, C.S., Lotter, A.F. 2008. Improved age modelling approaches as exemplified by the revised chronology for the Central European varved lake Soppensee. *Quaternary Science Reviews* 27, 61–71.

Blome, M.W., Cohen, A.S., Tryon, C.A., Brooks, A.S., Russell, J. 2012. The environmental context for the origins of modern human diversity: a synthesis of regional variability in African climate 150,000–30,000 years ago. *Journal of Human Evolution* 62, 563–592.

Bos, J.A.A., Dickson, J.H., Coope, G.R., Jardine, W.G. 2004. Flora, fauna and climate of Scotland during the Weichselian Middle Pleniglacial – palynological, macrofossil and coleopteran investigations. *Palaeogeography, Palaeoclimatology, Palaeoecology* 204, 65–100.

Boslough, M., Nicoll, K., Daulton, T.L., Scott, A.C., Claeys, P., Gill, J.L., Marlon, J.R., Bartlein, P.J. 2015. Incomplete Bayesian model rejects contradictory radiocarbon data for being contradictory. *Proceedings of the National Academy of Sciences* 112, E6722.

Bowen, D.Q., Phillips, F.M., McCabe, A.M., Knutz, P.C., Sykes, G.A. 2002. New data for the last Glacial Maximum in Great Britain and Ireland. *Quaternary Science Reviews* 21, 89–101.

Brauer, A., Hajdas, I., Blockley, S.P.E., Bronk Ramsey, C., Christl, M., Ivy-Ochs, S., Moseley, G.E., Nowaczyk, N.N., Rasmussen, S.O., Roberts, H.M., Spötl, C., Staff, R.A., Svensson, A. 2014. The importance of independent chronology in integrating records of past climate change for the 60–8 ka INTIMATE time interval. *Quaternary Science Reviews* 106, 47–66.

Brill, D., Jankaew, K., Brückner, H. 2015. Holocene evolution of Phra Thong's beach-ridge plain (Thailand) – Chronology, processes and driving factors. *Geomorphology* 245, 117–134.

Bristow, C.S., Armitage, S.J. 2016. Dune ages in the sand deserts of the southern Sahara and Sahel. *Quaternary International* 410, 46–57.

Bronk Ramsey, C. 2008. Deposition models for chronological records. *Quaternary Science Reviews* 27, 42–60.

Bronk Ramsey, C. (2009a). Bayesian analysis of radiocarbon dates. *Radiocarbon* 51, 337–360.

Bronk Ramsey, C. (2009b). Dealing with outliers and offsets in radiocarbon dating. *Radiocarbon* 51, 1023–1045.

Bronk Ramsey, C. 2017. *OxCal 4.3 Manual.* Version 4.3.

Bronk Ramsey, C., Lee, S. 2013. Recent and planned developments of the program OxCal. *Radiocarbon* 55, 720–730.

Bronk Ramsey, C., Albert, P., Blockley, S., Hardiman, M., Lane, C., Macleod, A., Matthews, I.P., Muscheler, R., Palmer, A., Staff, R.A. 2014. Integrating timescales with time-transfer functions: a practical approach for an INTIMATE database. *Quaternary Science Reviews* 106, 67–80.

Brown, A.G. 2008. Geoarchaeology, the four dimensional (4D) fluvial matrix and climatic causality. *Geomorphology* 101, 278–297.

Brown, E.J., Rose, J., Coope, R.G., Lowe, J.J. 2007. An MIS 3 age organic deposit from Balglass Burn, central Scotland: palaeoenvironmental significance and implications for the timing of the onset of the LGM ice sheet in the vicinity of the British Isles. *Journal of Quaternary Science* 22, 295–308.

Buck, C.E., Christen, J.A., James, G.N. 1999. Bcal: an on-line Bayesian radiocarbon calibration tool. *Internet Archaeology* 7.

Bueno, L., Feathers, J., De Blasis, P. 2013. The formation process of a odellingn open-air site in Central Brazil: integrating lithic analysis, radiocarbon and luminescence dating. *Journal of Archaeological Science* 40, 190–203.

Burrough, S.L., Thomas, D.S.G. 2013. Central southern Africa at the time of the African Humid Period: a new analysis of Holocene palaeoenvironmental and palaeoclimate data. *Quaternary Science Reviews* 80, 29–46.

Burrough, S.L., Thomas, D.S.G., Bailey, R.M. 2009. Mega-Lake in the Kalahari: a Late Pleistocene record of the Palaeolake Makgadikgadi system. *Quaternary Science Reviews* 28, 1392–1411.

Chiverrell, R.C., Foster, G.C., Thomas, G.S.P., Marshall, P., Hamilton, D. 2009. Robust chronologies for landform development. *Earth Surface Processes and Landforms* 34, 319–328.

Chiverrell, R.C., Thrasher, I.M., Thomas, G.S.P., Lang, A., Scourse, J.D., van Landeghem, K.J.J., McCarroll, D., Clark, C.D., Ó Cofaigh, C., Evans, D.J.A., Ballantyne, C.K. 2013. Bayesian modelling the retreat of the Irish Sea Ice Stream. *Journal of Quaternary Science* 28, 200–209.

Christen, J.A. 1994. Summarizing a set of radiocarbon determinations: a robust approach. . *Journal of the Royal Statistical Society, Series C (Applied Statistics)* 43, 489–503.

Christophe, C., Philippe, A., Guérin, G., Mercier, N., Guibert, P. 2018 Bayesian approach to OSL dating of poorly bleached sediment samples: Mixture Distribution Models for Dose (MD²). *Radiation Measurements* 108, 59–73.

Christophe, C., Philippe, A., Kreutzer, S., Guérin, G. 2017. BayLum: Chronological Bayesian Models Integrating Optically Stimulated Luminescence and Radiocarbon Age Dating. R Package version 0.1.1. https://CRAN.R-project.org/package=BayLum

Clark, C.D., Hughes, A.L.C., Greenwood, S.L., Jordan, C., Sejrup, H.P. 2012. Pattern and timing of retreat of the last British–Irish Ice Sheet. *Quaternary Science Reviews* 44, 112–146.

Clark-Balzan, L.A., Candy, I., Schwenninger, J.L., Bouzouggar, A., Blockley, S., Nathan, R., Barton, R.N.E. 2012. Coupled U-series and OSL dating of a Late Pleistocene cave sediment sequence, Morocco, North Africa: Significance for constructing Palaeolithic chronologies. *Quaternary Geochronology* 12, 53–64.

Clarke, M.L., Rendell, H.M., Wintle, A.G. 1999. Quality assurance in luminescence dating. *Geomorphology* 29, 173–185.

Clarkson, C., Smith, M., Marwick, B., Fullagar, R., Wallis, L.A., Faulkner, P., Manne, T., Hayes, E., Roberts, R.G., Jacobs, Z., Carah, X., Lowe, K.M., Matthews, J., Florin, S.A. 2015. The archaeology, chronology and stratigraphy of Madjedbebe (Malakunanja II): a site in northern Australia with early occupation. *Journal of Human Evolution* 83, 46–64.

Combès, B., Philippe, A. 2017. Bayesian analysis of individual and systematic multiplicative errors for estimating ages with stratigraphic constraints in optically stimulated luminescence dating. *Quaternary Geochronology* 39, 24–34.

Combès, B., Philippe, A., Lanos, P., Mercier, N., Tribolo, C., Guerin, G., Guibert, P., Lahaye, C. 2015. A Bayesian central equivalent dose model for optically stimulated luminescence dating. *Quaternary Geochronology* 28, 62–70.

Cunningham, A.C., DeVries, D.J., Schaart, D.R. 2012. Experimental and computational simulation of beta-dose heterogeneity in sediment. *Radiation Measurements* 47, 1060–1067.

Cunningham, A.C., Wallinga, J. 2012. Realizing the potential of fluvial archives using robust OSL chronologies. *Quaternary Geochronology* 12, 98–106.

Davies, L.J., Jensen, B.J.L., Froese, D.G., Wallace, K.L. 2016. Late Pleistocene and Holocene tephrostratigraphy of interior Alaska and Yukon: key beds and chronologies over the past 30,000 years. *Quaternary Science Reviews* 146, 28–53.

Dirks, P.H.G.M., Roberts, E.M., Hilbert-Wolf, H., Kramers, J.D., Hawks, J., Dosseto, A., Duval, M., Elliott, M., Evans, M., Grün, R., Hellstrom, J., Herries, A.I.R., Joannes-Boyau, R., Makhubela, T.V, Placzek, C.J., Robbins, J., Spandler, C., Wiersma, J., Woodhead, J., Berger, L.R. 2017. The age of *Homo naledi* and associated sediments in the Rising Star Cave, South Africa. *eLife* 6, e24231.

Discamps, E., Jaubert, J., Bachellerie, F. 2011. Human choices and environmental constraints: deciphering the variability of large game procurement from Mousterian to Aurignacian times (MIS 5–3) in southwestern France. *Quaternary Science Reviews* 30, 2755–2775.

Douka, K., Jacobs, Z., Lane, C., Grün, R., Farr, L., Hunt, C., Inglis, R.H., Reynolds, T., Albert, P., Aubert, M., Cullen, V., Hill, E., Kinsley, L., Roberts, R.G., Tomlinson, E.L., Wulf, S., Barker, G. 2014. The chronostratigraphy of the Haua Fteah cave (Cyrenaica, northeast Libya). *Journal of Human Evolution* 66, 39–63.

Duller, G.A.T. 2012. Improving the accuracy and precision of equivalent doses determined using the optically stimulated luminescence signal from single grains of quartz. *Radiation Measurements* 47, 770–777.

Duller, G.A.T., Bøtter-Jensen, L., Murray, A.S. 2000. Optical dating of single sand-sized grains of quartz: sources of variability. *Radiation Measurements* 32, 453–457.

Duller, G.A.T., Bøtter-Jensen, L., Murray, A.S., Truscott, A.J. 1999. Single grain laser luminescence (SGLL) measurements using a novel automated reader. *Nuclear Instruments and Methods in Physics Research B* 155, 506–514.

Duller, G.A.T. 2016. Challenges involved in obtaining luminescence ages for long records of aridity: examples from the Arabian Peninsula. *Quaternary International* 410, 69–74.

Durcan, J.A., King, G.E., Duller, G.A.T. 2015. DRAC: Dose Rate and Age Calculator for trapped charge dating. *Quaternary Geochronology* 28, 54–61.

Feathers, J.K., Bush, D.A. 2000. Luminescence dating of Middle Stone Age deposits at Die Kelders. *Journal of Human Evolution* 38, 91–119.

Feathers, J.K., Rhodes, E.J., Huot, S., Mcavoy, J.M. 2006. Luminescence dating of sand deposits related to late Pleistocene human occupation at the Cactus Hill Site, Virginia, USA. *Quaternary Geochronology* 1, 167–187.

Fienberg, S.E. 2006. When did Bayesian inference become 'Bayesian'? *Bayesian Analysis* 1, 1–40.

Flantua, S.G.A., Blaauw, M., Hooghiemstra, H. 2016. Geochronological database and classification system for age uncertainties in Neotropical pollen records. *Climate of the Past* 12, 387–414.

Fu, X., Cohen, T.J., Arnold, L.J. 2017. Extending the record of lacustrine phases beyond the last interglacial for Lake Eyre in central Australia using luminescence dating. *Quaternary Science Reviews* 162, 88–110.

Fullagar, R.L.K., Price, D.M., Head, L.M. 1996. Early human occupation of northern Australia: archaeology and thermoluminescence dating of Jinmium rock-shelter, Northern Territory. *Antiquity* 70, 751–773.

Galbraith, R. 2003. A simple homogeneity test for estimates of dose obtained using OSL. *Ancient TL* 21, 75–77.

Galbraith, R.F., Green, P.F. 1990. Estimating the component ages in a finite mixture. *International Journal of Radiation Applications and Instrumentation, Part D. Nuclear Tracks and Radiation Measurements* 17, 197–206.

Galbraith, R.F., Roberts, R.G. 2012. Statistical aspects of equivalent dose and error calculation and display in OSL dating: an overview and some recommendations. *Quaternary Geochronology* 11, 1–27.

Galbraith, R.F., Roberts, R.G., Laslett, G.M., Yoshida, H., Olley, J.M. 1999. Optical dating of single and multiple grains of quartz from Jinmium rock shelter, northern Australia: part I, Experimental design and statistical models. *Archaeometry* 41, 339–364.

Gelman, A., Carlin, J.B., Stern, H.S., Rubin, D.B. 2004. *Bayesian Data Analysis*. 2nd edition, Chapman and Hall/CRC Press, New York.

Glass, G.V. 1976. Primary, secondary, and meta-analysis of research. *Educational Researcher* 5 (10), 3–8.

Gliganic, L.A., Jacobs, Z., Roberts, R.G. 2012. Luminescence characteristics and dose distributions for quartz and feldspar grains from Mumba rockshelter, Tanzania. *Archaeological and Anthropological Sciences* 4, 115–135.

Greenbaum, N., Porat, N., Rhodes, E., Enzel, Y. 2006. Large floods during late Oxygen Isotope Stage 3, southern Negev desert, Israel. *Quaternary Science Reviews* 25, 704–719.

Grine, F.E., Bailey, R.M., Harvati, K., Nathan, R.P., Morris, A.G., Henderson, G.M., Ribot, I., Pike, A.W.G. 2007. Late Pleistocene human skull from Hofmeyr, South Africa, and modern human origins. *Science* 315, 226–229.

Groucutt, H.S., Petraglia, M.D., Bailey, G., Scerri, E.M.L., Parton, A., Clark-Balzan, L., Jennings, R.P., Lewis, L., Blinkhorn, J., Drake, N.A., Breeze, P.S., Inglis, R.H., Devès, M.H., Meredith-Williams, M., Boivin, N., Thomas, M.G., Scally, A. 2015. Rethinking the dispersal of *Homo sapiens* out of Africa. *Evolutionary Anthropology* 24, 149–164.

Guérin, G., Christophe, C., Philippe, A., Murray, A.S., Thomsen, K.J., Tribolo, C., Urbanova, P., Jain, M., Guibert, P., Mercier, N., Kreutzer, S., Lahaye, C. 2017a. Absorbed dose, equivalent dose, measured dose rates, and implications for OSL age estimates: introducing the Average Dose Model. *Quaternary Geochronology* 41, 163–173.

Guérin, G., Frouin, M., Tuquoi, J., Thomsen, K.J., Goldberg, P., Aldeias, V., Lahaye, C., Mercier, N., Guibert, P., Jain, M., Sandgathe, D., McPherron, S.J.P., Turq, A., Dibble, H.L. 2017b. The complementarity of luminescence dating methods illustrated on the Mousterian sequence of the Roc de Marsal: a series of reindeer-dominated, Quina Mousterian layers dated to MIS 3. *Quaternary International* 433, 102–115.

Guérin, G., Murray, A.S., Jain, M., Thomsen, K.J., Mercier, N. 2013. How confident are we in the chronology of the transition between Howieson's Poort and Still Bay? *Journal of Human Evolution* 64, 314–317.

Guibert, P., Bechtel, F., Bourguignon, L., Brenet, M., Couchoud, I., Delagnes, A., Delpech, F., Detrain, L., Duttine, M., Folgado, M., Jaubert, J., Lahaye, C., Lenoir, M., Maureille, B., Texier, J.-P., Turq, A., Vieillevigne, E., Villeneuve, G. 2008. *Une base de données pour la chronologie du Paléolithique moyen dans le Sud-Ouest de la France.* In Jaubert, J., Bordes, J.-G., Ortega, I. (eds) Les sociétés du Paléolithique dans un Grand Sud-Ouest de la France: nouveaux gisements, nouveaux résultats, odelling methodes. Société Préhistorique Française, Paris, 19–40.

Halfen, A.F., Lancaster, N., Wolfe, S. 2016. Interpretations and common challenges of aeolian records from North American dune fields. *Quaternary International* 410, 75–95.

Hamm, G., Mitchell, P., Arnold, L.J., Prideaux, G.J., Questiaux, D., Spooner, N.A., Levchenko, V.A., Foley, E.C., Worthy, T.H., Stephenson, B., Coulthard, V., Coulthard, C., Wilton, S., Johnston, D. 2016. Cultural innovation and megafauna interaction in the early settlement of arid Australia. *Nature* 539, 280–283.

Haslett, J., Parnell, A. 2008. A simple monotone process with application to radiocarbon-dated depth chronologies. *Journal of the Royal Statistical Society, Series C (Applied Statistics)* 57, 399–418.

Hesse, P.P. 2016. How do longitudinal dunes respond to climate forcing? Insights from 25 years of luminescence dating of the Australian desert dunefields. *Quaternary International* 410, 11–29.

Higham, T., Douka, K., Wood, R., Bronk Ramsey, C., Brock, F., Basell, L., Camps, M., Arrizabalaga, A., Baena, J., Barroso-Ruíz, C., Bergman, C., Boitard, C., Boscato, P., Caparrós, M., Conard, N.J., Draily, C., Froment, A., Galván, B., Gambassini, P., Garcia-Moreno, A., Grimaldi, S., Haesaerts, P., Holt, B., Iriarte-Chiapusso, M.-J., Jelinek, A., Jordá Pardo, J.F., Maíllo-Fernández, J.-M., Marom, A., Maroto, J., Menéndez, M., Metz, L., Morin, E., Moroni, A., Negrino, F., Panagopoulou, E., Peresani, M., Pirson, S., de la Rasilla, M., Riel-Salvatore, J., Ronchitelli, A., Santamaria, D., Semal, P., Slimak, L., Soler, J., Soler, N., Villaluenga, A., Pinhasi, R., Jacobi, R. 2014. The timing and spatiotemporal patterning of Neanderthal disappearance. *Nature* 512, 306–309.

Hobo, N., Makaske, B., Wallinga, J., Middelkoop, H. 2014. Reconstruction of eroded and deposited sediment volumes of the embanked River Waal, the Netherlands, for the period AD 1631–present. *Earth Surface Processes and Landforms* 39, 1301–1318.

Holliday, V.T. 2015. Problematic dating of claimed Younger Dryas boundary impact proxies. *Proceedings of the National Academy of Sciences* 112, E6721.

Huckleberry, G., Ferguson, T.J., Rittenour, T., Banet, C., Mahan, S. 2016. Identification and dating of indigenous water storage reservoirs along the Rio San José at Laguna Pueblo, western New Mexico, USA. *Journal of Arid Environments* 127, 171–186.

Hughes, A.L.C., Greenwood, S.L., Clark, C.D. 2011. Dating constraints on the last British–Irish Ice Sheet: a map and database. *Journal of Maps* 7, 156–184.

Hughes, A.L.C., Gyllencreutz, R., Lohne, Ø.S., Mangerud, J., Svendsen, J.I. 2016. The last Eurasian ice sheets – a chronological database and time-slice reconstruction, DATED-1. *Boreas* 45, 1–45.

Hughes, P.J., Sullivan, M.E., Hiscock, P. 2017. Palaeoclimate and human occupation in southeastern arid Australia. *Quaternary Science Reviews* 163, 72–83.

Hunt, C.O., Gilbertson, D.D., Hill, E.A., Simpson, D. 2015. Sedimentation, re-sedimentation and chronologies in archaeologically-important caves: problems and prospects. *Journal of Archaeological Science* 56, 109–116.

Huntley, D.J., Godfrey-Smith, D.I., Haskell, E.H. 1991. Light-induced emission spectra from some quartz and feldspars. *International Journal of Radiation Applications and Instrumentation, Part D. Nuclear Tracks and Radiation Measurements* 18, 127–131.

Huntley, D.J., Godfrey-Smith, D.I., Thewalt, M.L.W. 1985. Optical dating of sediments. *Nature* 313, 105–107.

Huntley, D.J., Lamothe, M. 2001. Ubiquity of anomalous fading in K-feldspars and the measurement and correction for it in optical dating. *Canadian Journal of Earth Sciences* 38, 1093–1106.

Jacobs, Z., Jankowski, N.R., Dibble, H.L., Goldberg, P., McPherron, S.J.P., Sandgathe, D., Soressi, M. 2016. The age of three Middle Palaeolithic sites: single-grain optically stimulated luminescence chronologies for Pech de l'Azé I, II and IV in France. *Journal of Human Evolution* 95, 80–103.

Jacobs, Z., Li, B., Jankowski, N., Soressi, M. 2015. Testing of a single grain OSL chronology across the Middle to Upper Palaeolithic transition at Les Cottés (France). *Journal of Archaeological Science* 54, 110–122.

Jacobs, Z., Wintle, A.G., Duller, G.A.T., Roberts, R.G., Wadley, L. 2008. New ages for the post-Howiesons Poort, late and final Middle Stone Age at Sibudu, South Africa. *Journal of Archaeological Science* 35, 1790–1807.

Jones, A.F., Macklin, M.G., Benito, G. 2015. Meta-analysis of Holocene fluvial sedimentary archives: a methodological primer. *Catena* 130, 3–12.

Kempf, P., Moernaut, J., Van Daele, M., Vandoorne, W., Pino, M., Urrutia, R., De Batist, M. 2017. Coastal lake sediments reveal 5500 years of tsunami history in south central Chile. *Quaternary Science Reviews* 161, 99–116.

Kennett, J.P., Kennett, D.J., Culleton, B.J., Tortosa, J.E.A., Bischoff, J.L., Bunch, T.E., Randolph Daniel Jr., I, Erlandson, J.M., Ferraro, D., Firestone, R.B., Goodyear, A.C., Israde-Alcántara, I., Johnson, J.R., Jordá Pardo, J.F., Kimbel, D.R., LeCompte, M.A., Lopinot, N.H., Mahaney, W.C., Moore, A.M.T., Moore, C.R., Ray, J.H., Stafford Jr., T.W., Tankersley, K.B., Wittke, J.H., Wolbach, W.S., West, A. (2015a). Bayesian chronological analyses consistent with synchronous age of 12,835–12,735 Cal B.P. for Younger Dryas boundary on four continents. *Proceedings of the National Academy of Sciences* 112, E4344–E4353.

Kennett, J.P., Kennett, D.J., Culleton, B.J., Tortosa, J.E.A., Bunch, T.E., Erlandson, J.M., Johnson, J.R., Jordá Pardo, J.F., LeCompte, M.A., Mahaney, W.C., Tankersley, K.B., Wittke, J.H., Wolbach, W.S., West, A. (2015b). Reply to Holliday and Boslough *et al.*: synchroneity of widespread Bayesian-modeled ages supports Younger Dryas impact hypothesis. *Proceedings of the National Academy of Sciences* 112, E6723–E6724.

Kingston, J.D., Harrison, T. 2007. Isotopic dietary reconstructions of Pliocene herbivores at Laetoli: implications for early hominin paleoecology. *Palaeogeography, Palaeoclimatology, Palaeoecology* 243, 272–306.

Lancaster, N., Wolfe, S., Thomas, D., Bristow, C., Bubenzer, O., Burrough, S., Duller, G., Halfen, A., Hesse, P., Roskin, J., Singhvi, A., Tsoar, H., Tripaldi, A., Yang, X., Zárate, M. 2016. The INQUA Dunes Atlas chronologic database. *Quaternary International* 410, 3–10.

Leighton, C.L., Bailey, R.M., Thomas, D.S.G. 2013. The utility of desert sand dunes as Quaternary chronostratigraphic archives: evidence from the northeast Rub' al Khali. *Quaternary Science Reviews* 78, 303–318.

Leighton, C.L., Thomas, D.S.G., Bailey, R.M. 2014. Reproducibility and utility of dune luminescence chronologies. *Earth-Science Reviews* 129, 24–39.

Li, G., Wen, L., Xia, D., Duan, Y., Rao, Z., Madsen, D.B., Wei, H., Li, F., Jia, J., Chen, F. 2015. Quartz OSL and K-feldspar pIRIR dating of a loess/paleosol sequence from arid central Asia, Tianshan Mountains, NW China. *Quaternary Geochronology* 28, 40–53.

Li, H., Yang, X. 2016. Spatial and temporal patterns of aeolian activities in the desert belt of northern China revealed by dune chronologies. *Quaternary International* 410, 58–68.

Livsey, D., Simms, A.R., Hangsterfer, A., Nisbet, R.A., DeWitt, R. 2016. Drought modulated by North Atlantic sea surface temperatures for the last 3,000 years along the northwestern Gulf of Mexico. *Quaternary Science Reviews* 135, 54–64.

Lunn, D.J., Jackson, C., Best, N., Thomas, A., Spiegelhalter, D. 2012. *The BUGS Book*. Chapman and Hall/CRC Press, New York.

Lunn, D.J., Thomas, A., Best, N., Spiegelhalter, D. 2000. WinBUGS – A Bayesian modelling framework: concepts, structure, and extensibility. *Statistics and Computing* 10, 325–337.

Macken, A.C., Staff, R.A., Reed, E.H. 2013. Bayesian age–depth modelling of Late Quaternary deposits from Wet and Blanche Caves, Naracoorte, South Australia: a framework for comparative faunal analyses. *Quaternary Geochronology* 17, 26–43.

Mayya, Y.S., Morthekai, P., Murari, M.K., Singhvi, A.K. 2006. Towards quantifying beta microdosimetric effects in single-grain quartz dose distribution. *Radiation Measurements* 41, 1032–1039.

McBrearty, S., Brooks, A.S. 2000. The revolution that wasn't: a new interpretation of the origin of modern human odelling. *Journal of Human Evolution* 39, 453–563.

McCabe, A.M., Clark, P.U., Clark, J., Dunlop, P. 2007. Radiocarbon constraints on readvances of the British–Irish Ice Sheet in the northern Irish Sea Basin during the last deglaciation. *Quaternary Science Reviews* 26, 1204–1211.

McCarroll, D., Stone, J.O., Ballantyne, C.K., Scourse, J.D., Fifield, L.K., Evans, D.J.A., Hiemstra, J.F. 2010. Exposure-age constraints on the extent, timing and rate of retreat of the last Irish Sea ice stream. *Quaternary Science Reviews* 29, 1844–1852.

Millard, A.R. 2008. A critique of the chronometric evidence for hominid fossils: I. Africa and the Near East 500–50 ka. *Journal of Human Evolution* 54, 848–874.

Millard, A.R. (2006a). Bayesian analysis of ESR dates, with application to Border Cave. *Quaternary Geochronology* 1, 159–166.

Millard, A.R. (2006b). Bayesian analysis of Pleistocene chronometric methods. *Archaeometry* 48, 359–375.

Mischke, S., Lai, Z., Aichner, B., Heinecke, L., Mahmoudov, Z., Kuessner, M., Herzschuh, U. 2017. Radiocarbon and optically stimulated luminescence dating of sediments from Lake Karakul, Tajikistan. *Quaternary Geochronology* 41, 51–61.

Munyikwa, K. 2005. The role of dune morphogenetic history in the interpretation of linear dune luminescence chronologies: a review of linear dune dynamics. *Progress in Physical Geography* 29, 317–336.

Murray, A.S., Olley, J.M. 2002. Precision and accuracy in the optically stimulated luminescence dating of sedimentary quartz: a status review. *Geochronometria* 21, 1–16.

Murray, A.S., Funder, S. 2003. Optically stimulated luminescence dating of a Danish Eemian coastal marine deposit: a test of accuracy. *Quaternary Science Reviews* 22, 1177–1183.

Murray, A.S., Wintle, A.G. 2003. The single aliquot regenerative dose protocol: potential for improvements in reliability. *Radiation Measurements* 37, 377–381.

Murray, A.S., Wintle, A.G. 2000. Luminescence dating of quartz using an improved single-aliquot regenerative-dose protocol. *Radiation Measurements* 32, 57–73.

Nathan, R.P., Thomas, P.J., Jain, M., Murray, A.S., Rhodes, E.J. 2003. Environmental dose rate heterogeneity of beta radiation and its implications for luminescence dating: Monte Carlo modelling and experimental validation. *Radiation Measurements* 37, 305–313.

Nespoulet, R., El Hajraoui, M.A., Amani, F., Ben Ncer, A., Debénath, A., El Idrissi, A., Lacombe, J.-P., Michel, P., Oujaa, A., Stoetzel, E. 2008. Palaeolithic and Neolithic occupations in the Témara region (Rabat, Morocco): recent data on hominin contexts and odelling. *African Archaeological Review* 25, 21–39.

O'Connell, J.F., Allen, J. 2015. The process, biotic impact, and global implications of the human colonization of Sahul about 47,000 years ago. *Journal of Archaeological Science* 56, 73–84.

O'Connell, J.F., Allen, J. 2004. Dating the colonization of Sahul (Pleistocene Australia–New Guinea): a review of recent research. *Journal of Archaeological Science* 31, 835–853.

Ó Cofaigh, C., Telfer, M.W., Bailey, R.M., Evans, D.J.A. 2012. Late Pleistocene chronostratigraphy and ice sheet limits, southern Ireland. *Quaternary Science Reviews* 44, 160–179.

Olley, J.M., Roberts, R.G., Yoshida, H., Bowler, J.M. 2006. Single-grain optical dating of grave-infill associated with human burials at Lake Mungo, Australia. *Quaternary Science Reviews* 25, 2469–2474.

Parnell, A.C., Haslett, J., Allen, J.R.M., Buck, C.E., Huntley, B. 2008. A flexible approach to assessing synchroneity of past events using Bayesian reconstructions of sedimentation history. *Quaternary Science Reviews* 27, 1872–1885.

Philippe, A., Guérin, G., Kreutzer, S. 2019. BayLum – An R package for Bayesian analysis of OSL ages: An introduction. *Quaternary Geochronology* 49, 16–24.

Plummer, M. 2003. *JAGS: a program for analysis of Bayesian graphical models using Gibbs sampling.* In Hornik, K., Leisch, F., Zeileis, A. (eds) Proceedings of the 3rd International Workshop on Distributed Statistical Computing (DSC 2003). Available online: https://www.r-project.org/conferences/DSC-2003/Proceedings/.

Rhodes, E.J., Bronk Ramsey, C., Outram, Z., Batt, C., Willis, L., Dockrill, S., Bond, J. 2003. Bayesian methods applied to the interpretation of multiple OSL dates: high precision sediment ages from Old Scatness Broch excavations, Shetland Isles. *Quaternary Science Reviews* 22, 1231–1244.

Richter, D., Grün, R., Joannes-Boyau, R., Steele, T.E., Amani, F., Rué, M., Fernandes, P., Raynal, J.-P., Geraads, D., Ben-Ncer, A., Hublin, J.-J., McPherron, S.P. 2017. The age of the hominin fossils from Jebel Irhoud, Morocco, and the origins of the Middle Stone Age. *Nature* 546, 293–296.

Rixhon, G., Briant, R.M., Cordier, S., Duval, M., Jones, A., Scholz, D. 2016. Revealing the pace of river landscape evolution during the Quaternary: recent developments in numerical dating methods. *Quaternary Science Reviews* 166, 91–113.

Roberts, D.H., Chiverrell, R.C., Innes, J.B., Horton, B.P., Brooks, A.J., Thomas, G.S.P., Turner, S., Gonzalez, S. 2006. Holocene sea levels, Last Glacial Maximum glaciomarine environments and geophysical models in the northern Irish Sea Basin, UK. *Marine Geology* 231, 113–128.

Roberts, R.G., Galbraith, R.F., Olley, J.M., Yoshida, H., Laslett, G.M. 1999. Optical dating of single and multiple grains of quartz from Jinmium rock shelter, Northern Australia: part II, Results and Implications. *Archaeometry* 41, 365–395.

Roberts, R.G., Galbraith, R.F., Yoshida, H., Laslett, G.M., Olley, J.M. 2000. Distinguishing dose populations in sediment mixtures: a test of single-grain optical dating procedures using mixtures of laboratory-dosed quartz. *Radiation Measurements* 32, 459–465.

Roberts, R.G., Jones, R., Smith, M.A. 1990. Thermoluminescence dating of a 50,000-year-old human occupation site in northern Australia. *Nature* 345, 153–156.

Rodnight, H., Duller, G.A.T., Wintle, A.G., Tooth, S. 2006. Assessing the reproducibility and accuracy of optical dating of fluvial deposits. *Quaternary Geochronology* 1, 109–120.

Rodríguez-Rey, M., Herrando-Pérez, S., Gillespie, R., Jacobs, Z., Saltré, F., Brook, B.W., Prideaux, G.J., Roberts, R.G., Cooper, A., Alroy, J., Miller, G.H., Bird, M.I., Johnson, C.N., Beeton, N., Turney, C.S.M., Bradshaw, C.J.A. 2015. Criteria for assessing the quality of Middle Pleistocene to Holocene vertebrate fossil ages. *Quaternary Geochronology* 30, 69–79.

Rosenberg, T.M., Preusser, F., Wintle, A.G. 2011. A comparison of single and multiple aliquot TT-OSL data sets for sand-sized quartz from the Arabian Peninsula. *Radiation Measurements* 46, 573–579.

Scourse, J.D., Evans, D.J., Hiemstra, J., McCarroll, D., Rhodes, E.J., Furze, M.F. 2006. Pleistocene stratigraphy, geomorphology and geochronology. In Scourse, J.D. (ed) *The Isles of Scilly: Field Guide*. Quaternary Research Association, London, 13–22.

Singarayer, J.S., Bailey, R.M., Ward, S., Stokes, S. 2005. Assessing the completeness of optical resetting of quartz OSL in the natural environment. *Radiation Measurements* 40, 13–25.

Small, D., Clark, C.D., Chiverrell, R.C., Smedley, R.K., Bateman, M.D., Duller, G.A.T., Ely, J.C., Fabel, D., Medialdea, A., Moreton, S.G. 2017. Devising quality assurance procedures for assessment of legacy geochronological data relating to deglaciation of the last British–Irish Ice Sheet. *Earth-Science Reviews* 164, 232–250.

Smedley, R.K., Scourse, J.D., Small, D., Hiemstra, J.F., Duller, G.A.T., Bateman, M.D., Burke, M.J., Chiverrell, R.C., Clark, C.D., Davies, S.M., Fabel, D., Gheorghiu, D.M., McCarroll, D., Medialdea, A., Xu, S. 2017. New age constraints for the limit of the British–Irish Ice Sheet on the Isles of Scilly. *Journal of Quaternary Science* 32, 48–62.

Spriggs, M. 1989. The dating of the Island Southeast Asian Neolithic: an attempt at chronometric hygiene and linguistic correlation. *Antiquity* 63, 587–613.

Stan Development Team. 2015. Stan modelling language: user's guide and reference manual. Version 2.8.0.

Steel, D. 2001. Bayesian statistics in radiocarbon calibration. *Philosophy of Science* 68, S153–S164.

Steffen, D., Preusser, F., Schlunegger, F. 2009. OSL quartz age underestimation due to unstable signal components. *Quaternary Geochronology* 4, 353–362.

Stokes, C.R., Tarasov, L., Blomdin, R., Cronin, T.M., Fisher, T.G., Gyllencreutz, R., Hättestrand, C., Heyman, J., Hindmarsh, R.C.A., Hughes, A.L.C., Jakobsson, M., Kirchner, N., Livingstone, S.J., Margold, M., Murton, J.B., Noormets, R., Peltier, W.R., Peteet, D.M., Piper, D.J.W., Preusser, F., Renssen, H., Roberts, D.H., Roche, D.M., Saint-Ange, F., Stroeven, A.P., Teller, J.T. 2015. On the reconstruction of palaeo-ice sheets: recent advances and future challenges. *Quaternary Science Reviews* 125, 15–49.

Thomas, D.S.G., Burrough, S.L. 2016. Luminescence-based dune chronologies in southern Africa: analysis and interpretation of dune database records across the subcontinent. *Quaternary International* 410, 30–45.

Thomsen, K.J., Murray, A.S., Bøtter-Jensen, L. 2005. Sources of variability in OSL dose measurements using single grains of quartz. *Radiation Measurements* 39, 47–61.

Thomsen, K.J., Murray, A.S., Jain, M., Bøtter-Jensen, L. 2008. Laboratory fading rates of various luminescence signals from feldspar-rich sediment extracts. *Radiation Measurements* 43, 1474–1486.

Thrasher, I. 2009. Optically stimulated luminescence dating of ice-marginal palaeosandar from the last Irish Sea Ice-Stream. PhD Thesis, Department of Geography, University of Liverpool.

Thrasher, I.M., Mauz, B., Chiverrell, R.C., Lang, A., Thomas, G.S.P. 2009. Testing an approach to OSL dating of Late Devensian glaciofluvial sediments of the British Isles. *Journal of Quaternary Science* 24, 785–801.

Tripaldi, A., Zárate, M.A. 2016. A review of Late Quaternary inland dune systems of South America east of the Andes. *Quaternary International* 410, 96–110.

Truscott, A.J., Duller, G.A.T., Bøtter-Jensen, L., Murray, A.S., Wintle, A.G. 2000. Reproducibility of optically stimulated luminescence measurements from single grains of Al_2O_3:C and annealed quartz. *Radiation Measurements* 32, 447–451.

Újvári, G., Molnár, M., Novothny, Á., Páll-Gergely, B., Kovács, J., Várhegyi, A. 2014. AMS ^{14}C and OSL/IRSL dating of the Dunaszekcső loess sequence (Hungary): chronology for 20 to 150 ka and implications for establishing reliable age–depth models for the last 40 ka. *Quaternary Science Reviews* 106, 140–154.

Vermeersch, P.M., Paulissen, E., Stokes, S., Charlier, C., van Peer, P., Stringer, C., Lindsay, W. 1998. A Middle Palaeolithic burial of a modern human at Taramsa Hill, Egypt. *Antiquity* 72, 475–484.

Veth, P., Ward, I., Manne, T., Ulm, S., Ditchfield, K., Dortch, J., Hook, F., Petchey, F., Hogg, A., Questiaux, D., Demuro, M., Arnold, L., Spooner, N., Levchenko, V., Skippington, J., Byrne, C., Basgall, M., Zeanah, D., Belton, D., Helmholz, P., Bajkan, S., Bailey, R., Placzek, C., Kendrick, P. 2017. Early human occupation of a maritime desert, Barrow Island, North-West Australia. *Quaternary Science Reviews* 168, 19–29.

Ward, G.K., Wilson, S.R. 1978. Procedures for comparing and combining radiocarbon age determinations: a critique. *Archaeometry* 20, 19–31.

Wintle, A.G., Murray, A.S. 2006. A review of quartz optically stimulated luminescence characteristics and their relevance in single-aliquot regeneration dating protocols. *Radiation Measurements* 41, 369–391.

Wohlfarth, B., Veres, D., Ampel, L., Lacourse, T., Blaauw, M., Preusser, F., Andrieu-Ponel, V., Kéravis, D., Lallier-Vergès, E., Björck, S., Davies, S.M., de Beaulieu, J.-L., Risberg, J., Hormes, A., Kasper, H.U., Possnert, G., Reille, M., Thouveny, N., Zander, A. 2008. Rapid ecosystem response to abrupt climate changes during the last glacial period in western Europe, 40–16 ka. *Geology* 36, 407–410.

Wood, R. 2015. From revolution to convention: the past, present and future of radiocarbon dating. *Journal of Archaeological Science* 56, 61–72.

Wood, R., Jacobs, Z., Vannieuwenhuyse, D., Balme, J., O'Connor, S., Whitau, R. 2016. Towards an accurate and precise chronology for the colonization of Australia: the example of Riwi, Kimberley, Western Australia. *PloS One* 11, e0160123.

Zeeden, C., Dietze, M., Kreutzer, S. 2018. Discriminating luminescence age uncertainty composition for a robust Bayesian modelling. Quaternary Geochronology 43, 30–39.

Zink, A., Porto, E. 2005. Luminescence dating of the Tanagra terracottas of the Louvre collections. *Geochronometria* 24, 21–26.

4 APPLICATIONS IN AEOLIAN ENVIRONMENTS

KATHRYN E. FITZSIMMONS

Research Group for Terrestrial Palaeoenvironments, Max Planck Institute for Chemistry, Hahn-Meitner-Weg 1, 55128 Mainz, Germany. Email: k.fitzsimmons@mpic.de

ABSTRACT: Aeolian, or wind-blown, deposits are the most widely dated category of sediment applicable to luminescence dating, and were instrumental in the development of the method. In this chapter is discussed the application of luminescence dating to coarse (>62 μm in diameter) aeolian deposits other than loess, which is covered in Chapter 5. Aeolian sediments are considered ideal for luminescence dating since they are assumed to have been exposed to substantial amounts of sunlight prior to deposition and are typically dominated by quartz. However, luminescence dating of aeolian sands is not without challenges. These include problems of pedogenic disturbance, characteristically low dose rates presenting problems for establishing reliable estimates of sediment moisture, effective burial depth in dynamic environments, as well as issues relating to inhomogeneity in β dose rates that lead to wide dose distributions. Despite this, the generation of large luminescence datasets has vastly improved our ability to extract meaningful palaeoenvironmental information from aeolian sands. This includes the possibility to build new quantitative accumulation models for linear dunes to understand potential climate forcing mechanisms, investigating how large transverse shoreline dunes have responded to changing wind regimes, and how sand sheets can be used to quantify rates of overturning and soil movement.

KEYWORDS: aeolian deposits, linear dunes, lunettes, sand sheets, sand ramps

4.1 INTRODUCTION

'Aeolian sands are ideal materials for the application of luminescence techniques.'

Ann Wintle (1993)

Aeolian deposits are wind-blown accumulations of sediment. Wind causes the accumulation and migration of sediment in many parts of the world, where it forms loess deposits, dunes, sand sheets and ramps (McKee 1979) – most particularly in deserts and desert margins at all latitudes, but also in regions with low vegetation cover and high wind strength, such as at the edges of glaciers and ice sheets. Particularly in desert areas of hydrologic stress, where the mean annual potential evaporation substantially exceeds annual precipitation, wind becomes the dominant geomorphic agent. Where sand-sized sediment supply is sufficient, dunes are created by the accumulation of sand through aeolian processes (Bagnold 1941). The form, dimensions and orientation of the resulting aeolian landforms is dependent upon a number of factors, including strength, uniformity and seasonality of wind regime, sediment supply, vegetation cover and type of sediment (Hesse 2010; Wasson and Hyde 1983).

This chapter discusses the applications of luminescence dating to aeolian deposits other than loess; the finer-grained loess is dealt with in Chapter 5. Consequently, the focus is on the dating of aeolian sands (62 μm–2 mm, although generally in the range 90–250 μm; Folk 1968) rather than silt-sized loessic material (Pécsi 1990) and almost exclusively on sand-sized quartz owing to the overwhelming dominance of the mineral in luminescence dating studies of aeolian sands.

Aeolian deposits represent dynamic geomorphic archives that respond to changes in environmental conditions, and therefore require a temporal framework for this change (Wintle 1993). Speculations as to the timing of aeolian activity in the past originally relied on numerical models, which predicted increases in continental albedo associated with increased aridity and desert dune activity during Quaternary glacial phases (CLIMAP 1976; Sarnthein 1978). Absolute dating of aeolian sands was hindered by the lack of suitable material with which to estimate age; organic material is poorly preserved in dryland regions, and soil carbonates and secondary gypsum were unreliable for radiocarbon and uranium-series dating, respectively (Callen 1984; Callen et al. 1983). The advent of luminescence dating techniques, and the recognition that sunlight exposure could effectively reset luminescence signals (Huntley et al. 1985; Wintle and Huntley 1979), provided the first opportunity to directly date the deposition of aeolian sands.

Luminescence dating determines the time elapsed since sediments were last exposed to sunlight. Since aeolian deposits are dominated by quartz and feldspar and exhibit a range of properties inherently suited to the luminescence method, these types of deposits have played a pivotal role in the development of luminescence dating techniques. Initially quartz thermoluminescence (TL), whereby the age of the sediment is effectively reset (or zeroed) by light as well as by heat (Aitken 1985; Wintle and Huntley 1979), was widely applied to a range of aeolian sediments, including linear dunes (e.g. Gardner et al. 1987; Nanson et al. 1992a; Prescott 1983). However, TL signals were often found to retain residual signal from prior to burial, resulting in overestimation of the true burial age (Stokes 1992, 1994) and poor dating precision (Aitken 1998). By that time, the more readily bleached optically stimulated luminescence (OSL) signal had been developed for dating applications based on experiments on Australian shoreline dunes (Huntley et al. 1985) and verified against independent age control (radiocarbon dating of organic material) in north American dunes (Stokes and Gaylord 1993). Infrared stimulated luminescence (IRSL) signals in aeolian feldspars also yielded reliable results compared with TL and OSL ages (Hütt et al. 1988), but gained minimal traction since quartz tends to dominate the aeolian sands of the world, is simpler to prepare in the laboratory, and is more easily bleached (Chapter 1; Wintle 1993).

Protocols for OSL dating of quartz were developed mostly using Australian aeolian samples (Murray and Roberts 1997b; Murray et al. 1997; Wintle 1997; Wintle and Murray 1997) – including (among others) the sample WIDG8, derived from the Widgingarri

I archaeological site in northwest Australia (Veth 1995). Perhaps most significantly for quartz OSL dating, aeolian sand, including WIDG8, was used to develop the now widely used single-aliquot regenerative dose (SAR) measurement protocol (Chapter 1; Murray and Wintle 2000, 2003; Wintle and Murray 2000). The first single-grain dating measurements were also made on aeolian sand (Murray and Roberts 1997a).

Particularly since the full range of luminescence methods has been applied to aeolian sands, care must be taken to assess the likely reliability of aeolian ages presented in a study, and the suitability of the protocols and methods used. Reliability is better assessed the more transparent the data reporting is (Hesse 2016; Lancaster *et al.* 2016), and this was often not considered necessary in earlier studies. Nevertheless, a degree of appreciation for the strengths and limitations of the method, and the inclusion of standard quality control checks or independent age control, can be sufficient. Despite recognised disadvantages of TL dating, various comparative studies have since argued that TL is often effectively bleached in aeolian deposits (e.g. Chase 2009; Cohen *et al.* 2012b). Ages obtained using protocols developed prior to the SAR standard (see Chapter 1) yielded substantially different results from SAR single-aliquot and single-grain chronologies (Duller and Augustinus 2006; Telfer and Thomas 2007; Thomas 2007), and cannot always be considered reliable unless compared against independent age control or remeasurement using SAR (Hesse 2016).

There are two main types of challenges for luminescence dating in aeolian sands. These form the basis for this chapter:

1. Despite Ann Wintle's optimistic assertion above, luminescence dating of aeolian sands is not without challenges, and the first challenge is methodological. For one thing, it cannot simply be assumed that aeolian sands are completely bleached prior to burial, remained at the same depth or contained consistent pore moisture, or that the sediments remained undisturbed following deposition. It is possible to identify in advance, based on geomorphic and sedimentological knowledge, if a particular deposit is likely to suffer from these sorts of problems, and to formulate appropriate sampling and measurement approaches to overcome these sorts of problems. Here the various methods are discussed by which it is possible to evaluate the degree of bleaching and post-depositional mixing in aeolian sands. Issues associated with low dose rates characteristic of aeolian sands are also discussed, along with their impact on the accuracy of the age, and the kinds of approaches available to account for these uncertainties. Finally, it is highlighted where problems exist for which there are as yet no simple methodological solutions, such as wide dose distributions associated with β dose rate heterogeneity.

2. The second challenge lies in the interpretation of the aeolian deposits as palaeoenvironmental archives. More precisely, what the luminescence ages which are generated represent, and how the combined chronological and geomorphic data can be used to generate meaningful information about past conditions or processes. The generation of large luminescence datasets from aeolian deposits of the world, particularly following the widespread adoption of the SAR protocol, has substantially improved our ability to do

so. Presented here are a number of case studies which go beyond simple chronological correlation and constructively fuse the two types of data to better understand palaeoenvironments and aeolian systems.

4.2 AEOLIAN DEPOSITS

4.2.1 Desert dunes

Desert dunes represent the most popular type of aeolian deposit to which luminescence dating is applied. Desert dunes take various forms, most common among which are linear (longitudinal) dunes and barchans. Since these are the features most often dated using luminescence, they are defined briefly here.

Linear dunes are the most regular and extensive landforms on the planet (McKee 1979). Linear, or longitudinal, dunes are elongate ridges of aeolian sand that typically form approximately parallel to the resultant vector of the sand-shifting winds (King 1960) (Fig. 4.1). Linear dune orientation typically lies within 30° of the overall wind vector (Brookfield 1970); however, orientational inertia may occur, so preserving evidence of prior wind regimes (Fitzsimmons 2007). The longitudinal orientation of such dunes may become established due to alternate (possibly seasonal) shifting of their steeper flanks (Bagnold 1941), or through net accumulation in cases where significant clay components prevent classic slip face development (Hesse 2011; Livingstone *et al.* 2007). Longitudinal migration of individual sand grains over long distances appears to be limited. Rather, direct transport of sand from adjacent interdune swales onto dunes, with a small longitudinal component, appears to be the dominant process (Hollands *et al.* 2006; Telfer 2011). Linear dunes are the type most likely to form in areas where wind regimes are relatively variable and sediment availability limited (Wasson and Hyde 1983). Vegetation cover may range from partial to non-existent, and biological soil crusts are common stabilisers of the land surface (Hesse and Simpson 2006).

Vertical growth of linear dunes may be episodic in nature, resulting in a compound stratigraphy of multiple units (Bristow *et al.* 2007; Fitzsimmons *et al.* 2007). These

Figure 4.1 A) Parallel linear dunes as seen from the air, Simpson Desert, Australia. B) Crest of a linear dune, looking downwind, and as surveyed in cross-section. This example is from the Madigan Line (Camp 8), Simpson Desert, Australia . Australian linear dunes are typically at least partially vegetated. Photographs taken by Paul Hesse.

have often been used to infer periods of increased aeolian activity (Chase and Thomas 2006, 2007; Fitzsimmons *et al.* 2007a; Singhvi *et al.* 2010; Stokes *et al.* 1997c; Telfer and Thomas 2007). Dune stratigraphic units may be separated by buried soils (paleosols), which develop during breaks in sediment accretion (Fitzsimmons *et al.* 2009). These depositional hiatuses have been linked to relatively humid climate phases resulting in reduced aeolian transport (Fitzsimmons *et al.* 2009), or to a reduction in sediment supply (Hollands *et al.* 2006). Not all dunes in a dunefield are active at the same time (Chase 2009; Hesse 2011; Hesse 2016). This state of partial activity is likely to have persisted in the past. Consequently, the degree to which dated episodes of dune activity relate to climatic conditions remains a matter of debate (Hesse 2016; Thomas and Bailey 2016, 2017) and will be discussed in Section 4.4.1.

Transverse dunes are aeolian deposits forming perpendicular to the prevailing winds (Fryberger 1979), typically under more uni-directional wind regimes than occur in linear dunefields (Bourke *et al.* 2010). Transverse ridges which are crescentic in form and individual crescent-shaped dunes, with horns pointing downwind and steep lee slopes, are called barchans (Bagnold 1941). Barchans are often highly mobile, dynamic landforms, generally smaller in scale than linear dunes (Bagnold 1941), although internal stratigraphy may be preserved (Wang *et al.* 2009). As such, barchans are rarely investigated from a chronological standpoint as palaeoenvironmental archives, although in some cases, chains of barchan dunes may develop and stabilise for sufficient periods of time to preserve episodes of aeolian activity (Bray and Stokes 2003, 2004). More often, luminescence dating has been used to shed light on rates of dune migration over very short time scales (Wang *et al.* 2009; Wolfe and Hugenholtz 2009), as will be discussed in Section 4.3.1.

4.2.2 Source-bordering lunette dunes

Source-bordering dunes are transverse aeolian landforms that occur adjacent to lakes or rivers. Clay-rich lunettes are a specific category that forms adjacent ephemeral lakes or playas (Bowler 1973, 1983). Source-bordering lunettes, occurring on the downwind margins of (ephemeral) lakes, are common on semi-arid desert margins (Hesse 2010). Lunettes gain their name from their crescentic morphology in plan view, and in contrast to the barchan develop limbs which extend upwind surrounding the lake, and have steeper slopes on the windward face (Bowler 1968).

Although the dominant formation process is aeolian, their formation and stratigraphy is genetically linked to the hydrology of the associated lake, which in the semi-arid zone is characterised by periodic filling and drying (Bowler 1983). The classical model of a lunette involves the aeolian transport of well-sorted clean sands onto the dune during lake-full phases (Fig. 4.2). Oscillating lake levels or drying conditions result in the exposure of the lake floor, and facilitate the formation of sand-sized aggregates of clay, which along with sand are blown onto the lunette (Bowler 1973, 1983). Periods of relative landscape stability, associated with regional drying or reduced sediment supply from the lake, facilitate pedogenesis on the lunette surface. Paleosols within lunettes in the semi-arid zone are characterised by clay illuviation, and carbonate precipitation (Bowler and Magee 1978).

Based on our understanding of the links between lake hydrology and lunette stratigraphy, luminescence dating of these landforms provides a chronological framework for hydrologic

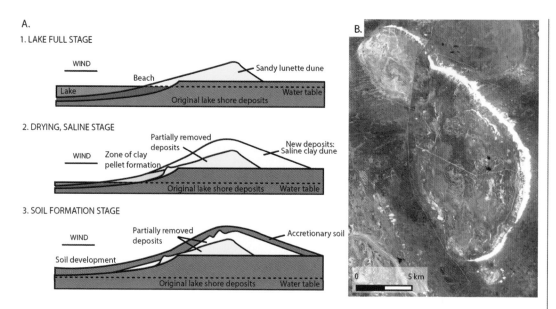

A.
1. LAKE FULL STAGE

WIND

Beach

Sandy lunette dune

Lake

Water table

Original lake shore deposits

2. DRYING, SALINE STAGE

Partially removed
deposits

WIND   Zone of clay
pellet formation

New deposits:
Saline clay dune

Original lake shore deposits   Water table

3. SOIL FORMATION STAGE

Partially removed
deposits

WIND

Accretionary soil

Soil development

Original lake shore deposits   Water table

B.

0   5 km

Figure 4.2 A) Schematic model showing the formation of a lunette dune with respect to lake hydrology (modified from Bowler 1973; Fitzsimmons 2017). B) Google Earth image of the Lake Mungo lunette, semi-arid southeastern Australia; the crescentic form of the lunette (white) is visible along the eastern (downwind) margins of the dry lake bed.

change in the lake basin (Bowler *et al.* 2003; Burrough and Thomas 2009). More recently, more nuanced palaeoclimatic information such as hydrologic balance, wind regime and precipitation has been generated from lunettes (Burrough *et al.* 2009a; Fitzsimmons 2017; Telfer and Thomas 2006), and will be discussed in Sections 4.3.2 and 4.3.3.

4.2.3 Sand ramps

Sand ramps are sloping sedimentary deposits that occur along mountain piedmonts (Bateman *et al.* 2012). They are rarely entirely aeolian in nature, but rather comprise interstratified aeolian sand with additional components that may include any or all of talus, colluvium, alluvial fan and fluvial material (Bateman *et al.* 2012; Lancaster and Tchakerian 1996; Turner and Makhlouf 2002). Consequently, these landforms result from a combination of upward (climbing) and downward (falling) aeolian transport (Livingstone and Warren 1996), as well as hydrological and gravitational processes relating to the adjacent hillslope (Bateman *et al.* 2012).

Rates of sediment accumulation calculated from luminescence ages provide insights into the possible causes of sediment accumulation, which may or may not be climatically driven (Bateman *et al.* 2012; Clarke and Rendell 1998).

4.2.4 Sand sheets (or coversands)

Sand sheets (or coversands), are low relief blankets of well-sorted sand up to several metres in thickness (Pye and Tsoar 1990). Topographic relief typically does not vary by more than

Figure 4.3 A) Sand ramps along the northern margins of the Drakensberg Mountains, southern Africa. Outlines of the ramps are shown; a car provides scale in the lower right corner. B) Sand ramp deposits in profile facilitate investigation of sedimentary structures, including evidence for aeolian and other processes. Figure modified from Telfer *et al.* (2012).

5 m vertically, and slope angles rarely exceed 6° (Pye and Tsoar 1990). Deposits of sand sheets may occur in either cold-climate or subtropical environments, and may form by a range of processes (Baillieul 1975; Boulter *et al.* 2007; Koster 1988).

Cold-climate sand sheets occur extensively across northwestern Europe (Koster 1988; Schwan 1988), northern United States and Canada (Pye and Tsoar 1990), and in high altitude regions subject to periodic periglacial activity such as the Drakensberg Mountain piedmont of southern Africa (Telfer *et al.* 2014). The proximity of these sand sheets to the extent of major ice sheets and glaciers has led researchers to believe that these deposits result from aeolian reworking of glacial outwash and other sources of glacial and periglacial sediment (Koster 1988). A combination of factors is thought to be required for cold-climate sand sheets, including a ready supply of sand-sized sediment; low vegetation cover; a low-relief landscape to facilitate aeolian drift; periodically reduced sand availability due to seasonally wet or frozen ground; and possibly also degradation of local permafrost and increased aridity within the soil (Kasse 1997). Sediments typically form horizontal or low angle beds which sometimes alternate between coarser and finer sands, possibly due to variable wind speed and sand transport potential (Schwan 1988).

Sand sheets in the subtropical and tropical zones result from various processes such as reworking and smoothing of ancient dunefields; mixing of *in situ* weathered sand and incoming aeolian material (Boulter *et al.* 2007); and *in situ* weathering of underlying sandstone bedrock (Baillieul 1975). Arid sand sheets are found on desert margins, such as in northern China (Zhou *et al.* 2009) and Botswana (Baillieul 1975); and in relatively stable subtropical regions dominated by regolith weathering processes (Boulter *et al.* 2007, 2010; Sanderson *et al.* 2001).

A genetic evolution from sand sheets to sandy loess and then to finer loess deposits, possibly corresponding to distance from source, has been observed in northwestern Europe (cold-climate cover sands) and China (desert-margin cover sands; Pye and Tsoar 1990; Fig. 4.4).

Luminescence dating of sand sheets has been applied to inform genetic links to glacial expansion (Singhvi *et al.* 2001), formation processes (Gliganic *et al.* 2016; Kristensen *et al.*

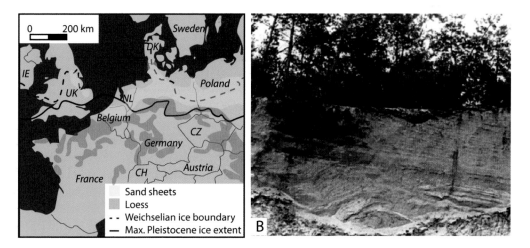

Figure 4.4 A) Spatial relationship between maximum extent of the Pleistocene ice sheets, sand sheet and loess distribution in northwestern Europe (modified from Koster 2005). B) Low angle bedding in sand sheets near Twente, Netherlands (modified from Schwan 1988).

2015), and in combination with nearby proxy records to reconstruct past environmental conditions (Boulter *et al.* 2010). Case studies of these applications, and the challenges inherent in working with these sorts of deposits, are presented in Sections 4.4 and 4.5.1.

4.3 METHODOLOGICAL CHALLENGES SPECIFIC TO AEOLIAN ENVIRONMENTS

4.3.1 OSL characteristics of aeolian sands

Since aeolian quartz sand was instrumental in the development of the luminescence dating technique and particularly the widely used SAR protocol, it is no great surprise that the OSL characteristics of this category of sediments are generally assumed to be well suited to the method (Fitzsimmons *et al.* 2010). Ultimately, the suitability of aeolian quartz sands to OSL dating is sample specific, and must always be individually assessed.

Aeolian sands are generally assumed to be completely bleached, or zeroed, at the time of deposition, since sunlight exposure of only a few seconds is required to reset the fast component of quartz OSL (Chapter 1; Aitken 1998). Incomplete bleaching is best assessed based on dose distribution, optimally using single grain rather than single-aliquot measurements (Duller 2008). For well bleached, undisturbed samples, aliquots should yield the same age as single grains (e.g. Fitzsimmons *et al.* 2014). The degree of bleaching should always be tested, particularly in high latitude environments experiencing extended periods of darkness during which aeolian transport may persist (Bristow *et al.* 2011b). The assumption of complete bleaching in Australian desert dune samples, which presumably experience substantial sunlight hours, was also called into question following observations of wide dose distributions both in aliquots and single grains (Lomax *et al.* 2007). It was proposed that the intense iron coatings on Australian dune quartz may play a role in sand grain bleachability.

Comparison of OSL signal decay of etched and non-etched quartz showed that although signal resetting for grains with iron coatings is slower (Fig. 4.5), complete bleaching takes place within 120 s, a period of time judged to be feasible within most aeolian environments, and therefore the dune sands were most likely completely bleached. Scatter in the dose distribution has other potential causes which are discussed in Sections 4.3.2 and 4.3.3.

Luminescence signal sensitivity, or 'brightness', is defined as the luminescence signal intensity per unit absorbed radiation dose, and relates to the efficiency with which absorbed radiation is transmitted as luminescence (Aitken 1998; Zimmerman 1971). This may increase with increasing exposure to temperature (Rhodes and Bailey 1997), bleaching (McKeever 1991) and irradiation (Zimmerman 1971). Ideally, OSL samples are highly sensitive and do not exhibit substantial sensitivity change over the SAR regeneration dose cycles. In real terms, this means that the T_x/T_N values measured after each regeneration dose through the SAR protocol would remain the same, resulting in $y = 1$ in a plot such as that shown in Figure 4.6. However, this is rarely the case.

One may reasonably expect that the high potential for bleaching and reworking in aeolian environments would result in sensitisation of the quartz grains (increased ability to effectively emit luminescence) and a corresponding 'bright' signal. This hypothesis is vindicated by studies on single aliquots (Fitzsimmons et al. 2010) from Australia, which yielded generally higher proportions of highly sensitive grains (>100 cts/s/gy) in aeolian samples compared with those from other depositional modes (Fig. 4.7A). Comparison between aeolian samples from Australian sediment basins of different sizes – inferring sediment residence time – suggests that quartz grains from the larger catchment, experiencing longer periods of time within the sediment system, are exposed to a greater chance of quartz sensitisation (Fig. 4.7B). However, at single-grain level the sensitivity of aeolian sediments from the smaller basin was indistinguishable from other depositional modes, indicating that sediment residence time plays a more important role than transport mode in sensitisation (Fitzsimmons 2011). Aeolian sediments from other parts of the world which have experienced relatively short sedimentary histories likewise exhibit low signal sensitivities (Bristow et al. 2011a, 2011b).

The SAR protocol incorporates a number of internal assessments to test the reliability of sample behaviour to the method regardless of sediment type. Checks include the OSL arising from a duplicate regenerative dose equal to one of the doses already given during SAR to test for recycling of the OSL signal, and a zero regenerative dose to assess thermal transfer of charge (see Chapter 1 for details; Murray and Wintle 2000, 2003). These are used as a criterion for rejection from further analysis in single-grain studies (Doerschner et al. in press; Fitzsimmons et al. 2014; Jacobs et al. 2006; Jacobs and Roberts 2007). IR depletion ratios (Duller 2003) are taken as a proxy for the degree of feldspar contamination of the quartz being measured, with high IR depletion ratios justifying additional more rigorous removal of feldspars from the quartz sample by etching or density separation. Occasionally even etching does not completely remove the feldspar signal, implying the presence of feldspar inclusions within the sand-sized grains (Hülle et al. 2010). In such instances it is preferable to avoid quartz OSL and apply feldspar IRSL measurement protocols in order to date the samples (Hülle et al. 2010).

Dose recovery tests are recommended as standard performance checks of quartz SAR dating, and involve the application of a known laboratory dose to either artificially bleached or modern analogue samples (Murray and Wintle 2003). The ability to recover dose within

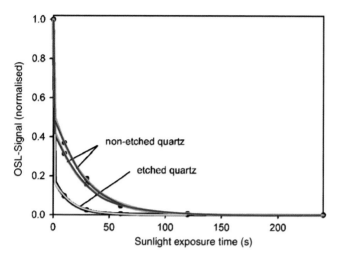

Figure 4.5 OSL signal decay of aliquots of etched and non-etched aeolian quartz from Australian desert dunes, following different sunlight exposure times (10, 30, 60, 120 and 240 s) using stimulation by blue LEDs for 0.1 s at 40 °C. Modified from Lomax et al. (2007).

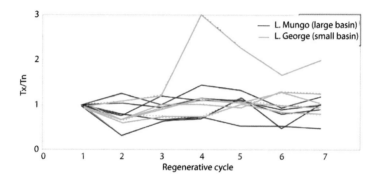

Figure 4.6 Test dose sensitivity change across regenerative dose cycles of the SAR protocol, normalised to the first test dose response (T_n) obtained after measurement of the natural signal, for five representative grains each from Australian aeolian sands (small and large sedimentary basins as shown). Variability between grains of the same sample can be observed. Modified from Fitzsimmons (2011).

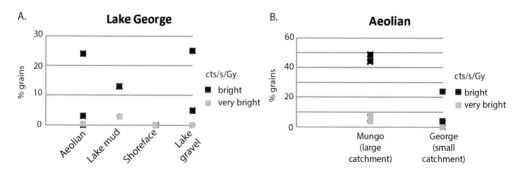

Figure 4.7 A) Proportions of bright (>100 cts/s/Gy) and very bright (>1000 cts/s/Gy) grains in quartz from different depositional contexts in the Lake George catchment, southeastern Australia. B) Proportions of bright aeolian grains from large and small sediment basins. Figure modified after Fitzsimmons (2011).

10% of unity is considered acceptable. The approach to dose recovery tests, however, is presently far from consistent. Rarely are all samples from a suite tested for dose recovery; rarely are the same grains or aliquots which were measured tested directly; and aliquots are generally measured rather than single grains (Arnold and Demuro 2015; Jacobs *et al.* 2008a). The optimal measurement and bleaching parameters for these measurements remain under discussion. Dose recovery has been shown to be dose dependent (Thomsen *et al.* 2012), and dependent on the method used to artificially bleach the samples (Choi *et al.* 2009; Doerschner *et al.* 2016; Wang *et al.* 2011). Although some comparative studies yield no difference in dose recovery ratios between single aliquots and single grains (Thomsen *et al.* 2016), others indicate a reliability on the measurement parameters that are of some concern (Doerschner *et al.* 2016). Studies have shown that there is no correlation between dose recovery ratios and accuracy of the OSL ages for samples with independent age control (Guerin *et al.* 2015). Dose recovery tests involving irradiation at much higher dose rates than those experienced in nature do not emulate the natural system and therefore cannot reliably be used to assess the suitability of a natural sample for OSL dating. At present, the utility of dose recovery tests as a useful assessment of the suitability of samples for OSL dating appears debatable. It nevertheless remains on the list of recommended quality control checks until a consistent set of measurement parameters is devised.

4.3.2 Scatter in dose distributions

Determination of the palaeodose (D_E) – the numerator of the luminescence age equation – requires a statistically significant number of aliquots or grains (often agreed to be >50 single grains; Rodnight 2008). The observed spread in grain or aliquot ages can be attributed to a range of causes, some more or less likely to apply to aeolian sands. These include:

- luminescence sensitivity variability and dose saturation characteristics (Duller *et al.* 2000)

- incomplete bleaching of the luminescence signal (Olley *et al.* 1998)

- thermal transfer of charge (Rhodes, 2000; Rhodes and Bailey 1997)

- analytical instrument consistency and variations in photon counting during measurement (Thomsen *et al.* 2005)

- post-depositional mixing (Section 4.1.4; Bateman *et al.* 2007a; Bateman *et al.* 2007b)

- variations in microdosimetry, particularly inhomogeneity in β radiation contributions to dose rates (Kalchgruber *et al.* 2003; Mayya *et al.* 2006; Thomsen *et al.* 2005; Vandenberghe *et al.* 2003).

In highly sensitive samples common in aeolian sands, the contribution from systematic counting statistics diminishes, and overdispersion (OD) dominates as a source of uncertainty (Galbraith *et al.* 2005).

In investigating the ages of linear and transverse parabolic dune deposition along the semi-arid desert margins of southeastern Australia (Murray-Darling Basin dunes; MDB), Lomax *et al.* (2007) observed substantial scatter in individual D_Es (aliquots containing 50 ± 25 grains; Fig. 4.8A). This was unexpected for what was thought to be well-bleached

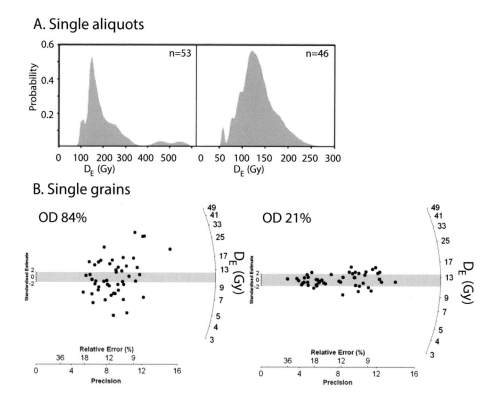

Figure 4.8 A) Weighted histograms of selected aliquot samples from MDB dunes, displayed as the sum of Gaussian curves for each D_E and error. B) Radial plots of single-grain dose distributions from MDB dunes, including (a) a linear dune sample and (b) a parabolic transverse (non-barchan) dune sample. The x-axis indicates relative error and precision of individual grains; the grey band indicates the 2-σ standard deviation of the standardised log dose. Figure modified after Lomax et al. (2007, 2011).

aeolian sands. OD was even larger (up to 70%) in subsequent single-grain measurements of the same sample set (Fig. 4.8B; Lomax *et al.* 2011). In the earlier study (Lomax *et al.* 2007), potential sources of OD were systematically checked and excluded from being the likely cause. Instrumental error was found to be minimal, and incomplete bleaching was excluded when even unetched grains with iron oxide coatings were shown to be zeroed (Section 4.3.1). The most likely causes were proposed to be either pedoturbation (Section 4.3.3) or β dose rate inhomogeneity. The latter phenomenon, also referred to as microdosimetry, may be caused by poor sorting of sediments, the occasional presence of high radiation minerals such as feldspars or zircons, or inhomogeneous distribution of carbonates (Olley *et al.* 1997; Singhvi *et al.* 1996).

Non-uniformity in microdosimetry is of particular concern in settings such as aeolian sands, which are often characterised by low dose rates (Section 4.4). In the case of the MDB dunes, dose rates averaged around 1 Gy/ka, and approximately 55% of the overall dose rate was derived from β radiation (Lomax *et al.* 2007). Individual mineral grains or pedogenic carbonates are most likely to be responsible for microdosimetry. Microdosimetry can be qualified, for example by autoradiography (Rufer and Preusser 2009), and mass spectrometry shows that dose rate 'hot spots' up to two times the average can occur (Schmidt *et al.* 2012). Statistical modelling has been used to reconstruct heterogeneous dose rates in three dimensions (Guérin and Mercier 2012; Guérin *et al.* 2012), assuming well-sorted sediment matrices.

Lomax *et al.* (2011) noted that scatter was most pronounced in the linear dune samples (OD >30%) and lower in samples from subparabolic (non-barchan, rounded transverse) dunes (OD *c.* 20%). According to models of β dose rate inhomogeneity (Mayya *et al.* 2006), the opposite trend ought to have been the case since subparabolic dunes yielded the lowest potassium concentrations. The authors concluded that the difference in sedimentation rates between the dune types, combined with pedoturbation effects, was most likely responsible for the scatter in dose distributions. Since subparabolic dunes record higher sedimentation rates than linear dunes, assuming consistent rates of pedoturbation, then sediment overturning in the subparabolic dunes was more likely to result in the mixing of grains of similar age, compared with linear dunes.

Regardless of cause, the question remains how best to calculate D_Es in cases of wide dose distributions. Where multiple dose populations are visible, the finite mixture model (FMM; Galbraith and Green 1990) is generally considered the best approach. OD values of 20% or even 35% have been used to justify the application of this age model (e.g. Cohen *et al.* 2012b; Fitzsimmons *et al.* 2014; Jacobs *et al.* 2008b). The FMM has also been used to argue for bimodality in the dose rate (Gliganic *et al.* 2012; Jacobs *et al.* 2008b, 2012), although this has yet to be demonstrated either in the environment or by simulation (Nathan *et al.* 2003). In cases of wide, yet log-normal dose distribution, either the central age model (CAM; Galbraith *et al.* 1999) or median value has been recommended (Lomax *et al.* 2007). In some extreme cases where OD values exceed *c.* 60%, samples may be rejected from age determination (Steele *et al.* 2016).

4.3.3 Shallow deposits and pedoturbation

Certain aeolian environments, particularly those with limited sediment supply, are at risk of pedoturbation and therefore mixing of the luminescence signal. This category of landscapes includes low-relief desert dune environments (e.g. Lomax *et al.* 2011) and shallow sand sheet deposits (e.g. Bateman *et al.* 2007a; Boulter *et al.* 2007; Gliganic *et al.* 2016). In this respect, the investigation of luminescence dose distributions, in particular at single-grain level (Gliganic *et al.* 2016; Kristensen *et al.* 2015), can elucidate the formation of shallow aeolian deposits at risk of sediment mixing and pedoturbation.

Mixed sediment samples can reasonably be expected to yield dose distributions reflecting the kind of mixing which has taken place based on the history of the individual grains that comprise the deposit (Gliganic *et al.* 2015, 2016). Figure 4.9A summarises the hypothetical effects of pedoturbation on dose distributions (Bateman *et al.* 2003), such as single well-defined peaks representative of well-bleached sediments of uniform age, dominant age peaks representing the true depositional age in-mixed with small proportions of older or younger grains skewing the resulting distribution, and thoroughly mixed sediment resulting in wide distributions or multiple peaks. On this basis, Bateman *et al.* (2003, 2007a) argued luminescence measurements are best made at single-grain level especially for potentially mixed sediments, since the true depositional age is masked by aliquot measurements (Duller 2008). Comparison of single-grain *vs.* single-aliquot dose

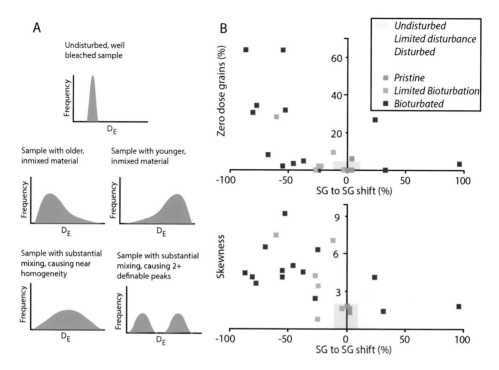

Figure 4.9 A) Hypothetical effects of pedoturbation on equivalent dose distribution (modified from Bateman *et al.* 2003). B) Bioturbation plots for North American sand sheet sites, ranging from undisturbed to heavily bioturbated, showing the shift between single-aliquot and single-grain measurements plotted against (top) zero dose grains and (bottom) skewness of the equivalent dose distribution (modified from Bateman *et al.* 2007a).

distributions from mixed sediments – such as proportions of zero dose grains and degree of skewness – can be used to elucidate pedoturbation processes and degree of mixing in aeolian deposits (Fig. 4.9B). More recent work using single-grain measurements has quantified the potential rates at which grains can be moved up or down the sediment column (Kristensen *et al.* 2015; Stockmann *et al.* 2013), for example upward transport of sand by ants (Rink *et al.* 2013), or the degree of downward mixing based on empirical observations and conceptual models (Gliganic *et al.* 2016).

The utility of single-grain dating for understanding potentially pedoturbated environments is well demonstrated by studies of subtropical sand sheet deposits (e.g. Bateman *et al.* 2007a; Boulter *et al.* 2007; Gliganic *et al.* 2015, 2016). Work on shallow sand sheets in subtropical Texas compared paired single grain and single-aliquot OSL dose distributions of deposits without internal structure, with those preserving buried paleosols, to assess how intact such deposits could be (Bateman *et al.* 2007b). In the case of the former, structureless, deposit, both single-grain and single-aliquot dose distributions were normally distributed, with minimal scatter, indicating an intact, undisturbed sand sheet that could be reliably dated. By contrast, the site containing buried paleosols yielded scattered single-grain dose distributions, facilitating identification of different types of pedoturbation down the profile. However, although apparent sediment age increased with depth, single-aliquot distributions did not match those of the single grains, and therefore the OSL chronology, even calculated from the finite mixture model, was unreliable (Bateman *et al.* 2007a). A similar study, also from Texan sand sheets, differentiated well-bleached aeolian grains with low OD from poorly bleached colluvial material with high OD within the same sample (Boulter *et al.* 2010). OD was highest in the samples closest to the present-day surface, indicating that the greatest degree of pedoturbation takes place at that level (Boulter *et al.* 2010).

More recent studies, combining single-grain dose distributions with a qualitative understanding of soil formation processes, have used single-grain quartz OSL to quantify rates of overturning and soil movement (Kristensen *et al.* 2015; Stockmann *et al.* 2013). Single-grain OSL has been used to quantify the upward movement of sand by ants (Rink *et al.* 2013). The minimum age model is recommended for age calculation in contexts where this is the dominant soil formation mechanism. In some instances, phases of enhanced mixing may be identified (Gliganic *et al.* 2015), although in others, mixing is so substantial as to completely obscure the true depositional age (Chazan *et al.* 2013). Distinction should be made between depositional and aggradational age determination in single-grain analyses of mixed soil profiles (Gliganic *et al.* 2016); the latter representing phases of increased sediment accretion and downward inmixing of sediment on a landscape scale.

4.4 LOW DOSE RATES

The natural radiation dose rate – the denominator of the luminescence age equation – derives from ionising α, β and γ radiation inherent within the sediment surrounding the sample, as well as contributions from cosmic ray dose rates (Aitken 1985, 1998). The dose rate of the present day is assumed to represent that prevailing throughout the burial history of the sample, although a number of studies challenge this assumption based on geomorphic knowledge and propose models to address potential change through time (e.g. Burrough *et al.* 2007; Telfer and Thomas 2007; Stone and Thomas 2008).

Aeolian sands often yield overall low dose rates compared with other types of sediments (Hesse 2016), ranging between 0.4–2.5 Gy/ka but often at the lower end of that range (<1.2 Gy/ka; e.g. Hesse 2016; Lomax *et al.* 2011; Telfer *et al.* 2017) (Fig. 4.10). This characteristic affects all types of aeolian sandy deposits and most likely reflects the dominance of quartz, and possibly also carbonates, combined with comparatively low proportions of clays

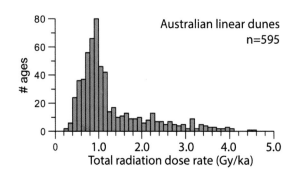

Figure 4.10 Frequency distribution of reported total radiation dose rates for luminescence age estimates in Australian dunes. Figure modified from Hesse (2016).

and feldspars within the mineral assemblage. This hypothesis is supported by positive correlations between dose rates and fine fractions (<63 μm) within linear dune sediments ($r^2 = 0.79$; Telfer *et al.* 2017). One notable exception to this pattern is the high dose rates of Mojave Desert sand ramps which range between 3.0–3.5 Gy/ka (Bateman *et al.* 2012) and differ in mineral assemblage from most of the case studies presented here.

The generally low dose rates and high saturation characteristics of aeolian quartz sand accounts for a number of very old published dune ages beyond MIS 6 (Fitzsimmons *et al.* 2007a; Hesse 2016; Sheard *et al.* 2006), particularly in Australia where dunes appear to have remained stable for long periods of time.

Low total dose rate values proportionately increase the contribution of components that are generally thought to be of minimal significance, thereby increasing their significance for the overall accuracy of the calculated age. Perhaps the most critical of these are moisture content, known to attenuate the β component of the dose rate (Section 4.4.1), and sample depth over the burial history as it impacts on cosmic dose rates (Section 4.4.2).

4.4.1 Approaches to calculating moisture content

Beta particle attenuation by moisture, present in the pore spaces of sediments, significantly affects the overall dose rate (Mejdahl 1979). Consequently, in aeolian systems characterised by low total dose rates, moisture contents play a significant role in the final age calculation. Ages increase by *c.* 1% for each 1% increase in moisture content (Cohen *et al.* 2012b).

Two approaches to determining moisture content are common in aeolian sands. Either the field moisture content of individual samples is used (e.g. Fitzsimmons *et al.* 2014), or the average for the sample set (e.g. Chase and Thomas 2007; Lomax *et al.* 2011; Telfer *et al.* 2017). It is unclear, however, how representative these values are of the long-term average water content.

Hesse (2016) reviewed the different values adopted for moisture content input from all studies in the Australian desert dunefields. Published moisture content values for Australian dunes are included in the INQUA dune chronology database (Hesse 2016; Lancaster *et al.* 2016). Most samples yielded water contents of <3%, ranging between 0.02–15.5% and with a mode of 1.5% (Fig. 4.11A). There is a weak correlation between moisture content and depth within individual dunes in the arid zone (Hesse 2016; Telfer 2011) (Fig. 4.11B). Hesse (2016) used these metadata, the modal value of 1.5% and log-transform mean of 1.7 ± 0.5% (Fig. 4.11A inset), to argue that the commonly used value of 5% could potentially overestimate the true age by 3%. He suggests that a long-term moisture content of 1.5 ± 1.5% might be the most parsimonious approach for aeolian sediments in arid regions, while dunes in more humid areas should typically yield an average log water content of 7.3 ± 3.0%.

In the case of aeolian deposits proximal to lakes, rivers or floodplains that have experienced flooding over the course of their burial history, it may be more realistic to adapt the moisture content to reflect periods of periodic flooding. Burrough *et al.* (2007) recommend taking the field moisture content value of each individual sample and incorporating the measured saturation moisture for a given period, adjusted according to sample depth relative to potential flooding height and degree of sediment sorting, to reflect variability in pore space.

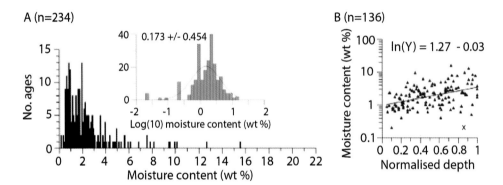

Figure 4.11 A) Frequency distribution of published water contents for Australian arid zone dunes; inset shows arid zone values converted to a log scale and fitted with Gaussian distribution. B) Measured moisture content against normalised depth (sample depth/dune thickness) for all arid zone dunes ($r^2 = 0.19$). Figure modified from Hesse (2016).

4.4.2 Approaches to calculating cosmic dose rates

The cosmic ray dose rate below the land surface is determined primarily by sample depth, is generally assumed to be constant through time, and is a function of latitude, altitude and burial depth (Prescott and Hutton 1994). It contributes a substantial proportion (up to 60%) to the total dose rate of aeolian sands with low concentrations of radioelements. Differences in sample depth input to the cosmic dose rate calculation may therefore substantially alter the final age.

Acknowledging that the depth of a sample may have varied throughout its burial history in dynamic aeolian environments, several studies in the Kalahari Desert undertook iterative modelling of overburden depth for individual samples (Burrough *et al.* 2007; Stone and Thomas 2008; Telfer and Thomas 2007). The overburden history was modelled based on the dose rate of a near surface sample, then calculated iteratively at various depths down the profile using the Prescott and Hutton (1994) algorithm. This dataset, implying gradual accumulation of overburden, was then compared with the unadjusted depositional ages that assume instantaneous deposition, and the final age model for an age–depth sequence was decided based on cluster analysis of the ages (Burrough *et al.* 2007). The gradual accumulation model was assumed unless samples adjacent one another down profile differed substantially from one another, indicating episodic deposition. This approach, while simplistic, is deemed more realistic than assuming instantaneous overburden deposition.

4.5 UNDERSTANDING LUMINESCENCE AGES WITHIN THE AEOLIAN GEOLOGICAL ARCHIVE

4.5.1 Aeolian deposits as climate archives

4.5.1.1 Linear dunes as climate archives

Sarnthein (1978) proposed that cold, arid conditions during the last glacial maximum (LGM) resulted in the expansion of sandy deserts worldwide. This hypothesis developed into the assumption that, globally, desert dunes became active during glacial phases, and inactive during interglacial periods (e.g. Nanson *et al.* 1992b). The advent of luminescence dating provided an optimal approach for testing this assumption.

Many of the early luminescence datasets from desert dunes around the world – generated by various protocols – were interpreted within the framework of Sarnthein's hypothesis (e.g. Nanson *et al.* 1992a, 1992b; Stokes *et al.* 1997a;, 1997b, 1997c). New datasets generated from SAR dating of desert linear dune quartz (e.g. Chase and Thomas 2007; Fitzsimmons *et al.* 2007a; Roskin *et al.* 2011; Stone and Thomas 2008; Yang *et al.* 2010a), likewise assumed that discrete phases of linear dune activity corresponded to periods of intensified aridity (e.g. Chase and Thomas 2007; Fitzsimmons *et al.* 2007a). This assumption persisted despite the fact that a number of aeolian pulses did not necessarily correlate with independent palaeoclimatic frameworks. In some instances an argument for enhanced aridity was supported by stratigraphic and sedimentological information (e.g. Fitzsimmons *et al.* 2009). In others, the datasets indicated a perpetual state of at least partial dune activity (e.g. Stone and Thomas 2008), or dune formation in temperate regions at times requiring unrealistically drier conditions (Hesse *et al.* 2003). This discord prompted a reassessment of our previously held assumptions linking desert dune activity with straightforward aridity (Chase 2009).

The Kalahari linear dunefields of southern Africa provide an excellent case study for the evolution of scientific thought with respect to understanding linear dunes as climate archives. Linear dunes in this region are extensive, but not continuous; broadly, they form a northern, northeastern and southwestern cluster, with an additional reticulate dunefield occurring along the western Namaqaland coast (Chase 2009). The first (non-SAR) quartz OSL chronologies, based on sampling at a number of sites both in the northern and southern dunefields, inferred that dunes preserved multiple episodes of activity, and therefore phases of aridity, over the last full glacial cycle (Stokes *et al.* 1997b, 1997c). Dunes in the more semi-arid northeastern Kalahari margin were interpreted to preserve episodic activity, whereas in the arid southwestern core of the desert, aeolian activity was more sustained (Stokes *et al.* 1997c). Dune building phases – constrained by quartz MAAD to 95–115, 41–46, 20–26, and 9–16 ka – were attributed to changes in the summer rainfall gradient and sea-surface temperatures in the southeastern Atlantic Ocean (Stokes *et al.* 1997c). The resulting OSL ages yielded uncertainties of the order of 15–20%, which were not accounted for in the interpretations of dune age clusters. Munyikwa (2005) compiled the dune dating database afresh and identified a different set of peak dune building phases at 57–60, 41–46 and 8–36 ka, leading to arguments that the earlier datasets should be excluded from future consideration (Telfer and Thomas 2007).

The advent of single-aliquot and single-grain quartz SAR dating resulted in a substantial revision of the existing Kalahari dune chronology, recognising not only the possible limitations of the older protocols (based on discrepancy in ages calculated from SAR and MAAD protocols; Duller and Augustinus 2006; argued by Thomas 2007). It was also recognised that earlier studies derived from non-representative sampling of the upper portions of dunes in pits and exposed profiles, and relatively small datasets (Telfer and Thomas 2007). The new wave of investigations involved high frequency sampling from dunes cored down to bedrock, and both single-aliquot and single-grain SAR analyses of representative samples (Stone and Thomas 2008; Telfer and Thomas 2007). The resulting chronologies – and interpretations regarding the palaeoclimatic drivers –

varied depending on region. Along the west coast of southern Africa, Chase and Thomas (2006) identified phases of aeolian activity at *c.* 63–73, 43–49, 30–33 and 16–24 ka coeval with marine proxy indications for intensified wind strength rather than aridity. A Holocene (4–8 ka) active dune phase was thought to correspond to a transitional period of increased temperatures and aridity but reduced wind strength. This study catalyzed a shift in thinking away from the simplistic assumption that dunes reflect arid conditions. It was supported by dune chronologies from the Witpan area in the southwestern Kalahari core (Telfer and Thomas 2007), clustering around 76–77, 52–57, 27–35 and 9–15 ka – out of phase with the western coastal dunes – which admitted that the climatic mechanisms driving aeolian activity in this region were unclear, and that the aeolian record here, despite high frequency sampling, was likely to be incomplete. Stone and Thomas (2008), working in the southern Kalahari, added that individual dunes recorded multiple phases of aridity, but that few phases were consistent between dunes. Artificial age clusters could be produced simply by reducing sampling frequency down a dune column. Taken as a whole, the Kalahari dunes appear to have remained partially active throughout the last full glacial cycle, so preventing meaningful assessment as to climatic drivers for dune reactivation. The apparently ambiguous link between dune activity and climate was reinforced by the application of dune activity indices to climate models for the LGM time slice, which yielded no resemblance to the actual distribution of sites within the Kalahari (Chase and Brewer 2009).

By the close of the decade, a large, state-of-the-art dataset of quartz SAR ages had been generated for sites across the Kalahari, yet no clear connection could be established between the timing of dune activity and climatic conditions (Chase 2009). It certainly could no longer be reliably assumed that aeolian activity equates to aridity. Although three main aeolian phases were identified across the Kalahari (c. 40–60, 20–35, 4–17 ka; Chase 2009), these periods of activity correspond to neither glacial nor interglacial events.

It became clear that linear dune systems, and our understanding of the various thresholds driving their formation and dynamics, needed to be reconsidered (Telfer and Hesse 2013; Thomas and Burrough 2016). It is reasonable to infer, based on observation of the partially active state of the present Kalahari dunefield, that sand deposition even during intensified reactivation episodes is likely to be associated with sand removal from another part of the dunefield. Detailed OSL sampling along the length of a single linear dune showed that activity is bimodal in nature (Telfer 2011). Growth by extension takes place under environmental conditions conducive to net sand accumulation (such as aridity, or strengthened wind regimes), while reworking along the length of the dune takes place at a local level irrespective of the prevailing climate (Telfer 2011).

The stalemate arising from the apparent lack of correlation between palaeoclimate frameworks and dated dune stratigraphies was driven largely by the lack of an adequate approach to understanding the production and interpretation of the dune records themselves. New quantitative models incorporating the effects of sampling and dating methods, landform processes and potential forcing conditions, overcome these individual issues by accounting for the thickness and preservation of sedimentary records as well as statistical uncertainties in luminescence ages (Bailey and Thomas 2014). When applied to the Kalahari dunefields, accumulation intensity peaks (excepting late Holocene accumulation in the south) correspond to cooler periods (Collins *et al.* 2014) and phases of decreasing February 25°S insolation (Fig. 4.12), together indicating drier climates (Thomas

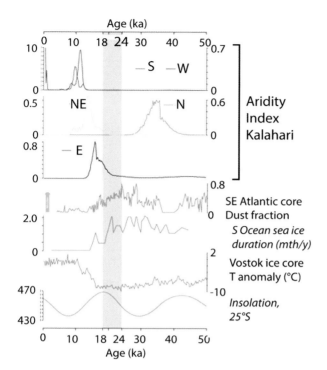

Figure 4.12 Plots of accumulation intensity for five Kalahari dunefields and independent climate proxies (modified from Thomas and Bailey 2017; data from Petit *et al.* 1999; Stute and Talma 1998; Stuut *et al.* 2002; Thomas and Bailey 2017). The grey shading corresponds to the LGM.

and Bailey 2017). Almost all linear dunefields yield peaks according to this approach during 16–10 ka. Since accumulation rates peaked after summer insolation maxima and corresponding periods of higher precipitation, and the dunefields lie downwind of major fluvial systems, there appears to be a direct link between fluvial pulses and aeolian sediment supply (Thomas and Bailey 2017). Peaks in dune accumulation are out of phase with peaks in aeolian dust input to the southeast Atlantic Ocean (Stuut *et al.* 2002), suggesting that the drivers for marine dust input may also need to be revised. The proposed accumulation rate variability model provides a more meaningful proxy for generating accumulation rates and palaeoenvironmental correlations from dune records across large dunefields (Bailey and Thomas 2014; Thomas and Bailey 2017).

4.5.1.2 Lunette dunes as hydrologic archives

The genetic link between lunette shorelines and palaeohydrology is well established (e.g. Bowler 1973, 1983, 1986; Burrough and Thomas 2009; Burrough *et al.* 2009b) (Fig. 4.2). Aeolian lunette shorelines may be preserved either as features forming at multiple levels representing lake stands of quantifiable volume (e.g. Lakes Frome-Callabonna and Eyre, central Australia, Fig. 4.13; DeVogel *et al.* 2004; Magee *et al.* 2004), or as multiple occupations of the same shoreline through time (e.g. Willandra Lakes, semi-arid southeastern Australia, Fig. 4.2; Bowler *et al.* 2012). Luminescence dating of lunettes therefore provides an unambiguous chronological framework for hydrologic change within a given lake basin (Burrough and Thomas 2008, 2009; Burrough *et al.* 2007; Fitzsimmons *et al.* 2014). Furthermore, when combined with understanding of the lake system, luminescence chronologies may be used as a powerful framework for understanding the influence of different climate subsystems on a catchment (e.g. Cohen *et al.* 2015).

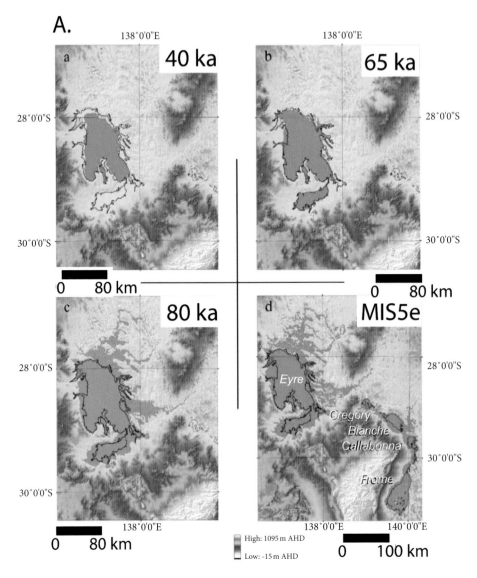

Figure 4.13 A) The outline of the Lake Eyre–Frome system, central Australia (modified from DeVogel *et al.* 2004), at palaeo-levels defined by luminescence dating (Magee *et al.* 2004). B) (facing page) History of lake filling and drying at Lakes Eyre and Frome, compared with other climatic records (labelled on the right). From top to bottom: January insolation at 30°S (Berger 1992) out of phase with the five Lake Eyre filling events dated using quartz OSL and TL by Magee *et al.* (2004); Combined single-grain quartz OSL and multi-grain TL chronology for the Lake Frome shorelines plotted against the Australian Height Datum, AHD, and showing periods of hydrologic connection with Lake Eyre (Cohen *et al.* 2012b); Histogram of summed individual U/Th speleothem ages from caves in southern Australia (Ayliffe *et al.* 1998); Global sea-level curve (Lambeck and Chappell 2001); Stacked sea-surface temperature (SST) record for the Southern Ocean (Barrows *et al.* 2007). The blue shaded zones correspond to high lake phases at Lakes Eyre (B, top) and Frome (B, second from top).

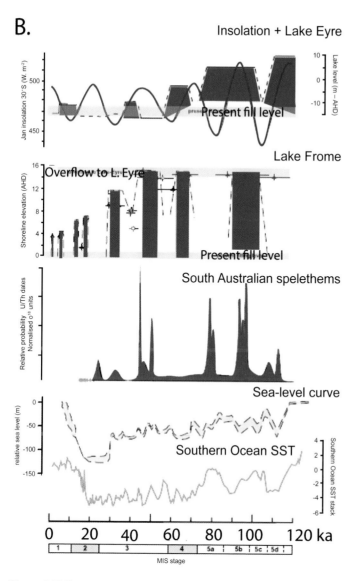

B.

Insolation + Lake Eyre

Lake Frome

South Australian spelethems

Sea-level curve

Southern Ocean SST

Figure 4.13 B

The example of the Lake Eyre–Frome system, arid Central Australia, provides a useful case study for the application of luminescence dating to lunette shorelines. The history of geochronological research in the region closely follows the evolution of the luminescence dating technique and the development of increasingly sophisticated approaches to refine the chronology.

The Lake Eyre basin is an endorheic system occupying approximately 1/6 of the Australian continental mainland. The catchment occupies the most arid central part of the continent. A string of presently unconnected terminal playa lakes (Fig. 4.13) are situated towards the southern margins of the basin. The catchment headwaters of Lake Eyre lie in the monsoon-watered latitudes of northern Australia (Magee *et al.* 2004). Under present-day conditions, Lake Eyre fills only during years of particularly intense monsoon rainfall, and does not connect with the Gregory-Frome chain of playas. By comparison, the Frome–Callabonna(–Gregory) system is filled by run-off from the adjacent Flinders Ranges to the west and Strzelecki Creek to the north, river systems which lie too far south to be influenced by the monsoon (Cohen *et al.* 2012b). The lunette shorelines of the Eyre–Frome system lakes have formed at multiple elevations, reflecting filling to different levels in the past (May *et al.* 2015). At the highest level of filling, Lake Eyre would have been hydrologically connected with the Frome lakes via the Warrawoocara overflow channel (Nanson *et al.* 1998). This constellation of climatic and hydrologic influences ensures that the lunette shorelines of the lakes in the basin provide an excellent opportunity to reconstruct long-term intensity of the Australian summer monsoon (Magee 2006), and the temperate-latitude westerly precipitation belt, across the continental inland.

The first applications of luminescence dating were made using coarse-grained quartz TL along key shoreline profiles of Lake Eyre, in combination with other dating methods including AMS radiocarbon on shell, humus, pollen and charcoal, and U/Th and amino acid racemisation (AAR) on eggshell (Magee *et al.* 1995). Unfortunately, however, no paired age estimates were produced to verify the accuracy of the TL ages. This was a cause of some controversy given a later shoreline TL dating study which appeared to contradict the earlier work (Nanson *et al.* 1998) as well as a subsequent chronology based on AAR (Magee and Miller 1998).

The first quartz OSL dating attempt of key Lake Eyre shorelines was made using the 'Australian slide' multiple aliquot method (Prescott *et al.* 1993), and produced a chronology of 30 age estimates for shorelines at five different levels over the last 150 ky (Magee *et al.* 2004). This chronologic framework was supported by independent age control based on additional quartz TL, U/Th and AMS radiocarbon dates that dated the intervening lake-dry phases that were not preserved within the lunette sediments. The five shoreline levels corresponded to five distinct periods extending from MIS 5e to the Holocene (Fig. 4.13B), and were interpreted to represent periods of intensified Australian monsoon (Magee *et al.* 2004). Correlation between the Lake Eyre palaeomonsoon record and insolation showed that Australian monsoon intensity corresponds more closely to northern hemisphere winter insolation minima, than to southern hemisphere summer insolation maxima (Fig. 4.13B). This pattern indicated that the northern hemisphere exerted a dominant control on the Australian monsoon, although dating uncertainties precluded correlation of monsoon peaks with specific insolation signals. The most recent re-dating of Lake Eyre shoreline sediments with single-grain quartz SAR (Cohen *et al.* 2015) reinforces the accuracy of the original quartz OSL chronology (Magee *et al.* 2004).

A recent research campaign directly targeting the lunettes of the entire system, primarily using single-grain quartz SAR and coupled with single-aliquot SAR and multi-grain quartz TL (Cohen *et al.* 2012b, 2015), provided an internally consistent chronology for the Frome–Callabonna system, yielding good agreement between ages from shorelines sharing the same elevations (Cohen *et al.* 2011). However, a number of samples yielded multiple single-grain populations, suggesting mixing or microdosimetry. The results, compared with the chronology for the Lake Eyre shorelines, indicated that the Frome system was last connected to Lake Eyre *c.* 50–47 ka (Cohen *et al.* 2011). Prior to this time, rainfall from both the northern monsoon and temperate latitude westerly systems was responsible for highstands at Lake Frome (Cohen *et al.* 2012b, 2011). After *c.* 47 ka, the Frome system filled solely in response to phases of increased precipitation associated with northward penetration of fronts from the Southern Ocean (Cohen *et al.* 2011, 2012b) (Fig. 4.13B). The Holocene hydrologic history at Lake Frome indicates four distinct highstands over the last 6 ky (Gliganic *et al.* 2014), correlating with increased effective precipitation in the adjacent Flinders Ranges to the west as determined by speleothem growth (Quigley *et al.* 2010).

Highly sensitive quartz grains enabled single-grain dating of particularly young grains from a shoreline at Lake Callabonna, corresponding to the Medieval Climate Anomaly (Cohen *et al.* 2012a). For the lake to fill to the volume calculated based on shoreline height implies a short-lived trough extending into Central Australia from sea-level high-pressure ridges over the central Indian Ocean and Tasman Sea. This is the first observed occurrence of anomalous climatic conditions occurring in Australia at the time of the northern hemisphere Medieval Climate Anomaly, although it is worth noting that the single-grain OSL ages did yield mixed populations, and associated TL samples overestimated the single-grain results.

Luminescence dating work in the Lake Eyre basin lunettes has so far yielded a number of significant discoveries informing palaeoclimate in continental inland Australia, and single-grain quartz SAR has crystallised a number of hypotheses that could not otherwise be answered due to a general lack of datable sediment and alternative proxies.

4.5.1.3 Reconstructing wind regimes from aeolian deposits

Transverse dunes are aeolian deposits where the longest axis forms perpendicular to the resultant sand-shifting wind vector (Bowler 1968). Consequently, the direction of sediment transport, irrespective of lake filling or drying, remains consistent with the prevailing wind regime. It is assumed that the thickest deposits of any given lunette-building phase will occur according to the orientation of the prevailing wind vector (Fitzsimmons 2017). Data on the distribution of stratigraphic thickness and verified by a luminescence chronology of the stratigraphic sequence at multiple points around the lunette feature, can be used as a proxy for reconstructing past wind regimes.

This approach was taken in a case study from the Lake Mungo lunette within the presently dry Willandra Lakes system of semi-arid southeastern Australia (Fitzsimmons 2017). During earlier studies, sedimentary characteristics provided the basis for correlation and were used to trace out distinctive stratigraphic units (Bowler 1998). Since stratigraphy appeared consistent between the localities in the southern half of the lunette, it was extrapolated across the lunette and used as a framework for the Willandra Lakes system (Bowler 1998, Bowler *et al.* 2012). Unit thickness was not measured. During more recent studies covering the central and northern parts of the lunette, a new, previously unrecognised stratigraphic unit (the Red Lunette) was identified (Fitzsimmons *et al.* 2015). This distinctive red unit was widespread in the northern part of the lunette, but was not preserved, and possibly not even deposited in the southern half. Comparative thicknesses were then investigated as a proxy for variability in prevailing wind directions through time (Fig. 4.14; Fitzsimmons 2017). Spatial visualisation of these data for two time slices revealed consistent patterns in the distribution of stratigraphic thickness between neighbouring lakes (Fig. 4.14B, C).

Prevailing wind direction appears to have shifted from a west-north-westerly regime over the period *c.* 54–32 ka to a more west-south-westerly vector just prior to the LGM (c. 24 ka), before shifting to a more direct westerly orientation during the

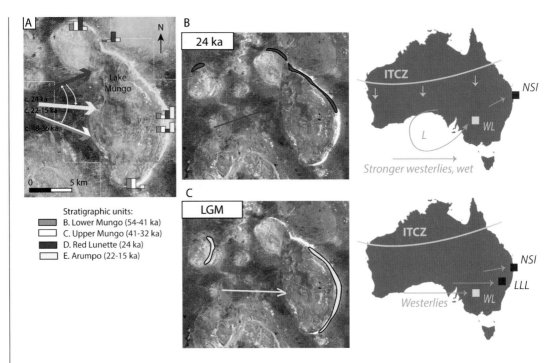

Figure 4.14 A) Relative thickness and OSL-based age of the major stratigraphic units along the Lake Mungo lunette, and proposed prevailing wind vectors through time. B) and C) Scenarios depicting climatic circulation changes at Lake Mungo/Willandra Lakes (WL) between (B) *c.* 24 ka just prior to the LGM and (C) *c.* 22–15 ka during and immediately following the LGM. Modified from Fitzsimmons (2017); eastern Australian sites of North Stradbroke Island (NSI; Petherick *et al.* 2009) and Lake Little Llangothlin lunette (LLL; Shulmeister *et al.* 2016) are also shown.

LGM (c. 22–15 ka) (Fitzsimmons 2017). Extrapolating these prevailing wind vectors into climate subsystem circulation, Fitzsimmons (2017) proposed that the west-north-westerly regime prevailing through *c.* 54–32 ka indicates a more southward penetration of the intertropical convergence zone (ITCZ) which facilitated relative weakening of the dominant westerly wind system. However, given the existing chronological control for this period (Bowler *et al.* 2003; Fitzsimmons 2017; Fitzsimmons *et al.* 2014, 2015) it is not possible to determine whether these conditions were sustained throughout this period, or prevailed in shorter-lived phases (Bayon *et al.* 2017). The short-lived megalake phase at *c.* 24 ka is much better constrained with single-grain quartz OSL dating (*n* = 5; average 23.7 ± 1.0 ka; Fitzsimmons 2017; Fitzsimmons *et al.* 2015) and therefore provides a higher resolution test case for linking changes in wind regime with climate circulation. The more southerly component to aeolian transport, coupled with a short pulse of high flow into the lakes at this time attributed to increased precipitation and run-off, is coeval with interpretations from the coastal site of North Stradbroke Island (NSI; Petherick *et al.* 2009; Fig. 4.14B). Millennial-scale southward shifts of the ITCZ may have initiated short-term intensification of north–south temperature gradients and stronger westerlies in the mid-latitudes (Whittaker *et al.* 2011), combined with increased penetration of summer rainfall into the northern parts of the Murray-Darling catchment (Bayon *et al.* 2017), within which the Willandra Lakes are located. Under such conditions, winds associated with the mid-latitude anticyclones could introduce a southerly component that would explain the Red Lunette deposition concentrating at the northern end of Lake Mungo, as well as the north-eastward dust transport to North Stradbroke Island. The later LGM-age lunette at Lake Mungo, and at Lake Little Llangothlin further northeast (Shulmeister *et al.* 2016), both with well-constrained single-grain quartz OSL chronologies (Fitzsimmons *et al.* 2014, 2015; Shulmeister *et al.* 2016), and independent age control (Bowler *et al.* 2012) – indicate a more sustained northward position of the ITCZ and northward penetration of westerly winds across inland Australia. The reconstruction of prevailing wind regimes from lunettes – fundamentally underpinned by robust luminescence chronologies – provides new insights into large-scale circulation patterns in semi-arid regions of the world.

4.5.1.4 Sand ramps as climate archives

Sand ramps are addressed within this chapter because it is generally the aeolian, rather than the non-aeolian, component which is dated with luminescence methods (Bateman *et al.* 2012; Kumar *et al.* 2017; Livingstone and Warren 1996). The combination of processes involved in the formation of sand ramps is precisely what makes them of interest as palaeoclimate archives. The challenge thus far, however, has been to understand the triggers for their formation, and to constrain rates of accumulation of the different phases (Bateman *et al.* 2012).

Sand ramps are common in desert piedmont areas, and have been documented in Africa (southern Namib, Bertram 2003; central Sahara, Busche 1998; Telfer *et al.* 2014; Drakensberg Range front, South Africa, Telfer *et al.* 2012), the Middle East and southern Central Asia (Ladakh, northwest India, Kumar *et al.* 2017; Iran, Thomas *et al.* 1997; Jordan, Turner and Makhlouf 2002), and north America (Mojave Desert, California, USA, Bateman *et al.* 2012; Clarke and Rendell 1998; Lancaster and Tchakerian 1996). The bulk of the geomorphic and geochronological data, however, derives from the Mojave Desert sand ramps of California, United States (Bateman *et al.* 2012 and references therein), and therefore this region forms our case study.

The first applications of luminescence dating to sand ramps in the Mojave Desert incorporated a range of techniques (Clarke 1994, 1996; Clarke and Rendell 1998; Clarke *et al.* 1996; Rendell *et al.* 1994; Rendell and Sheffer 1996). Rendell and Scheffer (1996) applied quartz TL, feldspar TL and additive dose feldspar IRSL to stratigraphic units across a number of sand ramp sites. The resulting chronologies were plagued by large uncertainties (up to 30%), high inter-aliquot scatter, age reversals at depth, and inadequate sampling strategies for the sites including lack of bracketing ages for individual stratigraphic units (Rendell and Sheffer 1996). Much of these limitations can be attributed to the now outdated methods, such as incorporation of residual signal (Lian and Roberts 2006) and inability to appropriately account for anomalous fading in the IRSL signal. The resulting quartz and feldspar TL chronologies nevertheless agreed reasonably well and indicated two major depositional phases in the aeolian units at *c.* 30–20 ka and 15–7 ka (Rendell and Sheffer 1996). In a subsequent study, Clarke and Rendell (1998) correlated phases of aeolian sediment accumulation with pluvial lake filling events, most likely associated with storms, which provided the sediment supply for aeolian transport.

The links between aeolian accumulation on sand ramps and potential climatic drivers are heavily dependent on reliable dating. To overcome the shortcomings of the outdated methods, Bateman *et al.* (2012) undertook quartz single-aliquot SAR dating of the stratigraphy of a single sand ramp (Soldier Mountain) at four different sections, including multiple samples from individual stratigraphic units. This study aimed to quantify accumulation rates of the aeolian layers, and to constrain the timing of interbedded stony horizons which have variously been interpreted as talus (Lancaster and Tchakerian 1996), desert pavement or slope wash (Bertram 2003), or rockfall (Turner and Makhlouf 2002), each with a range of implications as to timing of deposition and palaeoenvironmental conditions. While the authors conceded that single-grain measurements would have been more informative, they argued that the observed dose distributions overcame most of the problems of previous studies and provided a reliable chronology (Bateman *et al.* 2012). The new dataset was stratigraphically consistent, yielded smaller errors, and fitted well with the timing of independent palaeoenvironmental archives from the region. Aeolian accumulation rates indicated rapid accumulation of individual units over <5 ky, resulting in aeolian peaks around *c.* 13.8 ka and 12.4–7.8 ka (Bateman *et al.* 2012). These results are considerably younger than the previous work, most likely because the quartz OSL signal is more readily bleached than the TL.

Based on the new chronology from Soldier Mountain, Bateman *et al.* (2012) inferred that the nearby Lake Manix drained *c.* 15 ka and was associated with debris flows and alluvial fan aggradation. The first aeolian pulse took place *c.* 13.8 ka, after the lakes had dried, deriving sediment from shorelines and fluvial deltas. The final aeolian phase most likely involved free-form dunes moving across the sand ramp, either in response to increased aridity or simply due to the sand ramp occupying more space along the piedmont.

Arguably the Mojave Desert sand ramps would benefit from single-grain investigations to confirm the interpretations made by Bateman *et al.* (2012). As yet, while other sand ramps of the world (South Africa, Telfer *et al.* 2012; Ladakh, Kumar *et al.* 2017) have formed the focus of quartz SAR dating attempts, all have worked with aliquots – albeit of various sizes. Consequently, there is no confirmation from single-grain investigations to confirm interpretations of complete bleaching or lack of reworking. Such work remains a challenge for the future.

4.5.1.5 Aeolian sand sheets as climate archives

Luminescence dating of sand sheets provides some limited information relating to palaeoenvironmental conditions. In the case of the cold-climate sand belt of northern Europe, which approximates the maximum extent of the last glacial ice sheet (Fig. 4.4), a genetic link may be made between ice sheet expansion, aeolian sand supply, and reduced vegetation cover in a low relief landscape (Koster 1988). This genetic relationship appears to be borne out by luminescence chronologies of sand sheets across northwestern Europe (Koster 2005; Singhvi *et al.* 2001), whereby the timing of sand sheet accumulation coincides with the last ice age and deglacial period (Fig. 4.15). Sand sheet reactivation is also coeval with that of localised dune field activity (Fig. 4.15), such as in the Mainz–Gonsenheim area along the middle Rhine River valley (Radtke *et al.* 2001). Here, as is the case for the more widespread sand sheets, the feldspar IRSL ages underestimate the quartz estimates by as much as 25%.

One notable disadvantage of luminescence dating of cold-climate sand sheet deposits is that the ages lack the precision now available for high resolution palaeoclimate archives

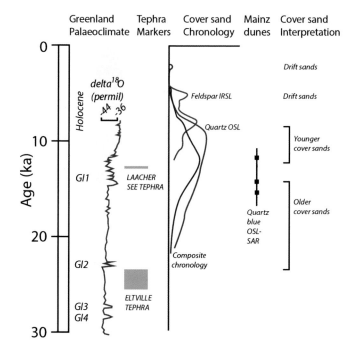

Figure 4.15 Correlation between the timing of cover sand reactivation in northwest Europe and palaeoclimate in the region. Left-right: Stratigraphic framework for climate changes during the Last Glacial period based on synchronised Greenland ice core records (Rasmussen *et al.* 2014); timing of tephra marker horizons in the region, the Laacher See tephra (van den Bogaard, 1995) and Eltville tephra (Zens *et al.* 2017); comparative probability density functions based on feldspar IRSL, quartz OSL and composite chronologies of European cover sands (Singhvi *et al.* 2001); quartz blue OSL-SAR dating of aeolian dunes at Mainz-Gonsenheim (BLSL-SAR; Radtke *et al.* 2001); and the geomorphic interpretation of the European cover sands (Koster 2005).

such as the Greenland ice cores (Rasmussen *et al.* 2014), and with which aeolian archives are correlated. In fact, where previously connections were made between sand sheet activity and the peak of the last ice age (older coversands) and final deglaciation (younger coversands; Koster 2005), the sand sheet chronology no longer precisely matches the most recent common time scale for climate change in the North Atlantic region (Rasmussen *et al.* 2014). Nevertheless, there does appear to be a link between sand sheet activity and cold climate phases coincident with ice sheet expansion. Older sand sheet activity has been dated to ice sheet expansions during MIS 4 and 6 (Frechen *et al.* 2001; Schokker *et al.* 2004), and as far back as MIS 10 (Sitzia *et al.* 2015).

While luminescence chronologies reaffirm the genetic link between sand sheet activation and ice sheet expansion, the limitations of dating precision prevent much more than that. This becomes evident when the chronology for the LGM and deglacial period are focused on (Fig. 4.15). Whilst it is possible that sand sheet activity did indeed peak around the transition to the Holocene (Singhvi *et al.* 2001), more likely the results reflect the methods used, their precision, and suitability for dating these kinds of deposits. In particular, the feldspar IRSL ages yield multiple age peaks and underestimate the quartz data; quartz ages use large aliquots and protocols preceding SAR and cannot adequately account for mixing or incomplete bleaching. Relatively few studies on European sand sheets applied the SAR protocol (Murton *et al.* 2003; Radtke *et al.* 2001; Sitzia *et al.* 2015; Vandenberghe *et al.* 2004). Arguably, then, potential does exist to improve the chronology for European sand sheet activity, simply by updating to the SAR protocol and using single-grain measurement.

4.5.2 Aeolian processes and dynamics

4.5.2.1 Desert dune dynamics

Various aspects of luminescence dating can also elucidate desert dune dynamics with respect to boundary conditions and more regional morphodynamic influences (Lancaster 2008; Telfer *et al.* 2017). Developments in the generation of luminescence datasets within desert dunes (see the INQUA Dunes Atlas chronologic database; Lancaster *et al.* 2016 and references therein) have produced a substantially more sophisticated understanding of the nature of desert dune dynamics (e.g. Ewing *et al.* 2015; Hesse 2016; Telfer *et al.* 2017). While these recent studies generate more questions than answers, reflections on dune formation mechanisms and even definitions – fundamentally underpinned by luminescence methods – clearly show more nuanced models for desert dune dynamics in comparison with earlier reviews (Lancaster 2008).

Migrating dunes preserve shorter records of aeolian activity due to constant reworking – the period required for complete reworking designated the reconstitution time (Lancaster 2008) – and luminescence dating provided the first opportunity to calculate reconstitution times and lateral migration rates. In the case of large migrating dunes in the Namib sand sea of southern Africa, this was constrained to *c.* 6 ky and 0.1 m/y (Bristow *et al.* 2007). In some instances, multiple orientations of linear dunes are overprinted on one another, and luminescence dating could be used to determine the timing of deposition of each orientation and potential links with climatic circulation (Lancaster *et al.* 2002). Conversely, linear dunes with relatively high fine-grained content appear to remain stable for long periods of time (Hesse 2011) – in the case of dunes in the Simpson Desert of Australia, stability could extend to *c.* 1 My (Fujioka *et al.* 2009). For linear dunes, net deposition was observed to occur in the part of the dune where wind speeds were reduced, implying that luminescence ages provide an indication of the time when dune activity ceased rather than peaked (Chase and Thomas 2007).

The generation of large desert dune chronologic datasets (Lancaster *et al.* 2016) provided an opportunity to reassess what the individual chronologies, as well as regional chronologies in aggregate, indicated about desert dune dynamics. Much of this reflection has focused on the extensive linear dunefields of Central Australia (Hesse 2010, 2011, 2016; Telfer *et al.* 2017). A new, detailed map of desert dunes across the Australian continent, ordered according to morphology, highlights the link between linear dune morphology, distribution and underlying topography (Hesse 2010). Underlying sediment basins or low-sediment craton plains appear to provide the fundamental morphodynamic influences for the establishment of linear dunefields (Hesse 2010). The stability of the Australian dunefields (e.g. Fujioka *et al.* 2009) was linked partly to the relatively dense vegetation cover which distinguishes the Australian desert from other dunefields of the world, and partly to a distinctive fine-grained component of dune sedimentology (Fitzsimmons *et al.* 2009; Hesse 2011) which may form under a broader range of climatic conditions than was previously thought.

Most recently, a chronostratigraphic and sedimentological study on two pairs of linear dunes in the Simpson Desert of central Australia, each with different spatial patterning, challenges our existing understanding of linear dune dynamics (Telfer *et al.* 2017). While

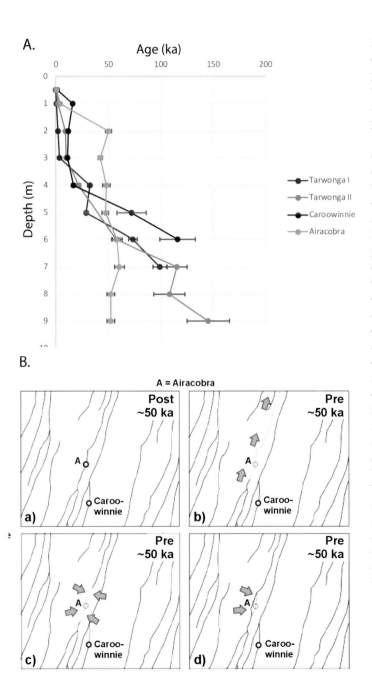

Figure 4.16 A) Age–depth profiles of four paired linear dunes in the Simpson Desert, Australia, with the pairs shown in red (Tarwonga I and II) and blue (Caroowinnie and Airacobra) tones. The distinctly different profile of Airacobra is apparent. B) The present dune plan morphology at Airacobra-Caroowinnie, established *c.* 50 ka, showing different dune building scenarios: b) if purely dune extensional mode assumed, the downwind dune would form subsequently; c) in the case of pattern interruption such as full-depth dune blowout, later infilled by accumulation at *c.* 50 ka; d) lateral migration of a small dune section by more than 50% of dune width. (modified from Telfer *et al.* 2017).

the OSL-based history of one pair of adjacent dunes was similar, yielding sporadic periods of net accumulation over the last 100–150 ka, another pair yielded quite different histories (Fig. 4.16). The results of this study indicate that small-scale, stochastic aeolian systems can overprint large-scale morphodynamic controls – namely, that the landscape cannot be dated by pattern analysis. The history of individual dunes cannot be used to represent an entire dunefield (Telfer *et al.* 2017). The mechanism for the radical difference in histories of the two adjacent dunes is unclear. Extensional growth parallel or oblique to the wind (Telfer 2011), vertical accretion (Craddock *et al.* 2015; Hollands *et al.* 2006), and lateral migration cannot be excluded. Linear dune formation is most likely to be the net result of multiple processes operating and possibly co-existing at different temporal and spatial scales (Ewing *et al.* 2015). The classical, non-vegetated seif dunes of hyperarid regions may be fundamentally different landforms to the vegetated linear features of the semi-arid zone (Telfer *et al.* 2017).

As actively migrating forms, barchans preserve relatively short sedimentary histories that limit their utility as palaeoenvironmental archives. Nevertheless luminescence dating provides a useful tool for understanding rates of migration of these features. One limitation of luminescence dating to this application is the reliance on sufficient signal to be able to date very young features reliably. This is highly variable depending on the sensitivity of the quartz or feldspar being targeted for dating. One attempt to date the migration rates of barchans in the Hexi Corridor of northern China was largely unsuccessful, since the quartz OSL signal was too dim for dating, and only one feldspar sample yielded sufficient signal (Wang *et al.* 2009). For that single sample, the resulting age of 28 ± 17 yr not only yielded a very high uncertainty, but a comparative date based on radiocarbon dating of underlying organic material provided a migration rate twice that of the luminescence result, calling both dating methods into question. By contrast, work investigating the development of stable parabolic dunes from active barchans in the Canadian prairies more reliably constrained the timing of last barchan migration to between 1810–1880 CE, which stabilised by 1910 (Wolfe and Hugenholtz 2009). Comparing the OSL chronology with historical information and LIDAR remote sensing data, the authors argued that active barchan migration took place under dry, cool climate conditions with low water tables, where sand transport exceeded the ability of existing vegetation cover to stabilise the surface.

Luminescence signal characteristics have also been used to test whether barchanoid forms on a lake bed – such as the Ntwetwe Pan within the Makgadigadi system in Botswana, southern Africa – were deposited by subaerial (aeolian) or subaqueous processes (Burrough *et al.* 2012). The OSL results were inconclusive in this case, although ultimately a case was made for aeolian deposition on the basis of morphology and sedimentology. Nevertheless, the proposal to compare the decay rates and bleachability of the different OSL signal components, since signal decay rates are dependent on incident light energy (Spooner 1994), and therefore that bleaching rate for both medium and fast components under high UV (subaerial) conditions should be comparable and less so under subaqueous conditions, is an interesting one worthy of further investigation.

4.5.2.2 Interplay between aeolian and non-aeolian processes

Luminescence dating of composite or interbedded landforms can elucidate the interplay between aeolian and non-aeolian processes. Perhaps the most common investigations of this kind are those which date interbedded aeolian and water-lain deposits, such as fluvial and lacustrine sediments (e.g. Burrough *et al.* 2007, 2012; Hollands *et al.* 2006; Yang *et al.* 2010b). In these cases, luminescence dating is a powerful tool when combined with stratigraphic and sedimentological investigations to understand alternating dominant processes.

It is not always possible to directly date non-aeolian units interbedded with aeolian material. One such example is interbedded stony horizons within sand ramps in mountain piedmonts. These stony horizons are of ambiguous provenance and have been variously described as talus (Lancaster and Tchakerian 1996), desert pavement or slope wash (Bertram 2003) or rock fall (Turner and Makhlouf 2002). Since they cannot be directly dated, bracketing their deposition can elucidate the nature and timing of these deposits, including whether they represent long or short breaks in the aeolian record, pedogenic or catastrophic depositional processes (Bateman *et al.* 2012). Detailed quartz OSL chronologies at Soldier Mountain sand ramp in the Mojave Desert provided the first opportunity to investigate this interplay (Bateman *et al.* 2012). This study calculated aeolian sedimentation rates of 1.74 m/ka and 3.3 m/ka for two discrete phases. This sequence is overlain by a stony armoured surface and followed by dissection. The OSL chronology constrained the earliest establishment of the stony surface to *c.* 1.8 ka, indicating that the processes involved in its formation are likely to be short term (Bateman *et al.* 2012).

4.6   SUMMARY AND FUTURE PERSPECTIVES

Luminescence dating of sandy aeolian deposits has had a long and productive history in the development of the method. Wind-blown sediments are eminently suitable to dating with a technique that determines the time elapsed since sediments were last exposed to sunlight. The likely exposure of these deposits to substantial amounts of sunlight during transport, the dominance of quartz within the mineral assemblage, the generally high degree of OSL signal sensitisation over repeated phases of bleaching and burial, and the relatively high saturation potential of aeolian quartz all inherently play in favour of application of luminescence dating. It therefore comes as no surprise that the number of studies applying luminescence dating to sandy aeolian deposits is vast and continues to increase. However, in part because of the long involvement of aeolian deposits in the development of luminescence dating protocols, care must be taken to assess the likely reliability of the aeolian ages presented in a study and the suitability of the protocols used. Reporting standards and data transparency are increasingly important (Hesse 2016; Lancaster *et al.* 2016).

Luminescence dating of aeolian sediments is not without challenges, although these are increasingly being addressed through advances in the method:

1. It cannot always be assumed that aeolian sands are completely bleached prior to burial, nor that the deposits remain undisturbed by pedoturbation. Partial or incomplete bleaching can occur anywhere, but may be a greater risk at high latitudes experiencing longer periods of darkness during which aeolian transport may occur (e.g. Bristow *et al.* 2011a, 2011b). Post-depositional disturbance is of particular concern in aeolian deposits with limited sediment supply; for example, the low relief

dunes of the Australian semi/arid zone (e.g. Fitzsimmons *et al.* 2007a; Lomax *et al.* 2011) or shallow sand sheet deposits (e.g. Bateman *et al.* 2007a; Boulter *et al.* 2010). These issues, while not exclusive to aeolian deposits, are increasingly well understood and quantified through single-grain measurements and statistical models that make sense of the data (Galbraith and Roberts 2012).

2. Single-grain dose distributions also have applications for understanding pedoturbated deposits beyond extracting depositional ages. Single-grain quartz OSL has been used to quantify rates of soil overturning (Kristensen *et al.* 2015; Stockmann *et al.* 2013), insect-based grain transport within sediment profiles (Rink *et al.* 2013), and constraint of phases of increased sediment accretion and downward mixing of sediment over landscape scales (Gliganic *et al.* 2016).

3. Low dose rates are strongly characteristic of aeolian deposits, which are often dominated by quartz and carbonates with low radiogenic contents. In such contexts, the importance of accurate and realistic moisture contents to account for dose rate attenuation can make a difference of several per cent in the calculated age. In general, a conservative approach, incorporating a moisture uncertainty that will realistically cover the full water saturation range of sandy sediment over the period during which it was buried, is recommended even though precision is reduced. Of additional concern is the contribution of variable sample depth to the cosmic dose rate component, especially in dynamic dune environments that may see variations in depth from the surface, and can likewise influence the resulting age estimate by several per cent. A range of approaches have been applied to aeolian environments, from simply taking the present-day depth to modelling a simplified burial depth history, iteratively adjusted down the sediment profile.

4. Many aeolian deposits yield wide dose distributions, which may be driven by a range of factors including sediment mixing and inhomogeneity of β dose rates within the sediment. Beta dose rate inhomogeneity is as yet only partly addressed by models which cannot completely account for differences at single-grain level.

The generation of large luminescence datasets, coupled with a better understanding of the nature of these geomorphic archives, has vastly improved our ability to extract meaningful palaeoenvironmental information from aeolian deposits. Particular highlights of recent years include:

1. The development of new quantitative models for accumulation intensity in linear dunes which provide direct correlation with potential climate forcing mechanisms (Bailey and Thomas, 2014; Thomas and Bailey, 2017). This new metric for aeolian deposits broke the stalemate arising from the apparent lack of correlation between palaeoclimatic proxy data and dune chronostratigraphies which had largely arisen from

the lack of an adequate approach to understanding the dune records themselves.

2. The generation of high resolution luminescence chronostratigraphies along the length of large transverse dunes to elucidate spatial shifts in deposition in response to changing wind regimes (Fitzsimmons 2017). The subsequent extrapolation of the prevailing wind vectors into climate subsystem circulation facilitates temporal reconstruction of continental-scale climate dynamics from aeolian archives.

3. Single-grain dose distributions in shallow aeolian sand sheet deposits can be used to quantify rates of overturning and soil movement (Gliganic *et al.* 2016; Kristensen *et al.* 2015), so bringing the application of luminescence beyond determining depositional ages and into a meaningful tool for quantifying soil processes.

These new approaches, fundamentally underpinned by experimental development of the luminescence method and advances in understanding the nature of these geomorphic archives, provide more meaningful, sophisticated perspectives from aeolian deposits as palaeoenvironmental records than previously possible.

REFERENCES

Aitken, M.J., 1985. *Thermoluminescence Dating*. Academic Press, London.

Aitken, M.J., 1998. *An Introduction to Optical Dating: The Dating of Quaternary Sediments by the use of Photon-Stimulated Luminescence*. Oxford University Press, New York.

Arnold, L.J., Demuro, M., 2015. Insights into TT-OSL signal stability from single-grain analyses of known-age deposits at Atapuerca, Spain. *Quaternary Geochronology* 30, 472–478.

Ayliffe, L.K., Marianelli, P.C., Moriarty, K.C., Wells, R.T., McCulloch, M.T., Mortimer, G.E., Hellstrom, J.C., 1998. 500 ka precipitation record from southeastern Australia: Evidence for interglacial relative aridity. *Geology* 26, 147–150.

Bagnold, R.A., 1941. *The Physics of Blown Sand and Desert Dunes*. Methuen and Co., London.

Bailey, R.M., Thomas, D.S.G., 2014. A quantitative approach to understanding dated dune stratigraphies. *Earth Surface Processes and Landforms* 39, 614–631.

Baillieul, T.A., 1975. A Reconnaissance Survey of the Cover Sands in the Republic of Botswana. *Journal of Sedimentary Petrology* 45, 494–503.

Barrows, T.T., Juggins, S., De Deckker, P., Calvo, E., Pelejero, C., 2007. Long-term sea-surface temperature and climate change in the Australian–New Zealand region. *Paleoceanography* 22, PA2215.

Bateman, M.D., Boulter, C.H., Carr, A.S., Frederick, C.D., Peter, D., Wilder, M., 2007a. Detecting post-depositional sediment disturbance in sandy deposits using optical luminescence. *Quaternary Geochronology* 2, 57–64.

Bateman, M.D., Boulter, C.H., Carr, A.S., Frederick, C.D., Peter, D., Wilder, M., 2007b. Preserving the palaeoenvironmental record in Drylands: Bioturbation and its significance for luminescence-derived chronologies. *Sedimentary Geology* 195, 5–19.

Bateman, M.D., Bryant, R.G., Foster, I.D.L., Livingstone, I., Parsons, A.J., 2012. On the formation of sand ramps: A case study from the Mojave Desert. *Geomorphology* 161–162, 93–109.

Bateman, M.D., Frederick, C.D., Jaiswal, M.K., Singhvi, A.K., 2003. Investigations into the potential effects of pedoturbation on luminescence dating. *Quaternary Science Reviews* 22, 1169–1176.

Bayon, G., De Deckker, P., Magee, J.W., Germain, Y., Bermell, S., Tachikawa, K., Norman, M.D., 2017. Extensive wet episodes in Late Glacial Australia resulting from high-latitude forcings. *Scientific Reports* 7, 44054.

Berger, A., 1992. Orbital variations and insolation database, in: Program, N.N.P. (Ed.), IGBP PAGES/World data centre for Paleoclimatologuy Data Contribution Series # 92–007.

Bertram, S., 2003. *Late Quaternary Sand Ramps in South-Western Namibia: Nature, Origin and Palaeoclimatological Significance.* Universität Würzburg, Würzburg.

Boulter, C., Bateman, M.D., Frederick, C.D., 2007. Developing a protocol for selecting and dating sandy sites in East Central Texas: Preliminary results. *Quaternary Geochronology* 2, 45–50.

Boulter, C., Bateman, M.D., Frederick, C.D., 2010. Understanding geomorphic responses to environmental change: a 19 000-year case study from semi-arid central Texas, USA. *Journal of Quaternary Science* 25, 889–902.

Bourke, M., Lancaster, N., Fenton, L., Parteli, E., Zimbelman, J., Radebaugh, J., 2010. Extraterrestrial dunes: An introduction to the special issue on planetary dune systems. *Geomorphology* 121, 1–14.

Bowler, J., Gillespie, R., Johnston, H., Boljkovac, K., 2012. Wind v water: Glacial maximum records from the Willandra Lakes, in: Haberle, S., David, B. (Eds.), *Peopled Landscapes: Archaeological and Biogeographic Approaches to Landscapes.* The Australian National University, Canberra, pp. 271–296.

Bowler, J.M., 1968. Australian landform example no.11: lunette. *Australian Geographer* 10, 402–404.

Bowler, J.M., 1973. Clay dunes: their occurrence, formation and environmental significance. *Earth-Science Reviews* 9, 315–338.

Bowler, J.M., 1983. Lunettes as indices of hydrologic change: A review of the Australian evidence. *Proceedings of the Royal Society of Victoria* 95, 147–168.

Bowler, J.M., 1986. Spatial variability and hydrologic evolution of Australian lake basins: Analogue for Pleistocene hydrologic change and evaporite formation. *Palaeogeography, Palaeoclimatology, Palaeoecology* 54, 21–41.

Bowler, J.M., 1998. Willandra Lakes revisited: environmental framework for human occupation. *Archaeology in Oceania* 33, 120–155.

Bowler, J.M., Magee, J.W., 1978. Geomorphology of the Mallee region in semi-arid northern Victoria and western New South Wales. *Proceedings of the Royal Society of Victoria* 90, 5–25.

Bowler, J.M., Johnston, H., Olley, J.M., Prescott, J.R., Roberts, R.G., Shawcross, W., Spooner, N.A., 2003. New ages for human occupation and climatic change at Lake Mungo, Australia. *Nature* 421, 837–840.

Bray, H.E., Stokes, S., 2003. Chronologies for Late Quaternary barchan dune reactivation in the southeastern Arabian Peninsula. *Quaternary Science Reviews* 22, 1027–1033.

Bray, H.E., Stokes, S., 2004. Temporal patterns of arid-humid transitions in the south-eastern Arabian Peninsula based on optical dating. *Geomorphology* 59, 271–280.

Bristow, C.S., Augustinus, P., Rhodes, E.J., Wallis, I.C., Jol, H.M., 2011a. Is climate change affecting rates of dune migration in Antarctica? *Geology* 39, 831–834.

Bristow, C.S., Augustinus, P.C., Wallis, I.C., Jol, H.M., Rhodes, E.J., 2011b. Investigation of the age and migration of reversing dunes in Antarctica using GPR and OSL, with implications for GPR on Mars. *Earth and Planetary Science Letters* 289, 30–42.

Bristow, C.S., Duller, G.A.T., Lancaster, N., 2007. Age and dynamics of linear dunes in the Namib Desert. *Geology* 35, 555–558.

Brookfield, M., 1970. Dune trends and wind regime in central Australia. *Zeitschrift fur Geomorphologie N. F. Supplementband* 10, 151–153.

Burrough, S.L., Thomas, D.S.G., 2008. Late Quaternary lake-level fluctuations in the Mababe Depression: Middle Kalahari palaeolakes and the role of Zambezi inflows. *Quaternary Research* 69, 388–403.

Burrough, S.L., Thomas, D.S.G., 2009. Geomorphological contributions to palaeolimnology on the African continent. *Geomorphology* 103, 285–298.

Burrough, S.L., Thomas, D.S.G., Shaw, P.A., Bailey, R.M., 2007. Multiphase Quaternary highstands at Lake Ngami, Kalahari, northern Botswana. *Palaeogeography, Palaeoclimatology, Palaeoecology* 253, 280–299.

Burrough, S.L., Thomas, D.S.G., Bailey, R.M., 2009a. Mega-Lake in the Kalahari: A Late Pleistocene record of the Palaeolake Makgadikgadi system. *Quaternary Science Reviews* 28, 1392–1411.

Burrough, S.L., Thomas, D.S.G., Singarayer, J.S., 2009b. Late Quaternary hydrological dynamics in the Middle Kalahari: Forcing and feedbacks. *Earth-Science Reviews* 96, 313–326.

Burrough, S.L., Thomas, D.S.G., Bailey, R.M., Davies, L., 2012. From landform to process:

Morphology and formation of lake-bed barchan dunes, Makgadikgadi, Botswana. *Geomorphology* 161–162, 1–14.

Busche, D., 1998. *Die zentrale Sahara – Oberflächenformen im Wandel*. Perthes Geographie im Bild. Justus Perthes, Gotha.

Callen, R.A., 1984. Quaternary climatic cycles, Lake Millyera region, southern Strzelecki Desert. *Transactions of the Royal Society of South Australia* 108, 163–173.

Callen, R.A., Wasson, R.J., Gillespie, R., 1983. Reliability of radiocarbon dating of pedogenic carbonate in the Australian arid zone. Sedimentary Geology 35, 1–14.

Chase, B., 2009. Evaluating the use of dune sediments as a proxy for palaeo-aridity: A southern African case study. *Earth-Science Reviews* 93, 31–45.

Chase, B., Thomas, D.S.G., 2007. Multiphase late Quaternary aeolian sediment accumulation in western South Africa: Timing and relationship to palaeoclimatic changes inferred from the marine record. *Quaternary International* 166, 29–41.

Chase, B.M., Brewer, S., 2009. Last Glacial Maximum dune activity in the Kalahari Desert of southern Africa: observations and simulations. *Quaternary Science Reviews* 28, 301–307.

Chase, B.M., Thomas, D.S.G., 2006. Late Quaternary dune accumulation along the western margin of South Africa: distinguishing forcing mechanisms through the analysis of migratory dune forms. *Earth and Planetary Science Letters* 251, 318–333.

Chazan, M., Porat, N., Sumner, T.A., Horwitz, L.K., 2013. The use of OSL dating in unstructured sands: the archaeology and chronology of the Hutton Sands at Canteen Kopje (Northern Cape Province, South Africa). *Archaeological and Anthropological Sciences* 5, 351–363.

Choi, J.H., Murray, A.S., Cheong, C.S., Hong, S.C., 2009. The dependence of dose recovery experiments on the bleaching of natural quartz OSL using different light sources. *Radiation Measurements* 44, 600–605.

Clarke, M.L., 1994. Infra-red stimulated luminescence ages from aeolian sand and alluvial fan deposits from the eastern Mojave Desert, California. *Quaternary Science Reviews* 13, 533–538.

Clarke, M.L., 1996. IRSL dating of sands: Bleaching characteristics at deposition inferred from the use of single aliquots. *Radiation Measurements* 26, 611–620.

Clarke, M.L., Rendell, H.M., 1998. Climate change impacts on sand supply and the formation of desert sand dunes in the south-west U.S.A. *Journal of Arid Environments* 39, 517–531.

Clarke, M.L., Wintle, A.G., Lancaster, N., 1996. Infra-red stimulated luminescence dating of sands from the Cronese Basins, Mojave Desert. *Geomorphology* 17, 199–205.

CLIMAP, 1976. The surface of the ice-age Earth. *Science* 191, 1131–1137.

Cohen, T.J., Nanson, G.C., Jansen, J.D., Jones, B.G., Jacobs, Z., Treble, P., Price, D.M., May, J.-H., Smith, A.M., Ayliffe, L.K., Hellstrom, J.C., 2011. Continental aridification and the vanishing of Australia's megalakes. *Geology* 39, 167–170.

Cohen, T.J., Nanson, G.C., Jansen, J.D., Gliganic, L.A., May, J.H., Larsen, J.R., Goodwin, I.D., Browning, S., Price, D.M., 2012a. A pluvial episode identified in arid Australia during the Medieval Climatic Anomaly. *Quaternary Science Reviews* 56, 167–171.

Cohen, T.J., Nanson, G.C., Jansen, J.D., Jones, B.G., Jacobs, Z., Larsen, J.R., May, J.-H., P.Treble, Price, D.M., Smith, A.M., 2012b. Late Quaternary mega-lakes fed by the northern and southern river systems of central Australia: varying moisture sources and increased continental aridity. *Palaeogeography, Palaeoclimatology, Palaeoecology* 356–357, 89–108.

Cohen, T.J., Jansen, J.D., Gliganic, L.A., Larsen, J.R., Nanson, G.C., May, J.-H., Jones, B.G., Price, D.M., 2015. Hydrological transformation coincided with megafaunal extinction in central Australia. *Geology* 43, 195–198.

Craddock, R.A., Tooth, S., Zimbelman, J.R., Wilson, S.A., Maxwell, T.A., Kling, C., 2015. Temporal observations of a linear sand dune in the Simpson Desert, central Australia: Testing models for dune formation on planetary surfaces. *Journal of Geophysical Research: Planets* 120, 1736–1750.

DeVogel, S.B., Magee, J.W., Manley, W.F., Miller, G.H., 2004. A GIS-based reconstruction of late Quaternary palaeohydrology: Lake Eyre, arid central Australia. *Palaeogeography, Palaeoclimatology, Palaeoecology* 204, 1–13.

Doerschner, N., Fitzsimmons, K.E., Blasco, R., Finlayson, G., Rodriguez-Vidal, J., Rosell, J., Hublin, J.-J., Finlayson, C., in press. Chronology of the late Pleistocene archaeological sequence at Vanguard Cave, Gibraltar: Insights from quartz single and multiple grain luminescence dating. *Quaternary International*.

Doerschner, N., Hernandez, M., Fitzsimmons, K.E., 2016. Sources of variability in single grain dose recovery experiments: Insights from Moroccan and Australian samples. *Ancient TL* 34, 14–25.

Duller, G., 2008. Single-grain optical dating of Quaternary sediments: Why aliquot size matters in luminescence dating. *Boreas* 37, 589–612.

Duller, G.A.T., 2003. Distinguishing quartz and feldspar in single grain luminescence measurements. *Radiation Measurements* 37, 161–165.

Duller, G.A.T., Augustinus, P.C., 2006. Reassessment of the record of linear dune activity in Tasmania using optical dating. *Quaternary Science Reviews* 25, 2608–2618.

Duller, G.A.T., Botter-Jensen, L., Murray, A.S., 2000. Optical dating of single sand-sized grains of quartz: sources of variability. *Radiation Measurements* 32, 453–457.

Ewing, R.C., McDonald, G.D., Hayes, A.G., 2015. Multi-spatial analysis of aeolian dune-field patterns. *Geomorphology* 240, 44–53.

Fitzsimmons, K.E., 2007. Morphological variability in the linear dunefields of the Strzelecki and Tirari Deserts, Australia. *Geomorphology* 91, 146–160.

Fitzsimmons, K.E., 2011. An assessment of the luminescence sensitivity of Australian quartz with respect to sediment history. *Geochronometria* 38, 199–208.

Fitzsimmons, K.E., 2017. Reconstructing palaeoenvironments on desert margins: New perspectives from Eurasian loess and Australian dry lake shorelines. *Quaternary Science Reviews* 171, 1–19.

Fitzsimmons, K.E., Rhodes, E.J., Magee, J.W., Barrows, T.T., 2007. The timing of linear dune activity in the Strzelecki and Tirari Deserts, Australia. *Quaternary Science Reviews* 26, 2598–2616.

Fitzsimmons, K.E., Magee, J.W., Amos, K., 2009. Characterisation of aeolian sediments from the Strzelecki and Tirari Deserts, Australia: Implications for reconstructing palaeoenvironmental conditions. *Sedimentary Geology* 218, 61–73.

Fitzsimmons, K.E., Rhodes, E.J., Barrows, T.T., 2010. OSL dating of southeast Australian quartz: A preliminary assessment of luminescence characteristics and behaviour. *Quaternary Geochronology* 5, 91–95.

Fitzsimmons, K.E., Stern, N., Murray-Wallace, C.V., 2014. Depositional history and archaeology of the central Lake Mungo lunette, Willandra Lakes, southeast Australia. *Journal of Archaeological Science* 41, 349–364.

Fitzsimmons, K.E., Stern, N., Murray-Wallace, C.V., Truscott, W., Pop, C., 2015. The Mungo mega-lake event, semi-arid Australia: Non-linear descent into the last ice age, implications for human behavior. *PLoS ONE* 10, e0127008.

Folk, R.L., 1968. *Petrology of Sedimentary Rocks*. Hemphill's, Austin.

Frechen, M., Vanneste, K., Verbeeck, K., Paulissen, E., Camelbeeck, T., 2001. The deposition history of the coversands along the Bree Fault Escarpment, NE Belgium. Geologie en Mijnbouw/ Netherlands *Journal of Geosciences* 80, 171–186.

Fryberger, S.G., 1979. Dune forms and wind regime, in: McKee, E.D. (Ed.), *A Study of Global Sand Seas*. United States Government Printing Office, Washington, pp. 137–170.

Fujioka, T., Chappell, J., Fifield, L.K., Rhodes, E.J., 2009. Australian desert dune fields initiated with Pliocene–Pleistocene global climatic shift. *Geology* 37, 51–54.

Galbraith, R.F. and Roberts, R.G. 2012. Statistical aspects of equivalent dose and error calculation and display in OSL dating: an overview and some recommendations. *Quaternary Geochronology* 11, 1–27.

Galbraith, R.F., Green, P.F., 1990. Estimating the component ages in a finite mixture. *Nuclear Tracks and Radiation Measurements* 17, 197–206.

Galbraith, R.F., Roberts, R.G., Laslett, G.M., Yoshida, H., Olley, J.M., 1999. Optical dating of single and multiple grains of quartz from Jinmium rock shelter, northern Australia. Part 1, Experimental design and statistical models. *Archaeometry* 41, 339–364.

Galbraith, R.F., Roberts, R.G., Yoshida, H., 2005. Error variation in OSL palaeodose estimates from single aliquots of quartz: A factorial experiment. *Radiation Measurements* 39, 289–307.

Gardner, G.J., Mortlock, A.J., Price, D.M., Readhead, M.L., Wasson, R.J., 1987. Thermoluminescence and radiocarbon dating of Australian desert dunes. *Australian Journal of Earth Sciences* 34, 343–357.

Gliganic, L.A., Cohen, T.J., May, J.-H., Jansen, J.D., Nanson, G.C., Dosseto, A., Larsen, J.R., Aubert, M., 2014. Late-Holocene climatic variability indicated by three natural archives in arid southern Australia. *The Holocene* 24, 104–117.

Gliganic, L.A., Cohen, T.J., Slack, M., Feathers, J.K., 2016. Sediment mixing in aeolian sandsheets identified and quantified using single-grain optically stimulated luminescence. *Quaternary Geochronology* 32, 53–66.

Gliganic, L.A., Jacobs, Z., Roberts, R.G., Domínguez-Rodrigo, M., Mabulla, A.Z.P., 2012. New ages for Middle and Later Stone Age deposits at Mumba rockshelter, Tanzania: Optically stimulated luminescence dating of quartz and feldspar grains. *Journal of Human Evolution* 62, 533–547.

Gliganic, L.A., May, J.H., Cohen, T.J., 2015. All mixed up: Using single-grain equivalent dose distributions to identify phases of pedogenic mixing on a dryland alluvial fan. *Quaternary International* 362, 23–33.

Guerin, G., Combes, B., Lahaye, C., Thomsen, K.J., Tribolo, C., Urbanova, P., Guibert, P., Mercier, N., Valladas, H., 2015. Testing the accuracy of a Bayesian central-dose model for single-grain OSL, using known-age samples. *Radiation Measurements* 81, 62–70.

Guérin, G., Mercier, N., 2012. Field gamma spectrometry, Monte Carlo simulations and potential of non-invasive measurements. *Geochronometria* 39, 40–47.

Guérin, G., Mercier, N., Nathan, R., Adamiec, G., Lefrais, Y., 2012. On the use of the infinite matrix assumption and associated concepts: A critical review. *Radiation Measurements* 47, 778–785.

Hesse, P.P., 2010. The Australian desert dunefields: formation and evolution in an old, flat, dry continent, in: Bishop, P., Pillans, B. (Eds.), *Australian Landscapes*. Geological Society, London, pp. 141–163.

Hesse, P., 2011. Sticky dunes in a wet desert: Formation, stabilisation and modification of the Australian desert dunefields. *Geomorphology* 134, 309–325.

Hesse, P.P., 2016. How do longitudinal dunes respond to climate forcing? Insights from 25 years of luminescence dating of the Australian desert dunefields. *Quaternary International* 410, 11–29.

Hesse, P.P., Simpson, R.L., 2006. Variable vegetation cover and episodic sand movement on longitudinal desert sand dunes. *Geomorphology* 81, 276–291.

Hesse, P.P., Humphreys, G.S., Selkirk, P.M., Adamson, D.A., Gore, G.B., Nobes, G.C., Price, D.M., Schwenninger, J.-L., Smith, B., Talau, M., Hemmings, F., 2003. Late Quaternary aeolian dunes on the presently humid Blue Mountains, Eastern Australia. *Quaternary International* 108, 13–22.

Hollands, C.B., Nanson, G.C., Jones, B.G., Bristow, C.S., Price, D.M., Pietsch, T.J., 2006. Aeolian–fluvial interaction: evidence for Late Quaternary channel change and wind-rift linear dune formation in the northwestern Simpson Desert, Australia. *Quaternary Science Reviews* 25, 142–162.

Hülle, D., Hilgers, A., Radtke, U., Stolz, C., Hempelmann, N., Grunert, J., Felauer, T., Lehmkuhl, F., 2010. OSL dating of sediments from the Gobi desert, Southern Mongolia. *Quaternary Geochronology* 5, 107–113.

Huntley, D.J., Godfrey-Smith, D.I., Thewalt, M.L.W., 1985. Optical dating of sediments. *Nature* 313, 105–107.

Hütt, G., Jaek, I., Tchonka, J., 1988. Optical dating: K-feldspars optical response stimulation spectra. *Quaternary Science Reviews* 7, 381–385.

Jacobs, Z., Duller, G.A.T., Wintle, A.G., 2006. Interpretation of single grain D_e distributions and calculation of D_e. *Radiation Measurements* 41, 264–277.

Jacobs, Z., Roberts, R.G., 2007. Advances in optically stimulated luminescence dating of individual grains of quartz from archeological deposits. *Evolutionary Anthropology: Issues, News, and Reviews* 16, 210–223.

Jacobs, Z., Roberts, R.G., Galbraith, R.F., Deacon, H.J., Grun, R., Mackay, A., Mitchell, P., Vogelsang, R., Wadley, L., 2008a. Ages for the Middle Stone Age of Southern Africa: Implications for Human Behavior and Dispersal. *Science* 322, 733–735.

Jacobs, Z., Roberts, R.G., Nespoulet, R., El Hajraoui, M.A., Débénath, A., 2012. Single-grain OSL chronologies for Middle Palaeolithic deposits at El Mnasra and El Harhoura 2, Morocco: Implications for Late Pleistocene human–environment interactions along the Atlantic coast of northwest Africa. *Journal of Human Evolution* 62, 377–394.

Jacobs, Z., Wintle, A.G., Duller, G.A.T., Roberts, R.G., Wadley, L., 2008b. New ages for the post-Howiesons Poort, late and final Middle Stone Age at Sibudu, South Africa. *Journal of Archaeological Science* 35, 1790–1807.

Kalchgruber, R., Fuchs, M., Murray, A.S., Wagner, G.A., 2003. Evaluating dose rate distributions in natural sediments using a-Al2O3: C grains. *Radiation Measurements* 37, 293–297.

Kasse, C., 1997. Cold-Climate Aeolian Sand-Sheet Formation in North-Western Europe (c. 14–12.4

ka); a Response to Permafrost Degradation and Increased Aridity. *Permafrost and Periglacial Processes* 8, 295–311.

King, D., 1960. The sand ridge deserts of South Australia and related aeolian landforms of the Quaternary arid cycles. *Transactions of the Royal Society of South Australia* 83, 99–108.

Koster, E.A., 1988. Ancient and modern cold-climate aeolian sand deposition: A review. *Journal of Quaternary Science* 3, 69–83.

Koster, E.A., 2005. Recent advances in luminescence dating of Late Pleistocene (cold-climate) aeolian sand and loess deposits in western Europe. *Permafrost and Periglacial Processes* 16, 131–143.

Kristensen, J.A., Thomsen, K.J., Murray, A.S., Buylaert, J.-P., Jain, M., Breuning-Madsen, H., 2015. Quantification of termite bioturbation in a savannah ecosystem: Application of OSL dating. *Quaternary Geochronology* 30, 334–341.

Kumar, A., Srivastava, P., Meena, N.K., 2017. Late Pleistocene aeolian activity in the cold desert of Ladakh: A record from sand ramps. *Quaternary International* 443, 13–28.

Lambeck, K., Chappell, J., 2001. Sea level change through the last glacial cycle. *Science* 292, 679–686.

Lancaster, N., 2008. Desert dune dynamics and development: insights from luminescence dating. *Boreas* 37, 559–573.

Lancaster, N., Tchakerian, V.P., 1996. Geomorphology and sediments of sand ramps in the Mojave desert. *Geomorphology* 17, 151–165.

Lancaster, N., Kocurek, G., Singhvi, A., Pandey, V., Deynoux, M., Ghienne, J., Lo, K., 2002. Late Pleistocene and Holocene dune activity and wind regimes in the western Sahara Desert of Mauritania. *Geology* 30, 991–994.

Lancaster, N., Wolfe, S., Thomas, D., Bristow, C., Bubenzer, O., Burrough, S., Duller, G., Halfen, A., Hesse, P., Roskin, J., Singhvi, A., Tsoar, H., Tripaldi, A., Yang, X., Zárate, M., 2016. The INQUA Dunes Atlas chronologic database. *Quaternary International* 410, 3–10.

Lian, O.B., Roberts, R.G., 2006. Dating the Quaternary: Progress in luminescence dating of sediments. *Quaternary Science Reviews* 25, 2449–2468.

Livingstone, I., Warren, A., 1996. *Aeolian Geomorphology: An Introduction*. Longman Singapore Publishers, Singapore.

Livingstone, I., Wiggs, G.F.S., Weaver, C.M., 2007. Geomorphology of desert sand dunes: A review of recent progress. *Earth-Science Reviews* 80, 239–257.

Lomax, J., Hilgers, A., Twidale, C.R., Bourne, J.A., Radtke, U., 2007. Treatment of broad palaeodose distributions in OSL dating of dune sands from the western Murray Basin, South Australia. *Quaternary Geochronology* 2, 51–56.

Lomax, J., Hilgers, A., Radtke, U., 2011. Palaeoenvironmental change recorded in the palaeodunefields of the western Murray Basin, South Australia – New data from single grain OSL-dating. *Quaternary Science Reviews* 30, 723–736.

Magee, J.W., 2006. Australian lake-level studies, in: Elias, S.A. (Ed.), *Encyclopedia of Quaternary Science*. Elsevier, Amsterdam, pp. 1359–1365.

Magee, J.W., Bowler, J.M., Miller, G.H., Williams, D.L.G., 1995. Stratigraphy, Sedimentology, Chronology and Paleohydrology of Quaternary Lacustrine Deposits at Madigan Gulf, Lake Eyre, South Australia. *Palaeogeography Palaeoclimatology Palaeoecology* 113, 3–42.

Magee, J.W., Miller, G.H., 1998. Lake Eyre palaeohydrology from 60 ka to the present: beach ridges and glacial maximum aridity. *Palaeogeography Palaeoclimatology Palaeoecology* 144, 307–329.

Magee, J.W., Miller, G.H., Spooner, N.A., Questiaux, D., 2004. Continuous 150 k.y. monsoon record from Lake Eyre, Australia: Insolation-forcing implications and unexpected Holocene failure. *Geology* 32, 885–888.

May, J.-H., Wells, S.G., Cohen, T.J., Marx, S.K., Nanson, G.C., Baker, S.E., 2015. A soil chronosequence on Lake Mega-Frome beach ridges and its implications for late Quaternary pedogenesis and paleoenvironmental conditions in the drylands of southern Australia. *Quaternary Research* 83, 150–165.

Mayya, Y.S., Morthekai, P., Murari, M.K., Singhvi, A.K., 2006. Towards quantifying beta microdosimetric effects in single-grain quartz dose distribution. *Radiation Measurements* 41, 1032–1039.

McKee, E.D., 1979. Introduction to a study of global sand seas, in: McKee, E.D. (Ed.), *A Study of Global Sand Seas*. United States Government Printing Office, Washington, pp. 1–20.

McKeever, S.W.S., 1991. Mechanisms of thermoluminescence production: some problems and a few answers? *Nuclear Tracks and Radiation Measurements* 18, 5–12.

Mejdahl, V., 1979. Thermoluminescence dating: beta-dose attenuation in quartz grains. *Archaeometry* 21, 61–72.

Munyikwa, K., 2005. Synchrony of Southern Hemisphere Late Pleistocene arid episodes: A review of luminescence chronologies from arid aeolian landscapes south of the Equator. *Quaternary Science Reviews* 24, 2555–2583.

Murray, A.S., Roberts, R.G., 1997a. Determining the burial time of single grains of quartz using optically stimulated luminescence. *Earth and Planetary Science Letters* 152, 163–180.

Murray, A.S., Roberts, R.G., 1997b. Determining the burial time of single grains of quartz using optically stimulated luminescence. *Earth and Planetary Science Letters* 152, 163–180.

Murray, A.S., Roberts, R.G., Wintle, A.G., 1997. Equivalent dose measurement using a single aliquot of quartz. *Radiation Measurements* 27, 171–184.

Murray, A.S., Wintle, A.G., 2000. Luminescence dating of quartz using an improved single-aliquot regenerative-dose protocol. *Radiation Measurements* 32, 57–73.

Murray, A.S., Wintle, A.G., 2003. The single aliquot regenerative dose protocol: potential for improvements in reliability. *Radiation Measurements* 37, 377–381.

Murton, J.B., Bateman, M.D., Baker, C.A., Knox, R., Whiteman, C.A., 2003. The Devensian periglacial record on Thanet, Kent, UK. *Permafrost and Periglacial Processes* 14, 217–246.

Nanson, G.C., Chen, X.Y., Price, D.M., 1992a. Lateral migration, thermoluminescence chronology and colour variation of longitudinal dunes near Birdsville in the Simpson Desert, central Australia. *Earth Surface Processes and Landforms* 17, 807–819.

Nanson, G.C., Price, D.M., Short, S.A., 1992b. Wetting and drying of Australia over the past 300 ka. *Geology* 20, 791–794.

Nanson, G.C., Callen, R.A., Price, D.M., 1998. Hydroclimatic interpretation of Quaternary shorelines on South Australian playas. *Palaeogeography, Palaeoclimatology, Palaeoecology* 144, 281–305.

Nathan, R.P., Thomas, P.J., Jain, M., Murray, A.S., Rhodes, E.J., 2003. Environmental dose rate heterogeneity of beta radiation and its implications for luminescence dating: Monte Carlo modelling and experimental validation. *Radiation Measurements* 37, 305–313.

Olley, J.M., Caitcheon, G.G., Murray, A.S., 1998. The distribution of apparent dose as determined by optically stimulated luminescence in small aliquots of fluvial quartz: Implications for dating young sediments. *Quaternary Geochronology* 17, 1033–1040.

Olley, J.M., Roberts, R.G., Murray, A., 1997. Disequilibria in the uranium decay series in sedimentary deposits at Allen's Cave, Nullarbor Plain, Australia: Implications for dose rate determinations. *Radiation Measurements* 27, 433–443.

Pécsi, M., 1990. Loess is not just the accumulation of dust. *Quaternary International* 7–8, 1–21.

Petherick, L.M., McGowan, H.A., Kamber, B.S., 2009. Reconstructing transport pathways for late Quaternary dust from eastern Australia using the composition of trace elements of long traveled dusts. *Geomorphology* 105, 67–79.

Petit, J.R., Jouzel, J., Raynaud, D., Baricov, N.I., Basil, I., Bender, M., Chappellaz, J., Davis, M., Delaygue, G., Delmott, M., Kotlyakov, V.M., Legrand, M., Lipenkov, V.Y., Lorius, C., Pepin, L., Ritz, C., Saltzman, E., Stievenard, M., 1999. Climate and atmospheric history of the past 420,000 years from the Vostok ice core, Antarctica. *Nature* 399, 429–436.

Prescott, J.R., 1983. Thermoluminescence dating of sand dunes at Roonka, South Australia. *PACT* 9, 505–512.

Prescott, J.R., Hutton, J.T., 1994. Cosmic ray contributions to dose rates for luminescence and ESR dating: Large depths and long term variations. *Radiation Measurements* 23, 497–500.

Prescott, J.R., Huntley, D.J., Hutton, J.T., 1993. Estimation of equivalent dose in thermoluminescence dating – the *Australian slide* method. *Ancient TL* 11, 1–5.

Pye, K., Tsoar, H., 1990. *Aeolian Sand and Sand Dunes*. Unwin Hyman, London.

Quigley, M.C., Horton, T., Hellstrom, J.C., Cupper, M.L., Sandiford, M., 2010. Holocene climate change in arid Australia from speleothem and alluvial records. *The Holocene* 20, 1093–1104.

Radtke, U., Janotta, A., Hilgers, A., Murray, A.S., 2001. The potential of OSL and TL for dating Lateglacial and Holocene dune sands tested with independent age control of the Laacher See tephra (12880 a) at the section `Mainz-Gonsenheim'. *Quaternary Science Reviews* 20, 719–724.

Rasmussen, S.O., Bigler, M., Blockley, S.P., Blunier, T., Buchardt, S.L., Clausen, H.B., Cvijanovic, I., Dahl-Jensen, D., Johnsen, S.J., Fischer, H., Gkinis, V., Guillevic, M., Hoek, W.Z., Lowe, J.J., Pedro, J.B., Popp, T., Seierstad, I.K., Steffensen, J.P., Svensson, A.M., Vallelonga, P., Vinther, B.M., Walker, M.J.C., Wheatley, J.J., Winstrup, M., 2014. A stratigraphic framework for abrupt climatic changes during the Last Glacial period based on three synchronized Greenland ice-core records: refining and extending the INTIMATE event stratigraphy. *Quaternary Science Reviews* 106, 14–28.

Rendell, H.M., Scheffer, N.L., 1996. Luminescence dating of sand ramps in the Eastern Mojave Desert. *Geomorphology* 17, 187–197.

Rendell, H.M., Lancaster, N., Tchakerian, V.P., 1994. Luminescence dating of late quaternary aeolian deposits at Dale Lake and Cronese Mountains, Mojave Desert, California. *Quaternary Science Reviews* 13, 417–422.

Rhodes, E.J., 2000. Observations of thermal transfer OSL signals in glacigenic quartz. *Radiation Measurements* 32, 595–602.

Rhodes, E.J., Bailey, R.M., 1997. The effect of thermal transfer on the zeroing of the luminescence of quartz from recent glaciofluvial sediments. *Quaternary Geochronology* 16, 291–298.

Rink, W.J., Dunbar, J.S., Tschinkel, W.R., Kwapich, C., Repp, A., Stanton, W., Thulman, D.K., 2013. Subterranean transport and deposition of quartz by ants in sandy sites relevant to age overestimation in optical luminescence dating. *Journal of Archaeological Science* 40, 2217–2226.

Rodnight, H., 2008. How many equivalent dose values are needed to obtain a reproducible distribution? *Ancient TL* 26, 3–10.

Roskin, J., Tsoar, H., Porat, N., Blumberg, D.G., 2011. Palaeoclimate interpretations of Late Pleistocene vegetated linear dune mobilization episodes: Evidence from the northwestern Negev dunefield, Israel. *Quaternary Science Reviews* 30, 3364–3380.

Rufer, D., Preusser, F., 2009. Potential of autoradiography to detect spatially resolved radiation patterns in the context of trapped charge dating. *Geochronometria* 24, 1–13.

Sanderson, D.C.W., Bishop, P., Houston, I., Boonsener, M., 2001. Luminescence characterisation of quartz-rich cover sands from NE Thailand. *Quaternary Science Reviews* 20, 893–900.

Sarnthein, M., 1978. Sand deserts during glacial maximum and climatic optimum. *Nature* 272, 43–44.

Schmidt, C., Pettke, T., Preusser, F., Rufer, D., Kasper, H.U., Hilgers, A., 2012. Quantification and spatial distribution of dose rate relevant elements in silex used for luminescence dating. *Quaternary Geochronology* 12, 65–73.

Schokker, J., Cleveringa, P., Murray, A.S., 2004. Palaeoenvironmental reconstruction and OSL dating of terrestrial Eemian deposits in the southeastern Netherlands. *Journal of Quaternary Science* 19, 193–202.

Schwan, J., 1988. The structure and genesis of Weichselian to early hologene aeolian sand sheets in western Europe. *Sedimentary Geology* 55, 197–232.

Sheard, M.J., Lintern, M.J., Prescott, J.R., Huntley, D.J., 2006. Great Victoria Desert: New dates for South Australia's ?oldest desert dune system. *MESA Journal* 42, 15–26.

Shulmeister, J., Kemp, J., Fitzsimmons, K.E., Gontz, A., 2016. Constant wind regimes during the Last Glacial Maximum and early Holocene: evidence from Little Llangothlin Lagoon, New England Tablelands, eastern Australia. *Clim. Past* 12, 1435–1444.

Singhvi, A.K., Banerjee, D., Ramesh, R., Rajaguru, S.N., Gogte, V., 1996. A luminescence method for dating 'dirty' pedogenic carbonates for paleoenvironmental reconstruction. *Earth and Planetary Science Letters* 139, 321–332.

Singhvi, A.K., Bluszcz, A., Bateman, M.D., Rao, M.S., 2001. Luminescence dating of loess–palaeosol sequences and coversands: methodological aspects and palaeoclimatic implications. *Earth-Science Reviews* 54, 193–211.

Singhvi, A.K., Williams, M.A.J., Rajaguru, S.N., Misra, V.N., Chawla, S., Stokes, S., Chauhan, N., Francis, T., Ganjoo, R.K., Humphreys, G.S., 2010. A ~200 ka record of climatic change and dune activity in the Thar Desert, India. *Quaternary Science Reviews* 29, 3095–3105.

Sitzia, L., Bertran, P., Bahain, J.-J., Bateman, M.D., Hernandez, M., Garon, H., de Lafontaine, G., Mercier, N., Leroyer, C., Queffelec, A., Voinchet, P., 2015. The Quaternary coversands of southwest France. *Quaternary Science Reviews* 124, 84–105.

Spooner, N.A., 1994. On the dating signal from quartz. *Radiation Measurements* 23, 593–600.

Steele, T.E., Mackay, A., Fitzsimmons, K.E., Igreja, M., Marwick, B., Orton, J., Schwortz, S., Stahlschmidt, M.C., 2016. Varsche Rivier 003: A Middle and Later Stone Age Site with Still Bay and Howieson's Poort Assemblages in Southern Namaqualand, South Africa. *PaleoAnthropology* 2016, 100–163.

Stockmann, U., Minasny, B., Pietsch, T.J., McBratney, A.B., 2013. Quantifying processes of pedogenesis using optically stimulated luminescence. *European Journal of Soil Science* 64, 145–160.

Stokes, S., 1992. Optical dating of young (modern) sediments using quartz: Results from a selection of depositional environments. *Quaternary Science Reviews* 11, 153–159.

Stokes, S., 1994. The timing of OSL sensitivity changes in a natural quartz. *Radiation Measurements* 23, 601–605.

Stokes, S., Gaylord, D.R., 1993. Optical dating of Holocene dune sands in the Ferris dune field, Wyoming. *Quaternary Research* 39, 274–281.

Stokes, S., Kocurek, G., Pye, K., Winspear, N.R., 1997a. New evidence for the timing of aeolian sand supply to the Algodones dunefield and East Mesa area, southeastern California, USA. *Palaeogeography, Palaeoclimatology, Palaeoecology* 128, 63–75.

Stokes, S., Thomas, D.S.G., Shaw, P.A., 1997b. New chronological evidence for the nature and timing of linear dune development in the southwest Kalahari Desert. *Geomorphology* 20, 81–93.

Stokes, S., Thomas, D.S.G., Washington, R., 1997c. Multiple episodes of aridity in southern Africa since the last interglacial period. *Nature* 388, 154–158.

Stone, A.E.C., Thomas, D.S.G., 2008. Linear dune accumulation chronologies from the southwest Kalahari, Namibia: challenges of reconstructing late Quaternary palaeoenvironments from aeolian landforms. *Quaternary Science Reviews* 27, 1667–1681.

Stuut, J.B.W., Prins, M.A., Schneider, R.R., Weltje, G.J., Jansen, J.H.F., Postma, G., 2002. A 300 kyr record of aridity and wind strength in southwestern Africa: inferences from grain-size distributions of sediments on Walvis Ridge, SE Atlantic. *Marine Geology* 180, 221–223.

Telfer, M.W., 2011. Growth by extension, and reworking, of a south-western Kalahari linear dune. *Earth Surface Processes and Landforms* 36, 1125–1135.

Telfer, M.W., Hesse, P.P., 2013. Palaeoenvironmental reconstructions from linear dunefields: recent progress, current challenges and future directions. *Quaternary Science Reviews* 78, 1–21.

Telfer, M.W., Thomas, D.S.G., 2006. Complex Holocene lunette dune development, South Africa: Implications for paleoclimate and models of pan development in arid regions. *Geology* 34, 853–856.

Telfer, M., Thomas, D., 2007. Late Quaternary linear dune accumulation and chronostratigraphy of the southwestern Kalahari: implications for aeolian palaeoclimatic reconstructions and predictions of future dynamics. *Quaternary Science Reviews* 26, 2617–2630.

Telfer, M.W., Thomas, Z.A., Breman, E., 2012. Sand ramps in the Golden Gate Highlands National Park, South Africa: Evidence of periglacial aeolian activity during the last glacial. *Palaeogeography, Palaeoclimatology, Palaeoecology* 313–314, 59–69.

Telfer, M.W., Mills, S.C., Mather, A.E., 2014. Extensive Quaternary aeolian deposits in the Drakensberg foothills, Rooiberge, South Africa. *Geomorphology* 219, 161–175.

Telfer, M.W., Hesse, P.P., Perez-Fernandez, M., Bailey, R.M., Bajkan, S., Lancaster, N., 2017. Morphodynamics, boundary conditions and pattern evolution within a vegetated linear dunefield. *Geomorphology* 290, 85–100.

Thomas, D.S.G., 2007. *Palaeoenvironmental Potentials of Detailed Late Quaternary Dune Construction Chronologies: Aridity Records or Aeolian Process Archives?* INQUA Quaternary International, Cairns, Australia.

Thomas, D.S.G., Bailey, R.M., 2016. Accumulation rate variability analysis of southern African Late Quaternary desert dune chronologies. *Quaternary International* 404, Part B, 193.

Thomas, D.S.G., Bailey, R.M., 2017. Is there evidence for global-scale forcing of Southern Hemisphere Quaternary desert dune accumulation? A quantitative method for testing hypotheses of dune system development. *Earth Surface Processes and Landforms* 42, 2284–2294.

Thomas, D.S.G., Burrough, S.L., 2016. Luminescence-based dune chronologies in southern Africa: Analysis and interpretation of dune database records across the subcontinent. *Quaternary International* 410, 30–45.

Thomas, D.S.G., Bateman, M.D., Mehrshahi, D., O'Hara, S.L., 1997. Development and Environmental Significance of an Eolian Sand Ramp of Last-Glacial Age, Central Iran. *Quaternary Research* 48, 155–161.

Thomsen, K.J., Murray, A., Jain, M., 2012. The dose dependency of the over-dispersion of quartz OSL single grain dose distributions. *Radiation Measurements* 47, 732–739.

Thomsen, K.J., Murray, A.S., Bøtter-Jensen, L., 2005. Sources of variability in OSL dose measurements using single grains of quartz. *Radiation Measurements* 39, 47–61.

Thomsen, K.J., Murray, A.S., Buylaert, J.P., Jain, M., Hansen, J.H., Aubry, T., 2016. Testing single-grain quartz OSL methods using sediment samples with independent age control from the

Bordes-Fitte rockshelter (Roches d'Abilly site, Central France). *Quaternary Geochronology* 31, 77–96.

Turner, B.R., Makhlouf, I., 2002. Recent colluvial sedimentation in Jordan: Fans evolving into sand ramps. *Sedimentology* 49, 1283–1298.

van den Bogaard, P., 1995. 40Ar/39Ar ages of sanidine phenocrysts from Laacher See Tephra (12,900 yr BP): Chronostratigraphic and petrological significance. *Earth and Planetary Science Letters* 133, 163–174.

Vandenberghe, D., Hossain, S.M., De Corte, F., Van den haute, P., 2003. Investigations on the origin of the equivalent dose distribution in a Dutch coversand. *Radiation Measurements* 37, 433–439.

Vandenberghe, D., Kasse, C., Hossain, S.M., De Corte, F., Van Den Haute, P., Fuchs, M., Murray, A.S., 2004. Exploring the method of optical dating and comparison of optical and 14C ages of Late Weichselian coversands in the southern Netherlands. *Journal of Quaternary Science* 19, 73–86.

Vandenberghe, D., De Corte, F., Buylaert, J.P., Kucera, J., Van den haute, P., 2008. On the internal radioactivity in quartz. *Radiation Measurements* 43, 771–775.

Veth, P., 1995. Aridity and settlement in northwest Australia. *Antiquity* 69, 733–746.

Wang, X.L., Wintle, A.G., Du, J.H., Kang, S.G., Lu, Y.C., 2011. Recovering laboratory doses using fine-grained quartz from Chinese loess. *Radiation Measurements* 46, 1073–1081.

Wang, Z., Zhao, H., Zhang, K., Ren, X., Chen, F., Wang, T., 2009. Barchans of Minqin: quantifying migration rate of a barchan. Sciences in Cold and Arid Regions 1, 0151-0156.

Wasson, R.J., Hyde, R., 1983. Factors determining desert dune type. *Nature* 304, 337–339.

Whittaker, T.E., Hendy, C.H., Hellstrom, J.C., 2011. Abrupt millennial-scale changes in intensity of Southern Hemisphere westerly winds during marine isotope stages 2–4. *Geology* 39, 455–458.

Wintle, A.G., 1993. Luminescence dating of aeolian sands: an overview, in: Pye, K. (Ed.), *The Dynamics and Environmental Context of Aeolian Sedimentary Systems*. Geological Society Special Publications, London, pp. 49–58.

Wintle, A.G., 1997. Luminescence dating: laboratory procedures and protocols. *Radiation Measurements* 27, 769–817.

Wintle, A.G., Huntley, D.J., 1979. Thermoluminescence dating of a deep-sea sediment core. *Nature* 289, 710–712.

Wintle, A.G., Murray, A.S., 1997. The relationship between quartz thermoluminescence, photo-transferred thermoluminescence, and optically stimulated luminescence. *Radiation Measurements* 27, 611–624.

Wintle, A.G., Murray, A.S., 2000. Quartz OSL: Effects of thermal treatment and their relevance to laboratory dating procedures. *Radiation Measurements* 32, 387–400.

Wolfe, S.A., Hugenholtz, C.H., 2009. Barchan dunes stabilized under recent climate warming on the northern Great Plains. *Geology* 37, 1039–1042.

Yang, L., Zhou, J., Lai, Z., Long, H., Zhang, J., 2010a. Lateglacial and Holocene dune evolution in the Horqin dunefield of northeastern China based on luminescence dating. *Palaeogeography, Palaeoclimatology, Palaeoecology* 296, 44–51.

Yang, X., Ma, N., Dong, J., Zhu, B., Xu, B., Ma, Z., Liu, J., 2010b. Recharge to the inter-dune lakes and Holocene climatic changes in the Badain Jaran Desert, western China. *Quaternary Research* 73, 10–19.

Zens, J., Zeeden, C., Römer, W., Fuchs, M., Klasen, N., Lehmkuhl, F., 2017. The Eltville Tephra (Western Europe) age revised: Integrating stratigraphic and dating information from different Last Glacial loess localities. *Palaeogeography, Palaeoclimatology, Palaeoecology* 466, 240–251.

Zhou, Y., Lu, H., Zhang, J., Mason, J.A., Zhou, L., 2009. Luminescence dating of sand-loess sequences and response of Mu Us and Otindag sand fields (north China) to climatic changes. *Journal of Quaternary Science* 24, 336–344.

Zimmerman, J., 1971. The radiation-induced increase of thermoluminescence sensitivity of fired quartz. *Journal of Physics C: Solid State Physics* 4, 3277–3291.

5 APPLICATIONS IN LOESSIC ENVIRONMENTS

THOMAS STEVENS

Department of Earth Sciences, Uppsala University, Villavägen 16, Uppsala, 75236, Sweden. Email thomas.stevens@geo.uu.se

ABSTRACT: Wind-blown, dominantly silt-sized terrestrial clastic deposits are called loess and are preserved on vast areas of the continents, providing excellent archives of past climate and dustiness. As loess is dominantly comprised of quartz and feldspar, and wind-blown transport generally ensures good exposure to sunlight, loess is ideal material for luminescence dating. Indeed, luminescence dating has become the chronological method of choice for much loess research. Luminescence dating of loess comes with a particular set of challenges and opportunities, and because loess has often been used as a medium to develop and test new luminescence protocols, judging the reliability of previously published ages and the best luminescence protocol to apply can be confusing. Furthermore, high dose rates, offsets between ages obtained from different quartz grain sizes, and early quartz saturation mean that the upper age limit for quartz OSL dating of loess is generally relatively young (potentially <30 ka). However, recently developed post-IR IRSL protocols using feldspars as dosimeters should enable dating of loess deposits well into the Middle Pleistocene. Identification and interpretation of reworked loess and palaeosols are major considerations when dating loess sequences, and their presence requires extra consideration when interpreting luminescence ages from these sediments. Furthermore, while soil stratigraphy presents opportunities to examine the accuracy of luminescence approaches, this can at times be misleading, and uncertainties over past water content and its effect on dose rates remain significant in loess deposits. A major opportunity with luminescence dating of loess lies in the application of luminescence protocols at high sampling resolution. This approach can highlight a range of syn- and post-depositional processes and allows derivation of much needed information on the rates of past dust mass accumulation, which in turn can provide insight into past atmospheric dustiness and help in improving models of past climate and dust activity.

KEYWORDS: loess, palaeosols, dust, mass accumulation rates

5.1 INTRODUCTION

Loess is a widespread aeolian-derived terrestrial sediment deposit, dominantly comprised of silt-sized clastic particles (Fig. 5.1), and covering around 10% of the land surface

153

(Pye 1987). Loess is often considered one of the key archives of climate change on land, recording detailed multi-millennial to millennial-scale climatic changes over the Quaternary (Porter 2001) and extending deeper into Cenozoic time in some locations (e.g. Guo *et al.* 2002). Recently, loess deposits have also received increased attention as an archive of large-scale landscape evolution (Nie *et al.* 2015) and atmospheric dustiness (Albani *et al.* 2015). As loess is formed from aeolian transport of dust (Smalley and Leach 1978; Smalley *et al.* 2009), loess sequences should be ideal candidates for direct numerical dating using luminescence methods as mineral grains are likely to be well-bleached prior to deposition. Indeed, many advances in luminescence methods have been achieved through testing and application of protocols in loess environments (Roberts 2008) and some of the earliest numerical dating studies on loess were performed using luminescence techniques (e.g. Wintle 1981). This connection between loess and luminescence technique development means that the wide range of luminescence dating protocols that have been applied to loess deposits can be bewildering to non-experts. Some of the important ones in use today are considered below from a practical point of view, but are outlined more generally in Chapter 1. The advantages and disadvantages in the use of these approaches in a loess context are also explored below. However, prior to considering the specific challenges and considerations in luminescence dating of loess, it is useful to first consider the characteristics of loess in more detail.

While the general definition of loess as a distinct sediment body of wind-blown, clastic, dominantly silt-sized material is relatively simple, this masks a degree of complexity in post-depositional alteration and reworking (Muhs 2013). The impacts of these post-depositional processes have a key influence on the practicalities of luminescence dating of loess deposits. Some aspects of loess sedimentology can broadly be considered as universal. For example, loess is dominantly comprised of silt-sized (~4–63 μm) quartz and feldspars, with secondary carbonates, clay minerals, micas and heavy minerals (Pye 1995). To varying extents, loess is also comprised of sand (~>63 μm) and clay-sized (~<4 μm) particles, although the proportions can vary considerably geographically, with source proximal deposits containing higher sand contents. Loess modal grain size therefore varies between ~20 μm to even 60 μm (Fig. 5.2), often decreasing downwind of source regions (Yang and Ding 2004). The typical grain size distribution of loess displays a clear single silt mode as part of a negatively skewed size distribution; luminescence dating making use of a range of grain-size fractions (Fig. 5.2). This skewed distribution has sometimes been interpreted as indicating the influence of both short-term (medium and coarse silt) and long-term (fine silt and clay) suspended dust on loess composition (Fig. 5.2), although the impact of grain-aggregate transport and post-depositional weathering on this distribution may be significant (Újvári *et al.* 2016). In any case, given that the majority of loess comprises medium to coarse silt, the sediment deposit can generally be considered as dominantly formed from source proximal wind-blown dust.

Loess deposits are widespread and drape or mantle the underlying landscape surfaces, covering pre-existing topography. Deposits may range from just cm to m through to hundreds of metres in thickness, and are relatively homogeneous. The majority of deposits show a classic alternation of massive, buff coloured loess (L) units and variously coloured soil/palaeosol (S) units showing varying degrees and types of weathering and alteration (Fig. 5.1). The type and intensity of soil formation in loess can have an impact on luminesce dating of these soil units and this subject is revisited later in this chapter. Sometimes loess and soil units may be intercalated with other sediments, such as marine and fluvial sediments (e.g. in the Lower Volga loess). While loess deposits generally lack primary structures, more weathered loess units and the soil units show some secondary structures, including burrows (krotovina), root (rhizoliths) and pseudomycelia casts and traces, carbonate nodules, oxidation and reduction streaks and soil aggregates. The alternation of loess and soil units in loess sequences, interpreted as representing the general dominance of cold and warm climates respectively, extends back in some regions throughout the Quaternary and records swings in glacial–interglacial climate (Fig. 5.3).

While many of these aspects of loess are relatively consistent between areas, other aspects are more variable, and this can impact luminescence sampling and interpretation. The degree and type of pedogenesis varies considerably in space and time, with some loess deposits showing only weak soil formation during interglacials, and almost no evidence of interstadial soil formation (e.g. NW Chinese Loess Plateau; Jeong *et al.* 2008), while others show extremely complex soil stratigraphy with multiple soil horizons within single loess units, possibly reflecting millennial-scale climate oscillations (e.g. W Europe; Haesaerts *et al.* 2016). The types of soils also vary considerably, with European loess deposits in particular exhibiting a range of soil types reflecting a gradient of climatic conditions, from wet maritime in the west to semi-arid continental and even Mediterranean in the east and south. This diversity makes traditional age models based on soil correlations rather

Figure 5.1 Classic loess deposits on the Chinese Loess Plateau. Left-hand: the lower Quaternary part of the Chinese loess type section at Luochuan. Right-hand: a high resolution sequence covering the last two glacial–interglacial cycles at Beiguoyuan (Huanxian) on the northwest of the plateau. Note the people for scale in both photos.

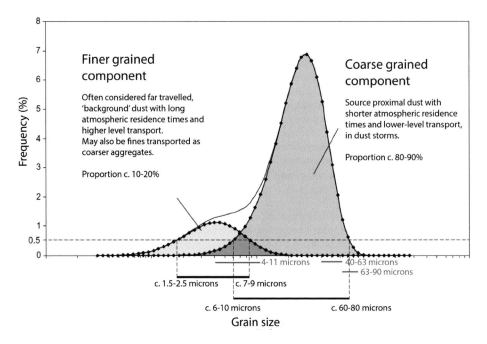

Figure 5.2 Typical loess grain-size distribution (example from Hungarian loess of the Carpathian Basin). The positive skew of typical loess can be seen with a tail of fine silt to clay-sized particles that are often considered as having undergone a separate, far travelled transport history from the main coarse grain silt mode. Also shown are the typical grain-size ranges used in luminescence dating of loess (in red). Figure modified from Varga *et al.* (2012). Note the $\log_{10}$ grain-size scale.

tricky to assign (Marković *et al.* 2015) and underscores the need to apply independent dating methods such as luminescence to such deposits. Indeed, new cross-continental correlations of loess in Europe have to a large extent been based on luminescence dating. This diversity in weathering in soils is also reflected in diversity of values in geochemical weathering indices such as the chemical index of alteration (Obreht *et al.* 2015).

In fact, to some extent post-depositional alteration may be considered a pre-requisite for all loess deposits, as clay and carbonate bonds formed post-depositionally allow loess-soil deposits to form vertical cliffs when incised by streams and rivers (Fig. 5.1). The combination of processes that allows this is considered by some researchers to be unique to loess deposits and is termed 'loessification' (Pécsi 1995). However, the degree to which loess can be considered a sediment, soil or rock is debated depending on the background and emphasis of loess researchers (Sprafke and Obreht 2016), and many argue that loess can essentially be considered as simply terrestrially accumulated wind-blown dust (Muhs 2013; Smalley *et al.* 2011). Perhaps more importantly from the perspective of the luminescence dating practitioner, the difference between loess and loess-derivatives must be considered. This distinction is not always agreed upon, although broadly the latter can be considered as aeolian dust deposits modified during or after deposition (often called reworked loess or 'loess-like' deposits). This modification can be via a number of processes, including solifluction, slope wash, gelifluction, colluvial or slope processes, other periglacial processes, or potentially changes in ground water (e.g. Lehmkuhl *et al.*

2016, Sprafke and Obreht 2016). Such processes can have an impact on luminescence ages taken from loess derivative deposits, as well as on their interpretation in terms of the timing of dust deposition. As such it is important to identify loess derivatives when in the field and during sampling. Furthermore, while loess generally is considered as geochemically and mineralogical quite homogeneous globally, being a good approximation of the upper continental crustal average in geochemical composition (Taylor *et al.* 1983), there is considerable variation in source area and pathway (Crouvi *et al.* 2008; Smalley and Leach 1978; Smalley *et al.* 2009), with consequent variation in source rock type (Stevens *et al.* 2013). This variation can have an impact on mineralogical and chemical composition (Obreht *et al.* 2016) and can also strongly influence the luminescence characteristics of quartz and feldspar (Schatz *et al.* 2012; Stevens *et al.* 2013). The impact of sediment source as well as loess syn- and post-depositional processes on luminescence dating is described in the following sections.

In spite of these complications, loess can generally be considered as an ideal sediment for the application of luminescence methods (Roberts 2008). While fossil gastropods, charcoal and other suitable material for ^{14}C dating are sometimes present in loess sequences (Pigati *et al.* 2013; Újvári *et al.* 2015; 2017), they are often only sporadically present or are absent in some areas, resultant ages can be mutually inconsistent (Novothny *et al.* 2009), and in any case are limited to the last *c.* 40 ka. In contrast, recent developments in luminescence dating (largely developed and tested on loess) allow independent dating back to the Middle Pleistocene (*c.* 250–300 ka; Buylaert *et al.* 2012; Li and Li 2012; Stevens *et al.* 2018) and luminescence techniques are unique in dating the timing of deposition of loess-constituent mineral grains themselves. Given the inherent assumptions and inaccuracies in age models derived from non-independent, stratigraphic/proxy correlation or tuning based approaches (Stevens *et al.* 2008) this makes luminescence dating an essential tool in much loess research.

Significant breakthroughs in loess research have come from luminescence dating and there are many opportunities for rich rewards in future applications. Some potential interesting areas are discussed in the sections below. Loess soil stratigraphy has also been correlated extensively with marine oxygen isotope stratigraphy, and so depending on the degree of security of these correlations the results of loess luminescence dating can be benchmarked against these expected ages, or vice versa can be used to test these proposed correlations. However, this approach can of course become circular in rationale if not carefully considered and independent age control may not always be as accurate as is hoped for (e.g. ^{14}C dates in central Asian loess; Li *et al.* 2016). Finally, in addition to providing age information for climate proxy or dust provenance reconstructions, luminescence dating of loess is a key tool to estimate past dust mass accumulation rates (MAR) (Kang *et al.* 2015; Kohfeld and Harrison 2003; Perić *et al.* 2018; Roberts *et al.* 2003; Stevens *et al.* 2016; Újvári *et al.* 2010). Given the critical role of atmospheric dust in the climate system, this further elevates the importance of good practice in luminescence dating of loess deposits so that dust MAR data can be properly incorporated into dust or climate modelling (Albani *et al.* 2015). These and other considerations are outlined in Section 5.2. First, the specific choice of luminescence protocol for loess is considered, then second, some general luminescence dating challenges are looked at within the context of loess deposits. Next, the aspects of luminescence dating rather more specific to loess environments are considered, and finally the importance of using high sampling resolution luminescence dating in the development

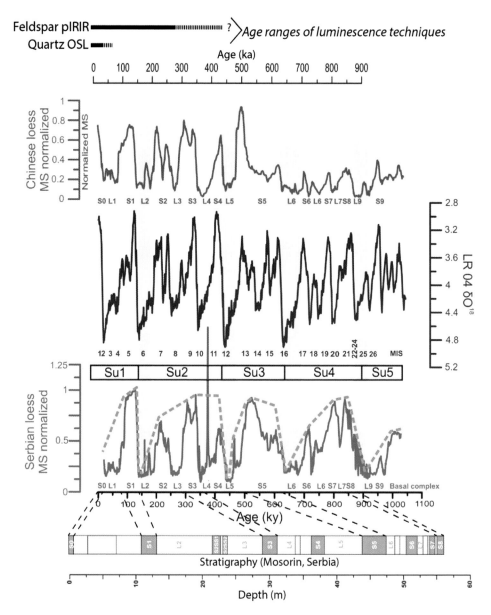

Figure 5.3 Generalised loess stratigraphy and age for Serbian and Chinese loess sequences, alongside approximate age ranges of quartz OSL and feldspar pIR-IRSL dating of typical loess. Normalised magnetic susceptibility (MS) curves for Serbian and Chinese loess composite profiles are shown alongside age based on correlation to the marine oxygen-isotope record of Lisiecki and Raymo (2005). Oxygen isotope stages and loess (L) and soil (S) units are labeled. Peaks in MS represent soil units. The Serbian composite sequence is correlated to a generalised soil stratigraphic log of the Mosorin site in Serbia, with soil and loess units labeled. SU labels and the orange dashed line shows so called 'super units' of Marković *et al.* (2015), which higher order stratigraphic units terminated by particularly deep glacials (grey). Figure modified from Marković *et al.* (2015).

of detailed age models and dust mass accumulation rate is outlined. This takes a practical rather than theoretical approach as much as possible, to encourage best practice in the application of luminescence dating to loess environments.

5.2 CHOICE OF PROTOCOL FOR LUMINESCENCE DATING OF LOESS

The long history of utilising loess deposits to test and develop luminescence methods, coupled with the issues (outlined below) surrounding the use of quartz and feldspar in standard single-aliquot regeneration (SAR) protocols, has led to a wide range of luminescence techniques being applied to loess deposits. Many early studies on loess used multiple aliquot additive OSL, IRSL and TL techniques (e.g. Singhvi *et al.* 1989; Watanuki and Tsukamoto 2001) although, as outlined in Chapter 1, it is generally considered that the application of SAR OSL (Murray and Wintle 2000) or IRSL protocols are now standard in routine luminescence dating of loess. However, the specific choice of protocol within that framework requires some consideration.

For loess samples anticipated to be Holocene or mid–late last glacial in age, the application of SAR OSL dating to separated quartz grains is recommended as the first-choice protocol. Standard internal tests should be applied to check the validity of the protocol for each loess section studied, preferably on multiple samples, including dose recovery and preheat plateau tests (Murray and Wintle 2000; 2003). Issues regarding choice of grain size are discussed below (Timar-Gabor *et al.* 2011), but historically, studies have used the fine silt ('fine grained'; 4–11 μm; Schmidt *et al.* 2010), course silt (40–63 μm; Stevens *et al.* 2008) or fine sand (63–90 μm; Timar-Gabor and Wintle 2013) fractions in quartz OSL dating. In addition, it is essential that the separation of quartz from other minerals is well achieved prior to dating. As feldspar is well known to suffer from anomalous fading, the decay of luminescence signal on laboratory timescales (Wintle 1973), contamination of quartz separates with feldspar is likely to lead to age underestimation (e.g. Little *et al.* 2002). Chemical treatment of grain size separates using H_2SiF_6 or HF is the first choice to remove feldspars. The success of this approach can be estimated by stimulation of aliquots of material using IR diodes to check for IRSL, via one of the proposed protocols outlined by Duller (2003). If feldspar contamination is still suspected, treatment should be repeated. However, some loess studies report samples where the IRSL signal (and presumably feldspar contamination) cannot be removed, and complete removal of feldspars from 4–11 μm fractions can be tricky. In such cases, it may be appropriate to apply a modified 'double SAR' protocol (Roberts and Wintle 2001; 2003; Zhang and Zhou 2007). Various modifications of this protocol have been proposed, but in essence, this approach makes use of the fact that quartz does not respond to IR stimulation so that, prior to measurement of the quartz OSL using blue light diodes, an IR stimulation step is added that should reduce or remove the signal from any feldspar grains in the aliquot but not affect the quartz OSL signal. This results in two equivalent dose (D_E) values; one for the IRSL step and one for the 'post-IR OSL' measurement. The latter should then be a good approximation of the pure quartz D_E. However, Roberts (2007) demonstrated on Nebraskan loess that even using this protocol the feldspar signal can still dominate the quartz signal, so for routine dating this should be considered a last resort after unsuccessful attempts to chemically isolate pure quartz grains.

A limiting factor in quartz luminescence dating of loess deposits is the relatively high dose rates in many loess deposits, and the relatively low equivalent doses before quartz grains become saturated and show limited signal growth with dose. Dose rates to quartz vary between c. 3–5 Gy ka^{-1} in typical loesses, with Chinese loess averages around 3.5 Gy ka^{-1} (Lai et al. 2006). This relatively high level is due to the mineralogical and grain-size composition of loess and imposes a lower upper limit to luminescence dating as compared to sandy sediments that are richer in low dose rate quartz. In addition, the quartz OSL signal has a relatively low upper limit of signal growth with dose, which further frustrates attempts to date loess sequences using quartz (Fig. 5.4). Estimates vary over the precise equivalent dose (D_E) value that can be considered as an upper limit for quartz dating, but it may be as low as 120–150 Gy in Chinese loess (Buylaert et al. 2007). Certainly, its applicability beyond c. 180–200 Gy (c. 60–70 ka) seems doubtful in Chinese (Lai 2010; Yi et al. 2015; Zhang et al. 2015; Stevens et al. 2016) and European (Moska et al. 2015; Stevens et al. 2011; Timar et al. 2010) loess and it is wise to expect this to be the case elsewhere too.

One way to detect this age underestimation is to compare quartz OSL ages to other chronometers such as feldspar or expected-age-based soil stratigraphy or radiocarbon dating. However, at times there are unexplained discrepancies between dating methods; for example, between radiocarbon and luminescence methods in New Zealand (Grapes et al. 2010) or Central Asian (Li et al. 2016) loess. If this is the case or no other chronometers are available and soil stratigraphy is poorly constrained, then some estimate of signal saturation can be made via examination of the luminescence growth curve or through trends in ages with depth. While there is uncertainty about how laboratory quartz OSL growth curves are best described mathematically (see Timar-Gabor and Wintle 2013 for more detail), if natural D_E values lie beyond the first exponential part of the growth curve then it seems likely that the resultant D_E will be an underestimate. Furthermore, lack of D_E growth with depth or increasing scatter in D_E values may indicate that quartz OSL D_E is being underestimated (Stevens et al. 2016). However, comparison of natural dose response curves with those generated in the laboratory show significant differences between these curves in both Romanian and Chinese loess (Chapot et al. 2012; Timar-Gabor and Wintle 2013). This deviation will drive age underestimation so Chapot et al. (2012) recommended an upper limit of c. 150 Gy for Luochuan loess. Timar-Gabor and Wintle (2013) showed that although quartz doses continued to grow beyond 300 Gy, the natural dose response curves start to saturate between 100–200 Gy (63–90 µm) and 200–300 (4–11 µm). Recently, as discussed below, Timar-Gabor et al. (2017) suggested that in loess deposits generally, this divergence in growth curve means that even measures such as position of D_E on the growth curve relative to saturation level may not sufficiently show the point where quartz OSL ages start to underestimate true age. As such, unless detailed analyses of laboratory versus natural dose response curves and multiple independent ages are obtained it seems wise to consider all quartz SAR OSL ages in loess beyond 150 Gy as potentially underestimating the true palaeodose, even if standard SAR protocol internal tests are passed. This is a major frustration in quartz OSL dating.

In addition to this low upper age limit, in some loess areas the quartz OSL signal may be weak, may show evidence of non-fast component dominated decay, or may consistently fail internal tests. For example, quartz from loess regions where there is a significant contribution from volcanic material, such as Japan, may show enhanced recuperation or dominance of components other than the fast component, which is ideally used for dating (Tsukamoto et al. 2003). Even on the Chinese Loess Plateau, which generally shows good fast component dominated quartz signals, there can be portions of sections where standard protocols may fail tests such as dose recovery tests (Stevens *et al.* 2013; part of Beiguoyuan section in China). In such cases, or in older loess where quartz is considered in saturation, it may be preferable to utilise the IRSL signal from feldspar dominated polymineral aliquots or separated feldspars. Although K-feldspar should be the first-choice dosimeter, it can be hard to isolate fractions of specific feldspar types, or indeed feldspars in general, if using 4–11 μm separates. However, given that quartz does not respond to IRSL (Banerjee *et al.* 2001), polymineral aliquots can also be used in IRSL dating. One major problem though is that when measured at 50 °C in conventional protocols, the feldspar IRSL signal suffers from anomalous fading of the luminescence signal over time. This is generally considered to be universal (Little *et al.* 2002), despite some studies on European loess not detecting significant fading (Frechen *et al.* 2001). This apparent contradiction may be a consequence of the lack of agreement over how to monitor and correct for this anomalous fading (Auclair *et al.* 2003; Huntley and Lamothe 2001; Kars *et al.* 2008). Furthermore, given that many of the proposed corrections for anomalous fading only work for the linear (young) part of the IRSL growth curve it is debatable how much use standard IRSL protocols are in dating loess beyond the limits of standard quartz OSL dating.

One potential way around this has been to try to isolate a non-fading feldspar signal. In recent years this has apparently been achieved through a double IR stimulation (pIR-IRSL) protocol (see Chapter 1), developed and tested on loess (Buylaert *et al.* 2009, 2012; Murray *et al.* 2014; Thiel *et al.* 2011; Thomsen *et al.* 2008;) (see Chapter 1 for details). The procedure has been successfully applied to Alaskan, European, central Asian and Chinese loess (Buylaert *et al.* 2015; Fitzsimmons *et al.* 2018; Klasen *et al.* 2017; Lauer *et al.* 2017; Li *et al.* 2016; Roberts 2012; Stevens *et al.* 2011, 2018). A slightly modified protocol (multi-elevated-temperature post-IR-IRSL MET or pIR-IRSL MET) was also developed on loess by Li and Li (2012) and a multiple aliquot regeneration (MAR) procedure of this protocol was applied by Chen *et al.* (2015) to Chinese Loess Plateau samples. Generally, the SAR protocols appear to be able to date loess deposits to *c.* 250–300 ka (Buylaert *et al.* 2012; Murray *et al.* 2014; Schmidt *et al.* 2014; Stevens *et al.* 2018) while Chen *et al.* (2015) argued their MAR approach allows successful dating to *c.* 480 ka (Fig. 5.3).

Although the pIR-IRSL methods are still being tested in many regions, and are still under development, they show considerable promise in extending the age range of

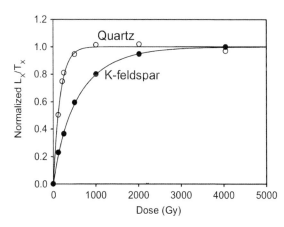

Figure 5.4 Dose response curves for coarse-grained quartz OSL and K-feldspar pIR-IRSL at 290 °C for a sample from the Niuyangzigou loess site in northeastern China. Data are fitted with a sum of two saturating exponential functions. Figure modified from Yi et al. (2016).

loess dating using a non-fading feldspar signal. However, this protocol should not yet be considered 'routine' for dating loess, and various specific tests must be applied prior to dating using pIR-IRSL protocols, in addition to the general standard SAR tests for recuperation, recycling and analysis of dose recovery (Wintle and Murray 2006). Some practical considerations are given here. A key question is the choice of temperatures used for the preheats and the first- and post-IR stimulations. While residual doses (see below) appear to decrease at lower pIR-IRSL stimulation temperatures, the measured rates of fading are apparently reduced at higher temperatures, leading to more stable signals (Buylaert *et al.* 2012; Roberts 2012). This has led to a great diversity of applied protocols under the pIR-IRSL umbrella, which can be highly confusing for the non-expert. For example, using Alaskan loess, Roberts (2012) showed that 225–270 °C pIR-IRSL stimulation temperatures resulted in agreement with independent tephra-based age control, but not pIR-IRSL at 290 °C. By contrast, in many studies on loess, pIR-IRSL stimulation temperatures of 290 °C have often been used and yield ages consistent with independent age control and quartz OSL dating (see compilation by Buylaert *et al.* 2012).

One approach to selecting temperatures is to conduct an experiment on multiple aliquots of a sample by changing the first-IR stimulation temperature and measuring D_E values on groups of these aliquots, ranging from 50–270 °C (a so-called 'first IR stimulation plateau), first proposed by Buylaert *et al.* (2012). Ideally, resultant D_E values obtained in this way should show no change with first IR stimulation temperature (Fig. 5.5), or a plateau of D_E values over a given temperature range. Such a result would give confidence that pIR-IRSL signals are stable (no significant fading) over a given first IR preheat range and would demonstrate whether any systematic differences in D_E with choice of protocol are likely to occur. Many studies report clear plateaus that suggest a wide range of first IR stimulation temperatures are appropriate, between 50 and 250 °C (Buylaert *et al.* 2015; Yi *et al.* 2015 on Chinese loess), although evidence from the pIR-IRSL MET protocol suggests higher first stimulation temperatures (>200 °C) may be more appropriate for older samples (>500 Gy). However, this comes at a cost as higher first IR stimulation temperatures reduce the pIR-IRSL signal intensity. In many samples this is not generally an issue so a first IR stimulation temperature of *c.* 200 °C is often chosen (Buylaert *et al.* 2015; Yi *et al.* 2015). If time or instrument availability permits, a detailed analysis of the pIR-IRSL signal can be undertaken through use of the pIR-IRSL MET protocol of Li and Li (2012). This has the advantage of being able to test a variety of first- and post-IR stimulation temperature combinations and identify the combination that allows greatest signal stability, intensity and any systematic differences in D_E. In general, greater first-IR and post-IR stimulation temperatures (up to 300 °C for the latter) show greatest signal stability for older samples. This is consistent with measurement of higher laboratory aliquot fading rates using lower temperature signals after aliquot dose and storage for varying time intervals (measurement of *g*-values; Buylaert *et al.* 2009; Thomsen *et al.* 2008). However, the pIR-IRSL MET approach is time consuming and, in most cases, demonstration of a first IR stimulation plateau may be sufficient for routine dating. Studies are also generally in agreement that raising the preheat temperature to >300 °C results in little erosion of the pIR-IRSL signal and was most likely to increase pIR-IRSL signal stability (Li and Li 2012; Murray *et al.* 2009). As such, many studies now use 320 °C preheats, as proposed by Thiel *et al.* (2011) for Austrian loess.

One issue that may complicate the choice of IR stimulation temperatures is the bleachability of the pIR-IRSL signal. Especially at higher temperatures, the pIR-IRSL signals have often been shown to be difficult to bleach, with a variable and sometimes large residual component even after prolonged laboratory or sunlight bleaching (Buylaert *et al.* 2012). The size of residual varies from just a few Gy (Li and Li 2011; Yi *et al.* 2015) to 10s of Gy (Buylaert *et al.* 2012; Stevens *et al.* 2011) and is dependent on bleaching time (Kars *et al.* 2014) as well as strongly correlated with burial dose (Buylaert *et al.* 2012). A practical approach was shown by Yi *et al.* (2015) who, using northeastern Chinese loess from Sanbahuo, conducted a series of bleaching experiments on samples using a SOL2 solar simulator. They found that the residual doses became constant after approximately 300 hours of exposure (6.2 ± 0.7 Gy). Alternatively, sunlight-bleaching experiments can be conducted as for Serbian loess in Stevens *et al.* (2011), although the experimental parameters are less easily controlled this way. If time or resource is not sufficient to determine residual doses after extended bleaching experiments, an estimate of the true residual dose at sunlight exposure can be obtained by bleaching multiple samples of varying D_E at a constant exposure time. Plotting D_E versus residual depth and tracing the intercept with the *y*-axis (Fig. 5.6) gives an estimate of the unbleachable (in nature) residual dose (Buylaert *et al.* 2012). Reassuringly, in most cases when these experiments are undertaken they result in residual D_E values being close to pIR-IRSL D_E values from modern polymineral fine grains from recently deposited Chinese dust (Buylaert *et al.* 2011). However, Kars *et al.* (2014) noted that in many samples the magnitude of the residual dose is highly dependent on bleaching protocol, which suggests every effort should be made to minimise the contribution of the residual dose (potentially by reducing pIR-IRSL stimulation temperatures), without compromising signal stability. Given these findings, it is therefore broadly accepted that derived pIR-IRSL residual dose values should be subtracted from sample pIR-IRSL D_E values. However, even after this there is some uncertainty over the reliability of such corrected pIR-IRSL ages in samples where residual dose is a significant fraction of D_E. In such samples the resultant corrected pIR-IRSL D_E values may still overestimate expected ages (for example, in Holocene samples; Stevens *et al.* 2011). Thus, quartz OSL is still the technique of choice for younger, Holocene to latest glacial age samples. In any case, given this persistent signal, a relatively long and high temperature IRSL stimulation clean out is required at the end of each SAR cycle (e.g. 325 °C for 200 s; Thiel *et al.* 2011).

A final consideration in pIR-IRSL dating of loess is the size of the test dose used. Yi *et al.* (2016) found a dependency of D_E on test dose when applying pIR-IRSL at 290 ºC to fine sand-sized K-feldspar grains extracted from northeastern Chinese loess at Niuyangziguo. Based on test dose D_E plateau tests (Fig. 5.7) and extensive dose recovery tests using differing test doses they suggested very small (<15% of D_E) or very large (>60% of D_E) test doses should be avoided and use of a test dose value corresponding to approximately 30% of the D_E is ideal (Yi *et al.* 2016). This dependency of D_E on test dose may be related to carryover of hard-to-bleach charge between SAR cycles in pIR-IRSL, although in any case this finding underscores the need to consider test dose size in any application of pIR-IRSL to loess or other material, potentially with analysis of test dose plateaus (Fig. 5.7).

An alternative to the pIR-IRSL methods to date loess beyond standard quartz SAR OSL is the thermally transferred optically stimulated luminescence (TT-OSL) method applied to quartz (see Chapter 1). This was developed at the Luochuan section in China (Wang *et al.* 2006) and initially showed great promise in dating even to 800 ka (2000 Gy). However, questions remain about the best procedure for test dose sensitivity change

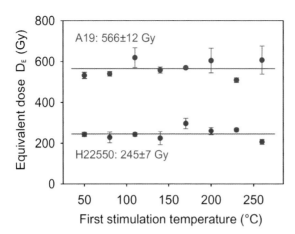

Figure 5.5 Dependence of D_E on first IR stimulation temperature for two samples under a pIR-IRSL protocol with second IR stimulation at 290 °C. Three aliquots were measured for each stimulation temperature. A19 is loess from Luochuan, Chinese Loess Plateau. Figure modified from Buylaert *et al.* (2012).

correction, given an apparent difficulty in removing the TT-OSL signal (Stevens *et al.* 2009), and a significant number of studies have cast doubt on the thermal stability of the signal (Adamiec *et al.* 2010; Brown and Forman 2012; Chapot *et al.* 2016). Violet-stimulated luminescence (VSL) of quartz has also been proposed as a means to extend the age range of luminescence dating (Jain 2009) and has also been tested on the Luochuan section in China (Ankjærgaard *et al.* 2016), showing potential for dating up until 600 ka. However, given the uncertainty surrounding TT-OSL and the relatively early stage of development of VSL it is recommended that these protocols be only applied as part of a project that aims to further develop and test these methods, not yet in routine dating applications. Currently then, the pIR-IRSL methods show some of the greatest promise for routine dating of loess deposits beyond the limits of quartz OSL dating. However, for younger loess, standard quartz SAR OSL remains the most robust method. It is worth noting here that this field of research is rapidly developing and in the near future there will no doubt be significant developments that help further improve the luminescence dating of older loess units. For now it seems sensible to obtain secure independent age control when possible (e.g. tephra in Alaskan loess; Roberts (2012)) and where this is absent ensure that both appropriate tests are applied (see above for pIR-IRSL) and that quartz SAR OSL ages are also present for the section (e.g. Stevens *et al.* (2011) for Serbian loess; Yi *et al.* (2015) for Chinese loess).

5.3 GENERAL LUMINESCENCE CHALLENGES WITHIN A LOESS CONTEXT

5.3.1 Testing the accuracy of luminescence ages using loess stratigraphic models

Arguably one of the main advantages of luminescence dating of loess deposits is that many loess areas around the world (but by no means all) have well-developed loess-soil stratigraphy that has been correlated to marine oxygen isotope stratigraphy (Porter 2001). While sub-Milankovic timescale climate proxy variations (<10,000–100,000 yr frequency) in loess are not well resolved under these age models (Stevens *et al.* 2007), the general

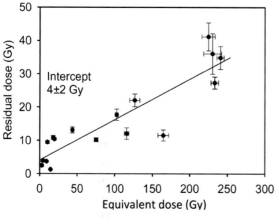

Figure 5.6 Residual doses observed after four hours of light exposure in a Hönle SOL2 solar simulator plotted against pIR-IRSL (second stimulation temperature at 290 °C) D_E values for 15 samples. Figure modified from Buylaert *et al.* (2012).

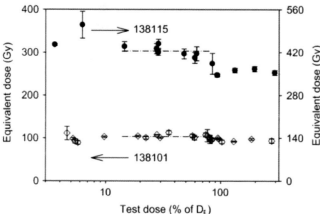

Figure 5.7: Dependence of D_E on test dose size for two samples of NE Chinese loess. The dash-dot line is the average of the D_E value for test doses ranging between 15 and 80% of the total (natural + added) dose. Figure modified from Yi *et al.* (2016).

loess-soil stratigraphy over multi-millennial timescales is well constrained in some regions, such as China (Ding *et al.* 2002), Europe (Marković *et al.* 2015) and some areas of the USA (Rutledge *et al.* 1996). This provides an opportunity to test the broad accuracy of the luminescence ages from a loess section with 'known age' marker horizons, such as soil unit boundaries. Numerous examples of this exist and have helped develop and test new methodologies in luminescence dating (Roberts 2008; Buylaert *et al.* 2012). As noted by Roberts (2008), an inevitable side effect of this testing of new methods and pioneering of new luminescence techniques on loess deposits is that many ages produced during these developmental phases should be viewed with caution. However, the range of luminescence techniques and protocols applied makes the assessment of what is reliable or not rather tricky for the non-specialist. Some aspects of this are covered in this chapter but for a more detailed review the reader is referred to Roberts (2008), and Buylaert *et al.* (2012).

Another note of caution should be sounded here. There is clearly an element of circularity in testing the validity of absolute numerical luminescence ages by comparison to age models defined by non-numerical means, such as correlation techniques. Although often conceptualised as rather simple, as Muhs (2013) highlights, loess stratigraphy is often rather complex. This is especially the case in loess areas that cover strong climatic gradients or experience the influence of multiple competing climatic influences. Loess soil stratigraphy is especially complex in European deposits, with numerous examples of mis-correlations based on incorrect age assignment of soils (see Marković *et al.* (2015) for an attempt to resolve this) due to missing soil units, spatial variations in soil expression due to local topographic or soil forming conditions, or changes in dominant soil-forming processes under strong environmental gradients. For example, Obreht *et al.* (2016) showed that over the Middle–Late Pleistocene the influence of continental versus Mediterranean climate has changed considerably over the Middle and Lower Danube Basin loess and that this shift in influence has been spatially variable. The result is that correlations based on similarity of soil type alone may be erroneous. A classic example of this is shown in cross-national European loess stratigraphic correlation, a highly complex problem involving numerous and changing soil unit nomenclature systems across national boundaries (Marković *et al.* 2015). After proposal of pedostratigraphic criteria for correlation at the 1961 6th INQUA Congress in Warsaw, Poland, the youngest Brown Forest Soil palaeosol exposed at loess sites was taken to be representative of the last interglacial (Fink 1962). However, this led to the correlation of last interglacial soils like Austrian Stillfried A with what we now know (thanks to multiple independent dating studies) are soils formed during marine oxygen-isotope stage 11, such as Hungarian Mende Base (MB) soil (Bronger 1976; Oches and McCoy 1995). As such, while the possibility to tie loess stratigraphy to marine oxygen-isotope stages has been a great help in the development and testing of luminescence methods, caution should be exercised in using this approach as a first order check on age reliability in many areas where such miscorrelations are possible.

5.3.2 Water content in loess sequences

One of the main uncertainties in luminescence dating is the estimation of past water content (see Chapter 1; Aitken 1985). Loess deposits are found in a wide range of geomorphological and climatic situations, ranging from semi-arid and arid plateau environments to humid river valleys and depressions, as well as waterlogged or permafrost areas, which means loess pore water content history can vary considerably. In addition, measurement of modern water content cannot be assumed as representative of the past water content integrated over the burial period. The range of soil types (waterlogged gley soils to open steppe soils) as well as the alternation of soil and loess unit stratigraphy underscores the fact that current conditions are unlikely to be representative of the range of conditions in the past. Even topography, and therefore proximity to the water table, can vary through time at any given loess site. This is especially the case in non-plateau loess landscapes such as in Europe where a range of loess landforms are found (Lehmkuhl *et al.* 2016), but also on classic stable loess plateau tableland such as on the Chinese Loess Plateau, where studies have shown a long history of gullying and infilling at a number of sites (Porter and An 2005). Since gully formation is likely to lower the local water table and dry out exposures, this would reduce the water content compared to adjacent deposits unaffected by this process. Under- or overestimation of the water content integrated over the burial period by failing to account for past changes or indeed inaccurate modern estimates will lead to erroneously under- and overestimates of the age respectively. In the case of loess with 'typical (for Chinese Loess Plateau)' U, Th, and

K values of 3–4 ppm, 9–11 ppm and *c.* 1.5–2% respectively, Stevens *et al.* (2013) showed that changing water content between 2.5 and 13% for a *c.* 8 ka old sample lead to a *c.* 1 ka difference in age (*c.* 6% change in age). While this may be within error limits of many luminescence ages, water contents chosen for loess vary dramatically in the literature. Over the Chinese Loess Plateau alone, values in the literature range from a few percent (Roberts *et al.* 2001) to over 25% or more (Lu *et al.* 2007). To some extent, this reflects variation in real water content due to the climatic gradient over the region, but such large differences will result in significantly different luminescence ages. Furthermore, approaches to error estimation vary from assigning a few percentage errors to >10% (Stevens *et al.* 2013), and some studies highlight substantial changes in water content at single sites (12–25%; Lu *et al.* 2007), likely to be due to changes between loess and soil stratigraphy. Despite this change in measured water content with stratigraphic depth and units, no clear correlation was seen between water content and grain size in Stevens *et al.* (2013). Estimated/assumed water contents also varied considerably in sampled Peoria loess of the continental USA, ranging from 10–20% (Muhs *et al.* 2013; Roberts *et al.* 2003; Rousseau *et al.* 2007) but extremely high water contents of up to 61% have been found in Alaskan loess, likely to be due to ice lensing due to permafrost conditions (Auclair *et al.* 2007). Extra caution is therefore required in areas that may have experienced or currently experience permafrost conditions, such as Alaskan and Siberian loess regions. Care should be taken to identify cryogenic features in loess sequences as these may indicate past permafrost conditions.

Given these uncertainties, some best practice can be recommended:

1. If possible, obtain *in situ* water content values from loess cores, or from fresh, newly dug exposures well away from old exposures (see Chapter 2 on sampling). If this is not possible and only old exposures are available, consider that it will be very unlikely that current *in situ* water contents will be indicative of even the most recent burial period.

2. Derive saturation water content values on each sample taken in the laboratory.

In the worst case, some attempt should be made, based on presumed burial history, landform, and proximity to the water table, to estimate the proportion of time the samples will have been below the water table and therefore saturated. As soils may show higher water contents than loess units (Stevens *et al.* 2013) it is important to obtain water content estimates from both loess and soil units, and not to assume that values from one unit are representative of both. A common approach is to measure water content on every sample taken, both *in situ* and saturation, and if no trends are seen then take the mean plus or minus the standard deviation for all samples, either the *in situ* value if believed to be representative, or some proportion of the saturated value (Stevens *et al.* 2008; 2018). If differences in units are seen then multiple averages are calculated or sample specific values can be used (e.g. Lu *et al.* 2007). In any case, large error terms are essential so as to account for the inherent uncertainty in estimating past water content, and these can be assigned based partly on variability in sample water content with depth. Even with this approach, and with careful measurement in fresh sections or cores, estimation of past water content in loess remains tricky and one of the major sources of age uncertainty.

5.3.3 Choice of grain size

One of the most fundamental questions regards the choice of the grain size to be dated (Fig. 5.2). Loess deposits constitute a wide range of grain sizes so loess provides an opportunity to measure luminescence signals from both fine (4–11 μm) and coarser fractions (40–63 μm or 63–90 μm). The decision concerning which grain size to focus on for dating purposes may partly be grounded in what the composition of loess is, with some finer loess (e.g. southern Chinese Loess Plateau) being more conducive to fine-grain dating and coarser loess deposits (e.g. lower Volga loess in Russia) providing plenty of material for coarser grain dating. Many loess studies have employed the fine-grain fraction for dating purposes (e.g. Watanuki and Tsukamoto 2001), largely due to the availability of material in this size fraction but also due to the fact that over 1 million grains are measured per aliquot, 3–5 times more than with coarse grain aliquots (Duller 2008; Roberts 2008). This significantly reduces aliquot-to-aliquot variability and reduces uncertainty estimates with fewer aliquots analysed, decreasing analysis time. However, for fine-grain quartz grains it is not possible to remove the outer *c.* 20 μm that has been alpha irradiated. This complicates the calculation of dose rate and introduces an assumption concerning alpha particle irradiation 'efficiency', as larger alpha particles result in lower luminescence signal per unit track length versus per unit absorbed beta particle dose (Aitken 1985). Estimates of this efficiency factor have varied but many adopted a value from Rees-Jones (1995) of 0.04, based on analysis of fluvial sediments. More recently, Lai *et al.* (2008) calculated an average alpha efficiency value of 0.035±0.003 based on fine grain quartz extracted from Chinese loess. While choice of alpha efficiency value around these limits does not result in large changes in age, it is sensible to assume a value that has been obtained from loess deposits (Lai *et al.* 2008).

The importance of grain-size choice in loess dating has been highlighted in recent work using SAR-OSL on quartz of different grain sizes from Romanian (Timar-Gabor *et al.* 2011), Serbian (Timar-Gabor *et al.* 2015) and Chinese loess (Timar-Gabor *et al.* 2017). These studies have shown that D_E values from coarse quartz (63–90 μm) are systematically higher than those on fine quartz (4–11 μm) for loess >~40 ka in age, despite the fact that that both grain fractions behave well in the SAR protocol (Fig. 5.8). Younger ages are in agreement and, reassuringly, match with independent age control (Anechitei-Deacu *et al.* 2014; Constantin *et al.* 2012; Trandafir *et al.* 2015). While this gives further confidence in standard quartz SAR OSL protocols, it also further underscores the fact that the quartz SAR OSL approach should be applied with caution beyond *c.* 40 ka (Fig. 5.3). Currently the mechanisms for this difference in age with grain size fraction analysed are not clear but it appears that quartz fine grains (4–11 μm) may underestimate the true age more than coarser quartz grains, and that this may be a world-wide phenomenon (Timar-Gabor *et al.* 2017). In any case, this indicates that continued growth in laboratory SAR dose response curves does not necessarily allow accurate recording of palaeodose at these levels, and further, that traditional means of estimating an age limit for quartz SAR OSL based on the shape of the dose response curve are not valid (Timar-Gabor *et al.* 2017). As noted above, this further requires that quartz SAR OSL dating of loess can only be assumed to be accurate for deposits up to 30–40 ka in age.

5.4 LUMINESCENCE DATING CONSIDERATIONS SPECIFIC TO LOESS AND PEDOGENIC ENVIRONMENTS

5.4.1 Loess, loess-like sediments and *in situ* disturbance

While loess is generally assumed to be relatively homogeneous, well bleached, and that luminescence ages are representative of the timing of aeolian deposition, there are some complicating factors that at certain stratigraphic depths and at certain sites can undermine these assumptions, notably with regard to reworked or 'loess-like' sediments. Here some practical consideration is given to this.

In many classical loess areas such as the Chinese Loess Plateau, loess is often deposited in a plateau environment, blanketing pre-existing relief with relatively constant topography over large distances (Pye 1987; 1995). However, while in general loess sequences in plateau areas can be considered as less likely to be influenced by post- or syn-depositional erosion or resedimentation, current topography is not necessarily an analogue for topography at the time of dust deposition (Lehmkuhl *et al.* 2016). Furthermore, in other areas, for example in North American and European deposits, loess may form distinct topographic features that are aligned with prevailing winds, such as dunes, pahas or gredas, for example in the Mississippi (Mason *et al.* 1999) or Rhine (Antoine *et al.* 2009) river valley loess areas. Luminescence dating is an excellent tool for understanding the formation of these features but this topographic heterogeneity can affect the degree and type of soil formation over short distances and increase the likelihood that loess is reworked via slope or other processes. In terms of the latter, while primary loess is highly likely to be well bleached on aeolian deposition, with ages representative of dust fall, reworked loess or loess derivatives may yield luminescence ages that are strongly affected by subsequent processes and may therefore yield age underestimates. Resultant ages may be representative of the secondary process affecting the loess, or mixed ages, as some grains are reworked or exposed. A good example of this comes from luminescence dating of deposits of what were considered primary loess in northwest England (Wilson *et al.* 2008). Fine silt and fine sand quartz OSL dating results yielded Holocene ages for these deposits, despite the fact no dust source exists for this material in the Holocene and that loess deposition elsewhere ceased well before the Holocene. As such, this was interpreted to indicate post-depositional reworking of the loess deposits. Through comparison of silt and sand ages as well as the presence of non-normal dose distributions, Wilson *et al.* (2008) suggested that reworking involved aeolian transport, overland flow and soil piping. In this case, luminescence dating helped demonstrate reworked loess deposits, in the absence of clear stratigraphic evidence at most of the sample points. However, in other scenarios where less background information is available, such anomalously young ages may go unnoticed and lead to error in interpretation. As such, it is important when dating loess to confirm as best as possible that the deposit is indeed primary loess, particularly in non-plateau depositional settings. Detailed stratigraphic, sedimentological and/or micromorphological analysis may be required. For example, the identification of gravel bands or faint banding structures in loess may be an indicator that the sediments are not primary loess (Fig. 5.9; Sprafke and Obreht 2016). Clear signs include fine laminations, apparent micro or macro fabric, incorporation of organic material or ceramic remains, dominance of grain-size fractions other than silt and the incorporation of coarser (coarse sand to gravel) material. However, many of these indicators may not be readily clear in the field, especially when trying to differentiate between reworked versus *in situ* soils (Lang 2003). It is always wise, therefore, to undertake detailed stratigraphic and sedimentological analysis simultaneously when taking luminescence samples.

Another consideration related to this is that loess may be prone to cracking if deposits lie close to a palaeogully, and also may undergo intense cracking and/or overturning during cryogenic processes (Velichko *et al.* 2006). These cracks will be infilled by younger material (Fig. 5.10) and so should be carefully avoided during sampling if the aim is to date the timing of loess deposition rather than subsequent cracking or cryoturbation. Related to this are problems caused by animal burrows or krotovina (Fig. 5.10). Many loess deposits are found in areas currently or formerly covered by grassland or steppe vegetation, and burrowing animals are common. Accidental sampling of an animal burrow will lead to age underestimation of the sediment level being dated, as shown in an experimental dating of a krotovina by Bateman *et al.* (2003). Burrowing and disturbance may be more common in soil units; this is discussed in a loess context below and more generally in Chapter 2 of this book. To avoid sampling these features inadvertently it is important to thoroughly clean the section and inspect it for signs of this disturbance.

5.4.2 Special considerations for dating soil units in loess

Another issue to consider is the sampling of soil and palaeosol units in loess deposits. Soils present problems for luminescence dating as the likelihood of reworking of sediment during soil-forming processes is large, which makes it difficult to interpret a luminescence age in terms of sediment deposition. However, many researchers consider loess in general to be a type of soil, given that to some extent soil-forming processes affect almost all loess deposits (Sprafke and Obreht 2016). Simplified loess stratigraphy, the alternation of palaeosols and loess units (Fig. 5.3), while very useful for generalised correlation and stratigraphic interpretation, can also be misleading, with many loess units, for example in moister regions, heavily affected by pedogenesis (Fig. 5.11). This aspect is highly variable geographically, with common loess soils ranging, for example, from highly rubified Mediterranean type soils, through to brown, humid forest soils, incipient cambisols, light brown to black steppe soils (chernozems) depicting dry continental climates, and tundra gleys indicating cold conditions with seasonally active layers in permafrost. On the Chinese Loess Plateau, loess units on the southern part of the plateau can be more weathered than soil units on the northwest of the plateau (Yang and Ding 2003). Clearly then, while general steps can be taken to avoid specific features like burrows and frost wedges (see above), avoiding soils when sampling loess deposits is seldom practical or indeed desirable. As such, it is critical to identify potential soil processes affecting loess deposits to aid interpretation of luminescence dates.

So how can soil processes in loess affect luminescence dates, and how can this be recognised? The primary way that soil-forming processes may affect luminescence ages is via re-exposure of pre-deposited grains or mixing of grains of different ages in the soil profile due to bioturbation. If soil-forming processes involve only small-scale movement of material, over only a few cm, this may have little noticeable impact on the luminescence ages. However, if mixing processes are more intense, then this can result in scatter in ages, age inversions or age underestimation (Bateman *et al.* 2003). For example, Stevens *et al.*

Figure 5.9 Photos of loess and reworked loess sediments from the Krems region in Lower Austria. A) Morphological situation of site (Krems Schießstätte; white rectangle). B) Outcrop view of the site. C) Reworked loess exposed in the northern part of the outcrop exhibiting partly discontinuous gravel bands and sporadic gravel (small holes); some marked. Figure modified from Sprafke and Obrecht (2016).

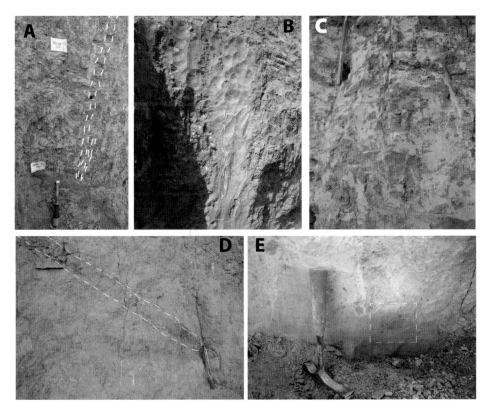

Figure 5.10 Photographs of disturbance in Lower Volga River loess, Russia. A) Infilled tension crack in loess (highlighted). B) Infilled cryogenic crack in loess (highlighted). C) Close up of cryogenic cracks in loess. D) Large section cut through krotovina in loess (highlighted). E) Krotovina exposed close to loess-soil boundary (in square).

(2007) argued that detailed quartz SAR OSL dating of late Quaternary Chinese Loess Plateau sites showed this effect clearly (Fig. 5.12A). Using a sampling resolution of between 10 and 20 cm the study showed that the large sections of early Holocene and late last glacial material produced inaccurate ages, as evidenced by multiple age inversions and scatter in ages. Interestingly, this effect impacted last glacial loess as well as Holocene soils, and this varied between sites depending on the prevailing climate. At more southerly sites with higher rainfall, much of the period between *c.* 7 and 20 ka could not be accurately dated (Fig. 5.12A), and, by implication, proxy records from this part of the section may be unreliable. The youngest dates in the disturbed part of the sequence may mark the maximum age of the cessation of this intense pedogenesis. This study utilised 40–60 μm grains, which made single-grain analysis difficult. However, if coarser grains can be used this raises the possibility of examining this effect on a single-grain level, perhaps allowing calculation of depositional ages. Stevens *et al.* (2007) argue that this effect would not be seen if sampling was conducted at a lower resolution (Fig. 5.12A) (the pros and cons of detailed, high sampling resolution dating of loess are discussed below), a hypothesis supported by recent evidence from compilations of ages from various loess sites on the Chinese Loess Plateau (Stevens *et al.* 2016). However, not all the sections analysed appear to show such variable and mutually inconsistent ages at the base of soils (Kang *et al.* 2015; Stevens *et al.* 2016) so it is important to consider how and if this effect may have impacted a study section.

One other potential complicating factor, but one that has seldom been investigated in loess, is the effect of soil-forming processes on dose rate. Soil formation in loess may involve carbonate dissolution and re-precipitation in carbonate horizons, redistribution of iron oxides (Kemp *et al.* 1995), changing soil-sediment water conditions, and clay formation and even eluviation/illuviation (translocation) of clays (He *et al.* 2008), all of which may impact the dose rate of a soil unit. Indeed, Kemp *et al.* (2006) suggested these effects may drive pedogenically induced variations in radionuclide composition that shift the apparent age of parent material when luminescence dating loess in Northern Pampa, Argentina. In light of these issues, Singhvi *et al.* (2001) recommended that knowledge of soil stratigraphy is vital in interpreting luminescence ages from loess, ideally augmented with micromorphological analyses. They suggested that due to bioturbation and churning of A-horizons in loess palaeosols, luminescence ages are more likely to reflect soil formation ages. In contrast, luminescence ages from B- and C- horizons are likely to reflect the age of loess deposition. However, as the latter will be impacted by clay and carbonate mobility, care must be taken to consider if these effects have been significant on dose rates. They also suggested comparing dose rates in pristine loess to those from B- and C-soil horizons. Significant differences may indicate post-depositional changes in dose rate, which undermines the principle of using current dose rates as a proxy for dose rates over the entire burial period. If these dose rate changes occurred rapidly after dust deposition then these processes may not cause an error in the integrated burial dose rate calculation. However, if they were slower, or a large proportion of the total burial time was prior to their occurrence, then measurement of modern dose rates through both *in situ* and laboratory methods would lead to values not representative of the burial dose. Again, detailed sampling may show up potentially anomalous ages that are affected by these processes, although if possible avoiding sampling from carbonate horizons or intensely weathered units is advisable. It is also worth keeping in mind that if large differences in dose rate are seen between loess and soil units, then the

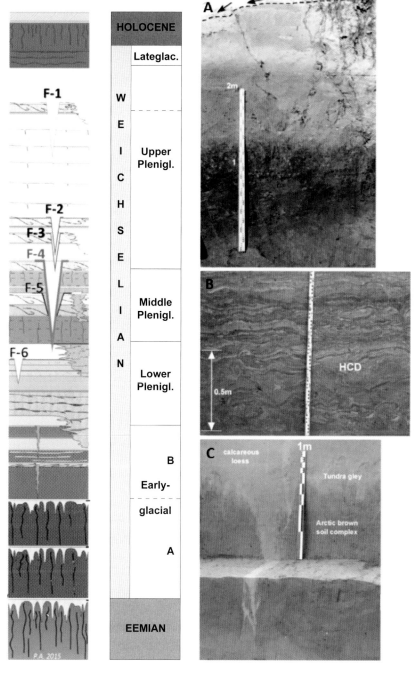

Figure 5.11 Summary of heavily weathered and altered loess pedo-lithostratigraphic sequence for northern France. F-numbers denote ice wedges. Photos on right hand side show: A) section of the Weichselian Early glacial at Saint-Sauflieu (Somme). **B)** Detailed view of the laminated colluvial deposits (HCD) including reworked lenses and nodules from Early glacial humic soils and showing periglacial deformations (cracks, faulting) at Hermies (North). **C)** Large ice-wedge casts preserved at the boundary between the Middle Pleniglacial brown soil complex and the Upper Pleniglacial calcareous loess at Havrincourt site (Northern France). Figure modified from Antoine *et al.* (2016), where a more detailed overview of the stratigraphy is presented.

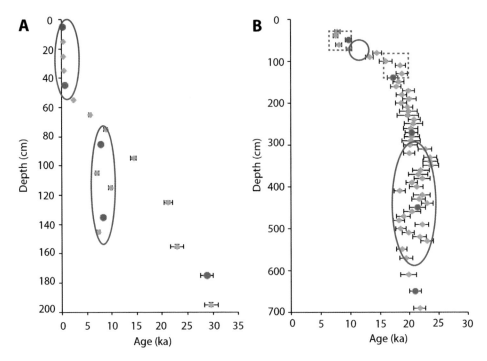

Figure 5.12 High sampling resolution quartz OSL ages at two sites on the Chinese Loess Plateau. A) Effect of agricultural disturbance and soil formation on ages from Shiguanzhi (Lantian) site on the southern Loess Plateau (highlighted in ovals). B) Changes in sedimentation rate, a hiatus and anomalously old ages shown by luminescence dates at Beiguoyuan (Huanxian) site on the northwest of the Chinese Loess Plateau (highlighted in ovals (hiatus and overestimated ages) and dashed squares (changes in sedimentation rate)). Luminescence ages shown in red are selected points to illustrate how lower sampling resolution analysis would probably miss much of the variation shown here, and possibly resulted in incorrect assessments of age change with depth. Data from Stevens *et al.* (2008).

calculated dose rate of samples taken close to these boundaries will need to be considered in more detail, as the gamma dose will have been impacted by the nearby sediments with different dose rate.

A further consideration touched on above relates to the thorny question of what process is actually being dated at the base of a soil, or at a soil loess boundary. In most loess areas, dust deposition will be greatly reduced or will even cease during soil formation periods, which means that soils form directly on loess deposited during cold stages prior to pedogenesis. Even in areas where dust deposition continues during soil formation, such as on the Chinese Loess Plateau, deposition rates can be greatly reduced over these intervals (but not always; Roberts *et al.* 2001). On the Chinese Loess Plateau, this effectively leads to a gradual shift from nearly fully accretionary soils in dust source proximal northwestern parts of the Loess Plateau, through to soils nearly entirely formed in pre-deposited loess in the far southeast where the soil-loess boundary has been superimposed on considerably older underlying loess. This fundamentally changes the interpretation of luminescence ages from these units: accretionary soils will likely (but not necessarily) yield dust depositional ages, while soils that have formed on underlying loess deposits may either yield soil formation ages, the timing of deposition of underlying the parent loess material, which may greatly precede soil formation, or a mixture of ages. This will vary depending on the type and intensity of soil formation, depending on the degree of soil formation and the specific soil horizon. The effect of this may be difficult to gauge in both the field and via examination of luminescence dates alone. Careful consideration of weathering proxies (e.g. rubification and chemical indices of weathering/alternation), soil micro and macromorphology, and ideally detailed, high sampling resolution luminescence sampling may help with interpretation.

These problems with understanding what is being dated also link to the difficulty in actually defining the boundary between loess and soil units in loess sequences. This was addressed by Lai and Wintle (2006) who examined the transition between last glacial loess (L1) and Holocene soil (S0) at Yuanbo section on the Chinese Loess Plateau in detail using quartz SAR OSL. Linear regression performed on ages from the section suggested an accumulation rate drop at around 13.5 ka, which could be considered the main Holocene-last glacial boundary. However, if colour changes in the field, or magnetic susceptibility methods, had been used to define the boundary (both widely used in the absence of independent ages), the boundary age would be placed at 15.1 or 9.8 ka respectively. This distinction is very important as soil-loess boundaries are often used as 'known age' tie points to correlate loess to other records in order to generate age models, under the assumption that these boundaries correspond in age to stratigraphic boundaries in marine oxygen isotope stratigraphy (see discussion in Stevens *et al.* (2007)). However, these results (uncertainty of >5 ka in boundary assignment) cast this approach into doubt and underscore the need for detailed dating of loess deposits to understand the nature of loess-soil boundaries, as well as loess accumulation generally. Understanding loess-soil unit boundaries, despite the complications associated, is therefore a major question that luminescence dating can address.

5.4.3 Heterogeneity or homogeneity of loess

Another aspect of loess that is significant for luminescence dating is that the material (at least unweathered) is often considered as homogeneous, both spatially and temporally, with a chemical composition that is a good approximation of the average composition of the upper continental crust (Taylor *et al.* 1983). This implies that loess globally is a mixture of eroded material from a wide range of continental rocks, and so should be geochemically similar both through space and time. From a luminescence dater's point of view, this would be an advantage as quartz and feldspar should therefore behave in a consistent and predictable way that would mean standard dating protocols would be easily developed and widely applied. This idea has been used to advantage by adopting standardised dose response or growth curves (SGCs) in an effort to speed up analyses and reduce analysis time, and indeed the technique was developed on luminescence deposits (Roberts and Duller 2004). The SGC technique relies on the principle that quartz luminescence signal intensity response to dose is similar, at least when the luminescence signal is derived from large numbers of grains. Roberts and Duller (2004) suggested that site-specific dose response curves should ideally be constructed and this approach was followed by Stevens *et al.* (2006; 2008) in high sampling resolution (10–20 cm) analysis of Chinese loess. The studies by Stevens *et al.* (2006; 2008) were the first such highly detailed luminescence dating analyses of loess deposits, an approach discussed below, and they were made possible by the use of SGCs. Indeed, the possibility of a universal luminescence dose response curve for loess quartz has also been demonstrated by Lai *et al.* (2007a), although Stevens *et al.* (2006) argued that site-specific differences could be seen in growth curves. Irrespective, this approach has great potential to speed up luminescence dating of loess deposits and allow high sampling resolution luminescence dating of loess.

However, although luminescence signal growth with dose does seem to be broadly similar in quartz grains, other key aspects may vary. While loess does generally closely reflect estimates of the average upper continental crustal composition, there are small differences in loess geochemistry in areas around the world (Gallet *et al.* 1998; Újvári, *et al.* 2015). Loess geochemistry can vary significantly locally, or even within a site; for example, due to differing sediment contributions from multiple but distinct river catchments, as illustrated at a loess site near the Balkan Mountains in Serbia (Obreht *et al.* 2016). While this often reflects variations in heavy minerals or clay mineralogy (which may impact dose rate), it is reasonable to ask the question as to how much this might also affect quartz and feldspar luminescence properties. Quartz from volcanic regions has been shown to suffer from high levels of signal recuperation and relatively weak fast component (Tsukamoto *et al.* 2003), which may require special analysis procedures in areas such as Japan and Argentina where loess deposits are strongly influenced by volcanic material. However, even in loess areas with apparently minimal input from active volcanic centres, standard SAR protocols and tests may need to be modified to obtain satisfactory quartz OSL dating results. For example, Schatz *et al.* (2012) showed that quartz OSL dating of the Tokaj section in Hungary could be appropriate, but satisfactory dose recovery test results could only be obtained when an added dose was given to samples prior to optical or thermal pre-treatment. In addition, some loess areas such as in the central Asian Xinjiang province of China show quartz luminescence signals that are relatively weak, yielding scatter in D_E values (Yang *et al.* 2014). This has led to controversy over the use of the technique in that region, exacerbated by mismatches between the OSL ages and gastropod ^{14}C ages (Feng *et al.* 2011). However, more recent studies have tended to show that quartz OSL dating of 63–90 μm grains yields results in line with stratigraphy and pIR-IRSL ages of K-feldspar, and that these also pass internal tests (Li *et al.* 2015). Contamination of gastropods at the limits of ^{14}C dating by modern carbon can lead to significant age underestimates, and this may account for some of this discrepancy. This appears to be a cautionary tale in placing too much faith in independent age control that may not always be warranted.

Even in loess areas with a long history of luminescence research and where reliable results are generally obtained via standard quartz SAR OSL approaches, it cannot always be assumed that all samples will behave consistently. At the northwestern Chinese Loess Plateau site Beiguoyuan, Stevens *et al.* (2013) showed that while much of the section could be dated using standard quartz SAR OSL approaches with a preheat combination that yielded good dose recovery results, a portion of the section with high accumulation rates and coarser grain size consistently failed dose recovery tests and yielded age overestimates. Testing of signal integration limits and deconvolution of luminescence decay curves into first order components demonstrated that quartz aliquots from these samples had a significant unstable ultrafast component that lead to internal test failures and inaccurate ages. This issue would only have been uncovered via high resolution sampling (see below) and furthermore, combining this information with mineralogy and geochemical data accessory to luminescence dating showed that a change in dust source occurred at this time. Thus, even in stable loess plateau areas with generally well-behaved quartz that can be dated using routine protocols, source changes can lead to changing quartz properties and problems with standard luminescence dating approaches. As such, it is crucial that performance of the SAR protocol is monitored frequently and that multiple representative samples are analysed throughout a loess sequence via dose recovery tests, rather than just one or two samples tested per sequence. It also suggests that even if universal quartz growth curves may be achievable, SGCs should perhaps initially be developed and checked on a section-by-section basis and using sufficiently representative samples from the whole section. In terms of feldspars, variation in fading rates and in the best choice of pIR-IRSL stimulation temperatures also suggests that variation in source asserts some control on the practicalities of dating loess. The best solution is to undertake the basic tests mentioned above to define the best SAR pIR-IRSL parameters for the samples in question.

5.4.4 High sampling resolution dating of loess and derivation of dust accumulation rates

One of the most promising aspects of luminescence dating of loess lies in the application of luminescence methods to loess profiles at high sampling resolution (small sampling interval of the order of 10 to a few 10s of cm) to generate detailed age models. As noted above, high sampling resolution dating can yield insight into source changes, erroneous ages, and the impacts of soil formation. In addition, the resulting detailed age models yield a number of benefits. Loess deposits are considered some of the best palaeoclimate archives available on land, with potential to derive high-resolution climatic records at a variety of timescales (Porter 2001). Luminescence dating at high sampling resolution has allowed detailed, independent age models to be developed for these valuable climate archives (Stevens *et al.* 2018; Sun *et al.* 2012). Furthermore, detailed age models from high sampling resolution dating allow for continental-scale chronostratigraphic correlation of loess deposits into the middle Pleistocene (Marković *et al.* 2015) and subsequent critical insight into continental-scale climate dynamics. In addition, loess records are also some of the widest spread dust deposits on the planet and represent an excellent but as yet underutilised archive of atmospheric dust deposition (Kohfeld and Harrison 2003). Atmospheric mineral dust is a fundamental component of the climate system (Knippertz and Stuut 2014) and dust drives climate change through multiple processes including direct atmospheric radiative forcing and oceanic productivity (Tegen 2013). Dust also responds to climate shifts in dust source regions via changes in dust production (Stevens *et al.* 2013), therefore comprising a complex feedback system for which the triggers dust and climate change are poorly understood. As well as regional palaeoclimate archives, loess deposits can be used as a near source dust archives and luminescence dating, through directly constraining timing of loess burial, allows the timing of dust deposition to be ascertained.

At high sampling resolution, luminescence age models can be developed to yield extremely detailed records of dust deposition, which often reveal highly variable dust dynamics. These can be related to fluctuations in regional ice sheet source areas (e.g. Wisconsin, USA loess, Schaetzl *et al.* 2014) or large-scale pulses of atmospheric dust over a wider scale (e.g. Chinese loess, Stevens and Lu 2009; Kang *et al.* 2015). One of the key ways to address this is via calculation of loess dust Mass Accumulation Rates (MAR, Kohfeld and

Harrison 2003). When coupled with grain-size distribution data, these MAR records are crucial in deriving information on the past loading of dust in the atmosphere (Albani *et al.* 2015), a major step in understanding dust–climate interactions. Dust MAR records show considerable variations (Újvári *et al.* 2010; 2017), and require detailed independent dating-based age–depth models as discrepancies between MAR values between independent and non-independent records are stark (Kohfeld and Harrison 2003). Luminescence dating is an ideal method to achieve this and high sampling resolution dating in particular can be used to understand short-term fluctuations in dust deposition (Roberts *et al.* 2003; Stevens and Lu 2009; Kang *et al.* 2015; Stevens *et al.* 2016). Some practical considerations when undertaking such analyses are outlined below.

The sampling resolution of application of luminescence methods has varied greatly, ranging from just a few samples over many metres, to sampling every 10–20 cm. Stevens *et al.* (2006, 2007, 2008) first used high sampling resolution luminescence dating using quartz OSL by taking advantage of the use of a SGC method (Roberts and Duller 2004). These studies and subsequent quartz OSL and pIR-IRSL work, focused especially on Chinese loess (Buylaert *et al.* 2008; 2015; Kang *et al.* 2015; Lai 2010; Lai *et al.* 2007b; Stevens *et al.* 2016, 2018; Sun *et al.* 2012), but also on European (Constantin *et al.* 2014; Stevens *et al.* 2011; Újvári *et al.* 2014) and American (Muhs *et al.* 2013) loess. They have provided critical insight into the nature of loess stratigraphy and sedimentation. Stevens *et al.* (2007; 2018) and Buylaert *et al.* (2008) suggested that gaps in the loess record exist on suborbital timescales on classic loess sites on the Chinese Loess Plateau and that soil formation on pre-deposited glacial loess obscured climate signals and invalidated the use of soil horizon bases as known age marker horizons. However, other studies have not uncovered such hiatuses at other sites (Kang *et al.* 2015, Stevens *et al.* 2016) and whether this is due to rarity of hiatuses or an artefact of much lower sampling resolution used in many studies is a key future focus. Based on analysis of changes in age with depth versus the size of error terms on luminescence ages in high sampling resolution dating studies on Chinese loess, Stevens *et al.* (2016) argued that hiatuses and pedogenesis-related disturbance were more common than has been reported. Such gaps or disturbance would be important to uncover as they would greatly impact the reliability of past climate and dust deposition reconstructions from parts of loess sequences, and indeed call into question non-independently dated age models for loess. An illustrative example is from the Jingbian desert marginal Chinese Loess Plateau site, used as an International Commission on Stratigraphy (ICS) type section for Quaternary climate change Cohen *et al.* 2013). Recent high sampling resolution post-IR IRSL dating of the sequence has drastically revised the chronology, and shown significant hiatuses at the site (Stevens *et al.* 2018). The age model also allowed detailed reconstruction of desert marginal processes and monsoon climate in this sensitive region.

Irrespective of the presence of hiatuses, luminescence dating of loess has highlighted significant variations in dust accumulation both between sites and through time and this is a significant future opportunity in luminescence dating of loess. Dust MAR values have been reviewed by Újvári *et al.* (2010) in Europe, and more recently by Kang *et al.* (2015) for the last glacial Chinese Loess Plateau. While independent dust MAR values remain few for European deposits, existing studies suggest significant variation between sites in Hungary and Serbia, although with potential peaks in accumulation rate during the last glacial maximum (Fuchs *et al.* 2008, Stevens *et al.* 2011, Újvári *et al.* 2014). In China, MAR data from loess is much more abundant. Stevens and Lu (2009) demonstrated considerable variation in loess rate of accumulation between sites using quartz OSL, although when compiling data from many luminescence dated sections, Kang *et al.* (2015) suggest that a peak in dust MAR of around 23–19 ka occurs across the Chinese Loess Plateau. Intriguingly, this appears to postdate a peak in dust accumulation identified in marine and ice cores (26–23 ka), notwithstanding age uncertainty and error in all these records. However, recent quartz OSL dating at the Xifeng section seems to suggest a dust MAR peak more in line with the ice and marine core estimates (Stevens *et al.* 2016). Perić *et al.* (2018) recently showed that the timing of last glacial peak dustiness also coincides between a site in Serbia and in China, hinting at wide-scale patterns in dust accumulation across Eurasia. The timing of dustiness peaks in loess is therefore a pressing research focus where luminescence dating can be used to great effect. At the Chinese loess sections, dust MAR values of 200–300 or more g m^{-2} a^{-1} are not uncommon. In the continental United States, however, extremely large pulses in accumulation of up to 10–16,000 g m^{-2} a^{-1} have been measured in late last glacial Peoria loess at sites in Nebraska and Iowa, along the Missouri and Mississippi Rivers (Roberts *et al.* 2003; Muhs *et al.* 2013). These represent extraordinary rates of loess accumulation over very short periods, likely to be related to the dynamics of the Laurentide ice sheet and outwash down major river systems. However, they raise problems with regard to calculation of MAR from luminescence dating as the rates of loess sedimentation become so large that the errors on luminescence ages prevent meaningful age increases with depth being identified. Muhs *et al.* (2013) utilised Bayesian modelling to refine their chronology, and this represents a promising path to reduce uncertainty and derive continuous age–depth functions. However, many studies to date have rather relied on regression analyses and confidence intervals to derive age–depth functions and measures of uncertainty for luminescence ages (Stevens *et al.* 2016). Perić *et al.* (2018) discuss the choice of age model in loess luminescence dating for MAR calculation in more detail.

The problems with taking discrete point age information with significant error margins and mixed systematic and random error components and turning these into an age depth model remain to be resolved (Perić *et al.* 2018; Zeeden *et al.* 2018). However, once an age model has been derived, the first derivative of the age model can be used to calculate the sedimentation rates of the loess sequence. In order to derive MAR values that give estimates of dust mass, and are therefore better representations of amounts of material deposited, measurements of bulk density should be made on the loess. Values in the literature vary considerably (e.g. 1.3–17 g cm^{-3}) so ideally these values should not be estimated but rather obtained directly from each loess sequence, and ideally at every sample point. To date, such studies have been relatively rare on loess and in a global compilation of Holocene MAR data from multiple dust archives, Albani *et al.* (2015) highlighted the paucity of data currently available from loess deposits. Given the importance of loess MAR estimates in reconstructing past dust flux, detailed, high sampling resolution luminescence dating of loess has enormous potential to contribute to major breakthroughs in understanding past dust activity, especially through addressing uncertainties in MAR calculation. It is important to note here that to ensure these datasets are of use to the dust and climate modelling community, in addition to luminescence ages and bulk density, it is also important to determine grain-size distribution data, so that MAR values for different particle sizes can be calculated. In addition, if the research question being tested concerns the nature of dust loading as reconstructed from loess, it is important to consider sampling location wisely, as dust MARs vary dramatically in different landform areas (Újvári et al. 2010). If close to a large river system this may skew loess MAR values to very high values that are not representative of wider atmospheric loading, while fluctuations in MAR may be more representative of river system dynamics rather than wider atmospheric dustiness. Certain landforms (e.g. greda) are likely to exhibit higher loess MAR values too, and local conditions may impart considerable variation on dust MAR, as reflected in highly variable and different loess sedimentation rates on the Chinese Loess Plateau (Stevens and Lu 2009). It should also go without saying that it is critical to avoid loess derivatives if the goal is to understand dust deposition from loess dating. Recent compilations (e.g. Kang et al. 2015) improve this situation by helping to identify wider, more widely representative trends, but ideally multiple sites should be analysed. In any case, as luminescence methods can also be used with increasing confidence well into the Middle Pleistocene, this now permits critical insight into the nature of loess deposition and preservation and dust activity over a much longer timescale than previously possible (Buylaert et al. 2015).

Crucially, much of this information is only available through high sampling resolution analysis. Take the hypothetical example presented in Figure 5.12B. The data in blue are coarse silt quartz SAR OSL luminescence ages taken at 10–20 cm intervals from the Beiguoyuan section on the northwestern part of the Chinese Loess Plateau, published in Stevens et al. (2008). The points in red are a selection of those dates taken at regular but lower sampling intervals of c. 2–3 m, which is typical for many loess luminescence studies. There are many features visible in the original dataset which disappear in the lower sampling resulution data. An apparent hiatus in the record around 10–14 ka, large swings in accumulation rate between 7 and 20 ka, and a set of ages between c. 3 and 5.3 m that overestimate the apparent true age (as represented by the bracketing intervals) are all not visible (Fig. 5.12B). These latter samples also fail dose recovery tests and are discussed above with reference to variability in quartz luminescence properties (Stevens et al. 2013). Without detailed, high sampling resolution dating, these issues and features would not be visible and loess accumulation at the site would be reconstructed as simply gradually reducing through time. Thus, high sampling resolution dating is essential for looking at shorter term, millennial-scale changes in dust accumulation and climate at loess sites. However, it is also noticable that there are lot of age points that do not explicitly add any extra information to this picture (Fig. 5.12B). As raised initially by Roberts (2008), this begs the question over at what level of detail does increased sampling resolution result in unnecessary age information. This is no trivial question as resources for dating are often limited. Unfortunately this is not an easy question to answer until after the reseracher has acquired excessive amounts of data as short-term changes in loess accumulation rates are not easily knowable prior to age determination via luminescence. Some rough estimate of what may be too much detail can be made if one considers the approximate age and thickness of a sequence, as well as the probable 6–10% errors on an average luminescence age. Clearly, if a 10 m interval of loess deposited somewhere around the last 40–50 ka is dated at 10 cm intervals this will result in an average age increase with depth of c. 100 yr per 10 cm. This is well under the 1σ age uncertainties of the dating method itself and therefore on average most ages will not yield any extra information. However, given that this loess deposition may not be evenly spread over this 10,000 yr period it is also possible that reducing this sampling interval too much will start to result in loss of information over any abrupt changes in accumulation rate. As such, a possible approach is to take samples at relatively high resolution and then analyse every 2nd, 3rd, or 4th sample as a way of examining the general trends in a sequence, before a decision is made concerning more detailed analysis. Fortunately, the growth curve information from this initial framework of ages can be used in construction of a quartz SGC that would greatly speed up more detailed analysis, as long as this is planned into the initial measurments (Roberts and Duller 2004).

There are also other benefits of high resolution luminescence dating. Recently, Stevens *et al.* (2018) used high sampling resolution post-IR IRSL at the Jingbian site in northern China to demonstrate a systematic lag of monsoon driven magnetic susceptibility behind orbital forcing over multiple precessional cycles. This provides crucial insight into a major debate over the nature of East Asian Monsoon climate forcing generally. Furthermore, many studies on loess deposits, for example on the Chinese Loess Plateau (Porter 2001), have suggested that variations in loess climate proxies such as grain size show variations that can be linked to short-term climate swings such as Heinrich events and Dansgaard-Oeschger cycles. However, while some luminescence dating studies have supported this assertion (Sun *et al.* 2012), others have suggested a more complex relationship where loess sedimentation or climate proxy swings are not always matched with such abrupt climate events, and vice versa (Stevens *et al.* 2008). This is another rich vein of future research for luminescence dating of loess deposits at high sampling resolution, at least during the late last glacial, and can reveal whether these proposed links are valid. It is tempting to overinterpret loess deposits due to the exceptional amount of climate and dust information contained in these sequences. This is especially true when no independent age model has been constructed, but is also true even with a detailed luminescence age model. It may be that the presence or absence of short-term climate proxy oscillations that match global climate events may simply be down to the error uncertainty on the luminescence ages and any resulting age model.

One final important aspect of high sampling resolution dating is the possibility to detect short-term changes in dust provenance from loess archives. Turning the issue with the poorly performing quartz SAR luminescence in samples from 3 to 5.3 m at Beiguoyuan (Chinese Loess Plateau) to their advantage, Stevens *et al.* (2013) demonstrated that this interval shows evidence for an abrupt change in dust source. This was confirmed through various other provenance indicators including heavy mineral analysis and subsequently via zircon U-Pb dating (Fenn *et al.* 2018) and suggested an abrupt change to a local dust source for that very arid and windy interval of the last glacial maximum. Interestingly, this source change was noticeable in the dose rate data, largely because of shifts in the abundance of radioisotopes, including the ratio of Th to U, a key indicator of source rock type. Thus, this accessory data to high sampling resolution luminescence dating can be used to identify general shifts in dust provenance, a significant added benefit to luminescence dating. Quartz sensitivity levels have also been used to identify shifts in dust provenance, predominantly again on the Chinese Loess Plateau. Lai and Wintle (2006) argued for a shift in dust source over the S0-L1 boundary at Yuanbao on the northwestern Chinese Loess Plateau, while Lü and Sun, (2011) showed that quartz sensitivity varies considerably across many of the possible source regions for Chinese loess. In summary, where resources permit, high sampling resolution luminescence dating of loess deposits should be undertaken to maximise information obtained from loess sequences.

5.5 SUMMARY AND RECOMMENDED APPROACH

The number of studies using luminescence dating to develop chronologies in loess deposits, or that use loess as a medium to develop and test luminescence techniques, is vast and at times rather bewildering. It is clear that loess deposits are near ideal deposits for application of luminescence methods and many past dating studies have yielded accurate results that have at times led to major corrections of previously incorrect chronostratigraphies. However, it is also true that a large number of past luminescence studies on loess have also yielded ages using methods that are likely to be inaccurate. There are clear opportunities with using luminescence in loess, not least in using high sampling resolution luminescence dating to uncover abrupt changes in climate or dust deposition rate. However, it is important that luminescence studies on loess carefully consider (1) the potential for reworking of loess deposits, (2) the possible influence of soil-forming processes on luminescence dates, (3) changes in water content stratigraphically, through time and between exposures and deeply buried loess, and (4) the possible influence of changing dust source on luminescence behaviour. The choice of grain size to be analysed is also now known to be a major consideration, especially towards the limits of quartz SAR OSL dating. It is recommended that caution is applied to the interpretation of any quartz OSL ages from loess older than *c.* 30–40 ka. However, although quartz SAR OSL is still the technique of choice for loess younger than this, it is now possible to obtain ages on Middle Pleistocene loess using the post-IR IRSL technique, which opens up great potential for future work.

As the goals of luminescence dating studies on loess are diverse and may for example be related to broad-scale approximate age finding, detailed high sampling resolution dating for dust accumulation rate calculation, analysis of Middle Pleistocene deposits, or even identification of provenance shifts, a single methodological and analytical approach cannot be completely recommended for dating loess overall. However, a checklist of key considerations can be stated:

1. The **choice of sampling location** and position of exposure should be carefully considered. Loess deposits are found comprising a range of landforms and certain landforms may be more prone to reworking loess sediments. Always consider the site/section position within the landform to minimise this. Even in loess plateau areas, episodic gullying in the past can disrupt classical stratigraphy and choosing positions that minimise the risk of this are ideal. Furthermore, to allow for more accurate estimate of water content, where possible, new or fresh sections should be sampled.

2. After cleaning the section (Chapter 2) a thorough assessment of the **site stratigraphy** should be conducted. This should assess where the site fits into regional stratigraphy (and therefore to inform on the best choice of luminescence technique) and to identify signs of reworked loess deposits and soil units (which may require specific consideration when analysing resultant ages). In addition, it is also desirable to obtain measurements from another stratigraphic tool or climate proxy, such as magnetic susceptibility. Ideally, this would be initially conducted in the field to better identify soil horizons, tephra layers or potentially

reworked deposits. Such proxy measurements can inform sampling, alert the investigator to potentially reworked loess, and help with analysis and interpretation of luminescence ages.

3. Plan for taking some **range finder ages** to get an approximate age of a deposit, or to at least determine if samples are in saturation and beyond the limit of luminescence (or require advanced techniques to measure palaeodose). Furthermore, it may be wise to plan to use multiple techniques (e.g. multiple-grain-size quartz OSL, and feldspar post-IR-IRSL), due to possible uncertainty about stratigraphy and due to the discrepancies in results from different methods.

4. While showing great promise for **dating of older loess deposits**, the pIR-IRSL techniques cannot yet be considered as routine for dating loess, and quartz SAR OSL techniques have a limited age-range of applicability that should be kept in mind.

5. **Dose rate data** should also not be considered only as an afterthought. Water content integrated through time is a major uncertainty in loess dating using luminescence and every effort should be made to analyse both *in situ* and saturation water content. The choice of water content should then reflect these measurements and an informed consideration of likely changes in the water table in the loess exposure. Dose rate data should be collected for every luminescence sample taken.

6. Where possible, **high sampling resolution** analyses should be conducted. Low sampling resolution analysis can provide misleading results due to inability to identify erroneous ages or reworking/hiatuses. Ideally, samples should be taken at between 10–50 cm intervals, although the complex issue of what is unnecessarily high resolution should be considered. High sampling resolution dating can be facilitated using standardised growth curves, if these are properly constructed and tested, and this has many advantages for improving understanding of loess stratigraphy, climate records, source, and not least dust accumulation rates, which is becoming a major strand of research into loess. If the latter is to be investigated, ensure that bulk density measurements are made in order to calculate mass accumulation rates (MAR), and ideally grain-size distribution analysis should be undertaken, to ensure that the data are of use to the dust-climate modelling community (Albani *et al.* 2015). In this way, luminescence dating of loess is making a large impact on analysis of the past dust cycle on a global scale.

REFERENCES

Adamiec, G., Duller, G.A.T., Roberts, H.M., Wintle, A.G. 2010. Improving the TT-OSL SAR protocol through source trap characterization. *Radiation Measurements* 4, 768–777.

Albani, S., Mahowald, N.M., Winckler, G., Anderson, R.F., Bradtmillers, L.I., Delmonte, B., Franşçois, R., Goman, M., Heavens, N.G., Hesse, P.P., Hovan, S.A., Kang, S.G., Kohfeld, K.E., Lu, H., Maggi, V., Mason, J.A., Mayewski, P.A., McGee, D., Miao, X., Otto-Bliesner, B.L., Perry, A.T., Pourmand, A., Roberts, H.M., Rosenbloom, N., Stevens, T., Sun, J. 2015. Twelve thousand years of dust: the Holocene global dust cycle constrained by natural archives. *Climate of the Past* 11, 869–903.

APPLICATIONS IN LOESSIC ENVIRONMENTS

Aitken, M.J. 1985. *Thermoluminescence dating*. Academic Press, London.

Anechitei-Deacu, V., Timar-Gabor, A., Fitzsimmons, K.E., Veres, D., Hambach, U. 2014. Multi-method luminescence investigations on quartz grains of different sizes extracted from a loess section in southeast Romania interbedding the Campanian Ignimbrite ash layer. *Geochronometria* 41, 1–14.

Ankjærgaard, C., Guralnik, B., Buylaert, J.-P., Reimann, T., Yi, S.W., Wallinga, J. 2016. Violet stimulated luminescence dating of quartz from Luochuan (Chinese loess plateau): Agreement with independent chronology up to ~600 ka. *Quaternary Geochronology* 34, 33–46.

Antoine, P., Rousseau, D-D., Moine, O., Kunesch, S., Hatté, C., Lang, A., Tissoux, H., Zöller, L. 2009. Rapid and cyclic aeolian deposition during the Last Glacial in European loess: a high resolution record from Nussloch, Germany. *Quaternary Science Reviews* 28, 2955–2973.

Antoine, P., Coutard, S., Guerin, G., Deschodt, L., Goval, E., Locht, J-L., Paris, C. 2016. Upper Pleistocene loess-palaeosol records from Northern France in the European context: Environmental background and dating of the Middle Palaeolithic. *Quaternary International* 411, 4–24.

Auclair, M., Lamothe, M., Huot, S. 2003. Measurement of anomalous fading for feldspar IRSL using SAR. *Radiation Measurements* 377, 487–492.

Auclair, M., Lamothe, M., Lagroix, F., Banerjee, S.K. 2007. Luminescence investigation of loess and tephra from Halfway House section, Central Alaska. *Quaternary Geochronology* 2, 34–38.

Banerjee, D., Murray, A.S., Bøtter-Jensen, L., Lang, A. 2001. Equivalent dose estimation using a single aliquot of polymineral fine grains. *Radiation Measurements* 33, 73–94.

Bateman, M.D., Frederick, C.D., Jaiswal, M.K., Singhvi, A.K. 2003. Investigations into the potential effects of pedogensis on luminescence dating. *Quaternary Science Reviews* 22, 1169–1176.

Bronger, A. 1976. Zur quarären Klima- und Landschaftsentwicklung des Karpatenbeckens auf (paläo-) pedologisher und bodengeogra- phischer Grundlage. Kieler Geographische Schriften – Band 45. Selbstverlag, Geographisches Institut der Universität Kiel. In German.

Brown, N.D., Forman, S.L. 2012. Evaluating a SAR TT-OSL protocol for dating fine-grained quartz within Late Pleistocene loess deposits in the Missouri and Mississippi river valleys, United States. *Quaternary Geochronology*, 12, 87–97.

Buylaert, J-P., Vandenberghe, D., Murray, A.S., Huot, S., Van den Haute, P. 2007. Luminescence dating of old (>70 ka) Chinese loess: A comparison of single-aliquot OSL and IRSL techniques. *Quaternary Geochronology* 2, 9–14.

Buylaert, J.P., Murray, A.S., Vandenberghe, D., Vriend, M., De Corte, F., Van den haute, P. 2008. Optical dating of Chinese loess using sand-sized quartz: Establishing a time frame for Late Pleistocene climate changes in the western part of the Chinese Loess Plateau. *Quaternary Geochronology* 3, 99–113.

Buylaert, J.P., Murray, A.S., Thomsen, K.J., Jain, M. 2009. Testing the potential of an elevated temperature IRSL signal from K-feldspar. *Radiation Measurements* 44, 560–565.

Buylaert, J.P., Thiel, C., Murray, A.S., Vandenberghe, D., Yi, S., Lu, H. 2011. IRSL and post-IR IRSL residual doses recorded in modern dust samples from the Chinese Loess Plateau. *Geochronometria* 38, 432–440.

Buylaert J-P., Jain, M., Murray, A.S., Thomsen, K.J., Thiel, C., Sohbati, R. 2012. A robust feldspar luminescence dating method for Middle and Late Pleistocene sediments. *Boreas* 41, 435–451.

Buylaert, J.P., Yeo, E-Y., Thiel, C., Yi, S., Stevens, T., Thompson, W., Frechen, M., Murray, A.S., Lu, H. 2015. A detailed post-IR IRSL chronology for the last glacial interglacial soil at the Jingbian loess site (northern China). *Quaternary Geochronology* 30, 194–199.

Chapot, M., Roberts, H.M., Duller, G.A.T., Lai, Z.P. 2012. A comparison of natural- and laboratory-generated dose response curves for quartz optically stimulated luminescence signals from Chinese Loess. *Radiation Measurements* 47, 1045–1052.

Chapot, M.S., Roberts, H.M., Duller, G.A.T., Lai, Z.P. 2016. Natural and laboratory TT-OSL dose response curves: Testing the lifetime of the TT-OSL signal in nature. *Radiation Measurements* 85, 41–50.

Chen, Y., Li, S.-H., Li, B., Hao, Q., Sun, J. 2015. Maximum age limitation in luminescence dating of Chinese loess using the multiple-aliquot MET-pIRIR signals from K-feldspar. *Quaternary Geochronology* 30, 207–212.

Cohen, K.M., Finney, S.C., Gibbard, P.L., Fan, J-X. 2013. The ICS international chronostratigraphic chart. *Episodes* 36, 199–204.

Constantin, D., Timar-Gabor, A., Veres, D., Begy, R., Cosma, C. 2012. SAR-OSL dating of different grain-sized quartz from a sedimentary section in southern Romania interbedding the Campanian Ignimbrite/Y5 ash layer. *Quaternary Geochronology* 10, 81–86.

Constantin, D., Begy, R., Vasiliniuc, S., Panaiotu, C., Necula, C., Codrea, V., Timar-Gabor, A. 2014. High-resolution OSL dating of the Costineşti section (Dobrogea, SE Romania) using fine and coarse quartz. *Quaternary International* 334–335, 20–29.

Crouvi, O., Amit, R., Enzel, Y., Porat, N., Sandler, A. 2008. Sand dunes as a major proximal dust source for late Pleistocene loess in the Negev desert, Israel. *Quaternary Research* 70, 275–282.

Ding, Z.L., Derbyshire, E., Yang, S.L., Yu, Z.W., Xiong, S.F., Liu, T.S. 2002. Stacked 2.6-Ma grain size record from the Chinese Loess based on five sections and correlation with deep-sea $\delta^{18}O$ record. *Paleoceanography* 17, PA000725.

Duller, G.A.T. 2003. Distinguishing quartz and feldspar in single-grain luminescence measurements. *Radiation Measurements* 37, 161–165.

Duller, G.A.T. 2008. Single-grain optical dating of Quaternary sediments: why aliquot size matters in luminescence dating. *Boreas* 37, 589–612.

Feng, Z.D., Ran, M., Yang, Q.L., Zhai, X.W., Wang, W., Zhang, X.S., Huang, C.Q. 2011. Stratigraphies and chronologies of late Quaternary loess-paleosol sequences in the core of the central Asian arid zone. *Quaternary International* 240, 156–166.

Fenn, K., Stevens, T., Bird, A., Limonta, M., Rittner, M., Vermeesch, P., Andò, S., Garzanti, E., Lu, H., Zhang, H., Lin, Z. 2018. Insights into the provenance of the Chinese Loess Plateau from joint zircon U-Pb and garnet geochemical analysis of last glacial loess. *Quaternary Research* 89, 645-659.

Fink, J. 1962. Studien zur absoluten und relativen chronologie der fossilen Böden in Österreich, II Wetzleinsdorf und Stillfried. *Archaeol. Austriaca* 31, 1–18.

Fitzsimmons, K.E., Sprafke, T., Zielhofer, C., Günter, C., Deom, J-M., Sala, R., Iovita, R. 2018. Loess accumulation in the Tian Shan piedmont: implications for palaeoenvironmental change in arid Central Asia. *Quaternary International* 469, 3043.

Frechen, M., van Vliet-Lanoe, B., van den Haute, P. 2001. The Upper Pleistocene loess record at Harmignies/Belgium – high resolution terrestrial archive of climate forcing. *Palaeoceanography, Palaeoclimatology, Palaeoecology* 173, 175–195.

Fuchs, M., Rousseau, D.D., Antoine, P., Hatté, C., Gauthier, C. 2008. Chronology of the Last Climatic Cycle (Upper Pleistocene) of the Surduk loess sequence, Vojvodina, Serbia. *Boreas* 37, 66–73.

Gallet, S., Jahn, B., Van Vliet Lanoë, B., Dia, A., Rossello, E. 1998. Loess geochemistry and its implications for particle origin and composition of the upper continental crust. *Earth and Planetary Science Letters* 156, 157–172.

Grapes, R., Rieser, U., Wang, N. 2010. Optical luminescence dating of a loess section containing a critical tephra marker horizon, SW North Island of New Zealand. *Quaternary Geochronology* 5, 164–169.

Guo, Z.T., Ruddiman, W.F., Hao, Q.Z., Wu, H.B., Qiao, Y.S., Zhu, R.X., Peng, S.Z., Wei, J.J., Yuan, B.Y., Liu, T.S. 2002. Onset of Asian desertification by 22 Myr ago inferred from loess deposits in China. *Nature* 416, 159–163.

Haesaerts, P., Damblon, F., Gerasimenko, N., Spagna, P., Pirson, S. 2016. The Late Pleistocene loess–palaeosol sequence of Middle Belgium. *Quaternary International* 411, 25–43.

He, X., Bao, Y., Hua, L., Tang, K. 2008. Clay illuviation in a Holocene palaeosol sequence in the Chinese Loess Plateau. In: *New Trends in Soil Micromorphology* (Eds. Kapur, S., Mermut, A., Stoops, G.), Springer, Berlin, pp. 237–252.

Huntley, D.J., Lamothe, M. 2001. Ubiquity of anomalous fading in K-feldspars and the measurement and correction for it in optical dating. *Canadian Journal of Earth Sciences* 38, 1093–1106.

Jain, M. 2009. Extending the dose range: probing deep traps in quartz with 3.06 eV photons. *Radiation Measurements* 44, 445–452.

Jeong, G.Y., Hiller, S., Kemo, R.A. 2008. Quantitative bulk and single-partucle mineraloty of a thick Chinese loess-paleosol section: implications for loess provenance and weathering. *Quaternary Science Reviews* 27, 1271–1287.

Kang, S., Roberts, H.M., Wang, X., An, Z., Wang, M. 2015. Mass accumulation rate changes in Chinese loess during MIS 2, and asynchrony with records from Greenland ice cores and North Pacific Ocean sediments during the Last Glacial Maximum. *Aeolian Research* 19, 251–258.

Kars, R.H., Wallinga, J., Cohen, K.M. 2008. A new approach towards anomalous fading correction for feldspar IRSL dating – tests on samples in field saturation. *Radiation Measurements* 43, 786–790.

Kars, R.H., Reimann, T.R., Ankjærgaard, K., Wallinga, J. 2014. Bleaching of the post-IR IRSL signal: new insights for feldspar luminescence dating. *Boreas* 43, 780–791.

Kemp, R.A., Derbyshire, E., Meng, X., Chen, F., Pan, B. 1995. Pedosedimentary reconstruction of a thick loess–paleosol sequence near Lanzhou in north-central China. *Quaternary Research* 43, 30–45.

Kemp, R.H., Zárete, M., Toms, P., King, M., Sanabria, J., Arguello, G. 2006. Late Quaternary paleosols, stratigraphy and landscape evolution in the Northern Pampa, Argentina. *Quaternary Research* 66, 119–132.

Klasen, N., Loibl, C., Rethemeyer, J., Lehmkuhl, F. 2017. Testing feldspar and quartz luminescence dating of sandy loess sediments from the Doroshivsty site (Ukraine) against radiocarbon dating. *Quaternary International* 432, 13–19.

Knippertz, P., Stuut, J-B. 2014. *Mineral Dust*. Springer.

Kohfeld, K.E., Harrison, S.P. 2003. Glacial–interglacial changes in dust deposition on the Chinese Loess Plateau. *Quaternary Science Reviews* 22, 1859–1878.

Lai, Z. 2010. Chronology and the upper dating limit for loess samples from Luochuan section in the Chinese Loess Plateau using quartz OSL SAR protocol. *Journal of Asian Earth Sciences* 37, 176–185.

Lai, Z., Wintle, A,.G. 2006. Locating the boundary between the Pleistocene and the Holocene in Chinese loess using luminescence. *The Holocene* 16, 893–899.

Lai, Z.P., Murray, A.S., Bailey, R.M., Huot, S., Bøtter-Jensen, L. 2006. Quartz red TL SAR equivalent dose overestimation for Chinese loess. *Radiation Measurements* 41, 114–119.

Lai, Z.P., Brückner, H., Zöller, L., Fülling, A. 2007a. Existance of a common growth curve for silt-sized quartz OSL of loess from different continents. *Radiation Measurements* 42, 1432–1440.

Lai Z.P., Wintle A.G., Thomas D.S.G. 2007b. Rates of dust deposition between 50 ka and 20 ka revealed by OSL dating at Yuanbo on the Chinese Loess Plateau. *Palaeogeography, Palaeoclimatology, Palaeoecology* 248, 431–439.

Lai, Z.P., Zöller, L., Fuchs, M., Brückner, H. 2008. Alpha efficiency determination for OSL of quartz extracted from Chinese Loess. *Radiation Measurements* 43, 767–770.

Lang, A. 2003. Phases of soil erosion-derived colluviation in the loess hills of South Germany. *Catena* 51, 209–221.

Lauer, T., Frechen, M., Vlaminck, S., Kehl, M., Lehndorff, E., Shahriari, A., Khormali, F. 2017. Luminescence-chronology of the loess palaeosol sequence Toshan, Northern Iran – A highly resolved climate archive for the last glacial–interglacial cycle. *Quaternary International* 429, 3–12.

Lehmkuhl, F., Zens, J., Krauβ, L., Schulte, P., Kels, H. 2016. Loess-paleosol sequences at the northern European loess belt in Germany: Distribution, geomorphology and stratigraphy. *Quaternary Science Reviews* 153, 11–30.

Li, B., Li, S.H. 2011. Luminescence dating of K-feldspar from sediments: A protocol without anomalous fading correction. *Quaternary Geochronology* 6, 468–479.

Li, B., Li, S-H. 2012. Luminescence dating of Chinese loess beyond 130 ka using the non-fading signal from K-feldspar. *Quaternary Geochronology* 10, 24–31.

Li, G., Wen, L., Duan, Y., Xia, D., Rao, Z., Madsen, D.B., Wei, H., Li, F., Jia, J., Chen, F. 2015. Quartz OSL and K-feldspar pIRIR dating of a loess/paleosol sequence from arid central Asia, Tianshan Mountains, NW China. *Quaternary Geochronology* 28, 40–53.

Li, G., Rao, Z., Duan, Y., Xia, D., Wang, L., Madsen, D.B., Jia, J., Wei, H., Qiang, M., Chen, J., Chen, F. 2016. Paleoenvironmental changes recorded in a luminescence dated loess/paleosol sequence from the Tianshan Mountains, arid central Asia, since the penultimate glaciation. *Earth and Planetary Science Letters* 448, 1–12.

Lisiecki, L.E., Raymo, M.E. 2005. A Plio-Pleistocene stack of 57 globally distributed benthic $\delta^{18}O$ records. *Paleoceanography* 20, PA1003.

Little, E.C., Lian, O.B., Velichko, A.A., Morozova, T.D., Nechaev, V.P., Dlussky, K.G., Rutterm N. 2002. Quaternary stratigraphy and optical dating of loess from the east European Plain (Russia). *Quaternary Science Reviews* 21, 1745–1762.

Lu, Y.C., Wang, X.L., Wintle, A.G. 2007. A new OSL chronology for dust accumulation in the last 130,000 yr for the Chinese Loess Plateau. *Quaternary Research* 67, 152–160.

Lü, T., Sun, J. 2011. Luminescence sensitivities of quartz grains from eolian deposits in northern

China and their implications for provenance. *Quaternary Research* 76, 181–189.

Marković, S.B., Stevens, T., Kukla, G.J., Hambach, U., Fitzsimmons, K.E., Gibbard, P., Buggle, B., Zech, M., Guo, Z., Hao, Q., Wu, H., O'Hara Dhand, K., Smalley, I.J., Újvári, G., Sümegi, P., Timar-Gabor, A., Veres, D., Sirocko, F., Vasiljević, D.A., Jary, Z., Svensson, A., Jović, V., Lehmkuhl, F., Kovács, J., Svirčev, Z. 2015. Danube loess stratigraphy – towards a pan-European loess stratigraphic model. *Earth-Science Reviews* 148, 228–258.

Mason, J.A., Natter, E.A., Zanner, C.W., Bell, J.C. 1999. A new model of topographic effects on the distribution of loess. *Geomorphology* 28, 223–236.

Moska, P., Jary, Z., Adamiec, G., Bluszcz, A. 2015. OSL chronostratigraphy of a loess-palaeosol sequence in Zlota using quartz and polymineral fine grains. *Radiation Measurements* 81, 23–31.

Muhs, D.R. 2013. Loess deposits: Origins and properties. In: *Encyclopedia of Quaternary Sciences, 3rd edition* (S. Ellias, Ed.). Elsevier, Amsterdam, pp. 573–584.

Muhs, D.R., Bettis III, E.A., Roberts, H.M., Harlan, S.S., Pace, J.B., Reynolds, R.L. 2013. Chronology and provenance of last-glacial (Peoria) loess in western Iowa and paleoclimatic implications. *Quaternary Research* 80, 468–481.

Murray, A.S., Wintle, A.G. 2000. Luminescence dating of quartz using an improved single-aliquot regenerative-dose protocol. *Radiation Measurements* 32, 57–73.

Murray, A.S., Wintle, A.G. 2003. The single aliquot regenerative dose protocol: potential for improvements in reliability. *Radiation Measurements* 37, 377–381

Murray, A.S., Buylaert, J.P., Thomsen, K.J., Jain, M. 2009. The effect of preheating on the IRSL signal from feldspar. *Radiation Measurements* 44, 554–559.

Murray, A.S., Schmidt, E.D., Stevens, T., Buylaert, J.-P., Marković, S.B., Tsukamoto, S., Frechen, M. 2014. Dating middle Pleistocene loess from Stari Slankamen (Vojvodina, Serbia) – limitations imposed by the saturation behaviour of an elevated temperature IRSL signal. *Catena* 117, 34–42.

Nie, J., Stevens, T., Rittner, M., Stockli, D., Garzanti, E., Limonta, M., Bird, A., Andò, S., Vermeesch, P., Saylor, J., Lu, H., Breecker, D., Hu, X., Liu, S., Resentini, A., Vezzoli, G., Peng, W., Carter, A., Ji, S., Pan, B. 2015. Loess Plateau storage of Northeastern Tibetan Plateau-derived Yellow River sediment. *Nature Communications* 6, 8511.

Novothny, A., Frechen, M., Horváth, E., Baláz, B., Oches, E.A., McCoy, W.D., Stevens, T. 2009. Luminescence and amino acid racemization chronology of the loess–paleosol sequence at Süttő, Hungary. *Quaternary International* 198, 62–76.

Obreht, I., Zeeden, C., Schulte, P., Hambach, U., Eckmeier, E., Tima-Gabor, A., Lehmkuhl, F. 2015. Aeolian dynamics of the Orlovat loess–palaeosol sequence, northern Serbia, based on detailed textural and geochemical evidence. *Aeolian Research* 18, 69–81.

Obreht, I., Zeeden, C., Hambach, U., Veres, D., Marković, S.B., Bösken, J., Svirčev, Z., Bačević, N., Gavrilov, M.B., Lehmkuhl, F. 2016. Tracing the influence of Mediterranean climate on Southeastern Europe during the past 350,000 years. *Scientific Reports* 6, 36334.

Oches, E.A., McCoy, W.D. 1995. Aminostratigraphic evaluation of conflicting age estimates for the 'Young Loess' of Hungary. *Quaternary Research* 43, 160–170.

Pécsi, M. 1995. The role of principles and methods in loess–paleosol investigations. *GeoJournal* 36, 117–131.

Perić, Z., Lagerbäck Adolphi, E., Stevens, T., Újvari, G., Zeeden, C., Buylaert, J-P., Marković, S.B., Hambach, U., Fischer, P., Schmidt, C., Schulte, P., Lu, H., Yi, S., Lehmkuhl, F., Obreht, I., Veres, D., Thiel, C., Frechen, M., Jain, M., Vött, A., Zöller, L., Gavrilov, M.B. 2018. Quartz OSL dating of late Quaternary Chinese and Serbian loess: A cross Eurasian comparison of dust mass accumulation rates. *Quaternary International* in press.

Pigati, J.S., McGeehin, J.P., Muhs, D.R., Bettis III, E.A. 2013. Radiocarbon dating late Quaternary loess deposits using small terrestrial gastropod shells. *Quaternary Science Reviews* 76, 114–128.

Porter, S.C. 2001. Chinese loess record of monsoon climate during the last glacial–interglacial cycle. *Earth-Science Reviews* 54, 115–128.

Porter, S.C., An, Z.S. 2005. Episodic gullying and paleomonsoon cycles on the Chinese Loess Plateau.. *Quaternary Research* 64, 234–241.

Pye, K. 1987. *Aeolian Dust and Dust Deposits*. Academic Press, London, 312 pp.

Pye, K. 1995. The nature, origin and accumulation of loess. *Quaternary Science Reviews* 14, 653–667.

Rees-Jones, J. 1995. Optical dating of young sediments using fine-grain quartz. *Ancient TL* 13, 9–14.

Roberts, H.M. 2007. Assessing the effectiveness of the double-SAR protocol in isolating a luminescence signal dominated by quartz. *Radiation Measurements* 42, 1627–1636.

Roberts, H.M. 2008. The development and application of luminescence dating to loess deposits: a perspective on the past, present and future. *Boreas* 37, 483–507.

Roberts, H.M. 2012. Testing Post-IR IRSL protocols for minimizing fading in feldspars, using Alaskan loess with independent chronological control. *Radiation Measurements* 47, 716–724.

Roberts, H.M., Duller, G.A.T. 2004. Standardized growth curves for optical dating of sediment using multiple-grain aliquots . *Radiation Measurements* 38, 241–252.

Roberts, H.M., Wintle, A.G. 2001. Equivalent dose determinations for polymineralic fine-grains using the SAR protocol: Application to a Holocene sequence of the Chinese Loess Plateau. *Quaternary Science Reviews* 20, 859–863.

Roberts, H.M., Wintle, A.G. 2003. Luminescence sensitivity changes of polymineral fine grains during IRSL and [post-IR] OSL measurements. *Radiation Measurements* 37, 661–671.

Roberts, H.M., Wintle, A.G., Mahar, B.A., Hu, M. 2001. Holocene sediment accumulation rates in the western Loess Plateau, China, and a 2500-year record of agricultural activity, revealed by OSL dating. *The Holocene* 11, 477–483.

Roberts, H.M., Muhs, D.M., Wintle, A.G., Duller, G.A.T., Bettis III, E.A. 2003. Unprecedented last-glacial mass accumulation rates determined by luminescence dating of loess from western Nebraska. *Quaternary Research* 59, 411–419.

Rousseau, D-D., Antoine, P., Kunesch, S., Hatté, C., Rossignol, J., Packman, S., Lang, A., Gauthier, C. 2007. Evidence of cyclic dust deposition in the US Great plains during the last deglaciation from the high-resolution analysis of the Peoria loess in the Eustis sequence (Nebraska, USA). *Earth and Planetary Science Letters* 262, 159–174.

Rutledge, E.M., Guccione, M.J., Markewich, H.W., Wysocki, D.A., Ward, L.B. 1996. Loess stratigraphy of the Lower Mississippi Valley. *Engineering Geology* 45, 167–183.

Schaetzl, R.J., Forman, S.L., Attig, J.W. 2014. Loess accumulation in the Tian Shan piedmont: Implications for palaeoenvironmental changes in arid Central Asia. *Quaternary Research* 81, 318–329.

Schatz, A-K., Buylaert, J-P., Murray, A., Stevens, T., Scholten, T. 2012. Establishing a luminescence chronology for a palaeosol-loess profile at Tokaj (Hungary): A comparison of quartz OSL and polymineral IRSL signals. *Quaternary Geochronology* 10, 68–74.

Schmidt, E., Machalett, B., Marković, S.B., Tsukamoto, S., Frechen, M. 2010. Luminescence chronology of the upper part of the Stari Slankamen loess sequence (Vojvodina, Serbia). *Quaternary Geochronology* 5, 137–142.

Schmidt, E.D., Tsukamoto, S., Frechen, M., Murray, A.S. 2014. Elevated temperature IRSL dating of loess sections in the East Eifel region of Germany. *Quaternary International* 334–335, 141–154.

Singhvi, A.K., Bronger, A., Sauer, W., Pant, R.K. 1989. Thermoluminescence dating of loess–paleosol sequences in the Carpathian basin (East–Central Europe): A suggestion for a revised chronology. *Chemical Geology: Isotope Geoscience section* 73, 307–317.

Singhvi, A., Bluszcz, A., Bateman, M.D., Someshwar Rao, M. 2001. Luminescence dating of loess–palaeosol sequences and coversands: methodological aspects and palaeoclimatic implications. *Earth-Science Reviews* 54, 193–211.

Smalley, I.J., Leach, J.A. 1978. The origin and distribution of loess in the Danube Basin and associated regions of East–Central Europe: a review. *Sedimentary Geology* 21, 1–26.

Smalley, I., O'Hara-Dhand, K., Wint, J., Machaelett, B., Jary, Z., Jefferson, I. 2009. Rivers and loess: The significance of long river transportation in the complex event-sequence approach to loess deposit formation. *Quaternary International* 198, 7–18.

Smalley, I., Marković, S., Svirčev, Z. 2011. Loess is (almost totally formed by) the accumulation of dust. *Quaternary International* 240, 4–11.

Sprafke, T., Obreht, I. 2016. Loess: Rock, sediment or soil – What is missing for its definition? *Aeolian Research* 399, 198–207.

Stevens T, Lu H. 2009. Optical dating as a tool for calculating sedimentation rates in Chinese loess: comparisons to grain-size records. *Sedimentology* 56, 911–934.

Stevens, T., Armitage, S.J., Lu, H., Thomas, D.S.G. 2006. Sedimentation and diagenesis of Chinese loess: implications for the preservation of continuous, high-resolution climate records. *Geology* 34, 849–852.

Stevens, T., Thomas, D.S.G., Armitage, S.J., Lunn, H.R., Lu, H. 2007. Reinterpreting climate proxy records from late Quaternary Chinese loess: A detailed OSL investigation. *Earth-Science Reviews* 80, 111–136.

Stevens T., Lu, H., Thomas, D.S.G., Armitage, S.J. 2008. Optical dating of abrupt shifts in the Late Pleistocene East Asian monsoon. *Geology* 36, 415–418.

Stevens, T., Buylaert, J.P., Murray, A. 2009. Towards development of a broadly-applicable SAR TT-OSL dating protocol for quartz. *Radiation Measurements* 44, 639–645.

Stevens T., Marković S.B., Zech M., Hambach H., Sümegi P. 2011. Dust deposition and climate in the Carpathian Basin over an independently dated last glacial–interglacial cycle. *Quaternary Science Reviews* 30, 662–681.

Stevens, T., Adamiec, G., Bird, A.F., Lu, H. 2013. An abrupt shift in dust source on the Chinese Loess Plateau revealed through high sampling resolution OSL dating. *Quaternary Science Reviews* 82, 121–132.

Stevens, T., Buylaert, J-P., Lu, H., Thiel, C., Murray, A., Frechen, M., Yi, S., Zeng, Z. 2016. Mass accumulation rate and monsoon records from Xifeng, Chinese Loess Plateau, based on a luminescence age model. *Journal of Quaternary Science* 31, 391–405.

Stevens, T., Buylaert, J-P., Thiel, C., Újvári, G., Yi, S., Murray, A.S., Frechen, M., Lu, H. 2018. Ice-volume-forced erosion of the Chinese Loess Plateau global Quaternary stratotype. *Nature Communications 9 art. No. 983.*

Sun Y.B., Clemens S.C., Morrill C., Lin, X., Wang, X., An, Z. 2012. Influence of Atlantic meridional overturning circulation on the East Asian winter monsoon. *Nature Geosciences* 5, 50–54.

Taylor, S.R., McLennan, S.M., McCulloch, M.T. 1983. Geochemistry of loess, continental crustal composition and crustal model ages. *Geochimica et Cosmochimica Acta* 47, 1897–1905.

Tegen, I. 2013. Glacial Climates: Effects of atmospheric dust. *Encyclopedia of Quaternary Science, 2nd edition* (ed. Elias), 729–736.

Thiel C., Buylaert J.P., Murray A.S., Terhorst B., Hofer I., Tsukamoto S., Frechen M. 2011. Luminescence dating of the Stratzing loess profile (Austria) – testing the potential of an elevated temperature post-IR IRSL protocol. *Quaternary International* 234, 23–31.

Thomsen, K.J., Murray, A.S., Jain, M., Botter-Jensen, L. 2008. Laboratory fading rates of various luminescence signals from feldspar-rich sediment extracts. *Radiation Measurements* 32, 1474–1486.

Timar, A., Vandenberghe, D., Panaiotu, E.C., Panaiotu, C.G., Necula, C., Cosma, C., van den Haute, P. 2010. Optical dating of Romanian loess using fine-grained quartz. *Quaternary Geochronology* 5, 143–148.

Timar-Gabor, A., Wintle, A.G. 2013. On natural and laboratory generated dose response curves for quartz of different grain sizes for Romanian loess. *Quaternary Geochronology* 18, 34–40.

Timar-Gabor, A., Vandenberghe, D.A.G., Vasiliniuc, S., Panaoitu, C.E., Panaiotu, C.G., Dimofte, D., Cosma, C. 2011. Optical dating of Romanian loess: A comparison between silt-sized and sand-sized quartz. *Quaternary International* 240, 62–70.

Timar-Gabor, A., Constantin, D., Marković, S.B., Jain, M. 2015. Extending the area of investigation of fine versus coarse quartz optical ages from the Lower Danube to the Carpathian Basin. *Quaternary International* 388, 168–176.

Timar-Gabor, A., Buylaert, J-P., Guralnik, B., Trandafir-Antohi, O., Constantin, D., Anechitei-Deacu, V., Jain, M., Murray, A.S., Porat, N., Hao, Q., Wintle, A.G. 2017. On the importance of grain size in luminescence dating using quartz. *Radiation Measurements* 106, 464–471.

Trandafir, O., Timar-Gabor, A., Schmidt, C., Veres, D., Anghelinu, M., Hambach, U., Simon, S. 2015. OSL dating of fine and coarse quartz from a Palaeolithic sequence on the Bistrița Valley (Northeastern Romania). *Quaternary Geochronology* 30, 487–492.

Tsukamoto, S., Rink, W.J., Watanuki, T. 2003. OSL of tephric loess and volcanic quartz in Japan and an alternative procedure for estimating De from a fast OSL component. *Radiation Measurements* 37, 459–465.

Újvári, G., Kovács, J., Varga, G., Raucsik, G., Marković, S.B. 2010. Dust flux estimates for the Last Glacial Period in East Central Europe based on terrestrial records of loess deposits: a review. *Quaternary Science Reviews* 29, 3157–3166.

Újvári, G., Molnar, D., Novothny, A., Pall-Gergely, B., Kovacs, J., Varhegyi, A. 2014. AMS 14C and OSL/IRSL dating of the Dunaszekcs☒ loess sequence (Hungary): chronology for 20 to 150 ka and implications for establishing reliable age–depth models for the last 40 ka. *Quaternary Science Reviews* 106, 140–154.

Újvári, G., Stevens, T., Svensson, A., Klötzli, U.S., Manning, C., Németh, T., Kovács, J., Sweeney, M.R., Gocke, M., Wiesenberg, G.L.B., Markovic, S.B., Zech, M. 2015. Two possible source regions for central Greenland last glacial dust. *Geophysical Research Letters* 42, 10,399–10,408.

Újvári, G., Kok, J.F., Varga, G., Kovács, J. 2016. The physics of wind-blown loess: implications for grain-size proxy interpretation in Quaternary paleoclimate studies. *Earth-Science Reviews* 154, 247–278.

Újvári, G., Stevens, T., Molnar, M., Demeny, A., Lambert, F., Varga, G., Jull, A.J.T., Pall-Gergely, B., Bulaert, J-P., Kovacs, J. 2017. Coupled European and Greenland last glacial dust activity driven by North Atlantic climate. *Proceedings of the National Academy of Sciences* 114, E10632–E10638.

Varga, G., Kovács, J., Újvári, G. 2012. Late Pleistocene variations of the background aeolian dust concentration in the Carpathian Basin: an estimate using decomposition of grain-size distribution curves of loess deposits. *Netherlands Journal of Geosciences – Geologie en Mijnbouw* 91, 159–171.

Velichko, AA., Morozova, T.D., Nechaev, V.P., Rutter, N.W., Dlusskii, K.G., Little, E.C., Catto, N.R., Semenov, V.V., Evans, M.E. 2006. Loess/paleosol/cryogenic formation and structure near the northern limit of loess deposition, East European Plain, Russia. *Quaternary International* 152–153, 14–30.

Wang, X.L., Wintle, A.G., Lu, Y.C. 2006. Thermally transferred luminescence in fine-grained quartz from Chinese loess: Basic observations. *Radiation Measurements* 41, 649–658.

Watanuki, T., Tsukamoto, S. 2001. A comparison of GLSL, IRSL and TL dating methods using loess deposits from Japan and China. *Quaternary Science Reviews* 20, 847–851.

Wilson, P., Vincent, P.J., Telfer, M.W., Lord, T.C. 2008. Optically Stimulated Luminescence (OSL) dating of loessic sediments and cemented scree in northwest England. *The Holocene* 18, 1101–1112.

Wintle, A.G. 1973. Anomalous fading of thermo-luminescence in mineral samples. *Nature* 245, 143–144.

Wintle, A.G. 1981. Thermoluminescence dating of late Devensian loesses in Southern England. *Nature* 289, 479–480.

Wintle, A.G., Murray, A.S. 2006. A review of quartz optically stimulated luminescence characteristics and their relevance in single aliquot regeneration dating protocols. *Radiation Measurements* 41, 369–391.

Yang, S., Ding, Z.L. 2003. Color reflectance of Chinese loess and its implications for climate gradient changes during the last two glacial–interglacial cycles. *Geophysical Research Letters* 30, 2058.

Yang, S., Ding, Z.L. 2004. Comparison of particle size characteristics of Teriary 'red clay' and Pleistocene loess in the Chinese Loess Plateau: implications for origin and sources of 'red clay'. *Sedimentology* 51, 77–93.

Yang, S.L., Forman, L.S., Song, Y.G., Pierson, J., Mazzacco, J., Li, X.X., Shi, Z.T., Fang, X.M. 2014. Evaluating OSL-SAR protocols for dating quartz grains from the loess in Ili Basin, Central Asia. *Quaternary Geochronology* 20, 78–88.

Yi, S., Buylaert, J-P., Murray, A.S., Thiel, C., Zeng, L., Lu, H. 2015. High resolution OSL and post-IR IRSL dating of the last interglacial cycle at the Sanbahuo loess site (northeastern China). *Quaternary Geochronology* 30, 200–206.

Yi, S., Buylaert, J.-P., Murray, A.S., Lu, H., Thiel, C., Zeng, L. 2016. A detailed post-IR IRSL dating study of the Niuyangziguo loess site in northeastern China. *Boreas* 45, 644–657.

Zeeden, C., Dietz, M., Kreutzer, S. 2018. Discriminating luminescence age uncertainty composition for a robust Bayesian modelling. *Quaternary Geochronology* 43, 30–39.

Zhang, J.F., Zhou, L.P. 2007. Optimization of the 'double SAR' procedure for polymineral fine grains. *Radiation Measurements* 42, 1475–1482.

Zhang, J., Nottebaum, V., Tsukamoto, S., Lehmkuhl, F., Frechen, M. 2015. Late Pleistocene and Holocene loess sedimentation in central and western Qilian Shan (China) revealed by OSL dating. *Quaternary International* 372, 120–129.

6 APPLICATIONS IN GLACIAL AND PERIGLACIAL ENVIRONMENTS

MARK D. BATEMAN

Geography Department, University of Sheffield, Winter St., Sheffield S10 2TN. Email: m.d.bateman@sheffield.ac.uk

ABSTRACT: Quartz and feldspars are almost ubiquitous within preserved glacial and periglacial sediments. Luminescence dating therefore, in theory, should be the method of choice for obtaining chronologies from these depositional settings. However, its application is not without challenges. These stem from potential poor luminescence resetting prior to burial and some poor luminescence characteristics of quartz in glacial environments. These challenges can be mitigated or overcome by careful targeting of sampling to collect the best reset sediments, optimising the luminescence signal used and by measuring stored palaeodoses at the single-grain or small aliquot level. Even where incomplete resetting has occurred, statistical models can be used to target and extract what are thought to be the well-reset component. In both glacial and periglacial environments, a strong understanding of processes and depositional context is needed to inform how this may affect sediments sampled in terms of both bleaching levels at burial and post-depositional mixing.

KEYWORDS: partial bleaching, till, sand wedges, single grain

6.1 INTRODUCTION

The Earth's recent geological past – the Quaternary period – saw in many parts of the world the growth and decay of multiple and extensive ice sheets and glaciers. It is therefore hardly surprising that many scientists want to reconstruct these and their relationship to climate through time (e.g. Clark *et al.* 2012; Dyke *et al.* 2001; Hughes *et al.* 2016; Toucanne *et al.* 2015). Unfortunately, this has not been without difficulties, in part due to chronometric challenges in glacial and periglacial environments. For example, organic materials are rarely preserved within glacial sediments, thereby limiting material that can be dated by radiocarbon, amino-acid racemisation or uranium series techniques. In the case of radiocarbon, even where it can be applied, it has an age range limit that covers only part of the last glacial–interglacial cycle. Additionally, both glacial and periglacial environments act like a deep freezer which means there is a high potential for carbon recycling/contamination (Briant and Bateman, 2009). As periglacial and glacial deposits

Figure 6.1 As exemplified by this photograph taken of the Spegegazzini glacier in Patagonia, both former and present-day glaciated landscapes present a number of logistical challenges both in terms of obtaining suitable sediments for luminescence dating and for that luminescence dating to reflect true burial ages. Photograph taken by Luca Galuzzi- www.galuzzi.it

rarely have contemporary carbonate or shell material, both uranium series and amino-acid racimisation dating are of limited utility. Whilst the application of cosmogenic dating to glacial deposits is growing, it is limited to boulders and bedrock exposed at surface and so can only date the time since first exposure. Given the dynamics of ice and the known multiple glaciations in some regions this limits its utility. In contrast, luminescence dating potentially can date events over the last the last two glacial–interglacial cycles (e.g. Bateman *et al.* 2011) and further back in low dose rate environments (e.g. Pawley *et al.* 2008). As outlined in Chapter 1, it is applicable to quartz and feldspars, which are almost ubiquitous within preserved glacial sediments. As such, the method is quite an attractive proposition to gain chronological frameworks for glacial and periglacial sequences. However, early studies were not promising, deriving age overestimates (e.g. Lamothe 1988, Duller *et al.* 1995) primarily from poor luminescence resetting prior to burial. However, recent advances in measurement, quartz/feldspar characterisation and analysis of data have led to an increasing number of Quaternary scientists reporting independently verified accurate luminescence ages of periglacial and glacial deposits (e.g. Evans *et al.* 2017; Houmark-Nielsen and Kjær 2003; Meyer *et al.* 2009; Smedley *et al.* 2016; Wysota *et al.* 2009).

This chapter outlines some of the challenges posed in applying luminescence dating techniques to glacial and periglacial deposits and how developments have either mitigated or overcome these challenges. The chapter then gives guidance on maximising the potential for gaining good luminescence chronologies from these environments before giving some case studies where luminescence has been successfully applied.

6.2 KEY LUMINESCENCE CHALLENGES

The main challenge for applying luminescence dating to glacial and periglacial sediments is incomplete resetting (also referred to as bleaching). The luminescence technique relies on sediments at some point between erosion, transport or deposition being exposed to sunlight for a sufficient duration to remove antecedent stored luminescence (Chapter 1). Unless this is met, the measured luminescence cannot be attributed only to the duration of burial. Quartz optically stimulated luminescence (OSL) signal has been shown to be reduced to <1% of its original level with only 10 seconds of exposure to sunlight (Godfrey-Smith *et al.* 1988). In theory, this criterion in most depositional environments should be possible to meet, even in glacial environments where sediment loads are high. King *et al.* (2014a) have shown sediment re-distribution in proglacial settings has a number of opportunities to reset the luminescence signal. However, glacial sediments have a number of pathways between erosion and final deposition and many of these make resetting of the luminescence signal either variable or unlikely. For example, sediment eroded at the base of an ice sheet hundreds of metres thick and redeposited as a sub-basal lodgement till will never have been exposed to sunlight so the luminescence signal will not related to the final burial. Alternatively, supraglacial sediment may get partially bleached during the mass-movement event which puts the sediment on top of the glacier in the first place. Some more of the grains may be reset as this material slumps down during transport down glacier and further partial resetting of some grains may take place as it is deposited at the glacier snout. In such instances, a proportion of grains will be reset at burial, other grains may have had their luminescence signal reduced but not fully reset, whilst other grains may be completely unreset, retaining a large luminescence signal from a previous burial or even the geological strata they were derived from. Unfortunately, with Quaternary research, many of the events/sediment requiring dating pertain to subglacial processes in which light exposure is unlikely (e.g. Lamothe 1988). In periglacial environments, the partial bleaching challenge is slightly more complex as sediments may be well reset on deposition, but due to subsequent active layer processes and cryoturbation may get mixed with older/younger sediments. As a result, replicate palaeodose measurements of a sample can give a wide range of values.

The challenge for both glacial and periglacial sediments is therefore to try to measure or extract the luminescence signal only from the reset grains to avoid age overestimation, ages in saturation and just highly variable ages (e.g. Houmark-Nielsen 2009; Thrasher *et al.* 2009). The other significant challenge in applying luminescence dating to glacial and periglacial environments has been poor luminescence characteristics of some quartz found in such environments, e.g. Switzerland, New Zealand and Chile. This so-called dim quartz fails to fulfil one of the other underpinning principles of luminescence dating – that of the mineral acting as a faithful dosimeter (see Chapter 1). For sediments, it is assumed the luminescence released during measurement is in proportion to the amount of stored background radioactivity. Dim quartz not only has a weak and harder to measure luminescence signal, but this signal also fails to be proportionate to the received background radioactivity.

6.3 GLACIAL ENVIRONMENTS

Former glaciated environments are often rich in landforms relating to when ice was eroding, transporting, deforming and depositing sediments which form them. In terms of depositional landforms, ice sheets and glaciers can form, for example, moraines, eskers and flutes, as well as extensive glacial sediments. The latter encompasses a range of types ranging from tills (diamicts), through to glaciofluvial, glaciolacustrine and mass-movement sediments. The glacial environment is complex. For example, glacial till can take a wide range of forms depending on whether it was deposited underneath the glacier (subglacial), within the glacier (englacial) or from the surface of the ice (supraglacial). It will also vary depending on the sediment load, particle size, water pressure, hydrology and ice thickness associated with the glacier at the time of deposition. Sediment forming till may have been eroded by the ice directly from bedrock or from previously deposited sediment. Alternatively, it may have come from debris falling onto the ice or been washed/blown on to the ice by water or wind. Adding to this complexity is the possibility that sediment can be transported by ice as large frozen rafts, e.g. the chalk rafts found in the Happisburgh Till in East Anglia, UK (Vaughan-Hirsch *et al.* 2013). Finally, glaciers and ice sheets appear adept at cannibalising their own deposits by depositing, eroded, transported and redeposited sediments many times. Readers are directed to Evans and Benn (2004) for details of the full range of sedimentary and bedding characteristics (lilthofacies) for each and every pathway within the glacial environment. Critical from a luminescence perspective is that glacial sediments are often deposited very rapidly as sediment packages, transported within or at the base of the glacier or formed through grinding under the glacier. As such, the previously built up luminescence signal may not be completely reset (bleached) prior to sediment burial. Even once beyond the ice limit, sediment resetting may be hampered. For example, resetting of glaciofluvial sediments is highly dependent on sediment load and turbidity as daylight is quickly reduced within water making bleaching inefficient (e.g. Berger 1990). However, because glaciofluvial sediments often have multiple phases of erosion, transportation and deposition before they finally get buried and preserved there are more opportunities for resetting than might be initially thought (King *et al.* 2014a, 2014b).

As outlined by Lüthgens and Böse (2010) there have been four main approaches taken to improve luminescence dating of glacial deposits and overcome the resetting challenges presented above.

1. Signal optimisation – utilising the fast to bleach signal in the most bleachable dosimeter
2. Avoid averaging effects – by avoiding measuring multiple grains on an aliquot
3. Extracting statistically the bleached D_E component
4. Targeted sampling – focusing on glacial sediments more likely to have been bleached prior to burial.

6.3.1 Signal optimisation

A number of early studies in applying both thermoluminescence (TL) and optical stimulated luminescence (OSL) to glacial sediments found the former gave much older ages. This was

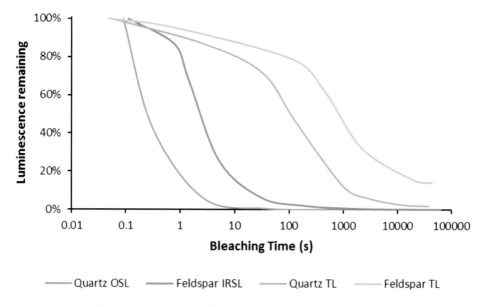

Figure 6.2 The relative bleachability of different minerals and luminescence techniques when sediment exposed to sunlight. The quartz OSL signal is bleached away quickest making it more likely to have been reset in glacial environments. Based on the data of Godfrey-Smith *et al.* (1988).

attributed to sufficient sunlight exposure to reset the OSL traps but not enough to reset the TL traps. As shown in Figure 6.2, the TL signal is slower to reset than the OSL signal taking ~1000 s of sunlight exposure to reduce it to below 10% of the original signal. To achieve the same with the OSL signal, only 2 s of light exposure is required. Given, as described above, the potential for glacial sediment to have only brief light exposure possibilities before burial this difference is critical. For example, Berger and Doran (2001) reported a 2 ka age differential between IRSL and TL ages from glacial lake sediment collected from Lake Hoare, Antarctica. Likewise, luminescence results from Swiss glaciofluvial sediments showed the TL results to be five times greater than the infrared stimulated luminescence (IRSL) results (Preusser 1999). As a result, most modern studies applying luminescence to glacial sediments opt for OSL, not TL.

As also shown in Figure 6.2 there is a difference in the speed in which quartz and feldspar grains are reset. Quartz is generally more bleachable, being reduced to 1% of the original luminescence with ~10 seconds of light exposure compared to the ~15 minutes required for feldspars to be reduced to the same level. Both minerals require ~60 minutes of sunlight exposure to reach the sample bleached level although this will vary if light has to penetrate through water first (Klasen *et al.* 2006). Spencer and Owen (2004) and Owen *et al.* (2002) compared OSL results from quartz to IRSL results from feldspars on glaciogenic sediments from the Karakoram Mountains, Pakistan. In both studies IRSL ages were much older than their OSL counterparts due to poorer bleaching of the IRSL signal. As reported by Bickel *et al.* (2015) in the context of glacial sediments from the Alps, whilst quartz resetting is generally better than that from feldspars, a much higher degree of confidence that full resetting of sediments did take place prior to burial can be placed on ages where both feldspar and quartz ages agree.

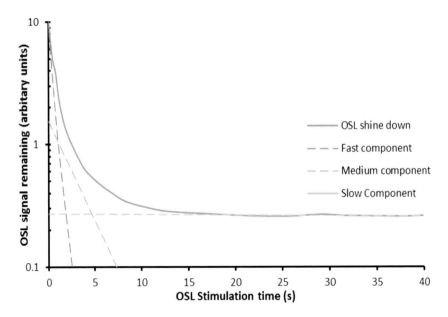

Figure 6.3 An example of a quartz OSL shine down curve with decaying amounts of luminescence with stimulation time. Also shown are the components making up this shine down curve. The fast component is both geologically stable and bleachable making it optimal for dating glacial sediment which may have been exposed to limited sunlight exposure prior to burial. Based on data from Bailey *et al.* (2011).

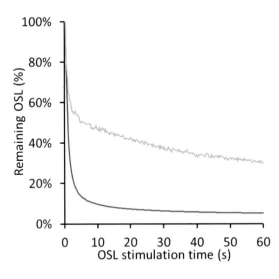

Figure 6.4 Comparison of a quartz OSL shine down curve dominated by the fast component (red) to one which is dominated by the medium–slow components (blue). The latter is sub-optimal for OSL dating purposes, especially in depositional environments where sunlight exposure prior to burial may be limited.

The final approach to signal optimisation is making sure that the luminescence emitted and measured from quartz is the geologically stable but easy to bleach component referred to in literature at the fast component. Generally, the quartz OSL signal consists of fast, medium and slow components (Fig. 6.3; e.g. Bailey *et al.* 2011; Singarayer *et al.* 2005). The medium and slow components, not only as the name suggests take longer to bleach, but are thought to be geologically unstable in part. This can lead to erroneous ages. For example, age overestimation of Scottish glaciogenic sediments was attributed by Lukas *et al.* (2007) to the fact that the quartz OSL signal had only a limited fast component and was dominated by the slow bleaching medium to slow components. Figure 6.4 shows an example of a sample dominated by medium-slow components. Where this is detected, switching away from quartz to feldspar-based measurements is one option to avoid problems but this might introduce other issues (see later). A different approach is to apply an early background subtraction (e.g. Ballarini *et al.* 2007). Usually, OSL measurements integrate the initial few seconds of luminescence as the signal and subtract from this the integral of the last few seconds of signal to remove the background. By integrating and subtracting all luminescence counts after the initial signal the early-background approach removes far more of the medium and slow components.

Alternatively, the different components within the OSL shine down can be directly measured and separated out using linear-modulated luminescence techniques (e.g. Singarayer *et al.* 2005). Whilst this allows an age to be based only on the stable and quick to bleach fast component OSL signal, linear-modulation measurements are time consuming and not straight forward to undertake routinely. Finally, it has been found that many glaciated regions have quartz that emits only limited OSL per unit of dose received, i.e. it is very insensitive. This has been shown to lead to age underestimates (Preusser *et al.* 2006). Duller (2006) encountered this when dating glaciofluvial sediments from the North Patagonian ice field in Chile, preventing in one case derivation of any single-grain results. Recent work by Smedley *et al.* (2016) as a result opted for dating similar sediments using feldspars and the pIRIR measurement protocol (for details see Chapter 1).

6.3.2 Avoiding averaging of D_E

With the development of measurement by the single-aliquot regenerative (SAR) protocol (Murray and Wintle 2000) it became possible to reliably make all the measurement needed to determine a D_E on a single subsample or aliquot. As a result, luminescence ages are now routinely based on the measurement of large numbers of replicate D_E values for each individual sample. Examination of the differences in D_E values between aliquots or single grains has been used to indicate whether a sample was totally reset prior to burial or not. Plotted replicate D_E values for well-bleached samples should appear as tight normal

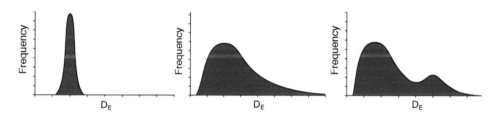

Figure 6.5 Replicate D_E distributions shown as combined probability plots for (left) a well-bleached sample and (middle and right) poorly bleached samples. In the latter two cases, the skewing indicates the majority of the D_E replicates are reset with some older D_E values (modified from Bateman *et al.* 2003).

distributions, whereas replicate D_E values for a poorly bleached sample will have a broad distribution which is skewed or possibly multi-modal (Fig. 6.5). How well the extent of partial bleaching is apparent in replicate D_E distributions is dependent in part on the size of the aliquot used (and therefore the number of grains on it) for luminescence measurements. Due to averaging, the more grains on an aliquot that are measured at the same time the less the variation in D_E that will be observed (Fig. 6.6; e.g. Wallinga 2002). As fine-grained luminescence measurements may have >100,000 grains on a standard 9.6 mm diameter aliquot averaging will be extreme. Also isolating a small number of fine grains is difficult. As a result, evaluation of D_E scatter for partial bleaching for fine-grained sediment (4–11 μm in diameter) is not possible.

For coarse grained samples, the averaging effect can be reduced by making aliquots with fewer grains mounted on them. Use of different-sized masks can restrict the surface area of the standard aliquot sample disc (9.6 mm diameter). Generally, a 'standard' aliquot will have the entire surface covered in a monolayer of grains and may therefore contain ~2300 grains. Masking to 5 mm to create a 'medium' aliquot reduces this to just over 600 grains. Masking the aliquot to 2 mm or 1 mm reduces the number of grains to ~100 and ~25 respectively, creating 'small' and 'ultra small' aliquots. For glacial environments, where incomplete resetting is likely it is recommended that aliquots should be 'small' or 'ultra small' in size and in many cases should be conducted at the single-grain level. Why the single-grain level? This is because even at small or ultra-small aliquot sizes, if all grains have the same luminescence sensitivity, it is still not an efficient way of detecting incomplete bleaching (Duller 2008) as averaging will be distorting the true D_E variability. In the worse cases it can cause phantom D_E components within the D_E distribution (Roberts *et al.* 2000). However, lots of studies have shown that for many sediment >90% of the measured luminescence signal is emitted from only 5–10% of the grains (Duller 2008). Therefore, in practice an 'ultra-small' aliquot containing 25 grains may only have one grain emitting most of the signal and therefore the D_E replicate data could be considered effectively as measurement at the signal grain level. This was the case for OSL work carried out on extracted quartz from glaciolacustrine sediment in the Vale of Pickering, UK by Evans *et al.* (2017). Here, a comparison of single-grain measurements to 'small' aliquots showed no difference in terms of the mean D_E or associated D_E variability (as shown by overdispersion in Fig. 6.7). Where dim quartz has been encountered, single-grain measurements, whilst optimal for determining levels of bleaching often fail to have any measurable grains. In such instances, increasing the aliquot size is necessary. This was the case for proglacial Swiss sediments measured by Preusser *et al.* (2007), forcing these authors to adopt 'small' aliquots for their measurements.

Key, whether a 'small' aliquot or single-grain approach to measurement is adopted, is to have a sufficient number of D_E replicates measured. For well-bleached samples, all giving very similar D_E values, whether 10 or 50 aliquots are measured will make little difference to the D_E distribution or the mean D_E calculated from it. Where partial bleaching is suspected to be an issue, seeing the true variability so that appropriate analysis can take place (see Section 6.3.3) is critical. Rodnight (2008) examined this issue and recommended a minimum of 50 replicates should be undertaken.

Bøe *et al.* (2007) measured large numbers (mostly > 100) of D_E replicates from quartz prepared from Norwegian glaciofluvial sediments. Measurements were both at the single-grain and 'ultra-small' aliquot level. In evaluating the D_E distributions, they concluded that,

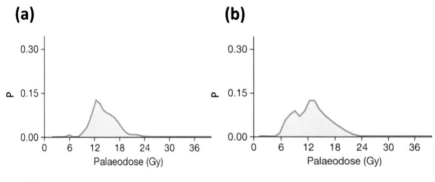

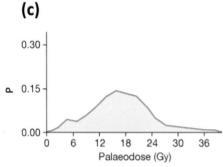

Figure 6.6 OSL data from quartz sample at different aliquot sizes showing how D_E distribution changes. a) standard 9.6 mm diameter aliquot containing ~2000 grains and therefore a high level of averaging; b) small 2 mm diameter aliquot containing ~ 100 grains. Note different D_E components not seen at 'standard aliquot' level now apparent, c) measurement of single grains showing true D_E variability of sample. Sediment shown is from a periglacial structure in East Anglia, UK (figure modified from Bateman *et al.* 2014).

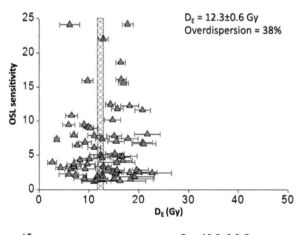

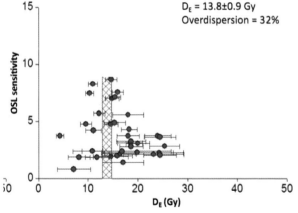

Figure 6.7 Replicate D_E data of quartz extracted from a glacial lacustrine sediment measured (left) at the 'small' aliquot and (right) single-grain levels. As few grains were contributing signal no discernible difference between the two measurement levels is apparent, either in terms of the mean (red bar) or overdispersion. Figure modified from Evans *et al.* (2017).

as both size ranges gave similar results and both sets of D_E distributions were normally distributed, the sediments had been well reset prior to burial. This was also found to be the case for glaciofluvial sediments from southern Sweden (Alexanderson and Murray 2007). Alexanderson and Murray (2012) went a step further, arguing that based on comparison of D_E distributions from large and small aliquots as well as single-grain measurements, even ice proximal sediments could be well bleached. As a caution, D_E distribution symmetry or asymmetry may in some cases not be taken as firm evidence of bleaching or partial bleaching (Fuchs and Owen 2008). Other luminescence tests such as dose recovery experiments, where samples are artificially bleached and given a known dose before D_E determination, have been recommended (e.g. Fuchs *et al.* 2007).

6.3.3 Extracting the bleached D_E component

Now it is possible to measure multiple replicate D_E values from a single sample at the single-grain level (or close to it), the true variety of palaeodose contained within the sample can be ascertained for poorly reset sediments. Whilst this is helpful in terms of warning that the sample has poor bleaching issues, it is not necessarily that helpful with age calculation. This requires a single D_E value from the sample, leading to the key question: which D_E value should be used? With a well-bleached sample (Fig. 6.5, left panel) a mean and standard deviation of the replicate D_E dataset describes the data well and would lead to a representative age being calculated for the sample. In the case of skewed or multi-modal D_E distributions (Fig. 6.5, middle and right panels) such an approach would not encapsulate the D_E variability. A mean in such circumstances would tend to overestimate true burial age by including data which was not fully reset at burial and therefore contained an antecedent signal from a previous burial. Added to this is the complexity that not all the D_E variability might be due to poor bleaching. Some could be due to post-depositional disturbance, e.g. ants or cryoturbation. If sediment has been post-depositional disturbed then older sediment could have been moved up or younger material moved down, resulting in a complex D_E distribution (see Bateman *et al.* 2007 for more details). This provides challenges as to what statistical model to analyse the data with (see Section below) as the lowest D_E component measured within a sample could be the best bleached and therefore closest to burial age or an artefact of an unrelated latter disturbance. Alternatively, some D_E variability could be due to inhomogeneity of the environmental beta dose rate (referred to as beta microdosimetry) and luminescence characteristics (e.g. Murray and Roberts 1997; Thomsen *et al.* 2005; Chauhan and Singhvi 2011). Single-grain replicate D_E data has been shown to be increasingly positively skewed as the number of potassium feldspar and zirconium grains in a bulk sediment reduces (Mayya *et al.* 2006). Zirconium grains may induce extremely high D_E values in adjacent quartz grains, whereas isolated (from sources of radiation) quartz grains may display very low D_E values. Chauhan and Singhvi (2011) developed a mathematical test to evaluate whether a given D_E distribution is significantly affected by microdosimetry.

The D_E variability often encountered in partially bleached glacial and associated sediments makes it difficult to objectively assess the true value of D_E that is appropriate to be used for the age calculation (Duller 2008). One approach for samples which, based on the D_E distributions, appear partially bleached, is to base an age estimate on the median D_E. This will be conservative and effectively be a maximum rather than true burial age.

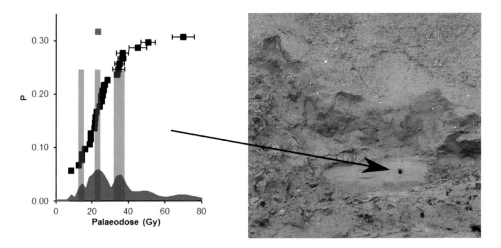

Figure 6.8 Example of a poorly bleached sand found within a glacial till at Barmston, UK. Shown in plot is combined probability of all measured single-grain OSL D_E values (blue), with individual D_E results (black points) and the average for the dataset (red point). Red bars indicate the D_E components extracted by use of the finite mixture model (FMM). In this instance the D_E dominant (middle) component was used to calculate and age of 21,500 ± 1100 years (Bateman *et al.* 2015).

If a true sediment burial age is required, then a multitude of statistical approaches have been developed to try and only select grains or aliquots which are believed to have been reset prior to burial from those which have not. The underlying assumption in all these approaches is that a sub-population of D_E values has been fully reset and as such most statistical models target the lower end of the D_E distribution (see Chapter 1 for more details). The statistical models most commonly used to evaluate D_E distributions are: (1) the minimum age model (MAM; Galbraith *et al.* 1999); the finite mixture model (FMM; Galbraith and Green 1990; Roberts *et al.* 2000); and (3) the internal–external uncertainty criterion (IEU) model (Medialdea *et al.* 2014; Thomsen *et al.* 2007). Both Bailey and Arnold (2006) and Boulter *et al.* (2007) tried to establish a decision tree to suggest which model is most appropriate for different types of D_E value distributions. However, neither includes all approaches used. Additionally, both rely on empirically based thresholds in supporting data such as over-dispersion and skewness calculations and so are not infallible.

The choice of model and how it is applied is largely a research decision made by the luminescence laboratory at the time of analysis. MAM targets the lowest grains/aliquots which are presumed to be the best reset. Critical to its successful application is the correct determination of the value for the expected over-dispersion of the sample if it was well bleached. Different values are reported in the literature ranging from 0.1 for single-aliquot measurements through to 0.3 for single-grain measurements. Sometimes this is estimated using a dose-recovery test on the sample in question or from modern well-bleached analogies. For example, Rittenour *et al.* (2015) in OSL dating poorly bleached ice proximal sediments at Lake Benson, USA determined an over-dispersion value of 0.15% from dose recovery experiments which they used in MAM to extract the final D_E values used to calculate ages for the samples. FMM has the advantage of

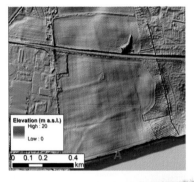

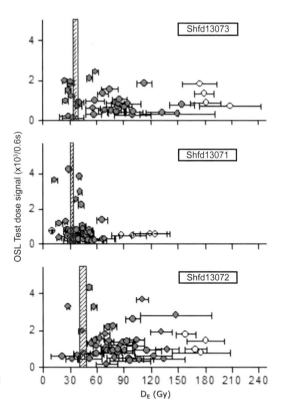

Figure 6.9 OSL dating the Ferriby moraine, UK. Top left: a digital elevation model of the moraine with the sample site denoted by 'A'. Bottom left: the exposure sampled showing a complex series of till, glaciotectonised lake and glaciofluvial sediments. Right: the D_E values determined from the ultra-small aliquots plotted against sensitivity with the blue bars indicating the IEU extracted final D_E used for age calculation purposes. Figure adapted from Bateman *et al.* (2018a).

its ability to identify and exclude any low D_E outliers that are measurement artefacts or result from beta heterogeneity that if included in age calculations would lead to age underestimations (Fig. 6.8; e.g. Smedley *et al.* 2017). The disadvantage of FMM is, depending on the exact D_E distribution (and the overdispersion used in the model – the σb-component), the lowest extracted component may include some aliquots/grains which were not fully bleached. Duller (2006) encountered some poorly bleached samples when he applied single-grain OSL to glaciofluvial sediments from Scotland and Chile. Using FMM, he was able to successfully determine ages in accordance with independent chronologies. Likewise, Bateman *et al.* (2015) applied FMM to single-grain OSL data from a sand lens found within a relict till to derive the first direct date of the last glacial till found at Barmston, UK. The IEU also targets only the lowest D_E values but is different to MAM in that the starting parameters are defined by artificially bleaching and giving a number of different irradiating doses to subsamples of the sediment under study. The relationship of over-dispersion to dose is then used to define what a well-bleached sample of this sediment should be (Medialdea *et al.* 2014). Bateman *et al.* (2018a) applied the IEU approach to ultra-small aliquot OSL measurements undertaken from a glacial moraine sequence formed by the last British and Irish Ice sheet at Ferriby, UK (Fig. 6.9).

As can been seen from the plots, the IEU utilises only the low end D_E values from which ages of 19.4 ± 1.8 ka, 21.7 ± 2.0 ka and 22.5 ± 1.6 ka were derived. These are within expectations from independent chronologies nearby.

Targeted sampling

As reviewed by Fuchs and Owen (Fig. 6.4, 2008), within glacial environments different landforms and sediments may have different probabilities of being reset prior to sediment burial (Figs. 6.10 and 6.11). These range from low for lateral moraines, subglacial lakes and subglacial tills to medium/low for frontal moraines and outwash fans. Glacial outwash is thought to have a medium probability of being reset whilst proglacial lacustrine and more distal glacio-aeolian deposits a high probability (Thrasher *et al.* 2009; Fig. 6.11). Within each of these landforms some sediment facies and types will also have higher or lower likelihoods of resetting. In poorly mixed sediments, usually indicative of mass-movement or direct deposition by ice, the probability of each sediment grain getting exposed to sunlight is low. Some will but probably not all. Clast size is often also used as a proxy for transportation energy levels. For a proglacial stream to move large cobbles requires more water velocity than to move sand. The relevance of this is that high energy environments, with lots of power, will transport a wide range of sediment sizes rapidly and have the potential to deposit large quantities in thicker packages. Predicting sediment bleaching from flow velocity though is difficult. Rendell *et al.* (1994) showed a three-hour light exposure was sufficient to zero the IRSL signals down to water depths of 10 m and 12–14 m for OSL. Higher sediment loads in fast-moving deep water may reduce light penetration and bleaching to only the top 10 s of centimetres of the water column but turbulence cycling sediment to the surface may still allow some bleaching. In low-energy deep-water environments, grains will mainly move at the base of the water column, as, for example, underflows or bedload, with only the finest grains getting to the surface and being reset. Flow regimes are often highly variable in glacial environments and sediment may go through a number of cycles of transportation, deposition and subaerial exposure as flow ceases. Resetting of sediment may occur at any point within these cycles, although in thicker sediment packages only surface sediment at deposition may get light exposure.

As a result, two approaches have been taken to try and improve the chances of getting robust luminescence-based chronologies from glacial environments. One is to look at the luminescence properties of contemporary glacial sediments, the other is to use detailed understanding of the depositional contexts (lithofacies) to inform where to sample. Bøe *et al.* (2007) collected samples from along present-day melt-water-fed rivers in Norway and compared them to older sediments from similar geomorphic contexts. They found that the contemporary glaciofluvial sediments had a low residual (D_E = 0.6 Gy), which they took to indicate the Last Glacial Maximum proglacial sediments should also have been reset. Modern fluvial sediment from Sweden also were found to show no significant incomplete bleaching (D_E = 0.5–2 Gy; Alexanderson and Murray 2007). A similar approach was taken by King *et al.* (2013) who studied modern glacial and glaciofluvial sediments from southern Norway. They found that whilst subglacial sediments had large inherited ages (of 0.81 ± 0.39 ka and 1.7 ± 0.77 ka) they were not as large as expected for completely unbleached sediments. Work by Bateman *et al.* (2012, 2018b) suggested that glacial grinding under the ice may lead to some OSL setting even in the absence of sunlight although this should not be relied upon to completely reset subglacial sediments. Paraglacial deposits surprisingly had smaller residuals than most of the glaciofluvial bar sediment sampled at 0.2 ± 0.07 ka for an avalanche sample and 0.89 ± 0.63 ka for a sheetwash sample (King *et al.* 2013). This they interpreted as indicating that paraglacial reworking maybe more effective at resetting OSL signal than previously thought. Glaciofluvial bar sediments were sampled up to 900 m from the glacial snout but show little increasing in resetting with distance and maintained a ~2.8 ± 1.1 ka inherited age. Also of note from this study is the fact that they showed that D_E variability (as determined by overdispersion values) was shown to initially increase as the amount of bleaching increased. This reflects that a totally unreset sample could have a low overdispersion and that with initial sunlight exposure some grains get reset, some partially reset whilst others remain completely unreset, leading to high variability. In a follow-up study (King *et al.* 2014a) using portable luminescence measurements on glaciofluvial sediment they demonstrate that sediments deposited by high-frequency but low-magnitude events were more likely to have been bleached. They also showed that within individual landforms – braided bars – bleaching varied, with bar tails and side-attached bar deposits being the best reset. They concluded that in glaciofluvial contexts these deposits would be the best to target for sampling. As a caveat to this, bar-head sediments appeared to be unbleached. Alexanderson and Murray (2012), in their study of modern subglacial to distal glaciofluvial sediment in Sweden, determined that significant bleaching occurs within 1 km of the ice front, although sediments still are not fully reset. They suggested that by 3–5 km glaciofluvial samples should be well reset. However, as different residual D_E levels were found when extended to other localities, it should not be assumed that all localities will be reset 3–5 km from the ice front. (King *et al.* 2014b). Specific depositional settings may still cause resetting issues even in features where the general sedimentation processes normally would cause good resetting (King *et al.* 2014b).

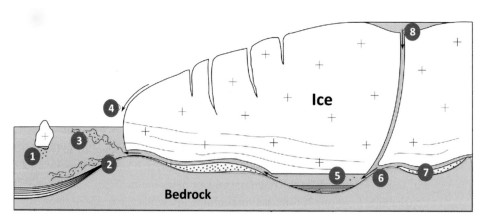

Figure 6.10 A schematic of pathways taken by sediments leading to the formation of both subglacial or proglacial deposits. Sediment finding its way to the proglacial environment is more likely to have been reset than that found under the ice. 1= sediment melting out from icebergs, 2 = turbidity currents, 3 = buoyant sediment plumes, 4 = supraglacial drainage, 5 = melt-out from debris rich ice, 6 = bedrock erosion, 7 = reworking of previously deposited sediments. Modified from Livingstone *et al.* (2015).

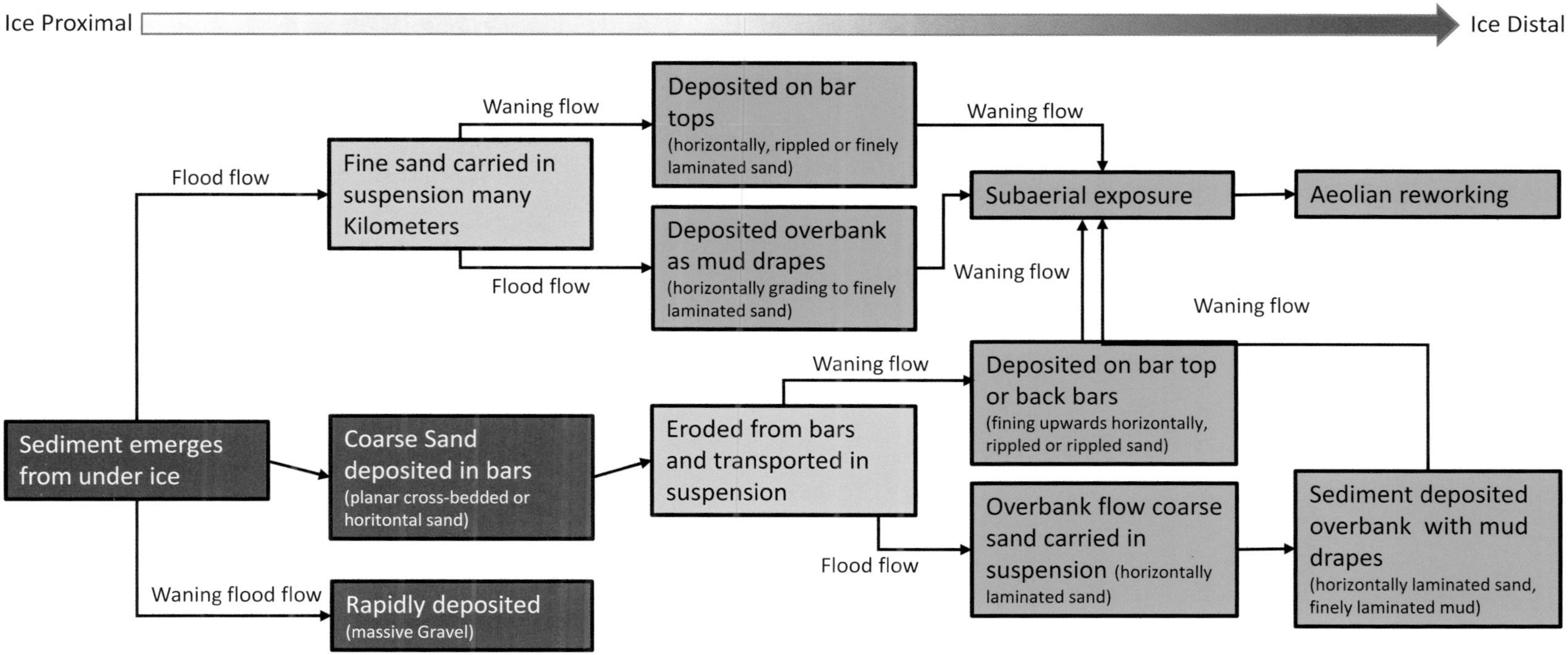

Figure 6.11 A schematic of the main pro-glacial sediment pathways within in a proglacial sandur and the potential for sediment bleaching. Red is likelihood of poor bleaching, orange medium potential for bleaching and green where there is a high potential for bleaching (modified from Thrasher *et al.* 2009).

APPLICATIONS IN GLACIAL AND PERIGLACIAL ENVIRONMENTS

The other strategy is to target only sediment facies where the probability of resetting prior to burial is higher (Fig. 6.11). This is the approach which Thrasher *et al.* (2009) advocated. In their study they targeted specific lithofacies in fossil glaciofluvial sediments on the Isle of Man by examining the sediment size, bedforms and unit configuration to determine their associated sub-environments (e.g. bars within a glaciofluvial system). They found that it was possible to identify sand-dominated sediments with bedforms characteristics of distal flow. These included finely laminated bar-top sands which are more likely to have been sub-aerially exposed or covered by a less turbid shallow water column during waning flow. Bar-top exposure on sandur plains can occur both on diurnal and annual cycles. Exposed bar-tops once flow has dropped further also have the potential for aeolian reworking thereby enhancing grain bleaching potential. They also identified rippled sand lithofacies formed in shallow water where light penetration through the water column was potentially sufficient to bleach sediment. In both instances these lithofacies yielded sufficient well-bleached quartz grains for OSL dating, producing ages in line with expectations. This is despite the samples still having positively skewed D_E distributions indicative of heterogeneous bleaching. In contrast when a deep sandur channel sediment was sampled this produced an age estimation ~10 ka older than expectations. This sediment was thought to reflect poor bleaching occurring in a deep, high-energy and sediment laden channel. Of note was the low overdispersion indicating that sediment in the channel may have been eroded, transported and deposited without bleaching or mixing with other age sediments. As a result of this targeted sampling approach they were able to produce an OSL-based chronology dating the sediments to between 17–14 ka which agreed with the expected retreat pattern of the British and Irish Ice Sheet through this area.

6.4 PERIGLACIAL ENVIRONMENTS

Periglacial environments are those formed under cold, non-glacial conditions (French 2007). Whilst in some areas they formed in front of expanding and contracting glaciers and ice sheets (e.g. the Alps), others have developed in the absence of glacial activity in areas where the climate is sufficiently cold for periglacial processes to dominate (e.g. North-western Europe during the Last Glacial Maximum ~21 ka). As through the Mid–Late Quaternary global climate and ice-volumes have gone through multiple cycles of warming/cooling and expanding/contracting so the amount of land affected by periglacial activity has also expanded and contracted. As current global climate is temperate, many former mid-latitude periglacial environments are now relict but are of interest to Quaternary Scientists if reliable chronologies for them can be established.

Many processes in periglacial environments involve the generation and movement of sediment (Fig. 6.12). Periglacial activity has been closely linked to the production and deposition of loess (see Chapter 5 for full details). Many periglacial deposits contain sand or are characterised by sand (e.g. sand wedges), thereby potentially allowing the application of luminescence dating to them. Additionally, still frozen periglacial sediments potentially have the advantage that they can have a very accurate and precise evaluation of their palaeomoisture, potassium and uranium levels. Potassium and uranium in frozen sediments cannot be leached or concentrated by water movement as happens in other environmental settings and water, which attenuates the background radiation, is less likely

Figure 6.12 Examples of periglacially derived sediments and features to which luminescence might be applied. Top left: loess on brecciated chalk at Pegwell Bay, Kent UK. Top right: sand filled involutions at Grenham Bay, Kent UK. Middle left: older coversand, Grubbenvorst, The Netherlands. Middle right: coversand-filled periglacial stripe exposed at Grimes Graves, UK. Bottom left: syngenetic sandwedge Liverpool Bay, Arctic Canada. Bottom middle: antisyngenetic sand wedge with darker vein bundles within it, Liverpool Bay, Arctic Canada. Bottom right: sediment slumped into former ice wedge, Barmston, UK.

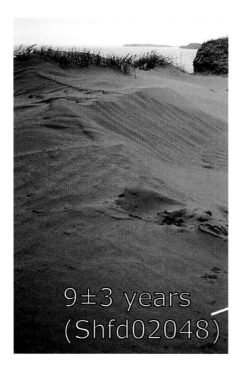

9±3 years
(Shfd02048)

Figure 6.13 Left: cliff-top dune found in a contemporary periglacial environment, Arctic Canada. Note the slip face and ripples indicating recent Aeolian movement. Right: luminescence sample collected from dune which showed good resetting of the luminescence signal had taken place in this environment (Modified from Bateman and Murton 2006).

to have varied if all pore spaces are frozen up. As a consequence, dose rates measured at the present day have a higher chance of reflecting the average dose rates since burial and are more likely to have been constant through time.

Just like glacial sediments however, understanding the formational processes associated with periglacial features is critical in terms of whether the luminescence signal in sediment is likely to have been reset at burial. It might be thought that sediment exposure to sunlight in high-latitude areas that experience seasonal winter darkness might be problematic for the application of luminescence dating. However, this is more than countered by the fact that summer sunlight hours are extended and most sediment transport in periglacial environments takes place from the thaw in spring through to autumn freeze up. Bateman and Murton (2006) demonstrated this when they reported fully reset sediment with an age of 9 ± 3 years from a cliff top dune in Arctic Canada whose surface showed fresh wind ripples and recent vegetation regrowth (Fig. 6.13). All relict periglacial sediments will have been part of the active layer either because they are near the surface and therefore part of the active layer or during permafrost degradation at the end of the periglacial period. As such, all will have been disturbed and mixed to a greater of lesser degree by ice melting, and seasonal freezing and thawing. Silt-sized material has the greatest frost susceptibility so could be expected to be affected by this more than sand and even more than gravel (which has a low frost susceptibility). Vandenberghe *et al.* (2009) showed this with their work on relict near-surface periglacial features which may have undergone post-depositional reworking leading to erroneously young dates. Post-depositional luminescence signal loss has also been reported due to repeated freeze–thaw cycles causing quartz grains to disintegrate (Dobrowolski and Fedorowicz 2007). The polygenetic and polycyclic nature of some periglacial features needs also to be understood so that the most appropriate luminescence measurement and analysis can be employed.

As reviewed by Bateman (2008), periglacial features can be usefully separated into four categories from the perspective of how easy or complex it is to apply luminescence to date them. This approach has been followed here with selected case studies given to illustrate issues and where possible, how they have been overcome or mitigated.

1. **Cold-climate aeolian** sediments such as dunes, coversands, syngenetic wedges, nival deposits should have been well exposed to sunlight during their wind-blown transport and therefore the luminescence time clock should have been reset prior to burial. Such deposits often have been preserved without post-depositional disturbance (e.g. active-layer cryoturbation), and have not been subject to multiple phases of activity or re-activation (Fig. 6.12). Generally such deposits have good potential for luminescence dating, and have no especial challenges so are not considered further here (see Chapter 4 for further details). The exception to this is where sediments and organic layers have been deposited and preserved in multiple thin layers. To illustrate this, Kolstrup et al. (2007) reported on OSL ages from contiguous samples taken from cold-climate aeolian sands (coversands) found in Jutland, Denmark and radiocarbon dates from associated organic horizons. Overall, the OSL ages increased with depth as would be expected but significantly there was a ~10% age underestimate when compared to the radiocarbon. This was attributed to problems in accurately determining the saturated palaeomoisture contents and density (sediment packing) both in the mineral and organic layers. These affected the cosmic dose rate and the attenuation of the beta and gamma dose rates from the sediments themselves (see Chapter 2). This study serves to exemplify how, even in what should be well reset aeolian sediments, a good understanding of both periglacial sediment/ depositional processes and their potential effects on luminescence dating (in this case the dose rate) is required.

2. **Patterned ground, sediment filled wedges and involutions** form by cracking processes induced by seasonal thawing and freezing of the sediments near the ground surface. They often incorporate into these cracks sandy sediments (often of aeolian origin) which provides reset material potentially suitable for luminescence dating (Fig. 6.12). However, such features can result from development over long periods of time and/or may relate to multiple periglacial events. This can cause problems in establishing a feature's true age. Well-preserved sand wedges may still have evidence of the vertically aligned sediment veins associated with individual thermal contraction events (Murton and Bateman 2007).

However, within individual relict wedges, multiple phases of cracking events (forming bundles of veins; Fig. 6.12) and rejuvenation of wedges (based on changes is wedge geometry) have been identified (e.g. Kolstrup 2004; Murton and Bateman 2007; Murton *et al.* 2000). This accords with present-day studies of wedges, which show cracks may form anywhere within a sand wedge (Mackay 1992). As such adjacent sediment veins maybe of very different antiquities. Additionally as near-surface features they may have undergone post-depositional active layer disturbance causing sand grains to move both laterally and vertically (e.g. Hallet and Waddington 1992; Peterson and Krantz 2003), during which exhumation and further exposure to sunlight could have taken place. In such cases luminescence ages may result in giving the age of when the periglacial feature was last active rather than when it started to form.

An early application of luminescence dating to sand wedges used thermoluminescence (TL) to date sand infills of three frost wedge casts in Jutland, Denmark (Kolstrup and Mejdahl 1986). This apparently worked well for two of the three, but issues were encountered for one wedge which returned an age of 39 ± 5 ka: well above the expected age of ~20 ka. This wedge, however, was interpreted as a composite wedge (the others had evidence of primary infilling) so the apparent age overestimation could have reflected both sediment mixing during secondary infilling as well as measurement at the multi-grain level with the harder to reset TL signal. TL was also applied to sediment infilling an ice-wedge at Wilczyce, Poland which dated to 47.0 ± 5.5 ka (Fiedorczuk 2001; cited in Kolstrup 2007). In this instance, the loessic host material that the wedge formed into was also dated, returning ages of 41.0 ± 3.5 ka and 40.0 ± 4.0 ka. As host material must pre-date the wedge, poor resetting of the wedge infill was suspected. This problem is not limited, however, to TL. OSL results from a large (8 m deep and 2 m wide at top) composite-wedge from Tjæreborg, Denmark also showed the wedge fill to be apparently older (230 ± 18 ka to 290 ± 20 ka) than the host material (133 ± 12 ka and 176 ± 16 ka; Kolstrup 2004). However, as this study had been undertaken with multiple single-aliquot replicates, the poor reproducibility of the palaeodose provided evidence that incomplete bleaching of the infilling sand or that, given its antiquity, multiple growth phases the wedge had taken place through time. Fine-grained ice-wedge cast fills from the Gobi Desert, Mongolia and sand from an ice wedge cast in the Qaidam Basin, Northern Tibet also, despite the use of easier to reset optical luminescence, showed poorer D_E reproducibility than would be expected for well-bleached aeolian-derived sediment (Owen *et al.* 1998, 2006). This led Bateman (2008) to speculate that when the ice wedges melted, sediment slumping had caused some mixing of different age sediments. More recently, Rémillard *et al.* (2015) applied multigrain aliquot OSL dating to ice-wedge pseudomorphs and composite-wedge casts on the Magdalen Islands, E Canada. The OSL chronology was corroborated by radiocarbon from organic material preserved in the some of the features. Based on the ages, they determined thermal contraction cracking occurred during the cold Younger Dryas period (12.9–11.5 cal. ka BP).

However, thermal contraction cracking may occur in winter, leading to incomplete bleaching in high Arctic settings. Additionally, crack locations within a sand wedge can be highly variable so that OSL samples may incorporate sediment from temporally divergent cracking events. Single-grain measurements offer the potential to look at the true D_E distribution within samples and to assess whether any scatter is due to natural processes, such as bioturbation (Bateman *et al.* 2007) or cryoturbation (e.g. Arnold *et al.* 2008). The first study to widely apply single-grain OSL measurements to sand wedge was the work of Buylaert *et al.* (2009) working in Flanders, Belgium. This study sampled both host and wedge sediments and showed at the single-grain level that the infill was well reset and the host material older. Where they found multiple phases of sand-wedge development in the stratigraphy, resultant OSL ages also reflected phased growth. Resultant ages clustered between 21.8 ± 1.2 ka to 13.8 ± 1.0 ka, falling in a period when NW Europe was known to have undergone periglacial conditions.

Bateman *et al.* (2010) encountered more complex sediments when trying to apply luminescence dating to an anti-syngenetic sand wedge in Arctic Canada (Figs. 6.12 and 6.14). Single-grain level measurements showed the three samples in the wedge to have poor D_E reproducibility leading to high age uncertainties, wedge ages older than the host material and in no stratigraphic order. The study found that this intra-sample D_E scatter was only to a minor amount attributable to OSL measurements issues relating to signal recuperation, some dim quartz grains not recovering known doses and some unaccounted sensitivity changes in the SAR protocol used. The majority of the D_E scatter was thought

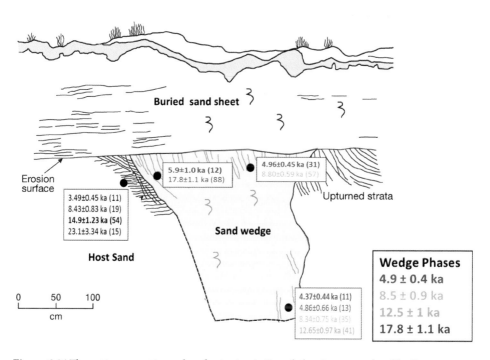

Figure 6.14 The anti-syngenetic sand wedge in Arctic Canada luminescence dated by Bateman *et al.* (2010). It is formed in the Cape Dalhousie Sands and which is overlain by an aeolian sand sheet. Also shown are the luminescence ages based on the single-grain D_E components extracted from the sample using finite mixture modelling with the amount of data in each component in parenthesis. These are colour coded to indicate their co-incidence with similar ages with other samples. Figure adapted from Bateman *et al.* (2010).

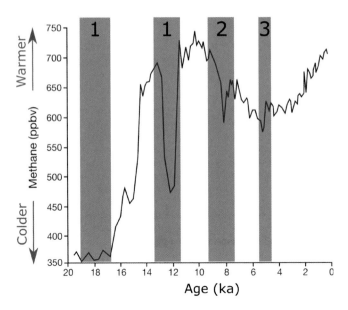

Figure 6.15 Coincidence of luminescence ages extracted from single-grain D_E components found in three samples within an anti-syngenetic sand wedge by Bateman *et al.* (2010) in Arctic Canada and cooling climatic phases. The proxy record for climate is taken from the GRIP methane record (after Blunier *et al.* 1995). Numbers shown in blue bars indicate number of component ages occurring in that phase. Figure adapted from Bateman *et al.* (2010).

to relate to thermal contraction cracking over thousands of years to form the sand wedges. When the FMM (see Chapter 1 for details) was applied to the data, all samples were found to have multiple D_E components within them. Ages calculated from these components showed similarities between samples from the wedge with all three samples having a component which dated to ~4.9 ka and two with a component dating to ~8.5 ka. Older components only occurred at the margins of the wedge and the ages relating to most of the grains from the host material were also older than the wedge material. As shown in Figure 6.15, the ages derived from the components within the samples also correlate well with known cooling events over the last 18 ka. The study of Bateman at al. (2010) therefore not only exemplifies the need to sample both host and wedge sediments, but also that multiple phases of activity require careful analysis of D_E data if accurate chronologies are to be obtained.

Luminescence dating has been applied to soft-sediment or cryoturbation involutions (Fig. 6.12) but they have their own issues to overcome. In most cases, limited if any light exposure may have taken place during involution formation so maximum ages only can be derived from the host sediment. If multiple cycles of involution activity have taken place, establishment of minimum ages may be difficult. Maximum fine-grained polymineral IRSL and OSL ages were reported from the deformed host sediment for involutions found in the Gobi Desert, Mongolia (Owen *et al.* 1998). In this case, minimum ages could only be derived from sediment deposited after the involutions were relict and periglacial activity stopped. Thus the involutions were thought to have formed between 22–15 ka and 13–10 ka. Likewise, bracketing ages between 23 and 18 ka were reported for soft-sediment deformation involutions in Lincolnshire, UK (Bateman *et al.* 2000). Both studies used multi-grained aliquots with no D_E replicates so could not establish the degree of partial resetting in these samples. Murton *et al.* (2003) applied OSL at the multi-grain aliquot level to quartz extracted from soft-sediment deformation involutions in Kent, UK (Fig. 6.12). They bracketed the development of involutions to between *c.* 21 and 18 ka based on the

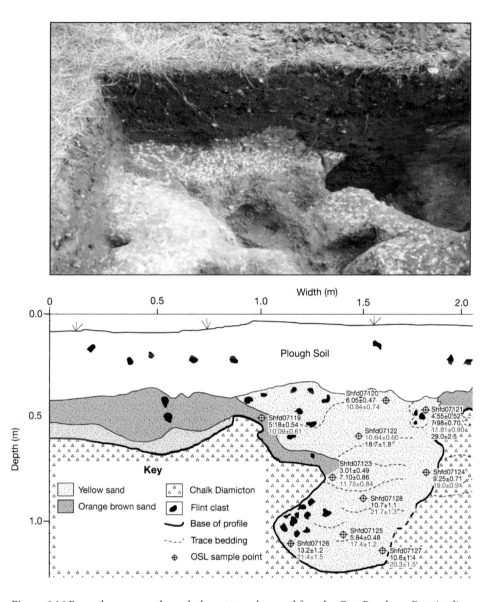

Figure 6.16 Example exposure through the patterned ground found at East Barnham, East Anglia. Top: photograph of the sediments. Bottom: stratigraphy of exposure, location of luminescence samples and ages derived from single-grained OSL dating of quartz as extracted using finite mixture modelling. Note the age component representing the largest proportion of the data is highlighted in red. Components representing <10% of OSL data within a sample are not shown. Figure modified from Bateman *et al.* (2014).

host sediment infilling the involutions and from overlying undisturbed loess.

In a major study, Bateman *et al.* (2014) applied OSL dating to the East Anglia region of the UK which contains abundant near-surface relict periglacial landforms including well-developed patterned ground, stripes and at least two generations of involutions.

Given the high potential for sand grains found in close proximity to one another having very different burial ages, sampling used narrow tubes (10 mm in diameter), extracted the easier to reset quartz and measurement was at the single-grain level to get the true range of D_E values present in samples. Intensive sampling was employed to try and understand periglacial sediment movement within features. Detailed statistical analysis of the resultant OSL D_E replicate data used finite mixture modelling to extract multiple D_E components from each sample as successfully employed by Bateman *et al.* (2010). Positive results were reported from this sampling, measurement and analysis approach. These showed that often the uppermost samples contained younger age components than samples at depth and a high degree of similar age components were found in different samples from the same site (Fig. 6.16). Results indicated that the East Anglian patterned ground experienced four main phases within the last 90–10 ka which coincide with colder climatic conditions but do not date back to earlier periglacial cycles (Fig. 6.17). Most sites were active during the cold of the Last Glacial Maximum ~21 ka and the cold climate oscillation of the Younger Dryas ~11–12 ka). Patterned ground forming polygons appeared to have a longer but more temporally and spatially varied record than the periglacial stripes sampled.

3. **Permafrost, massive ice-rich lenses and pingos** are essentially the products of concentration and freezing of water. During formation limited sediment movement takes place and what does is unlikely to get sufficient exposure to sunlight to reset the luminescence signal. No investigations to date have been made on the pressure effects of pore and segregated ice within permafrost on luminescence or how ground ice and organic matter within it might affect the concentration or leaching of radioactive compounds (Schirrmeister *et al.* 2016). Again, for relict features, some post-formational disturbance may also have occurred, e.g. sediment slumping during the melting of ice wedges. Where this has occurred poorly reset material of different ages could have been mixed together leading to erroneous ages being obtained. In such cases, an alternative sampling strategy is to sample under- and over-lying sediments for luminescence dating. This will get a maximum and minimum age for the feature. This approach was adopted by Murton *et al.* (2003) who applied luminescence to involutions (see above) and related their formation to permafrost aggradation and climatic cooling associated with Greenland Stadial 2c. French *et al.* (2007) reported OSL ages from Southern New Jersey, USA, which they also used to infer permafrost phases. Ages from epigenetic sand wedges returned minimum ages of 55–147 ka indicating the first phase of thermal contraction cracking. This was followed by fluvio-thermal erosion and gullying features (induced by climatic warming) which returned ages between 29.8 ± 2.9 and 34.8 ± 2.8 ka. Finally, a second phase of thermal contraction cracking and sand-wedge development was dated to 13.7 ± 1.6 and 16.8 ± 1.7 ka.

4. **Retrogressive thaw slumps, solifluction lobes, rock glaciers, screes and blockfields** are the most challenging from a luminescence dating perspective. They often fail to contain sediment of a suitable size and/or

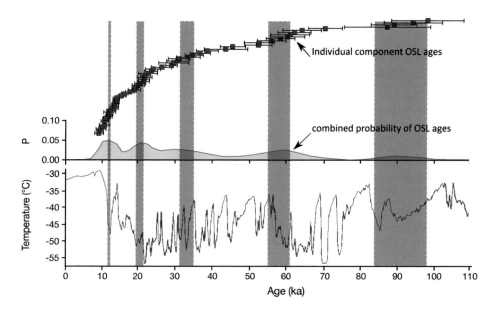

Figure 6.17 Single-grain OSL ages based on all D_E components extracted using finite mixture modelling for each sample from the periglacial structures of East Anglia presented in Bateman *et al.* (2014). Also shown the composite temperature record from Greenland (redrawn from Johnsen *et al.* 2001). Blue shaded rectangles indicate phases of periglacial activity. Figure modified from Bateman *et al.* 2014.

like many hillslope deposits have a low probability of sediment exposure to sunlight prior to burial (e.g. Fuchs and Lang 2008). An attempt to date Niveo-heterogenous fluvial sediments deposited by snow meltwater from Jutland, Denmark using OSL and TL produced ages of 91 ± 10 ka and 75 ± 7 ka (Christiansen (1998). However, the measurements made at the time (multigrain aliquots, no replicate D_E values) precluded an evaluation as to whether these are true burial ages or a result of poorly reset sediment. The resultant ages therefore should be treated with caution. Luminescence has also been applied to niveo-fluvial sediments in northeast Greenland (Christiansen et al. 2002). Based on skewed D_E distributions of measured replicates, poor sediment resetting prior to burial was encountered with only a minimum age approach getting close to independently derived chronologies. Whilst the sampled sediments were well sorted, snow re-distribution during winter darkness and mixing during the spring snow melt was attributed to have been the cause of this poor resetting. Luminescence ages have also been reported for a periglacial blockfield in the Falkland Islands (Hansom *et al.* 2008). Here, underlying sandy diamicton was sampled and a polymineral fraction measured using IRSL, OSL and TL. Although limited replicates were undertaken, some samples are reported as being reproducible and were used to indicate emplacement of the blockfield after 32–27 ka.

Hülle *et al.* (2009) attempted to apply luminescence to periglacial soliflucted slope deposits from Tanus, Germany using small multi-grained aliquots of feldspars and quartz. They encountered some poor resetting but also post-depositional mixing problems when the D_Es for the samples were measured which further work at the single-grain level may have helped to resolve. The mixing may relate to having sampled surficial sediment which had undergone both subsequent active layer processs but also been subject to pedogenesis and possible anthropogenic disturbance. An additional problem encountered with this study was the accurate determination of the background dose rate. The spatially heterogeneous nature of the sediments contained large differences in radionuclide composition and cause variable radiation flux attenuation. This necessitated measurement of the alpha and beta dose rate based on the fine sampled matrix and gamma dose rates based on *in situ* field measurements. Even then the authors recognised this was probably still problematic and recommended for solifluction deposits the use of *in situ* dosimeters at the point of sampling and left for a long-time to build up dose.

One study has attempted to evaluate the possibility of applying luminescence dating to rock glaciers (Fuchs *et al.* 2013). In this study of Alpine rock glaciers, dim and poor behavioural characteristics precluded the use of the faster to bleach quartz (see Section 6.3.1) but measurements on single feldspar grains were more promising although poorly bleached. Use of finite mixture model derived ages which are thought to reflect time elapsed since grains were incorporated into the rock glaciers and therefore minimum ages for the rock glaciers themselves. Resultant ages ranged from 3–8 ka, were consistent between sampling sites but different between rock glaciers. Whilst promising, further work is needed to follow this up, and on how the dose rates are affected by the high and variable ice contents found in rock glaciers.

Whilst as the above shows only limited luminescence dating has been carried out on these features up until the present, recent development of luminescence thermochronometric and rock-exposure techniques (see Chapter 11) may soon allow accurate and reliable ages to be routinely derived for some of this category of periglacial features.

6.5 SUMMARY

For glacial environments, luminescence dating, despite the low/variable probability of sediment bleaching prior to burial, has been shown to work in a surprising number of studies. This is more so where distal rather than proximal or subglacial sediments have been targeted for sampling. In terms of measurement, it is a requirement that steps should be taken to prove that partial bleaching is not an issue for samples and/or if it is, that the appropriate measurement level and analysis of D_E distributions has taken place.

For periglacial environments the development of optical rather than thermal stimulation, use of quartz rather than feldspar and the ability to measure individual grains has overcome many of the shortcomings raised by Kolstrup (2007) in the dating of periglacial sediments. This having been said, like glacial sediments, a strong understanding of periglacial processes and depositional context is needed to inform how this may affect sediments sampled in terms of both bleaching levels at burial and post-depositional mixing.

For both environments the following should be adhered to:

1. In most cases, careful thought about what to sample from where is needed, not only to make sure appropriate-sized material is sampled for luminescence but also to ensure that sediment most likely to have been reset prior to burial and not undergone disturbance is collected.

2. When sending samples for luminescence dating, it should be highlighted they are from glacial/periglacial environments and detailed and clear information about not only site stratigraphy but also lithofacies should be provided with the samples.

3. Given the potential for poor bleaching, optical rather than thermal measurements should be requested, and these should be made at the ultra-small aliquot or (preferably) the single-grain level with sufficient replicates to establish whether the D_E distribution is skewed or not

4. In areas prone to quartz with poor characteristics measurement should either be with feldspars and IRSL or extra measurement should be made to establish the quartz is working.

5. Where poor resetting is identified in the D_E replicate values from a sample, appropriate statistical analysis should have been undertaken to mitigate the effects of including poorly bleached results in age calculations.

Clearly some of the above are user-dependent (1–3) and other the luminescence laboratory used will need to do but the user should ascertain they have been done in order to have confidence in the results.

REFERENCES

Alexanderson, H. and Murray, A S. 2007. Was southern Sweden ice free at 19–25ka, or were the post LGM glacifluvial sediments incompletely bleached? *Quaternary Geochronology* 2, 229–236.

Alexanderson, H. and Murray, A.S. 2012. Problems and potential of OSL dating Weichselian and Holocene sediments in Sweden. *Quaternary Science Reviews*, 44, 37–60.

Arnold, L.J., Roberts, R.G., MacPhee, R.D.E., Willerslev, E., Tikhonov, A.N., Brock, F. 2008. Optical dating of perennially frozen deposits associated with preserved ancient plant and animal DNA in north-central Siberia. *Quaternary Geochronology* 3, 114-136.

Bailey, R.M. and Arnold, L.J. 2006. Statistical modelling of single grain quartz D_E distributions and an assessment of procedures for estimating burial dose. *Quaternary Science Reviews* 25, 2475–2502.

Bailey, R.M., Yukihara, E.G., McKeever, S.W.S. 2011. Separation of quartz optically stimulated luminescence components using green (525 nm) stimulation *Radiation Measurements* 46, 643-648.

Ballarini, M., Wallinga, J., Wintle, A.G., Bos, A.J.J. 2007. A modified SAR protocol for optical dating of individual grains from young quartz samples. *Radiation Measurements* 42, 360–369.

Bateman, M.D., Murton, J.B. 2006. Late Pleistocene glacial and periglacial aeolian activity in the Tuktoyaktuk Coastlands, NWT, Canada. *Quaternary Science Reviews* 25, 2552–2568.

Bateman, M.D. 2008. Luminescence dating of periglacial sediments and structures: A Review. *Boreas* 37, 574-588.

Bateman, M.D., Murton, J.B., Crowe, W. 2000. Reconstruction of the depositional environments associated with the Late Devensian and Holocene cover sand around Caistor, N. Lincolnshire, UK. *Boreas* 16, 1–16.

Bateman, M.D., Frederick, C.D., Jaiswal, M.K., Singhvi, A.K. 2003. Investigations into the potential effects of pedoturbation on luminescence dating. *Quaternary Science Reviews* 22, 1169–1176.

Bateman, M.D., Boulter, C.H., Carr, A.S., Frederick, C.D., Peter, D., Wilder, M. 2007. Detecting post-depositional sediment disturbance in sandy deposits using optical luminescence. *Quaternary Geochronology* 2, 57–64.

Bateman, M.D., Murton, J.B., Boulter, C., 2010. The source of De variability in periglacial sand wedges: depositional processes versus measurement issues. *Quaternary Geochronology* 5, 250–256.

Bateman, M.D., Carr, A.S., Dunajko, A.C., Holmes, P.J., Roberts, D.L., McLaren, S.J., Bryant, R.G., Marker, M.E., Murray-Wallace, C.V. 2011. The evolution of coastal barrier systems: a case study of the Middle-Late Pleistocene Wilderness barriers, South Africa. *Quaternary Science Reviews* 30, 63–81.

Bateman, M.D., Swift, D.A., Piotrowski, J.A. and Sanderson, D.C.W. 2012. Investigating the effects of glacial shearing of sediment on luminescence. *Quaternary Geochronology*, 10, 230–236.

Bateman, M.D., Hitchens, S., Murton, J.B., Lee, J.R. and Gibbard, P.L. 2014. The evolution of periglacial patterned ground in East Anglia, UK. *Journal of Quaternary Science* 29, 301–317.

Bateman, M.D., Evans, D.J.A., Buckland, P.C., Connell, E.R., Friend, R.J., Hartmann, D., Moxon, H., Fairburn,W.A., Panagiotakopulu, E., Ashurst,R.A. 2015:Last Glacial dynamicsof theVale of York and North Sea Lobes of the British and Irish Ice Sheet. *Proceedings of the Geologists' Association* 126, 712–730.

Bateman, M.D., Evans, D.J.A., Roberts, D.H., Medialdea, A., Ely, J.C. and Clark, C.D. (2018a). The timing and consequences of the blockage of the Humber Gap by the last British–Irish Ice Sheet. *Boreas* 47, 41–61.

Bateman, M.D., Swift, A., Piotrowski, J.A., Rhodes, E.J., Damsgaard, A. (2018b). Can glacial shearing of sediment reset the signal used for luminescence dating? *Geomorphology* 306, 90–101.

Berger, G.W. 1990. Effectiveness of natural zeroing of the thermoluminescence in sediments. *Journal of Geophysical Research* 95, 12375–12397.

Berger, G.W., Doran, P.T. 2001. Luminescence-dating zeroing tests in Lake Hoare, Taylor Valley, Antarctica. *Journal of Paleolimnology* 25, 519-529.

Bickel, L., Lüthgens, C., Lomax, J., Fiebig, M. 2015. Luminescence dating of glaciofluvial deposits linked to the penultimate glaciation in the Eastern Alps. *Quaternary International* 357, 110-124.

Bøe, A.-G., Murray, A. and Dahl, S.O. 2007. Resetting of sediments mobilised by the LGM ice-sheet in southern Norway. *Quaternary Geochronology* 2, 222–228.

Boulter, C., Bateman, M.D. and Frederick, C.D. 2007. Developing a protocol for selecting and dating sandy sites in East Central Texas: Preliminary results *Quaternary Geochronology* 2, 45–50.

Briant, R.M. and Bateman, M.D. 2009. Luminescence dating indicates radiocarbon age underestimation in Late Pleistocene fluvial deposits from eastern England. *Journal of Quaternary Science* 24, 916–927.

Blunier, T., Chappellaz, J.A., Schwander, J., Stauffer, B., Raynaud, D. 1995. Variations in atmospheric methane concentration during the Holocene epoch. *Nature* 374, 46–49.

Buylaert, J-P., Ghysels, G., Murray, A.S., Thomsen, K.J., Vandenberghe, D. de Corte, F., Heyse, I/, van den Haute, P. 2009 Optical dating of relict sand wedges and composite-wedge pseudomorphs in Flanders (Belgium). *Boreas* 36, 160–175.

Chauhan, N. and Singhvi, A.K. 2011. Distribution in SAR palaeodoses due to spatial heterogeniety of natural beta dose. *Geochronometria*, 38, 190–198.

Christiansen, H.H. 1998. Periglacial sediments in an Eemian–Weichselian succession at Emmerlev Klev, southwestern Jutland, Denmark. *Palaeogeography, Palaeoclimatology, Palaeoecology* 138, 245 258.

Christiansen, H.H., Bennike, O., Bocher, J., Elberling, B., Humlum, O. and Jakobsen, B.H. 2002. Holocene environmental reconstruction from deltaic deposits in northeast Greenland. *Journal of Quaternary Science 17,* 145–160.

Clark, C.D., Hughes, A.L.C., Greenwood, S.L., Jordan, C., Sejrup, H.S. 2012. Pattern and timing of retreat of the last British–Irish Ice Sheet. *Quaternary Science Reviews* 44, 112–146.

Dobrowolski, R., Fedorowicz, S. 2007. Glacial and periglacial transformation of palaeokarst in the Lublin–Volhynia region (se Poland, nw Ukraine) on the base of TL dating. *Geochronometria* 27, 41–46.

Duller, G.A.T. 2006. Single grain optical dating of glacigenic sediments. *Quaternary Geochronology* 1, 296–304.

Duller, G.A.T. 2008. Single-grain optical dating of Quaternary sediments: why aliquot size matters in luminescence dating. *Boreas* 37, 589–612

Duller, G.A.T., Wintle, A.G., Hall, A.M., 1995. Luminescence dating and its application to key pre-late Devensian sites in Scotland. *Quaternary Science Reviews* 14, 495-519.

Duller, G.A.T., Bøtter-Jensen, L., Kohsiek, P., Murray, A.S. 1999. A high-sensitivity optically stimulated luminescence scanning system for measurement of single sand-sized grains. *Radiation Protection Dosimetry* 84, 325–330.

Dyke, A.S. Andrews, J.T., Clark, P.U., England, J.H., Miller, G.H., Shaw, J., Veillette, J.J. 2001. The Laurentide and Innuitian ice sheets during the Last Glacial Maximum. *Quaternary Science Reviews* 21, 9–31.

Evans D.J.A. and Benn, D.I. 2004. *A Practical Guide to the Study of Glacial Sediments.* Routledge, London.

Evans, D.J.A., Bateman, M.D., Roberts, R.H., Medialdea, A., Hayes, L., Duller, G.A.T., Fabel D. and Clark, C.D. 2017. Glacial Lake Pickering: stratigraphy and chronology of a proglacial lake dammed by the North Sea Lobe of the British–Irish Ice Sheet *Journal of Quaternary Science* 32, 295–310.

French, H.M. Demitroff, M., Forman, S.L., Newell, W.L. 2007. A Chronology of Late-Pleistocene Permafrost Events in Southern New Jersey, Eastern USA Permafrost and Periglacial Processes 18, 49–59

French, H.M. 2007. *The Periglacial Environment* (3rd edition). Wiley, London.

Fuchs, M., and Lang A. 2008. Luminescence dating of hillslope deposits – A review. *Boreas* 37, 636–659.

Fuchs, M. and Owen, L.A. 2008. Luminescence dating of glacial and associated sediments: review, recommendations and future directions. *Boreas* 37, 636–659.

Fuchs, M., Woda, C. and Bürkert, A. 2007. Chronostratigraphy of a sedimentary record from the Hajar mountain range in north Oman: Implications for optical dating of insufficiently bleached sediments. *Quaternary Geochronology* 2, 202–207.

Fuchs, M.C. Böhlert, R., Krbetschek, M., Preusser, F., Egli, M. 2013. Exploring the potential of luminescence methods for dating Alpine rock glaciers. *Quaternary Geochronology* 18, 17–33.

Galbraith, R.F. and Green, P.F. 1990. Estimating the component ages in a finite mixture. *Radiation Measurements* 17,197-206.

Galbraith, R.F., Roberts, R.G., Laslett, G.M., Yoshida, H., Olley, J.M. 1999. Optical dating of single and multiple grains of quartz from Jinmium rock shelter, northern Australia, Part I: Experimental design and statistical models. *Archaeometry* 41, 339–364.

Godfrey-Smith, D.I., Huntley, D.J. and Chen, W.H. 1988. Optical dating studies of quartz and feldspar sediment extracts. *Quaternary Science Reviews* 7, 373–380.

Hallet B., Waddington E.D. 1992. Buoyancy forces induced by freeze–thaw in the active layer: implications for diapirism and soil circulation. In *Periglacial Geomorphology,* Dixon J.C. and Abrahams A.D. (eds.). John Wiley and Sons: Chichester, 251–279.

Hansom, J.D., Evans, D.A., Sanderson, C., Bingham, R.G., Bentley, M.J. 2008. Constraining the age and formation of stone runs in the Falkland Islands using optically stimulated luminescence. *Geomorphology* 94, 117–130.

Houmark-Nielsen, M. 2009. Testing OSL failures against a regional Weichselian glaciation chronology from southern Scandinavia. *Boreas,* 37, 660–677.

Houmark-Nielsen, M. and Kjær, K. H. 2003. Southwest Scandinavia, 40–15 ka BP: palaeogeography and environmental change. *Journal of Quaternary Science* 18, 769–786.

Hughes, A.L.C., Gyllencreutz, R., Lohne, O.S., Mangerud, J., Svendsen, J.I. 2016. The last Eurasian ice sheets – a chronological database and time-slice reconstruction, DATED-1. *Boreas* 45, 1–45.

Hülle, D., Hilgers, A., Kühn, P., Radtke, U. 2009. The potential of optically stimulated luminescence for dating periglacial slope deposits – A case study from the Taunus area, Germany. *Geomorphology* 109, 66–78.

Johnsen, S.J., Dahl-Jensen, D., Gundestrup, N., Steffensen, J.P., Clausen, H.P., Miller, H., Masson-Delmotte, V., Sveinbjörnsdottir, A.M. and White, J. 2001. Oxygen isotope and palaeotemperature records from six Greenland ice-core stations: Camp Century, Dye-3, GRIP, GISP2, Renland and NorthGRIP. *Journal of Quaternary Science* 16, 299–307.

King, G.E., Robinson R.A.J. and Finch, A.A. 2013. Apparent OSL ages of modern deposits from Fåbergstølsdalen, Norway: implications for sampling glacial sediments. *Journal of Quaternary Science* 28, 673–682.

King, G.E., Sanderson, D.C.W., Robinson R.A.J. and Finch, A.A. (2014a). Understanding processes of sediment bleaching in glacial settings using a portable OSL reader. *Boreas* 43, 955–972.

King, G.E., Robinson R.A.J. and Finch, A.A. (2014b). Towards successful OSL sampling strategies in glacial environments: deciphering the influence of depositional processes on bleaching of modern glacial sediments from Jostedalen, Southern Norway. *Quaternary Science Reviews*, 89, 94–107.

Klasen, N., Fiebig, M., Preusser, F. and Radtke, U. 2006. Luminescence properties of glaciofluvial sediments from the Bavarian Alpine Foreland. *Radiation Measurements* 41, 866–870.

Kolstrup, E. 2004. Stratigraphic and environmental implications of a large ice-wedge cast at Tjaereborg, Denmark. *Permafrost and Periglacial Processes 15,* 31–40.

Kolstrup, E., Murray, A., Possnert, G. 2007. Luminescence and radiocarbon ages from laminated Lateglacial aeolian sediments in western Jutland, Denmark. *Boreas 36,* 314–325.

Lamothe, M. 1988. Dating till using thermoluminescence. *Quaternary Science Reviews* 7, 273–276.

Livingstone, S.J., Piotrowski, J.P., Bateman, m.D., Ely, J.C., Clark, C.D. 2015. Discriminating between subglacial and proglacial lake sediments: an example from the D€anischer Wohld Peninsula, northern Germany. *Quaternary Science Reviews* 112, 86–108.

Lukas, S., Spencer, J.Q.G., Robinson, R.A.J. and Benn, D.I. 2007. Problems associated with luminescence dating of Late Quaternary glacial sediments in the NW Scottish Highlands. *Quaternary Geochronology* 2, 243–248.

Lüthgens C., Böse, M. 2010. From morphostratigraphy to geochronology e on the dating of ice marginal positions. *Quaternary Science Reviews* 44, 26-36.

Medialdea, A., Thomsen, K.J., Murray, A.S.G., Benito, G, 2014. Reliability of equivalent-dose determination and age-models in the OSL dating of historical and modern palaeoflood sediments. *Quaternary Geochronology* 22, 11–24.

Meyer, M.C., Hofmann ch.-ch., Gemmell, A.M.D., Haslinger, E. Häusler, H., Wangda, D. 2009. Holocene glacier fluctuations and migration of Neolithic yak pastoralists into the high valleys of northwest Bhutan. *Quaternary Science Reviews* 28, 1217–1237.

Mackay, J.R. 1992. The frequency of ice-wedge cracking (1967–1987) at Garry Island, western Arctic coast, Canada. *Canadian Journal of Earth Sciences 29,* 236–248.

Mayya, Y.S., Morthekai, P., Murari, M.K., Singhvi, A.K. 2006. Towards quantifying beta microdosimetric effects in single-grain quartz dose distribution *Radiation Measurements* 41, 1032–1039.

Murray, A S. and Roberts, R.G. 1997 Determining the burial time of single grains of quartz using optically stimulated luminescence. *Earth and Planetary Science Letters* 152, 163–180.

Murray, A.S. and Wintle, A.G. 2000. Luminescence dating of quartz using an improved single-aliquot regenerative-dose protocol. *Radiation Measurements* 32, 57–73.

Murton, J.B., Worsley, P. and Gozdzik, J. 2000. Sand veins and wedges in cold aeolian environments. *Quaternary Science Reviews* 19, 899–922.

Murton, J.B., Bateman, M.D., Baker, C.A., Knox, R., Whiteman, C.A. 2003. The Devensian periglacial record on Thanet, Kent, UK. *Permafrost and Periglacial Processes 14,* 217–246.

Murton, J.B. and Bateman, M.D. 2007. Syngenetic sand veins and anti-syngenetic sand wedges, Tuktoyaktuk Coastlands, western Arctic Canada. *Permafrost and Periglacial Processes 18,* 33–47.

Owen, L.A., Richards, B., Rhodes, E.J., Cunningham, W.D., Windley, B.F., Badamgarav, J., Dorjnamjaa, D. 1998. Relic permafrost structures in the Gobi of Mongolia: age and significance. *Journal of Quaternary Science 13,* 539–547.

Owen, L.A., Kamp, U., Spencer, J.Q. and Haserodt, K. 2002. Timing and style of Late Quaternary glaciation in the eastern Hindu Kush, Chitral, northern Pakistan: A review and revision of the glacial chronology based on new optically stimulated luminescence dating. *Quaternary International* 97–98, 41–55.

Pawley, S.M., Bailey, R.M., Rose, J., Moorlock, B.S.P., Hamblin, R.J.O., Booth, S.J., Lee, J.R. 2008. Age limits on Middle Pleistocene glacial sediments from OSL dating, north Norfolk, UK. *Quaternary Science Reviews* 27, 1363–1377.

Peterson, R.A., Krantz, W.B. 2003. A mechanism for differential frost heave and its implications for patterned-ground formation. *Journal of Glaciology* 49, 69–80.

Preusser, F. 1999. Luminescence dating of fluvial sediments and overbank deposits from Gossau, Switzerland: Fine grain dating. *Quaternary Geochronology* 18, 217–222.

Preusser, F., Ramseyer, K., Schlüchter, C. 2006. Characterization of low OSL intensity quartz from the New Zealand Alps. *Radiation Measurements* 41, 871–877.

Preusser, F., Blei, A., Graf, H.R. and Schlüchter, C. 2007. Luminescence dating of Wü¨rmian (Weichselian) proglacial sediments from Switzerland: Methodological aspects and stratigraphical conclusions. *Boreas* 36, 130–142.

Rémillard, A. M., Hétu, B., Bernatchez, P., Buylaert, J.-P., Murray, A. S., St-Onge, G. and Geach, M. 2015. Chronology and palaeoenvironmental implications of the ice-wedge pseudomorphs and composit-wedge casts on the Magdalen Islands (eastern Canada). *Boreas.* 44, 658–675.

Rendell, H.M., Webster, S.E., Sheffer, N.L. 1994. Underwater bleaching of signals from sediment grains: New experimental data. *Quaternary Geochronology* 13, 433–435.l 13

Rittenour, T.M., Cotter, J.F.P. and Arends, H.E 2015. Application of single-grain OSL dating to ice-proximal deposits, glacial Lake Benson, west-central Minnesota, USA. *Quaternary Geochronology* 30, 306–313.

Roberts, R.G., Galbraith, R.F., Yoshida, H., Laslett, G.M., Olley, J.M. 2000. Distinguishing dose populations in sediment mixtures: A test of single-grain optical dating procedures using mixtures of laboratory-dosed quartz. *Radiation Measurements* 32, 459–465.

Rodnight, H. 2008. How many equivalent dose values are needed to obtain a reproducible distribution? *Ancient TL* 26, 3–9.

Schirrmeister, L., Meyer, H., Andreev, A., Wetterich, S., Kienast, F., Bobrov, A., Fuchs, M., Sierralta, M., Herzschuh, U. 2016. Late Quaternary paleoenvironmental records from the Chatanika River valley near Fairbanks (Alaska). *Quaternary Science Reviews* 147, 259–278.

Singarayer, J.S., Bailey, R.M., Ward, S. and Stokes, S. 2005. Assessing the completeness of optical resetting of quartz OSL in the natural environment. *Radiation Measurements* 40, 13–25.

Smedley, R.K., Glasser, N.F. and Duller, G.A.T. 2016. Luminescence dating of glacial advances at Lago Buenos Aires (similar to 46 degrees S), Patagonia. *Quaternary Science Reviews* 134, 59–73.

Smedley, R.K., Scourse, J.D., Small, D., Hiemstra, J.F., Duller, G.A.T., Bateman, M.D., Burke, M.J., Chiverrell, R.C., Clark, C.D., Davies, S.M., Fabel, D., Gheorghiu, D.M., McCarroll, D., Medialdea, A., Xu, S 2017. New age constraints for the limit of the British–Irish Ice Sheet on the Isles of Scilly. *Journal of Quaternary Science* 32, 48–62.

Spencer, J.Q. and Owen, L.A. 2004. Optically stimulated luminescence dating of Late Quaternary glaciogenic sediments in the upper Hunza valley: Validating the timing of glaciation and assessing dating methods. *Quaternary Science Reviews* 23, 175–191.

Thomsen, K.J., Murray, A.S., Bøtter-Jensen, L. 2005. Sources of variability in OSL dose measurements using single grains of quartz. *Radiation Measurements* 39, 47-61.

Thomsen, K.J., Murray, A.S., Bøtter-Jensen, L. and Kinahan, J. 2007. Determination of burial dose in incompletely bleached fluvial samples using single grains of quartz. *Radiation Measurements*, 42, 370–379.

Thrasher, I.M., Mauz, B., Chiverrell, R.C., Lang, A. and Thomas, G.S.P 2009. Testing an approach to OSL dating of Late Devensian glaciofluvial sediments of the British Isles. *Journal of Quaternary Science*, 24, 785–801.

Toucanne, S., Soulet, G., Freslon, N., Silva Jacinto, R., Dennielou, B., Zaragosi, S., Eynaud, F., Bourillet, J.-F., Bayon, G. 2015. Millennial-scale fluctuations of the European Ice Sheet at the end of the last glacial, and their potential impact on global climate. *Quaternary Science Reviews* 123, 113–133.

Vandenberghe, D., Vanneste, K., Verbeeck, K., Paulissen, E., Buylaert, J-P. De Corte, F., van den Haute, P. 2009. Late Weichselian and Holocene earthquake events along the Geleen fault in NE Belgium: OSL age constraints. *Quaternary International* 199, 56–74.

Vaughan-Hirsch, D.P., Phillips, E., Lee, J.R. and Hart, J.K. 2013. Micromorphological analysis of poly-phase deformation associated with the transport and emplacement of glaciotectonic rafts at West Runton, north Norfolk, UK. *Boreas* 42, 376–394.

Wallinga, J. 2002. On the detection of OSL age overestimation using single-aliquot techniques. *Geochronometria* 21, 17–26.

Wysota, W., Molewski, P., Sokolowski, R.J. 2009. Record of the Vistula ice lobe advances in the Late Weichselian glacial sequence in north-central Poland. *Quaternary International* 207, 26–41, pp 433–435.

7 APPLICATIONS IN FLUVIAL AND HILLSLOPE ENVIRONMENTS

MARKUS FUCHS

Department of Geography, Justus-Liebig-University Giessen, Senckenbergstr. 1, 35390 Giessen, Germany. Email: Markus.Fuchs@geogr.uni-giessen.de

ABSTRACT: Fluvial and hillslope environments and their associated landforms and sediments represent one of the most important archives for reconstructing past environmental and climate conditions and for helping us determine the nature and magnitude of landscape evolution. Luminescence dating provides the necessary accurate timeframe and, in contrast to other dating methods such as radiocarbon dating, dates the sediment directly. It can be applied to a broad range of fluvial and hillslope environments, as quartz and feldspar are widely available as silt and sand grains. Furthermore, luminescence dating potentially covers an age range of 10–1,000,000 years, i.e. encompassing Holocene and Late to Middle Pleistocene time scales that are of most interest to Quaternary scientists and archaeologists. This chapter presents applications of luminescence dating for landforms and sediments from fluvial and hillslope environments and discusses specific challenges, including insufficient sunlight exposure and dose rate variability. To overcome these challenges, careful sampling and descriptions of the sampling site are essential.

KEYWORDS: hillslope, river, fluvial, alluvial, colluvial

7.1 INTRODUCTION

Fluvial and hillslope environments – specifically, their geomorphological processes, landforms and related sediments – are of fundamental interest in geosciences and their related fields. In most environments, fluvial processes are major agents shaping the Earth's surface, even in regions with infrequent rainfall and scarce surface run-off. Furthermore, since the beginning of human evolution, such environments have represented important areas of human colonisation and socioeconomic activities and have constituted major traffic routes. At the same time, they have been constantly exposed to changing climates and tectonic activities – as documented by the different landforms and sediments – providing valuable information for palaeoenvironmental reconstruction and evidence of paleoclimate change. This can be used to gain insight into landscape evolution.

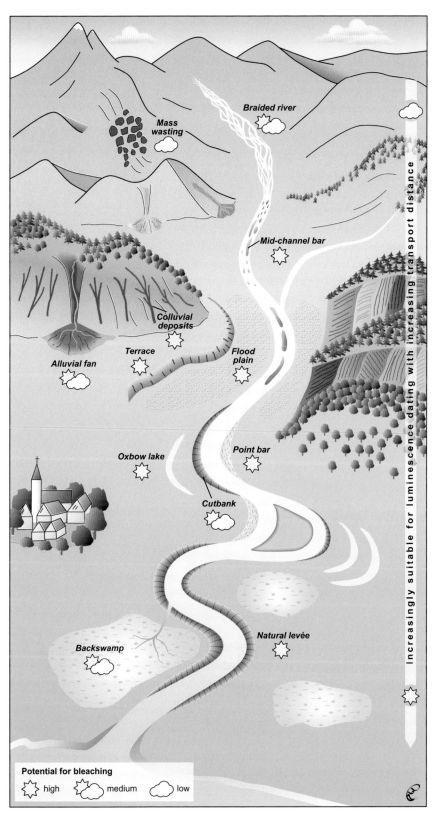

Figure 7.1 Hillslope and fluvial environments and their associated landforms and sediments, highlighting their suitability for luminescence dating. In fluvial systems, luminescence dating tends to become more suitable with increasing transport distance of the sediments.

Because of the overall importance of fluvial and hillslope environments, a basic understanding is necessary of these geosystems, their configuration, interaction and relation to other systems, as well as their system behaviour in response to internal (e.g. regolith thickening through weathering) or external (e.g. climate change) disturbances. Fluvial and hillslope environments display a large variety of landforms and sediment archives, resulting from different geomorphological processes of varying magnitudes and frequencies that commonly are dependent on climate and tectonic activity. Figure 7.1 provides a schematic overview of the different landforms and sediments associated with such environments.

The geomorphological processes operating on hillslopes are manifold, with fluvial processes, soil creep and mass movement as the major agents shaping the morphology of hillslopes by eroding sediment and transporting it downslope (Fig. 7.2). Major fluvial processes operating on hillslopes are sheet, rill and gully erosion, controlling the sediment flux to the valley floor and river system. Gully erosion may transport hillslope material in single events directly to the valley floor and river, but sheet and rill erosion is generally a more gradual transport process involving temporary on-slope sediment deposition. Sediments may also be deposited on the foot-slope, forming wedge-shaped colluvial deposits.

The valley floor and river not only receive sediments from the adjoining hillslopes but also from upstream reaches. Depending on the river transport capacity, sediments will be eroded, transported or deposited. Fluvial processes in river systems are diverse, complex and operate on various spatial and temporal scales, so fluvial environments possess a large variety of depositional and erosional landforms and units (Fig. 7.1), many with complex sedimentary architectures.

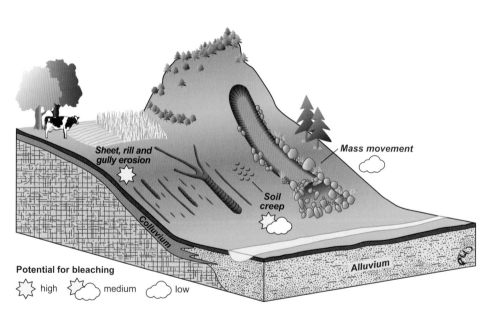

Figure 7.2 Soil creep, mass movements and fluvial processes (sheet, rill and gully erosion) operate on hillslopes, transporting sediments downslope to build colluvial deposits at footslope position. For luminescence dating, sediments from soil creep and mass movements are less favourable owing to lower bleaching potential.

To decipher the different information about past fluvial and hillslope systems, a robust time frame is crucial. Relative dating methods such as morpho- and lithostratigraphy provide preliminary information on the temporal succession of the sediments and landforms, but numerical dating methods are essential to establish reliable quantitative chronologies. Depending on sediment and material composition and the required accuracy, several numerical dating methods with their individual advantages and disadvantages are available for hillslope and fluvial environments. The fallout radionuclides ^{7}Be, ^{137}Cs and ^{210}Pb$_{ex}$ can be used for dating if investigations of short periods from days to years are needed, with ^{210}Pb$_{ex}$ reaching back to c. 100 years (e.g. Mabit et al. 2008). For investigating longer time scales in fluvial and hillslope environments, the most widely applied methods include radiocarbon, dendrochronology, ^{230}Th/U, terrestrial cosmogenic nuclide (TCN), electron spin resonance (ESR) and luminescence dating (Rixhon et al. 2016; Wagner 1998; Walker 2005).

Radiocarbon dating is a precise and widely used numerical dating method, with a half-life of ^{14}C of 5730 ± 40 a, making this method applicable for the last c. 50 ka (Hajdas 2008). In general, organic matter is used for radiocarbon dating, which is not always available in sufficient quantities and therefore limits its application in fluvial and hillslope environments. It also needs to be considered that radiocarbon dating does not date the sediment directly, but rather older organic matter that has been incorporated into the sediment. As a consequence of the former, radiocarbon ages often show significant age overestimation of the true sedimentation age (Fuchs and Wagner 2005; Lang and Hönscheidt 1999). An even more precise method than radiocarbon dating is provided by dendrochronology. However, because of its dependency on the availability of trees with annual tree-ring growth, the application of dendrochronology is spatially limited to dominantly the mid latitudes and temporally limited to the last c. 12 ka (Friedrich et al. 2004). Nevertheless, in Holocene fluvial environments, fossil trunks embedded in terraces are often the main source for the usage of dendrochronology (Radoane et al. 2015). Again, however, dendrochronology is only an indirect method that dates the incorporated wood and not the sediment itself, and thus often results in overestimation of the sedimentation age. Fluvial sediments sometimes host secondary carbonates, e.g. calcretes, which can be dated by ^{230}Th/U. Under favourable conditions, ages up to c. 600 ka can be produced (Bourdon et al. 2003). But because carbonate formation can only start after sedimentation of the fluvial material, ^{230}Th/U-dating can only provide minimum ages for the host fluvial sediments. TCN like ^{10}Be, ^{26}Al and ^{36}Cl are frequently used for surface-exposure dating (SED) of e.g. fluvial terraces or alluvial fans (Fuchs et al. 2015; Rixhon et al. 2011). The age range of SED is strongly dependent on the nuclides employed and on the setting of the sampling location, but Holocene to Middle Pleistocene age determinations are possible (Gosse and Phillips 2001; Ivy-Ochs and Kober 2008). However, SED ages are often associated with large errors or scattered ages, resulting from various uncertainties including erosion rate determination or inheritance of former nuclide accumulations. ESR with its use of widely available quartz as a datable mineral is increasing in importance as a dating method for fluvial sediments as it dates sedimentation directly (Tissoux et al. 2007; Toyoda 2015). The age range of ESR on quartz depends on the ESR signal used and on the sediment's mineralogy, but it is probably best applied to Middle to Early Pleistocene environments. A drawback of ESR as applied to quartz is the hard-to-bleach behaviour in daylight. This can result in incomplete bleaching of the ESR signal during sediment transport and deposition, and consequently to age overestimation (Toyoda *et al.* 2000). Therefore, ESR ages can often only be interpreted as maximum ages.

As outlined in Chapter 1, luminescence dating of sediments was introduced by Wintle and Huntley (1979), and since then its application to hillslope and fluvial systems in particular has contributed greatly to a better understanding of landscape evolution and paleoenvironmental conditions (e.g. Fuchs and Lang 2009; Rittenour 2008; Wallinga 2002). Because luminescence dating tends to use widely available quartz and feldspar minerals in the silt and sand grain size fraction, this dating method can be applied to many hillslope and fluvial sediments. Moreover, in contrast to ESR, luminescence ages are more accurate, because the luminescence signal bleaches much faster during sediment reworking.

In what follows, luminescence dating and specific issues regarding its application to hillslope and fluvial sediments are discussed (Section 2). Case studies are then presented, offering insight into the possibilities and challenges of luminescence dating in hillslope and fluvial environments (Section 3). The conclusion summarises this chapter on the application of luminescence dating to fluvial and hillslope environments (Section 4).

7.2 LUMINESCENCE DATING IN HILLSLOPE AND FLUVIAL ENVIRONMENTS

For hillslope environments, luminescence dating was reviewed by Fuchs and Lang (2009), for fluvial environments by Wallinga (2002), Rittenour (2008) and Rixhon *et al.* (2016). The following section discusses luminescence dating applied to both hillslope and fluvial environments and seeks to illustrate specific considerations needed to make sure the correct mineral and method is selected.

7.2.1 Mineral characteristics

The natural dosimeters of quartz and feldspar have different luminescence saturation levels because of their different numbers of traps available to store electrons, directly influencing the upper age range of luminescence dating (see Chapter 1). On average, the upper age range for quartz OSL is *c.* 150 ka, demonstrated by several studies on fluvial terraces (e.g. Cordier *et al.* 2012; Olszak and Adamiec 2016), though under favourable conditions (where background environmental dose rates are low) ages above 200 ka can be achieved (e.g. Wallinga *et al.* 2004). For older sediments, feldspars measured using IRSL should be chosen as feldspars have a much higher saturation level. Applied to fluvial sediments, the upper age range for feldspar IRSL is *c.* 300 ka (e.g. Cordier *et al.* 2012), but in favourable conditions ages up to *c.* 400 ka have been reported (e.g. Lowick *et al.* 2012; Roskosch *et al.* 2014).

The lower age range for luminescence dating is directly influenced by the brightness of the luminescence signal, which in turn depends on the sensitivity of the dosimeter and the dose rate. The general lower age limit of fluvial deposits is in the range of 1000 years (Fuchs *et al.* 2010; Preusser *et al.* 2016), and can be as low as in the range of 10 years (Olley *et al.* 1998; Wallinga *et al.* 2010). Luminescence signals from feldspar IRSL are brighter than signals from quartz OSL, but the sensitivity of quartz OSL signals is usually increased, when sediments experienced multiple reworking cycles of erosion, transportation and deposition during their geomorphological history (Pietsch *et al.* 2008). This is the case for sediments from middle or lower reaches of fluvial catchments, or at hillslopes with gradual downslope sediment transport including temporary on-slope sediment deposition. In contrast, for sediments from recently eroded bedrock with a very limited number of reworking cycles, quartz OSL sensitivity is often very low and therefore challenging for luminescence dating. This is especially the case in fluvial and hillslope systems of geomorphologically active high mountain environments or formerly glaciated terrains, where recently exposed bedrock is eroded and included into the process of erosion for the first time. Studies from the Himalayas (Blöthe *et al.* 2014; Jaiswal *et al.* 2008; Spencer and Owen 2004), the European Alps (Klasen *et al.* 2016) or formerly glaciated Fennoscandia (Alexanderson and Murray 2011; Bøe *et al.* 2007) illustrate the low quartz OSL sensitivity and its challenges for luminescence dating (see also Chapter 6).

As outlined in Chapter 1, a distinct difference between the mineral characteristics of quartz and feldspar is the phenomenon of anomalous fading, an unstable luminescence signal leading to signal loss and therefore age underestimation, frequently observed for feldspar (Wintle 1973). Anomalous fading is independent from the geomorphological history of the sediment and therefore, hillslope and fluvial processes have no positive influence on this phenomenon. However, there are suggestions to correct for anomalous fading, but these approaches to overcome anomalous fading are still under debate.

7.2.2 Signal resetting

Resetting of the luminescence signal by daylight exposure during the process of sediment reworking is a basic prerequisite for OSL and IRSL dating of sediments. If not sufficiently exposed to daylight, the 'luminescence clock' in the quartz and feldspar grains will not be reset to zero and the age of the sediment will be overestimated (see Chapter 1 for further details).

Under favourable conditions, luminescence signals from quartz and feldspar minerals are sufficiently bleached by daylight exposure within seconds and minutes, with quartz OSL bleaching faster than feldspar IRSL (Godfrey-Smith *et al.* 1988; and see Chapter 1 Section 1.3.2), demonstrated on modern fluvial deposits and their residual doses (Fuchs *et al.* 2005; Vandenberghe *et al.* 2007).

Sediments from hillslope and fluvial environments are commonly prone to insufficient bleaching (Jain *et al.* 2004), because various factors involved in the process of sediment reworking often hamper prolonged and direct daylight exposure. This is because:

1. **Mass movements, soil creep and fluvial processes** are dominantly responsible for sediment reworking on hillslopes. Downslope transport of material by mass movements (e.g. falling, sliding) often takes place as a rigid block (e.g. landslide) and only the outer layer of the block is exposed to daylight, whereas the inner sediment of the block is shielded from any daylight. In case of soil creep (e.g. solifluction), daylight exposure of the downslope-transported material is also limited to the surface and most of the sediment below the surface is shielded from daylight. In contrast, sheet, rill and gully erosion – including rainsplash as the preceding erosion process – commonly transport the material as disintegrated sediment downslope and individual mineral grains can be exposed

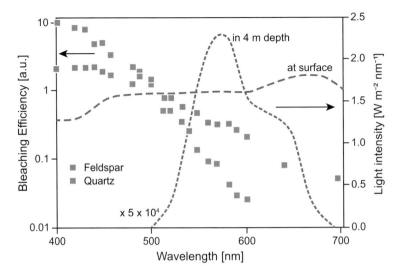

Figure 7.3 Bleaching efficiency (a.u. = arbitrary unit) of quartz and feldspar in turbid water. Dotted line indicates the light spectrum at the water surface, and the dashed line the spectrum at 4 m depth (latter data multiplied by 5 × 10⁴ due to scaling reasons). The data suggest that feldspar is bleaching better below a 4 m turbid water column, but due to the low light intensity in 4 m depth of turbid water, no significant bleaching will take place for either mineral (modified from Wallinga 2002).

to daylight. Especially, sheet and rill erosion commonly transport the sediment downslope gradually, with multiple erosion, transportation and deposition cycles, and each reworking process increases the probability of daylight exposure. However, gully erosion, for example, can limit the time of sediment exposure to daylight, because downslope sediment transport and foot-slope colluviation often takes place in short and single events.

2. The **water column** above the fluvially transported sediments attenuates the light and changes its spectral composition, reducing the bleaching efficiency of daylight exposure (Berger, 1990; Berger and Luternauer; 1987). Next to the depth of the water column, bleaching efficiency is reduced by suspended sediment caused by water turbidity (Ditlefsen, 1992; Fig. 7.3). Therefore, the mode of fluvial sediment transport and the transported grain sizes affect the bleaching efficiency of the mineral grains. Smaller grain sizes (silt) are usually transported in suspension and are therefore closer to the water surface, whereas coarser grain sizes (sand) are more likely to be transported as bedload or in saltation beneath a thicker water column. However, empirical studies indicate that the sand fraction is generally better bleached than the silt fraction (Fu *et al.* 2015; Schielein and Lomax, 2013; Vandenberghe *et al.* 2007) and within the sand fraction coarse sand is generally better bleached than fine sand (Olley *et al.* 1998; Truelsen and Wallinga, 2003; Wallinga, 2002). Reasons for this grains size dependency on the bleaching behaviour are not clear, but sand tends to be transported as single grains, whereas smaller grains are prone to coagulate and to

build aggregates, where the inner grains of an aggregate are shielded from daylight exposure. Independent from grain size, mineral coatings of iron, manganese or carbonate, for example, can also hamper mineral bleaching.

3. **Prolonged sediment transport** increases the probability of multiple sediment reworking and daylight exposure. For river systems, there is a positive correlation between bleaching efficiency and transport distance (Fig. 7.4), with insufficiently bleached sediment most likely in the upper reaches of a river and better bleached sediments further downstream (e.g. Schielein and Lomax, 2013; Stokes *et al.* 2001). However, at river confluences, tributaries with short-travelled sediment load can add insufficiently bleached sediment to the major trunk with its long-travelled and dominantly well-bleached sediments. For hillslope environments with generally short transport distances, sediments are prone to insufficient bleaching, but gradual downslope transport and multiple erosion, transportation and deposition cycles allow for daylight exposure, compensating for short-distance transport. In all cases of short-distance sediment transport the probability of transport during solely unfavourable conditions (e.g. at night) is increased.

4. **Fluvial processes vary in their mode, magnitude and frequency**, affecting the probability of mineral bleaching during sediment transport significantly. High-energy events like gully erosion on hillslopes and flash floods in upper reaches of fluvial catchments transport large quantities of sediment in turbid water in short periods of time, leading to limited daylight exposure. During floods, river bank erosion and undercutting destabilises the river bank, causing mass failure like e.g. slumping. The inner material of the collapsed sediment block is shielded from light and hampers resetting of the luminescence. This is also the case under periglacial conditions, when frozen chunks of sediment are eroded and transported without disintegrating.

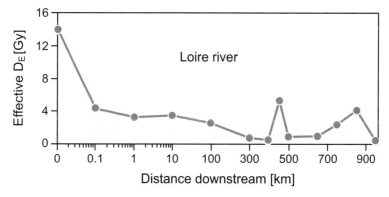

Figure 7.4 Bleaching efficiency for fluvial channel sediments from the Loire River, France. With increasing sediment transport distance, residual luminescence signals become smaller, thus the effective D_E from the unbleached luminescence signal is decreasing (modified from Stokes *et al.* 2001).

Even though sediments from hillslope and fluvial environments are prone to insufficient bleaching due to the above-mentioned reasons, luminescence dating has in many cases been successfully applied to these sediments. Besides fast bleaching of the OSL signal and multiple sediment reworking cycles increasing the probability of daylight exposure, mineral grains are frequently exposed to daylight by bioturbation and soil mechanical processes after sediment deposition, until the grains are covered through further sediments (Berger and Mahaney 1990; Fuchs and Lang 2009). However, to successfully apply luminescence dating to sediments from hillslope and fluvial environments, it is necessary to detect and possibly correct for insufficient bleaching, otherwise age overestimation of hundreds or thousands of years is possible (Jain *et al.* 2004; Wallinga 2002).

7.2.2.1 Detection of insufficiently bleached sediments

Due to the individual bleaching history of the mineral grains during the last process of sediment reworking, insufficiently bleached sediments commonly contain a mixture of well and poorly bleached mineral grains with different residual luminescence signals. Independent age control theoretically can identify insufficiently bleached sediments, but this assumes age control is available and correct. Alternatively, the measurement of residual luminescence signals from modern analogues provides first information on whether there is insufficient bleaching in a given environment. However, it is questionable if modern analogues represent the same mineral characteristics and bleaching history. More straightforward approaches to detect for insufficient bleaching are based on the analyses and comparison of (a) different luminescence signals, or (b) the degree of bleaching of individual mineral grains.

1. Because quartz OSL bleaches faster than feldspar IRSL, non-identical ages derived from both minerals indicate insufficient bleaching (Murray *et al.* 2012). This is demonstrated in a study on modern flood sediments from the Elbe River catchment in Germany, with up to one order of magnitude higher feldspar IRSL D_E values than derived from quartz OSL (Fuchs *et al.* 2005). In this study, the quartz OSL is also not completely zeroed, but the low luminescence residuals would only affect young fluvial deposits.

    Another indication for insufficiently bleached sediments is represented by non-identical ages derived from different grain sizes. Fluvial sediments from the Murrumbidgee River, Australia yielded different D_E values for different grain sizes from the sand fraction (Olley *et al.* 1998). Likewise, insufficiently bleached sediments from a flooding event of the Elbe River, Germany were also indicated by different D_E values derived from the silt and sand fractions (Fuchs *et al.* 2005). But again, residual luminescence signals are also evident for the better bleached coarser grain fraction. For colluvial sediments from a hillslope in Romania, Kadereit *et al.* (2006a) identified insufficient bleaching by non-identical D_E results from quartz sand and feldspar silt, but this discrepancy might also be caused by the different bleaching properties of the minerals quartz and feldspar.

    Different components of luminescence signals with their specific bleaching properties can also be used to identify insufficiently bleached

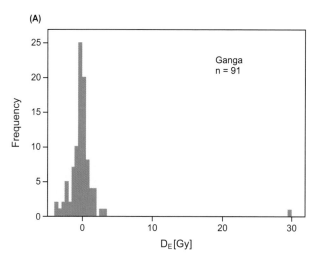

(A)

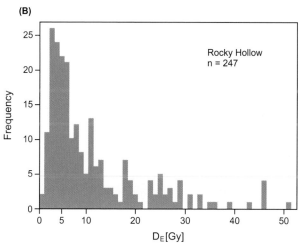

(B)

Figure 7.5 D_E estimated for quartz single grains on modern fluvial sediments from (A) the Ganga River, India, and (B) an ephemeral stream in Nebraska, USA. The tight D_E distribution of the Ganga sample indicates a well-bleached sample, whereas the sample from Nebraska is characterised by a wide D_E distribution, an indication of insufficient bleaching (modified from Jain *et al.* 2004 and Rittenour 2008, with data from Hanson 2006).

sediments (see Chapter 1). In the case of quartz OSL, the luminescence signal is composed of a fast, a medium, and several slow components, with the fast component bleaching most rapidly (Bailey, 2000; Bailey *et al.* 1997; Bulur, 1996). Thereafter, insufficient bleaching is indicated by non-identical ages derived from different luminescence components, demonstrated for sand-sized quartz on colluvial sediments from Oxfordshire, UK, as well as on fluvial sediments from various European rivers (Singarayer *et al.* 2005). Feldspar IRSL is also characterised by different bleaching properties of their signal components. Kadereit *et al.* (2010) used this fact to detect insufficient bleaching in the silt fraction of colluvial sediments from SW Germany.

2. Insufficiently bleached sediments are commonly characterised by a mixture of differently bleached mineral grains, with every individual grain showing a different D_E value, resulting in a wide and often positively

skewed D_E distribution. In contrast, well-bleached sediments with almost all mineral grains zeroed during the last process of sediment reworking are indicated by a tight D_E distribution. Figure 7.5 illustrates the D_E distribution of well-bleached fluvial sediments from a modern point bar of the Ganga River, India, indicated by a tight D_E distribution (Jain *et al.* 2004), whereas insufficiently bleached fluvial deposits from an ephemeral river in the Great Plains, USA are characterised by a wide D_E distribution (after Hanson, 2006).

The D_E measurement of single grains to detect insufficient bleaching is very time consuming and requires specialist luminescence equipment for measurement (Duller *et al.* 1999). As an alternative, small aliquots with a limited number of mineral grains (commonly less than 100–200 grains per aliquot) can be measured to detect insufficient bleaching. This is because for quartz OSL, only a limited number of grains (0.5–20%) emit sufficient luminescence for D_E determination and therefore small aliquots of *c.* 100–200 grains are nearly equivalent to single-grain measurements. However, the D_E measurement of single grains or small aliquots to detect insufficient bleaching is restricted to the sand fraction (Duller 2008). Even though insufficient bleaching is generally well documented in a wide D_E distribution (Fuchs *et al.* 2007), pedoturbation and microdosimetry issues also need to be considered as possible causes for a large D_E scatter (e.g. Bateman *et al.* 2003; Kalchgruber *et al.* 2003; Murray and Roberts 1997). Therefore, for identifying incomplete bleaching, other statistical parameters such as the skewness and the kurtosis of a D_E distribution may also be considered (e.g. Bailey and Arnold 2006).

Luminescence ages derived from insufficiently bleached sediments overestimate the age of the last process of luminescence signal resetting and must therefore be interpreted as maximum ages of sediment deposition. True depositional ages can be derived, if only well-bleached grains for D_E determination are considered.

7.2.2.2 D_E determination of insufficiently bleached sediments

Insufficiently bleached sediments are characterised by a wide and often positively skewed D_E distribution (Fig. 7.5b), with the lowest D_E values representing the best bleached grains. To extract the true D_E values, a number of statistical approaches are available, which all rely on the idea to isolate the best bleached D_E population from the lower end of the D_E distribution, generated by single grain or small aliquot measurements (see Chapter 1 Section 1.3.7).

Olley *et al.* (1998) used the lowest 5% of D_E values from a D_E distribution of a 70-year-old fluvial sediment from the Murrumbidgee River, Australia, receiving an age of *c.* 66 ± 7 years. The threshold of 5% is applicable for their specific site in Australia, but difficult to transfer to other study areas. In contrast, Fuchs and Lang (2001) suggested a sample specific threshold to separate well bleached from insufficiently bleached D_E values. They successfully applied their technique to Holocene fluvial sediments from the Assoposs River, Greece, as well as to colluvial sediments from adjacent hillslopes of the same region (Fuchs *et al.* 2004). Following their approach, the best achievable value for D_E precision is obtained from bleached and irradiated sample material and this sample specific value is then used to separate well-bleached from insufficiently bleached natural D_E values.

As outlined in Chapter 1, the minimum age (MAM) and finite mixture (FMM) models by Galbraith *et al.* (1999) and Galbraith and Green (1990) are widely used statistical approaches to derive D_E values from insufficiently bleached sediments, frequently applied to sediments from hillslope and fluvial environments (e.g. Bartz *et al.* 2017; Colarossi *et al.* 2015; Fuchs *et al.* 2014). Rodnight *et al.* (2006) applied MAM and FMM to insufficiently bleached sediments from paleochannels of the Klip River, South Africa. They rated the FMM as more reliable for their purpose, because MAM often underestimated the true D_E values due to its sensitivity to low D_E outliers. In comparison with radiocarbon ages the FMM model additionally gave more accurate results. Arnold *et al.* (2009) suggested a revised, un-logged approach of MAM for very young and modern-age samples and successfully applied it to young fluvial deposits from an arroyo system located in the southwest of the USA. In a further step, improving the precision and accuracy of luminescence-based chronologies, Bayesian modelling (see Chapter 3 for details) is increasingly applied to establish reliable age–depth profiles for fluvial deposits (Arnold and Roberts, 2009; Guerin *et al.* 2015), and in combination with bootstrap likelihoods, Cunningham and Wallinga (2012) were able to incorporate OSL data from insufficiently bleached samples into Bayesian modelling, which increased the coherence of their chronology established for floodplain sediments from the Waal River, The Netherlands.

7.2.3 Dose rate determination

The dose rate D_R builds up the luminescence signal and originates from natural low-level radioactivity (see Chapter 1 for details). Throughout the period of sediment storage, the dose rate D_R is generally considered as constant, but this is often a simplification. Especially changing water contents can lead to temporal variations of the D_R, because (a) water attenuates radiation, (b) water can change the chemical composition of the sediments and therefore may change the concentration of radionuclides and (c) may be responsible for radioactive disequilibria; especially for hillslope and fluvial environments, where water is a major geomorphic agent, variable water conditions and its consequences on the dose rate need to be considered.

1. The water content of the sediment body directly influences the dose rate: a 1% difference in water content results in *c.* 1% difference in luminescence age. Therefore, accurate estimates of the water content for the entire period of sediment storage are of crucial importance, noting that the present water content might not be representative for past conditions. This is particularly true for colluvial and fluvial sediments, where fluctuating water tables and variable fluvial activities of hillslope and fluvial environments directly affect the water content of the sediments. A profound understanding of the geomorphological setting and its paleoenvironmental history, as well as considering sedimentological-pedological aspects of the studied sediments (e.g. redoximorphic features, slickensides) is therefore essential for a sound estimation of former water contents. Based on grain size characteristics, Nelson and Rittenour (2015) suggested a model to calculate mean soil water contents especially for semi-arid environments. Their model used

grain-size dependent water retention curves in combination with regime maps of soil moisture in order to model average water contents. When applied to fluvial sediments from the Kanab Creek in Utah, USA, the model yielded OSL ages in agreement with radiocarbon ages. However, it is only applicable for regions, where the above-mentioned data and maps are available. For carbonate-rich sediments, Nathan and Mauz (2008) suggested a numerical model for estimating dose rate changes over time, which takes into account carbonate cementation and its impact on pore sizes of the sediment, which in turn directly affect the potential water content. The dose rate model assumes linear increase of carbonate mass and linear decrease of water mass in pores, resulting in differences between modelled and conventional OSL ages of up to 15%.

2. Disequilibria in the ^{238}U decay chain are well known and were described for fluvial and colluvial sediments by Krbetschek *et al.* (1994), who used alpha- and gamma spectrometry to detect the disequilibria. They explained the disequilibria by leaching and removal processes of the mobile nuclides ^{234}U, ^{226}Ra and ^{222}Rn from the ^{238}U decay chain. The consequences of radioactive disequilibria in the ^{238}U decay series on the dose rate of fluvial sediments are illustrated by Olley *et al.* (1996), measuring nuclide concentrations of modern fluvial sediments from rivers in southeastern New South Wales, Australia. Most of these sediments exhibited radioactive disequilibrium in the ^{238}U decay chain, but the deviation between true and calculated dose rate was less than 3% and therefore comparatively small. For carbonate-rich sediments, Nathan and Mauz (2008) provided a numerical model for dose rate estimation, taking into account geochemical sediment alteration and the associated radioactive disequilibria due to carbonate cementation. Even for sediments in present radioactive equilibrium, past disequilibria need to be considered, because under geochemical closed system conditions, a formerly existing radioactive disequilibrium will return to equilibrium after a specific time, depending on the initial degree and the half-life of the responsible nuclide. A typical scenario of radioactive disequilibria and the time required for re-equilibration is illustrated by Olley *et al.* (1996). In their specific example, no evidence of a former disequilibrium would be detected after *c.* 11 ka of initial disequilibrium and the resulting luminescence age calculated with a dose rate based on equilibrium conditions would be overestimated by *c.* 12%. In consequence, potential past equilibria should be considered in D_R estimation and should be modelled adequately (Preusser and Degering, 2007).

Next to temporal variations of the dose rate D_R, mainly controlled by changing sediment and soil moisture conditions, spatial variations of D_R are common phenomena due to sedimentological and mineralogical heterogeneities of the sediment body and its associated variation in radionuclide concentrations. In this context, the penetration range of alpha (*c.* 20 μm), beta (*c.* 2 mm) and gamma (*c.* 30 cm) radiation is of crucial importance

Figure 7.6 Example of stratified fluvial deposits with alternating layers of gravel, sand and loam. The sedimentological and mineralogical heterogeneity of the sediment body is associated with variations in radionuclide concentrations, resulting in different dose rates (D_R) for every sediment layer. To avoid dose rate heterogeneities for luminescence dating, the luminescence sample is taken in the centre of a homogeneous sediment layer. Photo shows a stream bank in the Eifel Mountains, Germany (modified from Stolz *et al.* 2012).

to understand the challenges associated with dosimetry and microdosimetry issues (Wagner 1998). This is especially true for stratified sediment bodies common for fluvial environments (Fig. 7.6), where the gamma radiation field of *c.* 30 cm radius can comprise different sediment layers with their individual mineralogy and radionuclide concentration, resulting in a heterogeneous gamma field (Kenworthy *et al.* 2014; Chapter 2 Section 2.2). On shorter penetration scales, heterogeneity of beta radiation frequently causes variation in microdosimetry (Kalchgruber *et al.* 2003; Mayya *et al.* 2006). This is especially important when dealing with single grain D_E analyses (Duller 2008; Mayya *et al.* 2006).

In addition to environmental low-level radioactivity, cosmic radiation contributes to the total dose rate. For fluvial deposits in high mountain ranges like the Himalayas or the Hajar Mountains in Oman, up to 11% of the total dose rate comes from the cosmic dose rate (Fuchs and Bürkert 2008; Owen *et al.* 2009). This may be exacerbated in situations where terrestrial radioactivity is low, e.g. in carbonate dominated environments. Fuchs *et al.* (2004) reported cosmic dose rates contributing up to 15% to the total dose rate for colluvial deposits in Greece. For cosmic dose rate estimation in accreting hillslope or fluvial contexts, the depth of sediment overburden can change significantly. Therefore, the average overburden depth during sediment storage needs to be estimated rather than using the overburden depth at the time of sampling. In practice, assuming e.g. constant sedimentation rates, the overburden depth at the time of sampling should be divided in half to account for increasing sediment overburden above the sampling site. If models other than linear sedimentation are available or more realistic then the cosmic dose rate calculations need to be adapted to take this into account.

As demonstrated, dose rate estimation can be challenging in hillslope and fluvial environments, because geomorphological and hydrological processes are particularly active in these environmental settings and therefore have a direct impact on dose rate estimations. Thus radioactive disequilibria, fluctuating water content or changes in sediment overburden need to be considered. An alternative approach for samples, where individual dose rate factors of water content, disequilibria or sediment overburden are difficult to estimate, is given by the time-consuming isochron method. Li *et al.* (2008) suggest an isochron method using K-feldspar extracts in a range of different grain size fractions to overcome changes in dose rate over time. This approach is based on the fact that K-feldspar has a significant internal dose rate, which is independent from possible external dose rate changes. Applying this method to Holocene fluvial deposits with a changing dose rate history from the Sala Us River in the Mu Us desert in China, stratigraphically correct ages in accordance with independent age control were obtained. However, a prerequisite for this approach are similar K concentrations in each mineral grain, or K concentrations need to be measured for every individual grain. The same is true for anomalous fading, where fading rates between the feldspar grains have to be the same or need to be measured individually.

7.2.4 Sampling strategies

The identification of appropriate sediments for luminescence dating while sampling in the field requires profound geoscientific knowledge of the different landforms, sediments and geomorphological processes associated with hillslope and fluvial environments. In this respect, the identification of sediments with the maximum likelihood of prolonged daylight exposure to ensure sufficient luminescence signal resetting is crucial. The bleaching potential of different landforms and their associated sediments from hillslope and fluvial environments is highlighted in Figures 7.1 and 7.2. As illustrated, deposits which experienced multiple reworking cycles and were transported over longer distances have the highest bleaching potential, while sediments transported in heavily sediment-laden turbid water, like e.g. slackwater deposits should be avoided. Furthermore, the sediments need to be checked for post-sedimentary disturbance and mixing, recognised by e.g. pedoturbation features and fluctuating water contents indicated by e.g. chemical sediment alteration like redoximorphic features should be documented adequately.

The above-mentioned challenges when dating fluvial and hillslope sediments require some considerations over and above those outlined in Chapter 2 with respect to the sampling strategy:

1. The mineral quartz is favoured over feldspar for D_E estimation, because the luminescence signal of quartz bleaches faster than the feldspar signal. In addition, quartz does not suffer from anomalous fading, a common problem of feldspar. In the case of dim quartz and sediments older than the quartz age range of *c.* 100–150 ka, feldspar is preferred.

2. For colluvial and fluvial sediments, the bleaching efficiency for sand size mineral grains is better than for grains from the silt fraction, thus larger grains are favoured over smaller grains. Due to dosimetry issues, the grain size range should be minimised (e.g. 90–125 μm) and grains larger than 300 μm should be avoided.

3. To avoid larger dose rate inhomogeneities often present in stratified fluvial deposits, luminescence samples should be taken from the centre of a sedimentological homogeneous layer of *c.* 60 cm thickness. In this context, sampling less than 30 cm below surface should be avoided because of a heterogeneous gamma radiation field (see Section 7.2.3), and

due to sediment mixing issues by pedoturbation. In the case of strong sediment heterogeneities, *in situ* dose rate measurements with e.g. a portable gamma spectrometer is recommended.

7.3 APPLICATION OF LUMINESCENCE DATING IN HILLSLOPE AND FLUVIAL CONTEXTS

Luminescence dating has been applied to a wide range of hillslope and fluvial environments, situated in different geographic settings and climate regions (e.g. Fuchs and Lang 2009; Rittenour 2008). These applications on various spatial and temporal scales have added immensely to our understanding of past environmental conditions and landscape evolution, answering questions about geomorphology, climate and human impact. The following case studies illustrate how luminescence dating has been applied in different hillslope and fluvial environments, highlighting the potential and challenge of this technique in fluvial research.

7.3.1 Hillslope environments (colluvium)

Hillslope deposits are called colluvium and represent the sediments produced by the geomorphological processes operating on hillslopes. Deposits generated by mass movements and soil creep are usually unsuitable for luminescence dating because of their dominantly non-bleached sediment character. Sheet, rill and gully deposits are usually transported as disintegrated sediment and individual mineral grains and are likely to be better bleached. In addition, the bleaching probability by daylight exposure is increased by the gradual downslope transport process with temporary on-slope sediment deposition, rendering colluvial sediments transported by sheet, rill and gully erosion commonly suitable for luminescence dating (Fuchs and Lang 2009).

7.3.1.1 Past climate change

Past climate change and its impact on hillslope processes and consequently on colluviation was the subject of luminescence dating in the early days of optical dating (Botha *et al.* 1994; Wintle *et al.* 1993, 1995a, 1995b). In these first studies on colluvial deposits in South Africa, Pleistocene periods of soil formation and colluviation were correlated with Oxygen Isotope Stages for the last 110 ka, using *inter alia* IRSL feldspar measurements. While fine-grain measurements did not allow to detect for insufficient bleaching, Wintle *et al.* (1995b) were able to explain IRSL age overestimations for some samples by measuring coarse grain feldspar extracts and identifying insufficient bleaching due to the broad D_E distribution.

A sequence of Pleistocene sandy colluvial deposits with intercalated paleosols was studied in KwaZulu-Natal, South Africa by Clarke *et al.* (2003), investigating periods of geomorphic stability and instability on hillslopes in relation to climate fluctuation of the last 100 ka. The chronology was established by a series of coarse grain IRSL feldspar measurements and their correctness was confirmed by ^{14}C ages up to *c.* 40 ka, which were derived from organic matter of the paleosols. It was thus demonstrated that colluvial sediments can be dated using feldspar grains and that incomplete bleaching is not necessarily an issue in these sediments.

The timing of sheet, rill and gully erosion was also investigated in tropical Tanzania by Eriksson *et al.* (2000), using colluvial deposits eroded from heavily degraded hillslopes. Based on OSL measurements of the coarse grain quartz fraction and on the analysis of D_E distributions, insufficient bleaching was observed in all samples. In order to avoid age overestimation, the best bleached D_E population from the lower end of the D_E distribution was chosen, applying MAM (see Chapter 1) after Galbraith and Laslett (1993). These measurements were performed on small aliquots with a limited number of grains per D_E measurement, avoiding strong averaging effects between well and insufficiently bleached grains. Finally, two major periods of erosion were identified, one representing the late Pleistocene, characterised by a climate shift from dry to wet conditions, and a more recent period of the last hundreds of years, most probably triggered by human activities and the clearance of natural vegetation cover.

Because in hillslope environments sediments are prone to insufficient bleaching, aliquot size and the number of mineral grains measured per aliquot is crucial. The dependency of D_E value on aliquot size was investigated by Duller (2004) on colluvial deposits from Tasmania. With decreasing aliquot size, the D_E scatter and skewness measured on coarse grain quartz increases and insufficient bleaching becomes detectable (Fig. 7.7). This would not be possible for large aliquots (*c.* 1000 grains per aliquot) due to the averaging effect of well and insufficiently bleached grains. When reducing the D_E determination to single-grain measurements, the averaging effect between well and insufficiently bleached grains can be excluded and the lowest D_E population from the D_E distribution might represent the best-bleached quartz grains. However, single-grain measurements are very time consuming and a large number of grains need to be measured, because of the generally low brightness of single quartz grains. Sediments from Australia are highly suitable for single-grain measurements due to their generally bright luminescence signals, but sediments from other regions around the world are often dim and therefore less favourable for single-grain

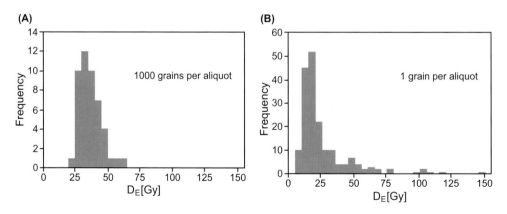

Figure 7.7 Equivalent dose (D_E) determination of quartz extracts from colluvial deposits from Tasmania: A) *c.* 1000 grains per aliquot, B) 1 grain per aliquot. With decreasing aliquot size, the D_E scatter and skewness increases and insufficient bleaching becomes detectable. This would not be possible for large aliquots, due to the averaging of well and insufficiently bleached grains. The lowest D_E population from the D_E distribution might represent the best-bleached quartz grains (modified from Duller 2004).

HANDBOOK OF LUMINESCENCE DATING

measurements and in most cases, less than 5–10% of the measured grains are sufficiently bright for D_E determination. Under these circumstances, small aliquots with a limited number of grains can be interpreted as single-grain measurements.

Even though hillslope sediments may be insufficiently bleached, in many cases the process of colluviation enables zeroing of the luminescence signal. This is demonstrated in a study on past hillslope processes and their sensitivity to Early Holocene climatic oscillations, where colluvial deposits from a small-scale hillslope environment in northeast Germany were investigated (Dreibrodt et al. 2010). In this study, D_E estimates are based on small aliquot OSL measurements of the coarse grain quartz fraction and due to the skewness and broadness of the D_E distribution, the colluvial deposits are characterised as well bleached. This interpretation is confirmed by stratigraphic information and independent age control, demonstrating that the process of colluviation on a slope length of less than 100 m is able to reset the luminescence signal to zero.

Early Holocene colluviation triggered by climate deterioration is also the conclusion of a study on loess-derived colluvial deposits in northwest England by Vincent et al. (2011). Even though the fine-grain quartz fraction (4–11 μm) and its associated difficulty to detect for insufficient bleaching was used for D_E determination, stratigraphic evidence and for some samples, D_E results from OSL coarse grain quartz indicate sufficient bleaching. Furthermore, a wide range of dose rates (1–3 Gy/ka) was observed in the study of Vincent et al. (2011), which demonstrates the importance of accurate dose rate measurements for every individual luminescence sample, even though field observations show homogeneous sediment characteristics and therefore suggest similar dose rates.

7.3.1.2 Past human-induced soil erosion

A major factor for Holocene soil erosion is the introduction of agriculture and associated deforestation to gain arable land. In Europe, this process started in the Neolithic period, when humans began building permanent settlements, growing crops and domesticating animals. The widespread deforestation triggered soil erosion and resulted in building up thick colluvial bodies at foot-slope positions. These sediments represent important sedimentary archives in geoarchaeological research for reconstructing past human agricultural activities. In this field of research, luminescence dating has intensively been used for establishing high-resolution chronologies to gain a better understanding of the temporal evolution of the human impact on the environment (e.g. Fuchs et al. 2004, 2010; Gerlach et al. 2012; Houben et al. 2012; Kadereit et al. 2006b; Lang 2003; Notebaert et al. 2011).

Colluvial deposits caused by man-induced soil erosion are hard to distinguish from colluviation caused by natural processes. Even though ceramics or other artefacts often imbedded in colluvial sediments indicate anthropogenic causes, detailed chronostratigraphic information is needed to correlate known cultural activities with periods of colluviation. In a detailed study about the history of Holocene soil erosion in Greece, Fuchs et al. (2004) investigated several colluvial profiles at foot-slope positions, establishing high-resolution OSL chronologies. Because of the short sediment transport distances and due to extreme rainfall events typical for the Mediterranean, the OSL samples were tested for insufficient bleaching by measuring coarse grain quartz on small aliquots and by analysing the D_E distributions (Fuchs and Wagner 2003). The resulting chronologies indicate a strong dependency between colluviation and periods of agricultural activities, starting already in the Early Neolithic of

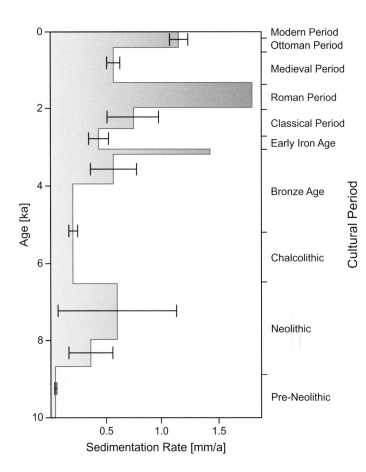

Figure 7.8

Sedimentation rates of colluvial deposits from Greece, derived from high-resolution OSL chronologies. There is a positive correlation between high sedimentation rates and agriculturally active periods, the latter leading to soil erosion and colluviation. These agriculturally active periods started in the Early Neolithic of the 7th millennium BCE (modified from Fuchs et al. 2004).

the 7th millennium BCE (Fig. 7.8). Man-induced soil erosion and colluviation has a strong impact on pedogenesis and topography, especially in European loess landscapes with a long lasting agricultural history. Kühn et al. (2017) demonstrate in their study from a loess landscape north of Frankfurt / Germany, how first pedogenesis was controlled by climate, then, with the onset of agriculture, dominantly controlled by human activities, leading to soil truncation and colluviation. This pedogenetic change was OSL dated to c. 4.7 ka and was accompanied by topographic levelling due to erosion at upslope and colluviation in foot-slope positions (Fig. 7.9), which was also reported from other loess landscapes with early human occupation (e.g. Kadereit et al. 2010). The temporal evolution of colluviation was established by a combination of OSL quartz coarse grain and ^{14}C ages, showing a strong overestimation of the ^{14}C ages in comparison to the OSL ages by 2–4 ka (Fig. 7.9). This highlights a common problem in ^{14}C sediment dating, where old organic material is incorporated in the sediment body and therefore does not represent the time of sedimentation (e.g. Lang and Hönscheidt 1999). In the study by Kühn et al. (2017), an independently dated tephra layer supports the correctness of the OSL ages. In addition, micromorphological analyses were carried out, indicating no signs of post-depositional pedoturbation (Bateman et al. 2003), further supporting the robustness of the OSL chronology. This example demonstrates the importance of combining analytical tools, such as micromorphology and OSL dating, for

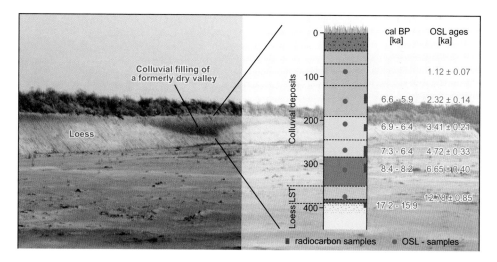

Figure 7.9 Colluvial filling of a formerly dry valley due to human-induced soil erosion, leading to topographic levelling in a loess landscape north of Frankfurt, Germany. The temporal evolution of colluviation was established by a combination of OSL and ^{14}C ages, with ^{14}C ages overestimating the process of colluviation due to the incorporation of old organic material into the reworked sediment. This is a common problem in ^{14}C dating (modified from Kühn et al. 2017). The correctness of the lowermost OSL age with 12.78 ± 0.85 ka is supported by independent age control of the tephra from the Laacher See eruption (LST), c. 13 ka cal BP.

identifying (post-) sedimentary processes, possibly influencing luminescence dating and its interpretation.

7.3.1.3 Colluvium associated with soil creep, rock falls and fault activities

As already stated, sediments associated with mass movements and soil creep are usually unsuitable for luminescence dating, because these geomorphic processes generally hinder sufficient daylight exposure of the sediment grains. Recent developments in rock surface dating by luminescence would theoretically allow to date e.g. boulders directly (see Chapter 11; Greilich and Wagner 2006; Sohbati et al. 2012), but so far the methods are still in an experimental stage of research. Until now, rockfalls have been dated indirectly, by dating the sediments associated with boulder deposition. Sohbati et al. (2016) follow this approach, indirectly dating boulders from a hillslope near Christchurch, New Zealand, to discuss possible seismic or human causes of these rockfall events. In this context, colluvial sediments underlying the boulders predate boulder accumulation, and the colluvial wedges upslope postdate the rockfall event (Fig. 7.10). The colluvial deposits were dated with quartz OSL and feldspar IRSL, using the 40–63 μm grain-size fraction. Because of the large number of grains per aliquot, no statistical analysis was possible to detect insufficient bleaching. Instead, the different bleaching characteristics for different types of luminescence signals (quartz OSL versus feldspar $pIRIR_{50}$ *vs.* $pIRIR_{290}$) were used to test for insufficient bleaching. Finally, the results from OSL and IRSL, as well as from both IRSL signals were in excellent agreement and consistent with the stratigraphy, indicating sufficient daylight exposure of the sediments before deposition. This study demonstrates that it is possible to

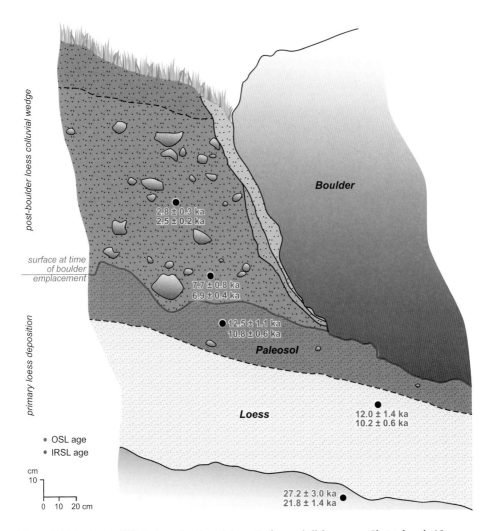

Figure 7.10 OSL and IRSL dating of colluvial deposits from a hillslope near Christchurch, New Zealand. Colluvial sediments underlying the boulders predate boulder accumulation but the colluvial wedges upslope postdate the rockfall event. Within errors, OSL and IRSL ages are in good agreement and are consistent with the stratigraphy, indicating sufficient bleaching of the sediments before deposition (modified from Sohbati *et al.* 2016).

date rockfalls by their associated sediments, under favourable conditions. However, this requires profound knowledge of both luminescence characteristics of different minerals and signals, and of stratigraphy and geomorphologic processes operating on hillslopes.

Periglacial slope deposits and their associated process of soil creep represent challenging hillslope deposits for luminescence dating because of their low probability for daylight exposure. A further problem is their generally heterogeneous sediment character, leading to possible dose rate heterogeneities. Periglacial slope deposits are characteristic for every periglacial environment, but their sediment structure is different between localities, depending upon various factors such as the lithology of the area. Hülle *et al.* (2009) investigated

Figure 7.11 Periglacial slope deposits in the Taunus low mountain range near Frankfurt, Germany. These are characterised by a heterogeneous sediment structure, grain sizes ranging from cobbles to clay, and a mixed mineralogy. To account for small-scale dose rate differences, *in situ* gamma spectrometer measurements are recommended as an additional method for dose rate determination (modified from Hülle *et al.* 2009).

periglacial slope deposits in the Taunus, a low mountain range near Frankfurt / Germany, taking samples from three different sections and depths. For D_E determination, two different grain size fractions from both quartz and feldspar were analysed. Surprisingly, the D_E scatter was relatively low, and also the skewness did not indicate insufficient bleaching of the samples, as well as the comparison between different mineral fractions, grain and aliquot sizes. In contrast, the sediment structure with a mixture of different grains sizes from cobbles to clay and a heterogeneous mixture in mineralogy, lead to small-scale dose rate differences between nearby samples. Under these circumstances, taking representative samples for dose rate measurements is problematic and *in situ* γ-spectrometer measurements are recommended as an additional method for dose rate determination, which better accounts for small-scale dose rate heterogeneities (Fig. 7.11). Hülle *et al.* (2009) also carried out mircromorphological analyses of the slope deposits, indicating post-depositional sediment mixing, which has to be taken into account when interpreting the obtained luminescence results. In conclusion, dating periglacial slope deposits is possible under certain circumstances (e.g. bright luminescence signals), but special efforts in detailed and spatially high-resolution dose rate determinations are needed. Including all uncertainties, luminescence ages derived from periglacial slope deposits often lack accuracy.

Colluvial deposits associated with past fault activities represent important archives for reconstructing paleoearthquakes. The vertical displacement of the topography leads to short distance hillslope processes and the resulting colluvial wedges are datable with luminescence techniques, post-dating the event of fault activity. Numerous studies from e.g. Iran (Fattahi *et al.* 2010), Israel (Porat *et al.* 2009), Greece (Tsodoulos *et al.* 2016) and Belgium (Vandenberghe *et al.* 2009) demonstrate successful dating of such deposits and despite the short transport distances, at least some of the mineral grains were sufficiently exposed to daylight. For more details about luminescence dating in tectonic settings, see Chapter 9.

APPLICATIONS IN FLUVIAL AND HILLSLOPE ENVIRONMENTS

7.3.2 Fluvial environments (alluvium)

Fluvial deposits associated with rivers and streams are called alluvium and represent the sediments transported and accumulated by streamflow. Within fluvial catchments, a large variety of fluvial sediments and landforms exist (Fig. 7.1), all representing their specific configuration of fluvial processes, and therefore have an impact on the suitability for luminescence dating. One major aspect when dealing with fluvial sediments for luminescence dating are their bleaching characteristics. Another major aspect is the often-heterogeneous sediment body, consisting of different grain sizes and a mixed mineralogy, which has a strong impact on the dose rate, as well as the hydrological situation of the sediments, with possible strong chemical sediment alterations over time. However, even though fluvial sediments are challenging for luminescence dating, this dating method is widely and successfully applied to sediments from fluvial environments (e.g. Rittenour, 2008; Rixhon *et al.* 2016; Wallinga, 2002), and its reliability was numerously checked with independent age control (Fig. 7.12).

7.3.2.1 Channels

Sediments from river channels are often dated by ^{14}C, but its use to establish accurate sedimentation ages and rates of geomorphological change is problematic. This might be due to incorporation of old organic material into the sediment, leading to ^{14}C age overestimation, and due to its vulnerability to ^{14}C contamination, especially for samples close to the upper ^{14}C age limit, leading to age over- or underestimation. In a study on fluvial channel fills from the Nene and Welland rivers in eastern England, Briant and Bateman (2009) present a set of paired AMS ^{14}C and OSL ages from sandy and organic rich

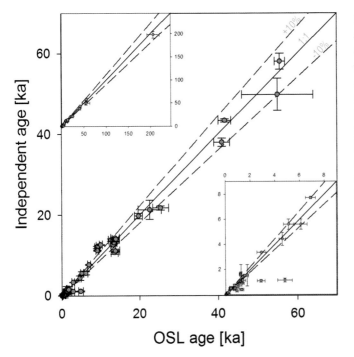

Figure 7.12 OSL ages from fluvial deposits compared with independent chronological control. Errors are stated as 1 σ and the solid line represents agreement between the OSL ages and the independent age control (data from Rittenour, 2008).

sediments, to test their reliability. In this study, significant younger ^{14}C ages were obtained in comparison to OSL ages derived from quartz coarse-grain extracts (90–125 µm), measured on small aliquots to account for insufficient bleaching. To explain the age difference from a luminescence dating perspective, possible OSL age overestimation can either result from D_E overestimation due to insufficient bleaching, or from dose rate underestimation. However, because of the general positive luminescence characteristics and the absence of any indication for insufficient bleaching with a tight and non-skewed D_E distribution, D_E overestimation seemed unlikely. The same was true for the dose rate, because only doubling the measured dose rate values would have allowed matching the OSL and ^{14}C ages, which would have been an unrealistic scenario for this study. Furthermore, the OSL ages were in good agreement with the stratigraphy, leading to the assumption that OSL dating did not overestimate, but ^{14}C dating underestimated the sedimentation ages of channel fillings due to ^{14}C contamination with younger organic material. In contrast, robust OSL age estimates from sandy channel fills could be established for the last glacial cycle.

Rittenour et al. (2005) OSL dated channel-belt deposits from the lower Mississippi valley, USA, to shed light on its fluvial landscape evolution during the last glacial cycle. Based on 69 OSL ages derived from the quartz sand fraction, OSL samples showed only little evidence for insufficient bleaching due to long distance sediment transport of up to 1000 km, as well as multiple sediment reworking. Therefore, age overestimation was only a minor problem for these samples, supported by available ^{14}C chronologies.

To gain a better understanding of river system dynamics, Rodnight et al. (2005) investigated the timing and rates of channel change and migration at the Klip River in South Africa. They sampled a sequence of scroll-bar ridges associated with an abandoned meander bend to obtain lateral migration rates for the last c. 1 ka (Fig. 7.13). Sand samples for quartz OSL dating were taken from the scroll-bar ridges and D_E measurements were carried out on small aliquots to detect insufficient bleaching. The scatter and skewness of the D_E distribution indeed indicated insufficient bleaching and a FMM (see Chapter 1) was applied to calculate ages from the best-bleach quartz grains. The obtained OSL ages were in correct chronological order for the sequence of scroll-bar ridges and an average lateral migration rate of 0.16 m/a over the last c. 1 ka was calculated.

Channel deposits and their luminescence dating also play an important role in reconstructing fault activities and their associated earthquakes. This is highlighted in a

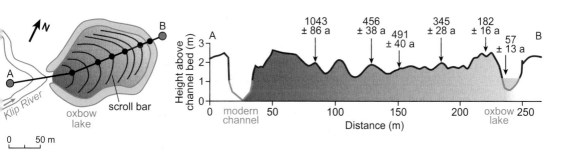

Figure 7.13 Abandoned meander and its associated scroll bars from the Klip River, South Africa, and transect across the scroll-bar sequence with OSL ages (modified from Rodnight et al. 2005).

245

study by Rockwell *et al.* (2009) at a strike-slip fault in northern Portugal, where Quaternary channels deflections and offsets were investigated in order to quantify the timing and magnitude of former fault activities. In this study, the earthquake-triggered channel deflection and offset was OSL dated using the quartz sand fraction from the channel deposits. Several metres of displacement for single rupture events were observed, with average slip rates of 0.3–0.5 mm/a for the Late Quaternary and earthquake magnitudes of up to M7. OSL dating of the channel deposits from this study was challenged by heterogeneous sediment structures, resulting in large differences in dose rate values even for closely spaced samples. To account for the small-scale dose rate variability, the dose rates from neighbouring samples were averaged and the correct stratigraphic order of the resulting OSL ages indicated that this approach was successful for this sedimentary setting.

7.3.2.2 Floodplains and terraces

Floodplains and river terraces are important fluvial landforms, with their formation dominantly controlled by tectonics or climate. The importance of these landforms for geomorphology and Quaternary research is shown in numerous studies, with luminescence dating as a major tool for establishing reliable age estimates (e.g. Blöthe *et al.* 2014; Cordier *et al.* 2012; Fuchs *et al.* 2015; Keen-Zebert *et al.* 2013; Kolb *et al.* 2016, 2017; Kolb and Fuchs 2018; Pederson *et al.* 2006; Winsemann *et al.* 2015).

For floodplain deposits from tributaries of the lower Rhine River in The Netherlands, Hobo *et al.* (2010) reconstructed sedimentation rates over the past decades to better understand floodplain dynamics in this region. Samples for luminescence dating were taken from sandy floodplain layers and, for D_E determination, quartz OSL measurements were applied to the sand fraction, using small aliquots. To account for insufficient bleaching and possible age overestimation, a measurement protocol after Wallinga *et al.* (2010) was applied. Ages as young as 8 ± 6 years could be derived, with maximum ages in 1.7 m depth of 565 ± 30 years. Sedimentation rates decreased with distance to the channel and sedimentation rates of 2–7 mm/a were calculated for distal channel areas, 3–9 mm/a for proximal areas and 9–25 mm/a on sand bars along natural levees. These ages and sedimentation rates were in good agreement with results from independent age control, demonstrating OSL dating as a reliable dating method for reconstructing floodplain sedimentation dynamics on decadal time scales. This is also true for the sand fraction deposited in the clayey distal area from the channel, where water turbidity and grain aggregation during sedimentation are thought to hinder sufficient daylight exposure, thus resetting of the luminescence signal. However, for very young OSL ages the relatively large age uncertainties need to be considered.

Even though floodplain deposits can be well bleached, in many cases these sediments show insufficient bleaching, which has to be considered especially for young sediments in the decadal age range. This is highlighted in a study by Sim *et al.* (2014), investigating sediments from a floodplain near Sydney, Australia, using single and multiple quartz grain OSL measurements. The results indicate that next to well-bleached samples, insufficiently bleached sediments show various degrees of bleaching and therefore luminescence residuals. For samples older than a few thousand years, the resulting age overestimation due to incomplete bleaching would be negligible, but for young floodplain deposits minimum age models need to be applied. Finally, results from single-grain OSL measurements, analysed with minimum age models, were in good agreement with independent age control and in stratigraphic order. Nevertheless, minimum age models applied to very young sediments should be used with care, because post-depositional sediment mixing might also be responsible for large D_E scatter.

In contrast to floodplain deposits, fluvial terraces are characterised by older alluvial deposits. Therefore, insufficient bleaching is of minor importance, because the percentage of residual luminescence signals is smaller compared to young sediments. Nevertheless, samples from fluvial terraces also need to be checked for insufficient bleaching, especially in cases where slower bleaching feldspar is used for D_E determination because of its higher age range. This is demonstrated in a study by Cordier *et al.* (2014), where fluvial terraces from the catchment of the river Moselle in Germany and France were investigated to gain a better understanding of fluvial response to climate change. To establish robust chronologies for the Middle and Late Pleistocene terraces, sandy layers were sampled and OSL ages were derived from the sand-sized quartz fraction. Due to the low D_E scatter, the samples were interpreted as well bleached and no application of minimum age models was needed, resulting in quartz OSL age estimates up to 120 ± 10 ka. Because these quartz OSL ages were already at their upper age limit, feldspar of older sediments was used, even though the application of feldspar is often hampered by anomalous fading, leading to age underestimation (see Chapter 1). Therefore, Cordier *et al.* (2014) applied an IRSL measurement procedure called post-IR IRSL, which better accounts for anomalous fading, but has the disadvantage that its luminescence signal is harder to bleach than the classical IRSL signal, making a test for a residual luminescence signal essential. Finally, after fading correction and identifying the unbleachable luminescence residuals, post-IR IRSL ages up to 155 ± 15 ka could be derived for terraces of the river Moselle, and 329 ± 35 ka for terraces of the river Sarre, a tributary to the river Moselle.

Based on late Pleistocene river terraces from the Panj River in the Pamir mountain range, Fuchs *et al.* (2014) calculated river incision rates using quartz OSL of the sand. As expected from fluvial deposits from alpine environments, the sediments were insufficiently bleached, indicated by a skewed and broad D_E distribution with the need to apply a MAM (see Chapter 1). Because of the challenging environmental setting for dose rate calculations, special emphasis was set on its determination. To check for possible radioactive disequilibrium, gamma spectrometry was applied and to avoid dose rate heterogeneities, OSL samples were taken away from sediment layer boundaries and 50 cm below the modern surface. To account for past variations in water content and to estimate realistic errors for this parameter, water saturation analyses were performed in addition to *in situ* water content measurements. This is especially important for regions with high seasonal differences in precipitation as the Pamir mountain range, and for considering regional paleoclimate differences. Based on detailed analyses of the most sensitive parameters for OSL age calculation, an average incision rate for the Panj River of 5.6 mm/a was estimated, with variations between 1.4–7.3 mm/a, depending on the structural situation of the Pamir mountain range.

7.3.2.3 Alluvial fans

Alluvial fans are mainly composed of poorly sorted sediments with a high content of clasts, representing high-energy fluvial transport systems. Luminescence dating of these sediments is therefore challenging, because the poorly sorted sediment structure leads to small-scale dose rate variations. Furthermore, due to the transport history, these sediments are prone to insufficient bleaching. Nevertheless, numerous studies from various environmental settings met these challenges and demonstrate the successful application of luminescence dating to these important fluvial landforms (e.g. Andreucci *et al.* 2014; Pope *et al.* 2008; Porat *et al.* 2010; Walker and Fattahi 2011).

Kenworthy *et al.* (2014) investigated several alluvial fans along the western front of the Lost River Range, USA, dominated by a mixture of unsorted gravels and sands, with a wide range of grain sizes and lithologies, and only thin to absent sand lenses. Because under these circumstances sampling with tubes was not possible, samples for luminescence dating were taken from the outcrops at night or below an opaque tarpaulin, sampling directly into opaque plastic bags, after removal the outer light-exposed sediments. When applying this procedure, contamination with loose and light-exposed sediments falling from above the sampling horizon needs to carefully be avoided. For OSL dating, quartz sand was extracted and measured on small aliquots. Little evidence for insufficient bleaching was observed and variations in D_E values were attributed to microdosimetry issues, a typical challenge when dealing with sediments of heterogeneous grain size distributions and lithologies. For dose rate analyses, the heterogeneity of the sediment body had to be taken into account and because *in situ* gamma spectrometry was not possible in this study, representative sample material was needed. This became all the more important after measuring individual dose rates for the sand (< 2 mm) and for the pebble fraction (1–5 cm). The results indicate a dependency of the dose rate from the grain size, with higher dose rates for the sand fraction than for the pebble fraction, the latter being mainly composed of low dose carbonates (Fig. 7.14). Finally, bulk samples with all grain sizes present in adequate proportion were taken for dose rate estimates and, as expected, the dose rates for the bulk samples ranged

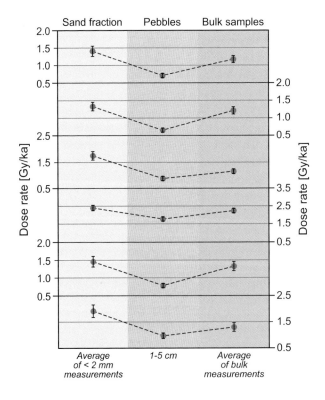

Figure 7.14 Illustration of dose rate dependency on grain size using samples taken from an alluvial fan in the Lost River Range, USA. The results indicate higher dose rates for the sand fraction than for the pebble fraction, and intermediate values for the bulk samples (modified from Kenworthy *et al.* 2014).

between the values for the sand and pebble fraction (Fig. 7.14). OSL age estimates from 4 to 120 ka were obtained and their correctness were supported by independent age control using U-series dating and a dated ash layer, as well as stratigraphic evidences. Nevertheless, the precision of the OSL age estimates was low, because of less precise dose rate and D_E estimates, due to the heterogeneity of the investigated sediment bodies.

7.3.2.4 Ephemeral river deposits

Low precipitation and infrequent surface run-off are characteristics of arid and semi-arid environments, resulting in ephemeral fluvial dynamics and their associated sediments. For luminescence dating, these deposits can be challenging, because high-magnitude low-frequency sediment transport associated with high sediment load and frequent short distance transport might lead to insufficient bleaching. In addition, determining the dose rate in arid environments can be challenging, because evaporates like calcretes or fluctuating water tables in playas can influence the dose rate over time. Despite these challenges, fluvial sediments from arid environments and associated playas (e.g. Bubenzer and Hilgers 2003; Fuchs and Bürkert 2008; May *et al.* 2015; Telfer *et al.* 2009), river end deposits (e.g. Eitel *et al.* 2006; Srivastava *et al.* 2006), wadis (e.g. Bartz *et al.* 2017) and arroyos (e.g. Arnold *et al.* 2007; Summa-Nelson and Rittenour 2012) were successfully dated with luminescence.

In a paleoenvironmental study from Fuchs and Bürkert (2008), thick playa-like sediments from the Hajar Mountain in Oman were investigated and based on OSL dating of quartz coarse grain extracts, a high-resolution chronology for the last 20 ka was established. Because of the limestone dominance of the small catchment, the mineral quartz in the sand fraction was scarce and large quantities of sediment had to be sampled

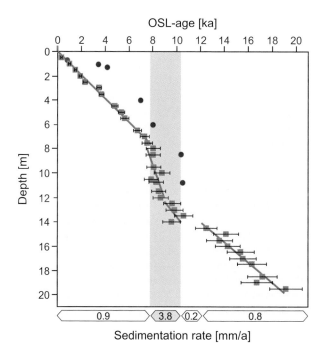

Figure 7.15 OSL age versus sampling depth for playa-like deposits from the Hajar Mountains, Oman. [14]C ages are given (circles), showing distinct age overestimations for this method due to the hard-water effect. Based on the OSL ages, increased sedimentation rates are evident *c.* 8–10 ka due to enhanced precipitation (modified from Fuchs *et al.* 2008).

(*c.* 1.5 kg per sample), compensating for the low quartz content. Therefore, sampling was carried out at night and, after carefully cleaning the light-exposed sediment profile, samples were taken directly into opaque plastic bags. Small aliquot OSL measurements of the coarse grain fraction showed a large D_E scatter, indicating insufficient bleaching for all of the samples. This was attributed to the short-term and extreme turbid surface run-off, resulting from low-frequency high-magnitude precipitation events. Therefore, a MAM was used to calculate D_E values from the best-bleached aliquots (Fuchs *et al.* 2007). Because sampling was undertaken during an exceptionally wet period, the measured water content for estimating the dose rate was not representative for the average water content since sediment deposition. Therefore, the porosity was estimated for every sediment sample to derive possible water content ranges and the mean of this range was used as average water content, including an error, which represented the possible water content range. Finally, OSL ages which were in stratigraphic order could be calculated and the resulting sedimentation rates were used as palaeorainfall proxies. The latter indicated periods of high and low precipitation, which were related to the position of the Intertropical Convergence Zone (ITCZ) and the associated SW monsoon pattern (Fig. 7.15).

From the Mu Us Desert in central north China, Li *et al.* (2008) calculated quartz OSL ages from Holocene fluvial deposits and observed significant age underestimation as well as an age reversal within the investigated sediment profile. These unreliable ages were the result of unrealistic dose rate values, derived from modern radionuclide concentrations and water content measurements. These values were not representative for the entire time since sediment deposition and past fluctuating water conditions and changes in the geochemical composition of the sediments, leading to dose rate changes over time. To consider changes in dose rate over time, Li *et al.* (2008) suggest an isochron approach using K-feldspar grains, and with this method, they were able to calculate stratigraphically correct ages and therefore highlighted that dose rate changes need to be considered.

7.3.2.5 Fluvial catchments

Fluvial catchments are important hydrological units, to better understand e.g. sediment dynamics and landscape evolution. To answer these questions, dating of the available landforms and their associated sediments is essential, with luminescence dating playing a crucial role in establishing the chronologies. This was demonstrated in numerous studies on small- (e.g. Dreibrodt *et al.* 2010; Fuchs *et al.* 2004; Rommens *et al.* 2007), meso- (e.g. Fuchs *et al.* 2011; Houben *et al.* 2012; Notebaert *et al.* 2011; Verstraeten *et al.* 2009) and large-scale (e.g. Hoffmann *et al.* 2009) fluvial catchments.

In a meso-scale fluvial catchment in Belgium, Notebaert *et al.* (2011) investigated Holocene floodplain and colluvial deposits to better understand geomorphological processes and their driving forces, and to reconstruct the catchment-wide sedimentation history. Reliable quartz OSL chronologies were established and resulting sediment mass accumulation rates as well as time-differentiated sediment budgets were used to identify periods of landscape stability and instability. A change in sedimentation dynamics was noted since Neolithic times and was attributed to an increase of agricultural activity. This lead to enhanced soil erosion, which culminated *c.* 1000 BCE (Fig. 7.16). Climate factors as sources for Holocene change in sediment dynamics could not be identified.

Fuchs *et al.* (2010) investigate the sediment dynamics and fluxes in a meso-scale catchment (97 km^2) in Bavaria, Germany. In their study, more than 70 OSL ages from

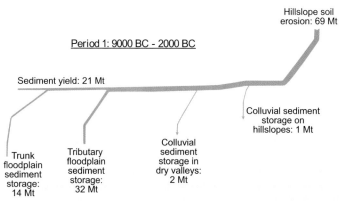

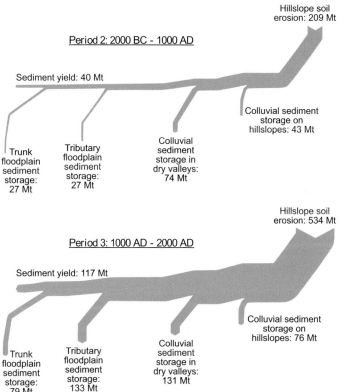

Figure 7.16 Based on OSL ages from colluvial and alluvial deposits, time-differentiated sediment budgets have been established for a meso-scale fluvial catchment (758 km²) in Belgium (modified from Notebaert *et al.* 2011).

different colluvial and alluvial deposits illustrate the evolutionary heterogeneity of individual sediment archives and their interaction, and demonstrate the importance of high-resolution OSL chronologies (Fuchs *et al.* 2011). In conclusion, investigating catchment-wide sediment dynamics asks for a comprehensive OSL dating approach including different sediment archives.

7.4 SUMMARY

Sediments from fluvial and hillslope environments can successfully be dated by luminescence techniques, as has been demonstrated by numerous studies from various climate regimes. Quartz OSL from small aliquot or single-grain measurements is the most widely applied luminescence dating technique for waterlain sediments, because it caters for the problem of insufficient bleaching best and shows negligible luminescence residuals and no fading, so it is relatively accurate. For luminescence dating beyond the age range of quartz, and for quartz samples with less favourable luminescence characteristics such as insufficient signal brightness, feldspar minerals represent a credible alternative.

To successfully apply luminescence dating to sediments from fluvial and hillslope environments, the challenges of insufficient bleaching and dose rate variability need to be considered but can be addressed by careful sampling and descriptions of the sampling site. A comprehensive understanding of the nature of fluvial and hillslope environments and their associated sediments is therefore essential to select the most appropriate samples for dating.

REFERENCES

Alexanderson, H., Murray, A. 2011. Problems and potential of OSL dating Weichselian and Holocene sediments in Sweden. *Quaternary Science Reviews*, 44, 37–50.

Andreucci, S., Panzeri, L., Martini, I.P., Maspero, F., Martini, M., Pascucci, V. 2014. Evolution and architecture of a West Mediterranean Upper Pleistocene to Holocene coastal apron-fan system. *Sedimentology*, 61, 333–361.

Arnold, L. J., Roberts, R.G. 2009. Stochastic modelling of multi-grain equivalent dose (D_E) distributions: implications for OSL dating of mixed sediment samples. *Quaternary Geochronology*, 4, 204–230.

Arnold, L.J., Bailey, R.M., Tucker, G.E. 2007. Statistical treatment of fluvial dose distributions from southern Colorado arroyo deposits. *Quaternary Geochronology*, 2, 162–167.

Arnold, L.J., Roberts, R.G., Galbraith, R.F., DeLong, S.B. 2009. A revised burial dose estimation procedure for optical dating of young and modern-age sediments *Quaternary Geochronology*, 4, 306–325.

Bailey, R.M. 2000. The interpretation of quartz optically stimulated luminescence equivalent dose versus time plots. *Radiation Measurements*, 32, 129–140.

Bailey, R.M., Arnold, L.J. 2006. Statistical modelling of single grain quartz D_E distributions and an assessment of procedures for estimating burial dose. *Quaternary Science Reviews*, 25, 2475–2502.

Bailey, R.M., Smith, B.W., Rhodes, E.J. 1997. Partial bleaching and the decay form characteristics of quartz OSL. *Radiation Measurements*, 27, 123–136.

Bartz, M., Rixhon, G., Kehl, M., El Ouahabi, M., Klasen, N., Brill, D., Weniger, G.C., Mikdad, A., Brückner, H. 2017. Unravelling fluvial deposition and pedogenesis in ephemeral stream deposits in the vicinity of the prehistoric rock shelter of Ifri n'Ammar (NE Morocco) during the last 100 ka. *Catena*, 152, 115–134.

Bateman, M.D., Frederick, C.D., Jaiswal, M.K., Singhvi, A.K. 2003. Investigations into the potential effects of pedoturbation on luminescence dating. *Quaternary Science Reviews*, 22, 1169–1176.

Berger, G.W. 1990. Effectiveness of natural zeroing of the thermoluminescence in sediments. *Journal of Geophysical Research*, 95, 12375–12397.

Berger, G.W., Luternauer, J.J. 1987. Preliminary fieldwork for thermoluminescence dating studies at the Fraser River delta, British Columbia. *Geological Survey of Canada*, 87/IA, 901– 904.

Berger, G.W., Mahaney, W.C. 1990. Test of thermoluminescence dating of buried soils from Mt. Kenya, Kenya. *Sedimentary Geology*, 66, 45–56.

Blöthe, J.H., Munack, H., Korup, O., Fülling, A., Garzanti, E., Resentini, A., Kubik, P.W. 2014. Late Quaternary valley infill and dissection in the Indus River, western Tibetan Plateau margin. *Quaternary Science Reviews*, 94, 102–119.

Bøe, A.-G., Murray, A., Dahl, S.O. 2007. Resetting of sediments mobilised by the LGM ice-sheet in southern Norway. *Quaternary Geochronology*, 2, 222–228.

Botha, G.A., Wintle, A G., Vogel, J.C. 1994. Episodic late Quaternary paleogully erosion in northern KwaZulu-Natal, South Africa. *Catena*, 23, 327–340.

Bourdon, B., Henderson, G.M., Lundstrom, C.C., Turner, S.P. 2003. *Uranium-Series Geochemistry*. Mineralogical Society of America. Washington DC.

Briant, R.M., Bateman, M.D. 2009. Luminescence dating indicates radiocarbon age underestimation in late Pleistocene fluvial deposits from eastern England. *Journal of Quaternary Science*, 24, 916–927.

Bubenzer, O., Hilgers, A. 2003. Luminescence dating of Holocene playa sediments of the Egyptian Plateau Western Desert, Egypt. *Quaternary Science Reviews*, 22, 1077–1084.

Bulur, E. 1996. An alternative technique for optically stimulated luminescence (OSL) experiment. *Radiation Measurements*, 26, 701–709.

Clarke, M.L., Vogel, J.C., Botha, G.A., Wintle, A.G. 2003. Late Quaternary hillslope evolution recorded in eastern South African colluvial badlands. *Palaeogeography, Palaeoclimatology, Palaeoecology,* 197, 199–212.

Colarossi, D., Duller, G.A.T., Roberts, H.M., Tooth, S., Lyons, R. 2015. Comparison of paired quartz OSL and feldspar post-IR IRSL dose distributions in poorly bleached fluvial sediments from South Africa. *Quaternary Geochronology*, 30, 233–238.

Cordier, S., Frechen, M., Harmand, D. 2014. Dating fluvial erosion: fluvial response to climate change in the Moselle catchment (France, Germany) since the Late Saalian. *Boreas*, 43, 450–468.

Cordier, S., Harmand, D., Lauer, T., Voinchet, P., Bahain, J.-J., Frechen, M. 2012. Geochronological reconstruction of the Pleistocene evolution of the Sarre valley (France and Germany) using OSL and ESR dating techniques. *Geomorphology*, 165–166, 91–106.

Cunningham, A.C., Wallinga, J. 2012. Realizing the potential of fluvial archives using robust OSL chronologies. *Quaternary Geochronology*, 12, 98–106.

Ditlefsen, C. 1992. Bleaching of K-feldspars in turbid water suspensions: A comparison of photo- and thermoluminescence signals. *Quaternary Science Reviews*, 11, 33–38.

Dreibrodt, S., Lomax, J., Nelle, O., Lubos, C., Fischer, P., Mitusov, A., Reiss, S., Radtke, U., Nadeau, M., Grootes, P.M., Bork, H.-R. 2010. Are mid-latitude slopes sensitive to climatic oscillations? Implications from an Early Holocene sequence of slope deposits and buried soils from eastern Germany. *Geomorphology*, 122, 351–369.

Duller, G.A.T. 2004. Luminescence dating of Quaternary sediments: recent developments. *Journal of Quaternary Science*, 19, 183–192.

Duller, G.A.T. 2008. Single-grain optical dating of Quaternary sediments: why aliquot size matters in luminescence dating. *Boreas*, 37, 589–612.

Duller, G.A.T., Bøtter-Jensen, L., Kohsiek, P., Murray, A.S. 1999. A high-sensitivity optically stimulated luminescence scanning system for measurement of single sand-sized grains. *Radiation Protection Dosimetry*, 84, 325–330.

Eitel, B., Kadereit, A., Blümel, W.D., Hüser, K., Lomax, J., Hilgers, A. 2006. Environmental changes at the eastern Namib Desert margin before and after the Last Glacial Maximum: New evidence from fluvial deposits in the upper Hoanib River catchment, northwestern Namibia. *Palaeogeography, Palaeoclimatology, Palaeoecology*, 234, 201–222.

Eriksson, M.G., Olley, J.R., Payton, R.W. 2000. Soil erosion history in central Tanzania based on OSL dating of colluvial and alluvial hillslope deposits. *Geomorphology*, 36, 107–128.

Fattahi, M., Nazari, H., Bateman, M.D., Meyer, B., Sébrier, M., Talebian, M., Dortz, K. Le, Foroutan, M., Ahmadi Givi, F., Ghorashi, M. 2010. Refining the OSL age of the last earthquake on the Dheshir fault, Central Iran. *Quaternary Geochronology*, 5, 286–292.

Friedrich, M., Remmele, S., Kromer, B., Hofmann, J., Spurk, M., Kaiser, K.F., Orcel, C. and Küppers, M. 2004. The 12,460-year Hohenheim oak and pine tree-ring chronology from central Europe – a unique annual record for radiocarbon calibration and paleoenvironmental reconstructions. *Radiocarbon*, 46, 1111–1122.

Fu, X., Li, S.-H., Li, B. 2015. Optical dating of aeolian and fluvial sediments in north Tian Shan range, China: Luminescence characteristics and methodological aspects. *Quaternary Geochronology*, 30, 161–167.

Fuchs, M., Lang, A. 2001. OSL dating of coarse-grain fluvial quartz using single-aliquot protocols on sediments from NE Peloponnese, Greece. *Quaternary Science Reviews*, 20, 783–787.

Fuchs, M., Lang, A. 2009. Luminescence dating of hillslope deposits – a review. *Geomorphology*, 109, 17–26.

Fuchs, M., Wagner, G.A. 2003. Optical dating of sediments: Recognition of insufficient bleaching by small aliquots of quartz for reconstructing soil erosion in Greece. *Quaternary Science Reviews*, 22, 1161–1167.

Fuchs, M., Wagner, G.A. 2005. Chronostratigraphy and geoarchaeological significance of an alluvial geoarchive: comparative OSL and AMS ¹⁴C dating from Greece. *Archaeometry*, 47, 849–860.

Fuchs, M., Lang, A., Wagner, G.A. 2004. The History of Holocene soil erosion in the Phlious Basin, NE-Peloponnese, Greece, provided by optical dating. *The Holocene*, 14, 334–345.

Fuchs, M., Straub, J., Zöller, L. 2005. Residual Luminescence signals of recent river flood sediments: A Comparison between quartz and feldspar of fine- and coarse-grain sediments. *Ancient TL*, 23 (1), 25–30.

Fuchs, M., Woda, C., Bürkert, A. 2007. Chronostratigraphy of a sedimentary record from the Hajar mountain range, N-Oman. *Quaternary Geochronology*, 2, 202–207.

Fuchs, M., Bürkert, A. 2008. A 20 ka fluvial sediment record from the Hajar Mountain range, N-Oman, and its implication for detecting arid–humid periods on the southeastern Arabian Peninsula. *Earth and Planetary Science Letters*, 265, 546–558.

Fuchs, M., Fischer, M., Reverman, R. 2010. Colluvial and alluvial sediment archives temporally resolved by OSL dating: Implications for reconstructing soil erosion. *Quaternary Geochronology*, 5, 269–273.

Fuchs, M., Will, M., Kunert, E., Kreutzer, S., Fischer, M., Reverman, R. 2011. The temporal and spatial quantification of Holocene sediment dynamics in a meso-scale catchment in northern Bavaria / Germany. *The Holocene*, 21, 1093–1104.

Fuchs, M. C., Gloaguen, R., Krbetschek, M., Szulc, A. 2014. Rates of river incision across the main tectonic units of the Pamir identified using optically stimulated luminescence dating of fluvial terraces. *Geomorphology*, 216, 79–92.

Fuchs, M., Reverman, R., Owen, L.A., Frankel, K. 2015. Reconstructing the timing of flash floods using terrestrial cosmogenic nuclide ¹⁰Be surface exposure dating: A case study from the Leidy Creek alluvial fan and valley, White Mountains, CaliforniaNevada, USA. *Quaternary Research*, 83, 178–187.

Galbraith, R.F., Green, P.F. 1990. Estimating the component ages in a finite mixture. *Nuclear Tracks and Radiation Measurements*, 17, 197–206.

Galbraith, R., Laslett, G. 1993. Statistical models for mixed fission track ages. *Radiation Measurements*, 21, 459–470.

Galbraith, R.F., Roberts, R.G., Laslett, G.M., Yoshida, H., Olley, J.M. 1999 Optical dating of single and multiple grains of quartz from Jinmium Rock Shelter, Northern Australia: Part I, Experimental design and statistical models. *Archaeometry*, 41, 339–364.

Gerlach, R., Fischer, P., Eckmeier, E., Hilgers, A. 2012. Buried dark soil horizons and archaeological features in the Neolithic settlement region of the Lower Rhine area, NW Germany: Formation, geochemistry and chronostratigraphy. *Quaternary International*, 265, 191–204.

Godfrey-Smith D.I., Huntley, D.J., Chen, W.-H. 1988. Optical dating studies of quartz and feldspar sediment extracts. *Quaternary Science Reviews*, 7, 373–380.

Gosse, J. C., Phillips, F.M. 2001. Terrestrial *in situ* cosmogenic nuclides: theory and application. *Quaternary Science Reviews*, 20, 1475–1560.

Greilich, S., Wagner, G.A. 2006. Development of a spatially resolved dating technique using HR-OSLOriginal Research Article. *Radiation Measurements*, 41, 738–743.

Guerin, G., Combès, B., Lahaye, C., Thomsen, K.J., Tribolo, C., Urbanova, P., Guibert, P., Mercier, N., Valladas, H. 2015. Testing the accuracy of a Bayesian central-dose model for single-grain OSL, using known-age samples. *Radiation Measurements*, 81, 62–70.

Hajdas, I. 2008. Radiocarbon dating and its applications in Quaternary studies. *Quaternary Science Journal*, 57, 2–24.

Hanson, P.R. 2006. Dating ephemeral stream and alluvial fan deposits on the central Great Plains: Comparing multiple-grain OSL, single-grain OSL, and radiocarbon ages. United States Geological Survey Open File Report 2006–1351, p. 14. Available at: http://pub- s.usgs.gov/of/2006/1351/pdf/of06–1351_508.pdf.

Hobo, N., Makaske, B., Middelkoop, H., Wallinga, J. 2010. Reconstruction of floodplain sedimentation rates: a combination of methods to optimize estimates. *Earth Surface Processes and Landforms*, 35, 1499–1515.

Hoffmann, T., Erkens, G., Gerlach, R., Lóostermann, J., Lang, A. 2009. Trends and controls of Holocene floodplain sedimentation in the Rhine catchment. *Catena*, 77, 96–106.

Houben, P., Schmidt, M., Mauz, B., Stobbe, A., Lang, A. 2012. Asynchronous Holocene colluvial and alluvial aggradation: A matter of hydrosedimentary connectivity. *The Holocene*, 23, 544–555
.

Hülle, D., Hilgers, A., Kühn, P., Radtke, U. 2009. The potential of optically stimulated luminescence for dating periglacial slope deposits – A case study from the Taunus area, Germany. *Geomorphology*, 109, 66–78.

Ivy-Ochs, S., Kober, F. 2008. Surface exposure dating with cosmogenic nuclides. *Quaternary Science Journal*, 57, 179–209.

Jain, M., Murray, A. S., Bøtter-Jensen, L. 2004. Optically stimulated luminescence dating: How significant is incomplete light exposure in fluvial environments? *Quaternaire*, 15, 143–157.

Jaiswal, M., Srivastava, P., Tripathi, J., Islam, R. 2008. Feasibility of the SAR technique on quartz sand of terraces of NW Himalaya: A Case Study from Devprayag. *Geochronometria*, 31, 45–52.

Kadereit, A., Sponholz, B., Rösch, M., Schier, W., Kromer, B., Wagner, G.A. 2006a. Chronology of Holocene environmental changes at the tell site of Uivar, Romania, and its significance for late Neolithic tell evolution in the temperate Balkans. *Zeitschrift für Geomorphologie N.F.*, 142, 19–45.

Kadereit, A., Dehner, U., Hansen, L., Pare, Ch., Wagner, G. A. 2006b. Geoarchaeological studies of man-environment interaction at the Glauberg, Wetterau, Germany. *Zeitschrift für Geomorphologie N.F.*, 142, 109–133.

Kadereit, A., Kühn, P., Wagner, G.A. 2010. Holocene relief and soil changes in loess-covered areas of south-western Germany: The pedosedimentary archives of Bretten-Bauerbach (Kraichgau). *Quaternary International*, 222, 96–119.

Kalchgruber, R., Fuchs, M., Murray, A.S., Wagner, G.A. 2003. Evaluating dose rate distributions in natural sediments using a-Al2O3:C. *Radiation Measurements*, 37, 293–297.

Keen-Zebert, A., Tooth, S., Rodnight, H., Duller, G.A.T., Roberts, H.M., Grenfell, M. 2013. Late Quaternary floodplain reworking and the preservation of alluvial sedimentary archives in unconfined and confined river valleys in the eastern interior of South Africa. *Geomorphology*, 185, 54–66.

Kenworthy, M.K., Rittenour, T.M., Pierce, J.L., Sutfin, N.A., Sharp W.D. 2014. Luminescence dating without sand lenses: An application of OSL to coarse-grained alluvial fan deposits of the Lost River Range, Idaho, USA. *Quaternary Geochronology*, 23, 9–25.

Klasen, N., Fiebig, M., Preusser, F. 2016. Applying luminescence methodology to key sites of Alpine glaciations in Southern Germany. *Quaternary International*, 420, 249–258.

Kolb, T. and Fuchs, M. 2018. Luminescence dating of pre-Eemian (pre-MIS 5e) fluvial terraces in Northern Bavaria (Germany) – Benefits and limitations of applying a pIRIR225-approach. *Geomorphology*, 321, 16–32.

Kolb, T., Fuchs, M., Zöller, L. 2016. Deciphering fluvial landscape evolution by luminescence dating of river terrace formation: a case study from Northern Bavaria, Germany. *Z. Geomorph. Suppl. N.F.*, 60, 29–48.

Kolb, T., Fuchs, M., Moine, O., Zöller, L. 2017. Quaternary river terraces as archives for paleoenvironmental reconstruction: new insights from the headwaters of the Main River, Germany. *Z. Geomorph. Suppl. N.F.*, 61, 53–76.

Krbetschek, M.R., Rieser, U., Zöller, L., Heinicke, J. 1994. Radioactive disequilibria in palaeodosimetric dating of sediments. *Radiation Measurements*, 23, 485–489.

Kühn, M., Lehndorff, E., Fuchs, M. 2017. A type locality for Late Pleniglacial to Holocene pedogenesis and colluviation in Central European Loess (Gambach, Germany). *Catena*, 154, 118–135.

Lang, A. 2003. Phases of soil erosion-derived colluviation in the loess hills of Southern Germany. *Catena*, 51, 209–221.

Lang, A., Hönscheidt, S. 1999. Age and source of colluvial sediments at Vaihingen-Enz, Germany. *Catena*, 38, 89–107.

Li, B., Li, S.-H., Wintle, A.G. 2008. Overcoming environmental dose rate changes in luminescence dating of waterlain deposits. *Geochronometria*, 30, 33–40.

Lowick, S.E., Trauerstein, M., Preusser, F. 2012. Testing the application of post IR-IRSL dating to fine grain waterlain sediments. *Quaternary Geochronology*, 8, 33–40.

Mabit, L., Benmansour, M., Walling, D.E. 2008. Comparative advantages and limitations of the fallout radionuclides ^{137}Cs, $^{210}Pb_{ex}$, and ^7Be for assessing soil erosion and sedimentation. *Journal of Environmental Radioactivity*, 99, 1799–1807.

May, J.-H., Barrett, A., Cohen, T.J., Jones, B.G., Price, D., Gliganic, L.A. 2015. Late Quaternary evolution of a playa margin at Lake Frome, South Australia. *Journal of Arid Environments*, 122, 93–108.

Mayya, Y.S., Morthekai, P., Murari, M.K., Singhvi, A.K. 2006. Towards quantifying beta microdosimetric effects in single-grain quartz dose distribution. *Radiation Measurements*, 41, 1032–1039.

Murray, A.S., Roberts, G.R. 1997. Determining the burial time of single grains of quartz using optically stimulated luminescence. *Earth and Planetary Science Letters*, 152, 163–180.

Murray, A.S., Thomsen, K.J., Masuda, N., Buylaert, J. P., Jain, M. 2012. Identifying well-bleached quartz using the different bleaching rates of quartz and feldspar luminescence signals. *Radiation Measurements*, 47, 688–695.

Nathan, R. P., Mauz, B. 2008. On the dose rate estimate of carbonate-rich sediments for trapped charge dating. *Radiation Measurement*, 43, 14–25.

Nelson, M. S., Rittenour, T.M. 2015. Using grain-size characteristics to model soil water content: Application to dose rate calculation for luminescence dating. *Radiation Measurements*, 81, 142–149.

Notebaert, B., Verstraeten, G., Vandenberghe, D., Marinova, E., Poesen, J., Govers, G. 2011. Changing hillslope and fluvial Holocene sediment dynamics in a Belgian loess catchment. *Journal of Quaternary Science*, 26, 44–58.

Olley, J.M., Murray, A., Roberts, R. 1996. The effects of disequilibria in the uranium and thorium decay chains on burial dose rates in fluvial sediments. *Quaternary Science Reviews*, 15, 751–760.

Olley, J.M., Caitcheon, G., Murray, A. 1998. The distribution of apparent dose determined by optically stimulated luminescence in small aliquots of fluvial quartz: implications for dating young sediments. *Quaternary Geochronology*, 17, 1033–1040.

Olszak, J., Adamiec, G. 2016. OSL-based chronostratigraphy of river terraces in mountainous areas, Dunajec basin, West Carpathians: A revision of the climatostratigraphical approach. *Boreas*, 45, 483–493.

Owen, L.A., Robinson, R., Benn, D.I., Finkel, R.C., Davis, N.K., Yi, C., Putkonen, J., Li, D., Murray, A.S. 2009. Quaternary glaciation of Mount Everest. *Quaternary Science Reviews*, 28, 1412–1433.

Pederson, J.L., Anders, M.D., Rittenhour, T.M., Sharp, W.D., Gosse, J.C. and Karlstrom, K.E. 2006. Using fill terraces to understand incision rates and evolution of the Colorado River in eastern Grand Canyon, Arizona. Journal of Geophysical Research, 111, 1–10.

Pietsch, T.J., Olley, J.M., Nanson, G.C. 2008. Fluvial transport as a natural luminescence sensitiser of quartz. *Quaternary Geochronology*, 3, 365–376.

Pope, R., Wilkinson, K., Skourtsos, E., Triantaphyllou, M., Ferrier, G. 2008. Clarifying stages of alluvial fan evolution along the Sfakian piedmont, southern Crete: New evidence from analysis of post-incisive soils and OSL dating. *Geomorphology*, 94, 206–225.

Porat, N., Duller, G.A.T., Amit, R., Zilberman, E., Enzel, Y. 2009. Recent faulting in the southern Arava, Dead Sea Transform: Evidence from single grain luminescence dating. *Quaternary International*, 199, 34–44.

Porat, N., Amit, R., Enzel, Y., Zilberman, E., Avni, Y., Ginat, H., Gluck, D. 2010. Abandonment ages of alluvial landforms in the hyperarid Negev determined by luminescence dating. *Journal of Arid Environments*, 74, 861–869.

Preusser, F., Degering, D. 2007. Luminescence dating of the Niederweningen mammoth site, Switzerland. *Quaternary International*, 164–165, 106–112.

Preusser, F., May, J.-H., Eschbach, D., Trauerstein, M., Schmitt, L. 2016. Infrared Stimulated Luminescence dating of 19th century fluvial deposits from the Upper Rhine River. *Geochronometria*, 43, 131–142.

Radoane, M., Nechita, C., Chiriloaei, F., Radoane, N., Popa, I., Roibu, C., Robu, D. 2015. Late Holocene fluvial activity and correlations with dendrochronology of subfossil trunks: Case study of northeastern Romania. *Geomorphology*, 239, 142–159.

Rittenour, T.M. 2008. Luminescence dating of fluvial deposits: application to geomorphic, palaeoseismic and archaeological research. *Boreas*, 37, 613–635.

Rittenour, T.M., Goble, R.J., Blum, M.D. 2005. Development of an OSL chronology for Late Pleistocene channel belts in the lower Mississippi valley, USA. *Quaternary Science Reviews*, 24, 2539–2554.

Rixhon, G., Braucher, R., Bourlès, D., Siame, L., Bovy, B., Demoulin, A. 2011. Quaternary river incision in NE Ardennes (Belgium) – insight from 10Be/26Al dating of river terraces. *Quaternary Geochronology*, 6, 273–284.

Rixhon, G., Briant, R.M., Cordier, S., Duval, M., Jones, A., Scholz, D. 2016. Revealing the pace of river landscape evolution during the Quaternary: recent developments in numerical dating methods. *Quaternary Science Reviews*, http://dx.doi.org/10.1016/j.quascirev.2016.08.016

Rockwell, T., Fonseca, J., Madden, C., Dawson, T., Owen, L.A., Vilanova, S., Figueiredo, P. 2009. Palaeoseismology of the Vilariça Segment of the Manteigas-Bragança Fault in northeastern Portugal. In Reicherter, K., Michetti, A.M., Silva, P.G. (eds) *Palaeoseismology: Historical and Prehistorical Records of Earthquake Ground Effects for Seismic Hazard Assessment*. The Geological Society, London, Special Publications, 316, 237–258.

Rodnight, H., Duller, G.A.T., Tooth, S., Wintle, A.G. 2005. Optical dating of a scroll-bar sequence on the Klip River, South Africa, to derive the lateral migration rate of a meander bend . *The Holocene*, 15, 802–811.

Rodnight, H., Duller, G.A.T., Wintle, A.G. and Tooth, S. 2006. Assessing the reproducibility and accuracy of optical dating of fluvial deposits. *Quaternary Geochronology*, 1, 109–120.

Rommens, T., Verstraeten, G., Peeters, I., Poesen, J., Govers, G., Van Rompaey, A., Mauz, B., Packman, S., Lang, A. 2007 Reconstruction of late-Holocene slope and dry valley sediment dynamics in a Belgian loess environment. *The Holocene*, 17, 777–788.

Roskosch, J., Winsemann, J., Polom, U., Brandes, C., Tsukamoto, S., Weitkamp, A., Bartholomäus, W. A.,Henningsen, D., Frechen, M. 2014. Luminescence dating of ice-marginal deposits in northern Germany: evidence for repeated glaciations during the Middle Pleistocene (MIS 12 to MIS 6). *Boreas*, 44, 103–126.

Schielein, P., Lomax, J. 2013. The effect of fluvial environments on sediment bleaching and Holocene luminescence ages – A case study from the German Alpine Foreland. *Geochronometria*, 40, 283–293.

Sim, A.K., Thomsen, K.J., Murray, A.S., Jacobsen, G., Drysdale, R., Erskine, W. 2014. Dating recent floodplain sediments in the Hawkesbury–Nepean River system, eastern Australia using single-grain quartz OSL. *Boreas*, 43, 1–21.

Singarayer, J.S., Bailey, R.M., Ward, S., Stokes, S. 2005. Assessing the completeness of optical resetting of quartz OSL in the natural environment. *Radiation Measurements*, 40, 13–25.

Sohbati, R., Murray, A., Chapot, M.S., Jain, M., Pederson, J. 2012. Optically stimulated luminescence (OSL) as a chronometer for surface exposure dating. *Journal of Geophysical Research*, 117, 1–7.

Sohbati, R., Borella, J., Murray, A., Quigley, M., Buylaert, J.-P. 2016. Optical dating of loessic hillslope sediments constrains timing of prehistoric rockfalls, Christchurch, New Zealand. *Journal of Quaternary Science*, 31, 678–690.

Spencer, J. Q., Owen, L.A. 2004. Optically stimulated luminescence dating of Late Quaternary glaciogenic sediments in the upper Hunza valley: validating the timing of glaciation and assessing dating methods. *Quaternary Science Reviews*, 23, 175–191.

Srivastava, P., Brook, G.A., Marais, E., Morthekai, P., Singhvi, A.K. 2006. Depositional environment and OSL chronology of the Homeb silt deposits, Kuiseb River, Namibia. *Quaternary Research*, 65, 478–491.

Stokes, S., Bray, H.E., Blum, M.D. 2001. Optical resetting in large drainage basins: tests of zeroing assumptions using single-aliquot procedures. *Quaternary Science Reviews*, 20, 879–885.

Stolz, C., Grunert, J. and Fülling, A. 2012. The formation of alluvial fans and young floodplain deposits in the Lieser catchment, Eifel Mountains, western German Uplands: A study of soil erosion budgeting. *The Holocene*, 22, 267–280.

Summa-Nelson, M.C., Rittenour, T.M. 2012. Application of OSL dating to middle to late Holocene arroyo sediments in Kanab Creek, southern Utah, USA. *Quaternary Geochronology*, 10, 167–174.

Telfer, M.W., Thomas, D.S G., Parker, A.G., Walkington, H., Finch, A.A. 2009. Optically

Stimulated Luminescence (OSL) dating and palaeoenvironmental studies of pan (playa) sediment from Witpan, South Africa. *Palaeogeography, Palaeoclimatology, Palaeoecology*, 273, 50–60.

Tissoux, H., Falguères, C., Voinchet, P., Toyoda, S., Bahain, J.-J., Despriée, J. 2007. Potential use of Ti-center in ESR dating of fluvial sediment. *Quaternary Geochronology*, 2, 367–372.

Toyoda, S. 2015. Paramagnetic lattice defects in quartz for applications to ESR dating. *Quaternary Geochronology*, 30, 498–505.

Toyoda, S., Voinchet, P., Falguères, C., Dolo, J.M., Laurent, M. 2000. Bleaching of ESR signals by the sunlight: a laboratory experiment for establishing the ESR dating of sediment. *Applied Radiation Isotopes*, 52, 1357–1362.

Truelsen, J. L., Wallinga, J. 2003. Zeroing of the OSL signal as a function of grain size: Investigating bleaching and thermal transfer for a young fluvial sample. Geochronometria, 22, 1–8.

Tsodoulos, I.M., Stamoulis, K., Caputo, R., Koukouvelas, I., Chatzipetros, A., Pavlides, S., Gallousi, C., Papachristodoulou, C., Ioannides, K. 2016. Middle–Late Holocene earthquake history of the Gyrtoni Fault, Central Greece: Insight from optically stimulated luminescence (OSL) dating and paleoseismology. *Tectonophysics*, 687, 14–27.

Vandenberghe, D., Derese, C., Houbrechts, G. 2007. Residual doses in recent alluvial sediments from the Ardenne (S Belgium). *Geochronometria*, 28, 1–8.

Vandenberghe, D., Vanneste, K., Verbeeck, K., Paulissen, E., Buylaert, J.-P., Corte, F. De, Van den Haute, P. 2009. Late Weichselian and Holocene earthquake events along the Geleen fault in NE Belgium: OSL age constraints. *Quaternary International*, 199, 56–74.

Verstraeten, G., Rommens, T., Peeters, I., Poesen, J., Govers, G., Lang, A. 2009. A temporarily changing Holocene sediment budget for a loess-covered catchment (central Belgium). *Geomorphology*, 108, 24–34.

Vincent, P.J., Lord, T.C., Telfer, M.W. and Wilson, P. 2011. Early Holocene loessic colluviation in northwest England: new evidence for the 8.2 ka event in the terrestrial record? *Boreas*, 40, 105–115.

Wagner, G.A. 1998. Age Determination of Young Rocks and Artefacts. Springer, Heidelberg.

Walker, M. 2005. *Quaternary dating methods*. Wiley.

Walker, R.T., Fattahi, M. 2011. A framework of Holocene and Late Pleistocene environmental change in eastern Iran inferred from the dating of periods of alluvial fan abandonment, river terracing, and lake deposition. *Quaternary Science Reviews*, 30, 1256–1271.

Wallinga, J. 2002. Optically stimulated luminescence dating of fluvial deposits: a review. *Boreas*, 31, 303–322.

Wallinga, J., Törnqvist, T., Busschers, F., Weerts, H. 2004. Allogenic forcing of the late Quaternary Rhine-Meuse fluvial record: the interplay of sea-level change, climate change and crustal movements. *Basin Research*, 16, 535–547.

Wallinga, J., Hobo, N., Cunningham, A.C., Versendaal, A.J., Makaske, B., Middelkoop, H. 2010. Sedimentation rates on embanked floodplains determined through quartz optical dating. *Quaternary Geochronology*, 5, 170–175.

Wintle, A.G. 1973. Anomalous fading of thermoluminescence in mineral samples. *Nature*, 245, 143–44.

Wintle, A.G., Huntley, D. 1979. Thermoluminescence dating of a deep-sea sediment core. *Nature*, 279, 710–712.

Wintle, A.G., Li, S.H., Botha, G.A. 1993. Luminescence dating of colluvial deposits. *South African Journal of Science*, 89, 77–82.

Wintle, A.G., Botha, G.A., Li, S.H., Vogel, J.C. 1995a. A chronological framework for colluviation during the last 110 kyr in KwaZulu/Natal. *South African Journal of Science*, 91, 134–139.

Wintle, A.G., Li, S.H., Botha, G.A., Vogel, J.C. 1995b. Evaluation of luminescence dating applied to Late Holocene colluvium near St Paul's Mission, Natal, South Africa. *The Holocene*, 5, 97–102.

Winsemann, J., Lang, J., Roskosch, J., Polom, U., Böhner, U., Brandes, C., Glotzbach, C., and Frechen, M. 2015. Terrace styles and timing of terrace formation in the Weser and Leine valleys, northern Germany: Response of a fluvial system to climate change and glaciation. *Quaternary Science Reviews*, 123, 31–57.

8 APPLICATIONS TO COASTAL AND MARINE ENVIRONMENTS

ALASTAIR C. CUNNINGHAM,[1] TORU TAMURA[2] AND SIMON J. ARMITAGE[3]

[1]Nordic Laboratory for Luminescence Dating, Department of Geoscience, Aarhus University, Risø Campus, DK-4000 Roskilde, Denmark. Email: alacun@dtu.dk
Centre for Nuclear Technologies, Technical University of Denmark, DTU Risø Campus, Denmark

[2]Geological Survey of Japan, AIST, Central 7, 1-1-1 Higashi, Tsukuba, Ibaraki 305-8567, Japan

[3]Centre for Quaternary Research, Department of Geography, Royal Holloway, University of London, Egham, Surrey, TW20 0EX. Also SFF Centre for Early Sapiens Behaviour (SapienCE), University of Bergen, Post Box 7805, 5020, Bergen, Norway.

ABSTRACT: Coastal sedimentary environments are well suited for the application of luminescence dating. The usually high proportion of quartz grains in clastic sediment, and the high degree of bleaching that occurs in the coastal setting, may provide close-to-ideal conditions. The favourable conditions have prompted applications that stretch the boundaries of the typical luminescence age range. The low residual (unbleached) signal in quartz permits the dating of coastal sediment deposited in the last few decades, and enables luminescence methods to be applied to questions of contemporary coastal geomorphology. At the upper end of the age range, bioclastic coastal barriers and raised beaches provide low-dose rate environments for mineral grains, allowing the dating of coastal barriers deposited over the last glacial period, and sometimes beyond. Shallow marine and deltaic sediments present more of a challenge. The necessity of using finer grain sizes makes identification of a well-bleached signal more difficult, and errors in water content measurement can have a large effect on the estimated dose rate. The dose rate calculation for deep-sea sediments is complicated further, because insoluble U-series radionuclides produced in the water column create a time-dependency in the sediment dose rate. These challenges are not insurmountable, however, and shallow marine settings, in particular, offer a promising area of application for future study.

KEYWORDS: Beach ridges, tsunami, delataic sediments, deep-sea sediments, disequilibrium, carbonates

8.1 INTRODUCTION

Coastal environments are particularly suitable for luminescence dating, and have been the setting for an enormous number of luminescence studies over the last 30 years. Marine sediments are more challenging, but have nonetheless been the focus of prominent TL and OSL developments (e.g. Stokes *et al.* 2003; Wintle and Huntley 1979). The suitability of coastal sediments for OSL has also had a long history of being utilised for the development and testing of OSL methods, from the original demonstration by Huntley *et al.* (1985), to more recent attempts to test the accuracy and inter-laboratory reproducibility of the most common techniques (Buylaert *et al.* 2008; Murray *et al.* 2015).

The principal advantage of coastal sediments comes through the quality of bleaching in the coastal environment. Sunlight bleaching at – or prior to – deposition of the sediment is the basic requirement of the OSL dating method (see Chapter 1). Sediment reworking on beaches and coastal dunes allows plentiful opportunity for the bleaching of the quartz fast component – the OSL signal most commonly used for dating. Moreover, coastal sediments are often derived from deposits that have experienced multiple cycles of subaerial exposure immediately prior to deposition, e.g. fluvial outwash, and so may have been bleached even before arriving at the coast. Even inter-tidal sediment may have a well-bleached quartz signal (Madsen *et al.* 2005), thanks to subaerial exposure of sediment on tidal flats and the lateral migration of tidal channels (Fruergaard *et al.* 2015a, b).

Besides the advantages of bleaching, quartz from the coastal zone also tends to be more sensitive than from other settings. In laboratory measurements and simulations, the quartz signal increases ('sensitises') with repeated irradiation, heating and bleaching (McKeever *et al.* 1996). It is likely that cycles of irradiation and bleaching during transport have a similar effect (e.g. Pietsch *et al.* 2008). Coastal and marine sediment is generally far-travelled, so the quartz ought to be more sensitive; some limited evidence from natural sediment is consistent with that idea (Sawakuchi *et al.* 2011). Sandy coastal sediment is also advantageous for the dose rate estimation. Dunes are generally homogeneous in grain size and radionuclide distribution (at the OSL sampling scale), so that standard assumptions hold for dose rate calculations. Furthermore, the saturation water content in sandy sediment is strictly limited by the porosity to about 25% of dry weight, which means that any uncertainty in the water content history can have only a limited effect on the age estimate.

The advantageous bleaching conditions of coastal sediment have prompted research that pushes the boundaries of OSL in other dimensions. A prominent line of research since the early 2000s has sought to date coastal sediment deposited over the last tens to hundreds of years. Applications include the study of beach-barrier progradation (Ballarini *et al.* 2003; Nielsen *et al.* 2006), dune scarp records of storm surges and sea level (Buynevich *et al.* 2007; van Heteren *et al.* 2000), and back-barrier overwash sediments from storm-surges and tsunamis (e.g. Tamura *et al.* 2015). The extension of the topic to very young sediment has brought OSL dating into the awareness of coastal scientists, planners and engineers. OSL is now a powerful tool for understanding modern coastal processes that operate over decades – bridging a gap between records of past environmental change and modern instrumental records.

Nevertheless, there are a number of potential problems that are characteristic of coastal and marine environments. In the absence of poor bleaching, the largest sources of error come through the measurement of radionuclide content and the calculation of dose rate. The key challenges in the dose rate term are introduced in Section 8.2. Subsequent sections explore the use of OSL dating in a range of coastal and marine settings – Holocene barriers, event records, deltaic and inter-tidal sediments, marine sediment cores, and relict

Pleistocene shorelines. The majority of OSL studies on coastal sediment will fall under one of these categories. Problems that are unique to each setting are covered within the subsection. For more general problems, like sampling strategies and the treatment of poorly bleached dose distributions, see Chapters 1 and 2.

8.2 KEY CHALLENGES

8.2.1 Young sediment

The high degree of bleaching observed in coastal sediments has prompted the application of OSL dating towards very young sediment (i.e. a few hundred years or younger). Significantly, this has allowed OSL dating to be applied to questions of modern coastal processes, such as morphodynamics of beach-barriers and estuaries, or event return frequencies. For dating young sediment, it is crucial to extract a well-bleached OSL signal, because any residual dose may be disproportionately large compared to the natural signal. In consequence, applications to very young sediment almost always use quartz – the feldspar signals are more difficult to bleach, and may have large residual doses even under ideal bleaching conditions (Reimann *et al.* 2011). Even so, some modifications to the single-aliquot regenerative dose (SAR) measurement protocol are warranted for young samples, so that the negative effects of poor bleaching, thermal transfer and weak signals can be minimised.

As with all samples for luminescence dating, preheating is necessary to empty unstable electron traps before each OSL measurement, especially the 110 °C TL trap (see Chapter 1). However, excessive preheating can induce an OSL signal that is unrelated to the burial dose; when combined with variability in thermal transfer between measurements it could be confused with poor bleaching, and/or lead to an age overestimate. Tests on young sediment have found the pre-heat should be kept relatively low, so that thermal transfer does not become a significant source of the OSL signal (see Madsen and Murray 2009). The pre-heat dependence of the OSL signal is usually tested as part of quality control procedures for each sample or site; for young samples this is more practical with a thermal transfer test (see Fig. 8.1), because of the confounding effects of poor bleaching. Thermal transfer tests are performed by first bleaching the natural signal, then incrementally pre-heating and measuring the OSL (e.g. Truelsen and Wallinga 2003; Nielsen *et al.* 2006). Commonly, preheats for young samples are kept in the range 180–200 °C.

Palaeodose (D_E) reconstruction for very young samples must use the most bleachable part of the quartz OSL signal – the fast component. Attempts have been made to isolate the fast component through curve fitting, but this is not very practical for weak signals (Cunningham and Wallinga 2009; Tamura *et al.* 2015). Instead, the choice of signal and background integrals can be selected to maximise the proportion of fast component in the net signal. Ballarini *et al.* (2007) established that an 'early' background integral is more suitable for this purpose, but this reduced the signal-to-noise ratio. However, by increasing the length of the initial signal, and using an early background ~2.5 times longer, the signal-to-noise ratio can be kept high while maximising the fast component signal (Fig. 8.1; Cunningham and Wallinga 2010). The signal-to-noise ratio may also be improved by reducing the thickness of the detection filter (Ballarini *et al.* 2005)

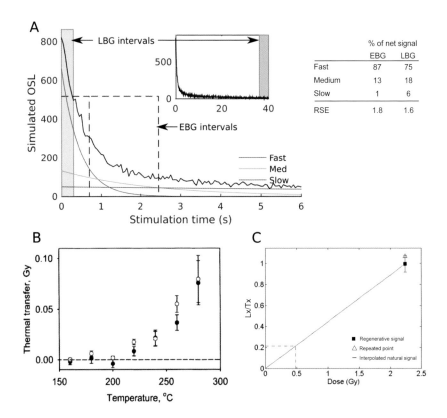

Figure 8.1 Adaptations of the quartz measurement and analysis protocols for the dating of young coastal sediment. A) Use of an 'early' background subtraction when defining the net signal from the OSL decay curve, in order to increase the proportion of the fast component in the net signal (Cunningham and Wallinga, 2010). The composition of the net signal under each choice of background interval is shown in the adjacent table. B) Use of lower preheat temperatures of ~200 °C to avoid thermal transfer from deep traps; the appropriate preheat temperature can be assessed using a thermal-transfer test. The two samples shown here are from young coastal sediments on Rømø barrier island (Madsen *et al.* 2007a). C) Use of a single regenerative dose (with a relatively high test dose) to define the dose response; this reduces the measurement time, allowing more aliquots to be measured (Modified from Cunningham and Wallinga (2009)).

The susceptibility of young samples to bleaching also requires the aliquot size to be kept as small as possible – i.e. the fewest number of grains in each aliquot that will still provide a measurable signal (see Chapter 1). This strategy means that many aliquots must be measured to produce statistically sound D_E distributions. However, because the burial dose for young samples is very low (< 1 Gy), there is also no need to measure a full dose response curve for each aliquot; a single regenerative point is sufficient. A linear fit between zero and the regenerative dose point is adequate to define the dose response function, and so many more aliquots can be measured within experimental time limits. A relatively high test dose (e.g. equal to the regenerative dose) is desirable, so as not to jeopardise the signal-to-noise ratio of the test-dose OSL (Ballarini *et al.* 2007)

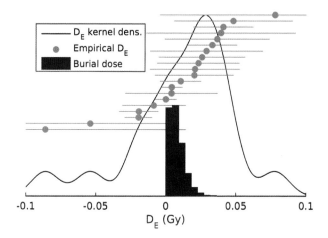

Figure 8.2 Simulated D_E distribution and burial dose analysis for very recent sediment, where some aliquots indicate a negative D_E. The D_E distribution is plotted in ascending order, and also by a kernel density estimate. The burial dose is estimated here using a Bayesian estimate of the central dose (Tamura *et al.* 2018a). The negative D_E values must be included in the dose model. However, the burial dose estimate is strictly limited to be >0 Gy, so its probability distribution may be non-gaussian.

When the natural OSL signal is very weak or absent, then background subtraction will sometimes lead to D_E values that are negative. As this is a random process, and dose models seek to account for random processes, these values must be included when fitting a dose model to the data. However, the widely used central and minimum age models (CAM and MAM, Galbraith *et al.* 1999) include a log transformation, so cannot be used on such datasets. 'Unlogged' versions of these models are available, and alternative Bayesian dose models have no need for log transformation (e.g. Guérin *et al.* 2017). It should be noted that while negative D_E values are possible, the true burial dose must still be positive. In consequence, the burial dose estimate may have a probability distribution truncated at zero (Fig. 8.2), and cannot be easily summarised by 'central estimate ± uncertainty' in the usual manner.

Dating of very young sediment is now fairly routine, and usually performs well against independent age control (Ballarini *et al.* 2003; Madsen *et al.* 2005; Tamura *et al.* 2015). However, it has been noted that zero-age sediment can display a residual dose > 0 Gy, equivalent to an age of at least a few years (Madsen and Murray 2009). When sampling young sediments, it is therefore advisable to collect and measure the luminescence of a modern analogue of the target sediment, so that the practical lower age limit of the dating method can be estimated.

8.2.2 Dose rates

8.2.2.1 Measurement and uncertainty

The dose rate term is dominated by the external beta and gamma contributions (see Chapter 1 for further details). These are calculated from measured concentrations of K, U and Th in the bulk sediment via spectroscopic methods, or by holistic estimates of each component via integral particle- or photon-counting instruments (e.g. thick-source beta counting and field gamma spectrometry). These measurements are likely to be standardised within the OSL laboratory, or obtained from an external laboratory. Unfortunately, the available evidence suggests that the accuracy of such measurements is often poor. Murray *et al.*'s (2015) laboratory inter-comparison found that the dry dose rate estimate, measured at multiple laboratories around the world, had a relative standard deviation of 12% – much too large, considering that the laboratories involved are likely to have been among the more rigorous. For most sediments then, it is very likely that the largest single source of dating error lies in the dry dose rate estimate.

To avoid significant errors in the bulk dose rate estimate, evidence is needed that the instruments and methods in question are likely to produce accurate measurements. This evidence might come through the submission of a reference material as part of a batch of samples (e.g. when using an external ICP-MS laboratory), or a published validation of internal procedures from an OSL laboratory. The 'reference' sample should be a natural sediment with typical radionuclide concentrations – not a high-activity standard. An estimate of the random measurement uncertainty can be obtained through replicate measurements single sample. Typically, mass or activity estimates of K, U and Th – when performing well – might provide a random uncertainty in the dry dose rate of ~6%. Integral counting methods can provide much higher degrees of precision (Cunningham *et al.* 2018), but are not yet in widespread use.

8.2.2.2 Non-uniform radiation field

Standard dose rate calculations require that the sediment is homogeneous and extends to infinity in all directions; this is known as the 'infinite matrix (IM)' assumption. The IM assumption is essentially met if the grain size is much smaller than the range of beta particles in sediment (~2 mm), and the sediment composition is consistent within the range of gamma photons (~30 cm). These assumptions are often met for coastal sediment, when targeting sandy sediment away from stratigraphic boundaries, or through measurement or modelling of gamma dose rates in different sedimentary beds (Chapter 2). However, where the concentrations of different minerals are non-uniform at the millimetre scale, the beta dose rate received by the grains may differ from the sediment average. Micro-bedding of coastal sediment provides one such scenario; for example, through concentrations of heavy minerals (lags) in beach-dune sediment. The ebb and flow of the tide also leads to a striking sand-silt microstratigraphy, which can be preserved in inter-tidal sediments if they are not subsequently obliterated through bioturbation (Fig. 8.3a). The sand component tends to be quartz rich, with relatively low radionuclide concentrations and many dosimeters (the quartz grains); the silt beds will have higher radionuclide concentrations but fewer dosimeters. In consequence, the beta dose rate derived from the bulk sediment will overestimate the dose rate received by the quartz grains, because the grains are preferentially located in low-dose rate zones (Fig. 8.3b; see Martin *et al.* 2015). To complicate matters further, the saturation water content will be lower in the sandy beds, so the water-content estimate derived from the bulk sediment may not be the most appropriate beta dose rate calculations. The beta dose rate calculation may then require a Monte Carlo transport model, combined with radionuclide measurements on the different beds. Thankfully, the Dosivox interface (Martin *et al.* 2015) makes such modelling available to the geoscience community, allowing users to avoid the long learning process previously necessary for Monte Carlo transport codes.

Shells present similar difficulties for the beta dose rate calculation. Shell carbonate is denser than sediment – because of the absence of pore space – making the shells a better absorber of radiation by volume. Shells are also likely to have a much lower radionuclide content than the sediment (mostly through the absence of K). The IM beta dose rate estimated from the bulk sediment is then not applicable to the quartz grains, as the grains are located

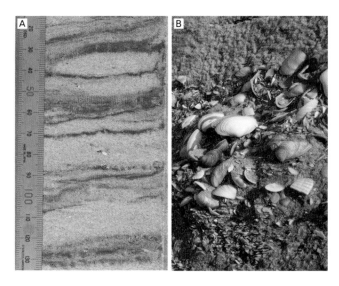

Figure 8.3 Examples of sediment for which the mean beta dose rate to dosimeter grains differs from the mean beta dose rate in the bulk sediment. A) Core photograph showing flaser bedding composed of fine sand and mud drapes (modified from Fruergaard *et al.* 2015). B) Shell-rich storm-surge sediment within a coastal dune (Photographs M Frvergaard and A. Cunningham).

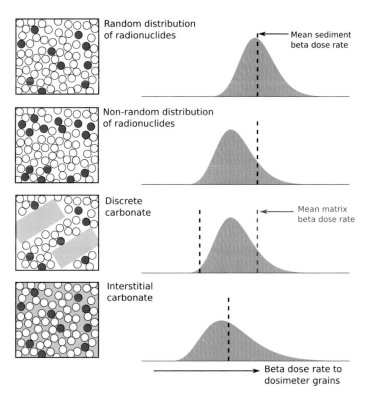

Figure 8.4 Schematic illustration showing some influences on the beta dose rate distribution to quartz grains in sandy sediment. A) With a random distribution of radioactive sources (e.g. grains of K-feldspars, zircons, monazite), the beta dose rate distribution is approximately log-normal, and the mean dose rate received by the quartz dosimeters equals the mean for the bulk sediment. B) Micro-bedding of the sediment causes a non-random distribution of source grains. The dose rate received by the quartz grains is less than the sediment average. C) Sediment with discrete carbonate zones of (shells or shell hash, with very-low radionuclide concentrations). The mean dose rate to the quartz grains is greater than the sediment average, but less than that of the non-carbonate matrix. D) The presence of interstitial carbonate reduces the mean beta dose rate, and increases the variance.

in the higher dose rate sediment matrix. A simple solution to this problem is to remove the shells before radionuclide measurements, then use the matrix-only beta dose rate. However, the dose rate to quartz grains is also affected by proximity to the shells: grains lying next to a shell receive a lower beta dose rate than other grains (Fig. 8.4). The strength of these two effects depends on the size and shape of the shell material. Cunningham (2016) modelled this dependency using a Monte Carlo transport code, and devised a simple strategy to follow once the size and mass of the shell material have been measured. When the shells are large they can be removed, and the beta dose rate calculated from measurements on the matrix material. Fine shell fragments should be included in the radionuclide measurements, with the bulk beta dose rate calculated as normal. Shell material of intermediate size – small shells and shell hash – presents the most difficulty. Neither the bulk-sediment nor matrix-only dose rates are accurate, and a model-based correction must be applied.

More commonly, carbonate in coastal sediment is interstitial, present as beach-rock or aeolianite. Beachrock forms in the intertidal zone, through the mixing of salt- and freshwater, and microbe activity (Mauz *et al.* 2015). Coastal aeolianite is also common in warmer climates, forming through the cementation of bioclastic, skeletal carbonate sand, and leading to the long-term preservation of coastal barrier dunes (see Section 8.7). Interstitial carbonate absorbs radiation, reducing the dose rate to minerals grains (and by shortening the range of beta particles, should also increase the heterogeneity in beta dose rates). However, as the carbonate is distributed uniformly, the infinite-matrix assumption still holds, and the bulk sediment can be used for the calculating the beta and gamma dose rates. This reasoning presumes that the carbonate precipitation is effectively contemporaneous with sediment deposition (relative to the burial age). If the cementation process is more gradual, then the ratio of the dose rate to the mineral grains decreases over time as the carbonate fills the available pore space. In such cases, the dose rate (both beta and gamma) must be modelled as a function of time, and of the duration of the cementation process (Mauz and Hoffmann 2014; Nathan and Mauz 2008).

8.2.2.3 U-series disequilibria

Dose rate calculations require assumptions on the present and former degree of equilibrium in the activities of the U-series radionuclides. For samples which have high a U concentration relative to K and Th, there is potential for a significant error in the estimated dose rate (and age) if these assumptions are not valid. Dose rates derived from top-of-chain mass-concentration estimates of K, U and Th, via conversion factors, presume that the decay chains of U and Th are in secular equilibrium; that is, the activity of all the daughter radioisotopes in the series is the same as the parent. This is not relevant for K, which has no radioactive daughters, or for the ^{232}Th series, which has no long-lived daughter isotopes and for which no significant disequilibria have been detected (Olley *et al.* 1996). For the ^{238}U series, the different chemical and physical properties of the isotopes provide several mechanisms through which disequilibria can be created. Through the weathering of rocks and soil, river water is enriched in ^{234}U relative to the parent ^{238}U, with the sediment correspondingly depleted of ^{234}U. Additionally, both the ^{238}U and ^{235}U series contain isotopes that are insoluble, and are deposited in excess of their parent in marine sediments following their production in the ocean water (see Section 8.6). U and Ra in groundwater and seawater can also be adsorbed by both biogenic and inorganic calcite; and the gaseous and non-reactive radon is liable to either escape or concentrate in terrestrial sediment.

Detection of the present-day state of disequilibrium is only possible with spectroscopic methods (principally high-resolution Ge-crystal gamma spectrometry (HRGS)). However, in a closed system (i.e. no mobility of U-series isotopes), an initial disequilibrium at deposition will equilibrate over time; thus, even if a sample is determined through HRGS to be in secular equilibrium, there is no guarantee that this was true for the entire burial period. In the case of U uptake by carbonate, the rate and timing of uptake can only be guessed at, so attempts to account for all uptake scenarios can lead to large uncertainties in the dose rate estimate (e.g. Zander et al. 2007).

Fortunately, for terrestrial sediment, the potential dose rate error resulting from U-series disequilibrium is limited, because the proportion of the total dose rate derived from the U series is usually low. Typically, the largest contribution to the dose rate comes from ^{40}K, due to its concentration in K-feldspar and some organic matter. For example, the beach-ridge in the Skagen peninsular, Denmark, sampled by Murray et al. (2015), has dose rate contributions from K, U and Th of roughly 86%, 8% and 6% respectively. If we suppose a 50% loss of radon throughout the burial period, then the presumption of equilibrium would cause an error of <3% in the total dose rate – barely detectable given other sources of uncertainty. The potential error is larger for sediments with low K content, such as coastal sites in Australia and South Africa with grains derived from old and weathered sedimentary rock. In such cases, the error can be minimised by presuming that the present-day state of disequilibrium has persisted throughout the burial period (Olley et al. 1996).

8.3 BEACH-RIDGE SYSTEMS

Beach-ridge systems occur on depositional coasts, where the shoreline progrades seawards and the beach-to-foredune morphology is successively abandoned inland as an elongated mound – the beach ridge (Hesp 1984; Tamura 2012; Taylor and Stone 1996). The application of OSL dating to beach-ridge systems has been motivated by their preservation of a time-ordered sequence of events, but usually lacking the datable material for radiocarbon dating (Huntley et al. 1993; Isla and Bujalesky 2000). Beach ridges are formed primarily by waves and winds acting on the beachface and foredune, where mineral grains are well bleached by sunlight. As such, they are ideal sedimentary landforms for OSL dating. In an early application of the SAR protocol to young coastal sediment, Ballarini et al. (2003) obtained OSL ages of <250 yr for a sequence of beach ridges on the island of Texel, formed by the shoreward migration of sand shoals. The OSL dates were consistent with historical maps for 18 samples out of 20. With the youngest sample indicating an age of 7 yr, Ballarini et al. (2003) concluded that the potential age-offset due to incomplete bleaching was generally < 5 yr, while a few samples showed offsets of several tens of years (Ballarini et al. 2007). Non-fading post-IR signals from K-feldspar, although harder to bleach, have also been applied successfully to Holocene beach-ridge systems. Reimann and Tsukamoto (2012) assessed a post-IR protocol for young coastal samples, and found that a relatively low stimulation temperature of 150 °C was required to limit size of the unbleachable residual. A beach-ridge sequence in Ruhnu Island, Estonia, showed post-IR IRSL ages that were consistent with sporadic radiocarbon ages and regional sea-level history (Preusser et al. 2014).

For sandy beach ridges, samples for OSL dating can be collected both from the aeolian and beach facies (Fig. 8.5). Strictly speaking, dates from aeolian and beach facies must be interpreted differently: the date of beach sand indicates the last time that the shoreline occurred at that location, while a date for the foredune indicates the time when the shoreline occurred further seawards – most likely at the next ridge (López and Rink 2007). However, the difference between the aeolian and beach ages in a single ridge would be negligible in most cases, as ridge intervals are generally multi-decadal (Tamura 2012) and well within the dating uncertainty for samples >1 ka. In contrast, an internal chronology is recognisable in young (< 500 yr) ridges if the ridge-forming process is relatively slow (Fig. 8.6; Tamura *et al.* 2017). Storm erosion (Buynevich *et al.* 2007) and shoreline re-orientation (Rodriguez and Meyer 2006) may cause a hiatus of decades to millennia, even in a single ridge or between neighbouring ridges.

Beach ridges composed of gravel or sand-and-gravel provide different options for the sampling strategy (Fig. 8.5). Samples can be collected from the aeolian facies (e.g. Orford *et al.* 2003; van Heteren *et al.* 2000) and subtidal shoreface facies (Roberts and Plater 2007) above and below beach gravel, and perhaps the gravel and cobbles if rock-surface dating is to be attempted. Gravel beach deposits are often overlain by aeolian sand, causing more pronounced relief of the beach ridges. OSL samples can be collected from the aeolian facies, with their ages constraining the time when the gravel beach was abandoned. The bleaching of subtidal shoreface sand is not guaranteed, because seawater strongly absorbs the ultraviolet component of sunlight. Rink and Pieper (2001), for example, detected

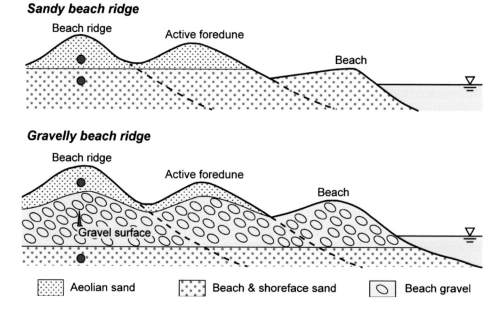

Figure 8.5 Profiles showing simplified stratigraphy of sandy and gravelly beach ridges (modified from Tamura 2012) with sampling strategies. Red dots show locations of OSL samples. OSL dating of the gravel surface is also feasible if gravels are shielded from sunlight during sampling. Dashed lines define isochrones, broadly showing the successive seaward accretion of beach ridges – the detailed chrono-stratigraphy can be more complex, depending on the formative processes.

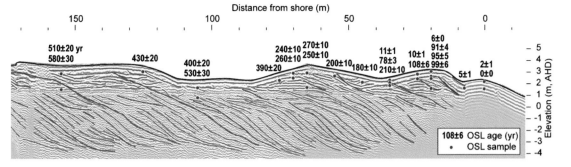

Figure 8.6 Ground-penetrating radar profile and OSL chronology of beach ridges in Cowley Beach, northeastern Australia. Blue and red lines trace reflection and truncation surfaces, respectively. The dashed green line indicates the groundwater table. Red dots show locations of OSL samples. OSL ages are expressed relative to AD 2015. The average beach-ridge interval in Cowley Beach is 250–270 years, and decadal-scale internal age architecture is recognisable in the first and second ridges from the shore. Modified from Tamura *et al.* (2018b).

pronounced residual TL in the underwater beach sand, although their TL signal needed several hours to bleach completely (Rink 1999). In contrast, Roberts and Plater (2007) confirmed that the quartz OSL of two modern sand samples in the upper shoreface showed negligible residual dose – equivalent to 15 or 40 years. They then systematically applied the quartz SAR protocol to shoreface sand facies found below gravel ridges in the Dungeness foreland, southern England. When combined with radiocarbon dates of inter-ridge swales, the chronology of the ridge sequence could be used for characterising episodic coastal changes following the mid Holocene (Plater *et al.* 2009). Mineral grains on the surface of beach cobbles can be expected to be well bleached, and advances in rock-surface dating methods have made such cobbles feasible to date (see Chapter 11). Simms *et al.* (2011) have applied the SAR protocol to quartz grains disaggregated from cobbles of raised beach ridges in Antarctica and obtained dates consistent with radiocarbon.

Beach-ridge sand is usually homogeneous, making the dose rate calculations relatively simple. Nevertheless, the presence of heavy mineral lags or carbonate should be considered (see Section 8.2). The time-averaged overburden thickness is also required for calculation of the cosmic dose rate. For beach ridges, this can usually be presumed identical to the sampling depth, because the topography of beach ridges – once isolated from coastal processes – tends to get preserved. However, if the overlying sediment has been affected by blowouts, transgressive dune deposition, or anthropogenic changes, then the current sample depth may not be a good estimate of the mean depth over the burial period. OSL ages are especially sensitive to errors in the assumed overburden thickness if the sample depth is very close to the surface (<1.5 m), and/or the radioactivity of the sediment is particularly low. The water-content term may also need consideration; a clear groundwater table is likely to develop around the boundary between the aeolian and beach facies (e.g. Bristow and Pucillo 2006), and the seasonal changes in its level may affect the water content for samples taken from within the range of fluctuation. A practical assumption of the water content could be based on the average of measured or assumed values in wet and dry seasons, with appropriate uncertainty added.

While OSL dating of beach-ridge sequences often provide reasonable results, there can be apparent age reversals or outliers in the sequence – most likely caused by random errors in the dose rate. However, a beach-ridge sequence is a stratigraphic succession, with the depositional age increasing landwards. If this knowledge is used as prior information for a Bayesian chronological model, then random uncertainties in OSL ages can be reduced and outliers can be identified. Brill *et al.* (2015) used this approach to estimate changes in accretion rate for a beach-ridge plain on Phra Thon Island, Thailand, after creating an age-distance model of shore-normal transects using the OxCal software. If each beach-ridge is sampled at multiple depths, then a two-dimensional profile of deposition age can be created using a similar rationale (Tamura *et al.* 2018a). However, outliers might also be explained through the processes of ridge formation (e.g. slumping). On the Mekong delta, Tamura *et al.* (2010) identified hierarchical sets of beach ridges with some inconsistent ages within the ridge sets. After inspection of the ridge architecture, these age reversals are inferred to result from local reworking of the ridges (Tamura *et al.* 2012a).

8.4 STORM-SURGE AND TSUNAMI SEDIMENT

Storm surges and tsunamis may deposit sediment beds by washover processes, with preservation of the sediment in back-barrier marshes (Fig. 8.7). Washover sediment may contain plant fragments and shells which can be used for radiocarbon dating, but as these materials are reworked they may not provide an accurate date for the event. The intervening marsh sediment is generally preferred as the source material for radiocarbon dating of washover events (e.g. Donnelly *et al.* 2001; Sawai *et al.* 2012). The principal advantage of OSL for dating washover sediment is in the direct use of the event deposit– especially useful if the over- and underlying sediment differ greatly in age. Furthermore, the timescale of the last few hundred years is of most interest in event chronologies, which is a period difficult to date with radiocarbon due to a large calibration plateau. OSL can provide dates from decades to centuries, and beyond, by a single method, which would otherwise require the composite application of radiocarbon, Pb[210] and Cs[137] dating. Applied in a number of locations worldwide, OSL dating of washover deposits has been reasonably successful (Madsen *et al.* 2009; Davids *et al.* 2010a; Brill *et al.* 2012, 2017; Prendergast *et al.* 2012; Nentwig *et al.* 2015; May *et al.* 2017).

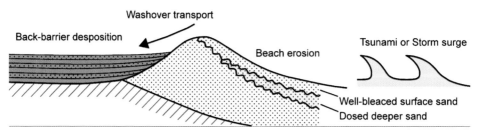

Figure 8.7 Schematic cross-section illustrating the phases of washover sedimentation, including erosion of beach sand, transport, and back-barrier deposition. Eroded beach-surface sand is generally well bleached, but unbleached deeper and older sand may also be eroded and transported depending on the depth of truncation. Bleaching during the short washover phase itself is probably very limited. The washover sand captured in the back-barrier marsh sediment is likely to include both well-bleached and unbleached grains.

Tsunamis and storms transport sediments in turbid water during a short period of time, ranging from seconds to hours. The bleaching of grains during transport is far from guaranteed, so there is clear potential for OSL dating to overestimate the depositional age. Nevertheless, comparisons with historical events (Banerjee *et al.* 2001; Cunningham *et al.* 2011; Tamura *et al.* 2015) and other numerical chronologies (Davids *et al.* 2010a; Huntley and Clague 1996; Madsen *et al.* 2009; Nentwig *et al.* 2015; Spiske *et al.* 2013) show that OSL ages of washover deposits rarely suffer obvious age overestimation. The primary reason for this is that the beach surface sand, from which washover sand is mainly derived, has already been bleached by sunlight and prior to the washover event (see Fig. 8.6). Several studies have explicitly assessed the residual dose of modern beach sand and deposits of modern coastal washover. Banerjee *et al.* (2001) observed a very low quartz OSL residual dose of beach surface sand, equivalent to 6 ± 3 yr in the Isles of Scilly, UK. Davids *et al.* (2010a) determined residual doses of quartz OSL and K-feldspar IRSL of beach sand in New England as 0.036 ± 0.016 Gy and 0.068 ± 0.026 Gy, respectively, which are equivalent to 20–40 yr of natural irradiation, and support the observation of Madsen *et al.* (2009) in the same region. Davids *et al.* (2010a) inferred that the negligible residual dose of the beach-surface sand was the reason that their OSL and IRSL chronologies of Holocene back-barrier washover deposits were consistent with radiocarbon dates.

The 2004 Indian Ocean tsunami provided an opportunity to assess the residual dose of modern tsunami deposits (Bishop *et al.* 2005). Using a portable OSL reader, Sanderson and Murphy (2010) measured the residual dose of 250 samples collected from the tsunami deposit and assessed the relationship with sediment source. The residual dose of samples of nearshore origin was equivalent to less than a few decades, while reworked material from terrigenous or anthropogenic deposits showed a residual dose equivalent to centuries-to-millennia. For tsunami deposits of nearshore origin, a residual dose equivalent to several 10s to 100 years has also been measured in fully processed quartz (Brill *et al.* 2012; Murari *et al.* 2007; Prendergast *et al.* 2012). Regardless of depositional context, a residual dose of this level is insignificant for deposits several thousand years old (e.g. Cunha *et al.* 2010). In contrast, the residual signal cannot be ignored for younger deposits, and a few studies have pointed out the possibility that some washover events erode the beach and foredune more deeply, and transport grains with a higher residual signal (Brill *et al.* 2017; Madsen *et al.* 2009; Ollerhead *et al.* 2001). However, provided that at least some of the grains have a negligible residual signal, a well-bleached population of grains should be identifiable in the equivalent-dose distribution (of single grains) using a suitable statistical model. Brill *et al.* (2012) and Prendergast *et al.* (2012) examined the residual dose of the Indian Ocean tsunami deposits on Phra Thong Island, Thailand, and found that using the minimum age model (MAM) of Galbraith *et al.* (1999) led to comparable OSL ages for three layers of tsunami deposits. Nentwig *et al.* (2015) obtained similar results for a set of tsunami beds deposited over the last 1000 years in central Chile. They found evidence of poor bleaching, with reasonable dates produced using the MAM applied to multi-grain aliquot data.

Washover sand units in back-barrier environments are typically preserved as intercalations in marsh peat. The sand layers are generally several centimetres to a few tens of centimetres thick, and are likely to receive a gamma dose from the over- and underlying marsh sediment. An accurate estimate of the gamma dose rate to the washover bed then requires that the gamma-dose contribution from the marsh sediment be included. For example, Madsen *et al.* (2009) studied a 1.2 m thick sequence of washover sand and

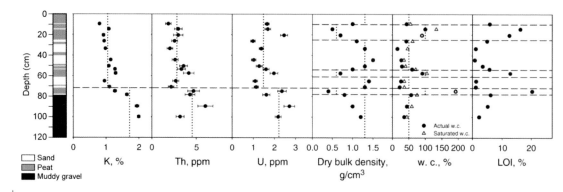

Figure 8.8 Variations in radionuclide concentrations, bulk density, water content and loss-on-ignition, plotted against depth in a sediment core obtained in the Little Sippewisset Marsh in Massachusetts, USA (modified from Madsen *et al.* 2009). The sediment succession shows alternations of washover sand and marsh peat, overlying muddy gravel. Dashed lines show boundaries of units defined for a multi-layer modelling of the gamma dose rate to individual samples.

salt-marsh peat in New England (Fig. 8.8). Sampling at high resolution, they measured the changes in radionuclide concentrations and water content throughout the profile. By constructing a simplified gamma dose rate model based on seven broad units, they estimated the gamma dose rate to each sample using the equations given in Appendix H of Aitken (1985). The resulting gamma dose rate differed by −20 to +10% to that estimated directly using the elemental composition of each sample, and total dose rate differed by −7 to +3%. The importance of modelling the gamma dose rate in this way will depend on the site characteristics: the thickness of the sand and peat layers, the difference in the radionuclide concentrations between layers, and the relative contribution of the gamma dose compared with other components of the dose rate (alpha, beta and cosmic).

In addition to back-barrier deposits, evidence of past storms may be found within the barrier sediment, in the form of erosional storm scarps, barrier-breach sediment and backwash deposits (Chaumillon *et al.* 2017). These archives preserve information on the magnitude of the event that is not available from back-barrier sediment. For example, Clemmensen *et al.* (2016) found that the elevation of a gravel berm crest formed during a 2013 storm in Denmark was closely related to storm flood level – a relationship that might be exploited as a proxy indicator of storm magnitude on micro-tidal coasts. Nichol *et al.* (2003) identified a gravelly tsunami deposit lying 14.3 m above sea level in the foredune of Whangapoua barrier, New Zealand, and by dating the underlying sediment constrained the depositional age to less than 4.7 ka. Similarly, cliff-top boulders in nothern Scotland, inferred to have been emplaced by high-energy storms, have been dated through quartz OSL on the bracketing aeolian units (Hall *et al.* 2006; Sommerville *et al.* 2003).

The use of barrier (rather than back-barrier) sediment has distinct advantages from an OSL perspective. Due to favourable conditions for bleaching, and for the dose rate estimation, Holocene barrier sediments are about the most ideal material for conventional quartz OSL dating. Barrier sediments are also well suited for profiling with ground penetrating radar (GPR), and the combination of GPR profiling and OSL dating has enabled storm-surge records to be obtained from barrier sediment. On the New England coast, Buynevich

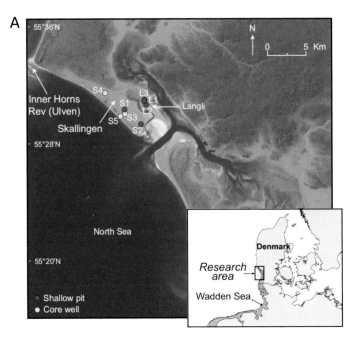

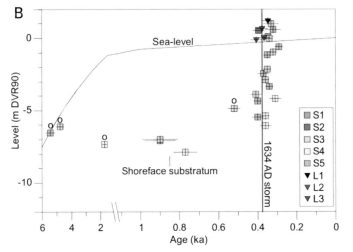

Figure 8.9 (A) Skallingen barrier-spit on the Danish Wadden Sea coast, with locations of sediment cores extracted by Fruergaard *et al.* (2013). (B) OSL dates from several sediment cores. Around 7 m of subtidal sediment was deposited around the time of a catastrophic storm in 1634 AD, or in the recovery phase in the following decades. Prior to the storm, historical maps show the coastline was 3–4 km landwards of its current location. Modified from Fruergaard *et al.* (2013).

et al. (2007) identified hurricane-induced storm scarps in a prograding barrier, on the basis of GPR profiles and heavy mineral lags in sediment cores. Their OSL chronology was based on the buried beach sediment immediately overlying the scarps, which can be expected to build up during the fairweather recovery phase following the storm (and so deposited under good bleaching conditions). Cunningham *et al.* (2011) identified a storm-surge deposit lying up to 6.5 m above sea level in a temporarily exposed dune barrier in the Netherlands. A prominent shell bed, dipping gently landwards, was tracked with GPR from the beach exposure to ~1 km inland. The shell bed was interpreted as a perched fan, deposited as the storm-surge overtopped low dunes or breached the first dune barrier. A large number of OSL dates on the shell bed and bracketing sediment constrained the date of the storm to the late 18th century, which was consistent with documentary evidence of large storm surges in 1775 and 1776 AD.

The geomorphological impact of a catastrophic storm of 1634 AD in the Danish Wadden Sea has been assessed through OSL dating of a barrier-island system (Fruergaard *et al.* 2013; Fruergaard and Kroon 2016). Changes in the barrier morphology are apparent from historical maps produced before and after the storm. Fruergaard *et al.* (2013) summarised the OSL dates from a number of sediment cores extracted from the Skallingen spit (Fig. 8.9). With high-resolution sampling, they could show that ~7 m of subtidal shoal sediment was deposited during or shortly after the storm. Assessed with the map evidence, the storm shoal is thought to represent a rapid coastal recovery (shoreface healing) within 30 to 40 years of the barrier breaching. The present configuration of the barrier-island system is strongly influenced by the 1634 storm and long-term recovery.

8.5 ESTUARY AND DELTA SEDIMENTS

As with other clastic coastal sediment, OSL dating of deltaic and intertidal sediments has been spurred by the paucity of other applicable dating methods. A long-standing question for OSL dating has been over the likelihood of intertidal sediments being bleached before deposition. Light intensity drops rapidly with water depth, and the high suspended sediment load in estuaries and deltas reduces the intensity further. Sanderson *et al.* (2007) measured the light intensity and spectrum in relatively still water in the Mekong delta. The overall light intensity was reduced to 5% of the surface level by 1.5 m of water. Importantly, the attenuation was most severe for the UV–blue spectral range, i.e. the range most effective at bleaching OSL signals. As sand-sized grains are transported mostly by rolling or saltation, the prospective light exposure under the full water column may be limited.

On the other hand, intertidal sediments are constantly reworked through migration of the tidal channels, and the subaerial exposure on mudflats and sandflats should provide ample opportunity for bleaching. Instrumental measurements of bed level from the micro-tidal Wadden Sea show that the gross movement of sediment far exceeds the net accumulation (Andersen *et al.* 2006; Fruergaard *et al.* 2015a, 2015b), confirming that reworking of sediment is significant. As deposition of channel sediment is primarily on the channel sides, it is possible that even deep channels will deposit well-bleached sediment (Fruergaard *et al.* 2015a 2015b). However, these processes may depend somewhat on site-specific variables, such as the sedimentation rate and suspended sediment load, meaning that between-site variation in the degree of bleaching can be expected (Mauz *et al.* 2010).

The work of Anni Madsen and colleagues has thoroughly examined the applicability of quartz OSL dating to estuarine tidal flats in the backbarrier of Skallingen, Denmark. Madsen *et al.* (2005) dated sand grains of 90–180 μm extracted from a core from the mudflat, and obtained very young ages of < 80 years, consistent with a ^{210}Pb and ^{137}Cs chronology. They also found the negligible residual dose equivalent to 7 yr in the surface sample. Madsen *et al.* (2007a) examined three further cores and found reproducible OSL results for a range of estuary environments from supratidal to subtidal, over an age range of 1000 years. A thicker and older sediment succession observed in a salt-marsh cliff was examined by Madsen *et al.* (2007b), who obtained OSL ages that are younger than 2000 years and consistent with radiocarbon dates. The uppermost two samples in this succession were found to have been incompletely bleached, attributed to contamination from nearby

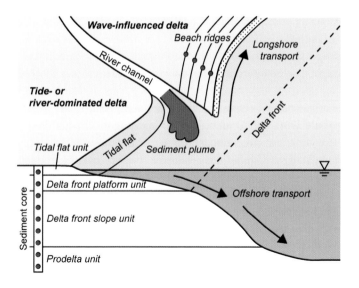

Figure 8.10 Depiction of two types of deltas: a tide- or river-dominated delta, and a wave-influenced delta, with a typical sediment succession resulting from the delta progradation. In tide-dominated deltas, sediment supplied from the river is bleached by sunlight while being transported in suspension, and/or temporarily stored in the tidal flat; the sediment is ultimately deposited in the subaqueous part of the delta, such as delta front and prodelta. Wave-influenced deltas commonly have a sandy beach and associated beach ridges. OSL dating of deltaic beach ridges is just as feasible as for other beach ridge systems. Red dots show the OSL sampling points for typical sampling strategies.

construction works. More recently, Chamberlain *et al.* (2017) investigated tidal sediments in the Ganges-Brahmaputra-Meghna delta, finding a limited degree of poor bleaching in fine-grained quartz signals that could be accounted for using a burial dose model. These encouraging results show that OSL is generally well suited for dating estuarine, salt-marsh, and subaerial delta sediment, and future applications should be encouraged.

Deltaic systems include the sub-environments of estuary and tidal flat, and if wave-influenced, they are commonly characterised by the development of sandy beaches and beach ridges (Fig. 8.10). Beach ridges that have developed on the delta plain have been the subject of OSL dating for the reconstruction of deltaic shoreline changes (Tamura *et al.* 2012b; Maselli and Trincardi 2013; Vespremeanu-Stroe *et al.* 2017), and sea-level changes (Giosan *et al.* 2006), and benefit from similarly advantageous bleaching conditions as beach-barriers. In contrast, the bleaching conditions in subaqueous delta environments, such as the delta front and prodelta, are less clear-cut, and OSL dating in these settings has yet to be properly tested. The grain sizes available typically range from fine to coarse silt, so aliquots contain many thousands of grains each. The D_E distribution does not provide much information on the degree of bleaching, because between-aliquot variation in D_E is obscured by within-aliquot signal averaging. An alternative approach to assess bleaching is to exploit the differences in bleaching rates between different OSL signals. Of the main signals used in dating protocols, the quartz fast-component signal is most easily bleached in sunlight, followed by the feldspar IR50 (infrared-stimulated luminescence, measured at 50 °C), followed by post-IR signals (e.g. Kars *et al.* 2014). If the age obtained from a feldspar protocol (either a fading-corrected IRSL, or a stable post-IR signal) are consistent with the quartz age, the implication is that the signals are well bleached; a well-bleached quartz signal might also be inferred if the feldspar ages are only slightly older (Murray *et al.* 2012).

A few studies in East Asia have shown encouraging results after assessing bleaching in this way. Sugisaki *et al.* (2015) assessed the bleaching of fine grains (4–11 μm) of 40 samples from two sediment cores obtained at water depths of 36 m and 38 m, near the mouth of the Yangtze River. All but a few samples showed similar age estimates from the quartz and feldspar signals, indicating that most of the samples were well-bleached prior to deposition in the subaqueous delta. Sugisaki *et al.* (2015) also measured the residual doses in modern suspended fine grains from the turbid Yangtze River. While they found a significant residual in the K-feldspar post-IR IRSL, the quartz residual was only 0.1–0.2 Gy. Gao *et al.* (2017) also compared equivalent doses of quartz OSL and K-feldspar post-IR IRSL from fine grains, in this case a 40 m thick Holocene subaqueous delta succession in the northern Yangtze delta plain. They found little difference in the quartz and feldspar ages, giving a good indication that the bleaching was sufficient. They then determined a reasonable OSL chronology of the sedimentary succession, inferring that the incised-valley fill and prograding delta wedge formed in response to post-glacial sea-level rise and the subsequent highstand. A similar quartz OSL chronology was also given by Kim *et al.* (2015) for both the fine grains and coarse grains from a 55 m thick sediment core taken in the delta plain of the Nakdong River, southeastern Korea.

These attempts, although restricted to East Asian deltas, show that OSL is viable in subaqueous delta sediments – at least for the age range of centuries to millennia. For such sediments, a reliable OSL protocol would prove particularly useful. At present, shallow-marine sediment cores are usually dated using radiocarbon, but the sporadic occurrence of carbon-datable material makes it difficult to identify changes in sedimentation rate. The advantage of OSL is that the sampling interval can be chosen at will, allowing changes in sedimentation rate to become apparent (Kim *et al.* 2015; Sugisaki *et al.* 2015). In this setting, the combination of OSL and radiocarbon dating becomes particularly useful. Nonetheless, there is room for discussion as to whether the apparent changes in sedimentation rates reflect changes in terrigenous sediment supply or subaqueous sediment reworking, or are artefacts caused by dating errors (Sugisaki *et al.* 2015).

8.6 DEEP-SEA MARINE SETTINGS

Luminescence dating of deep-sea sediments has a long history, with the first thermoluminescence ages being reported in 1979 (Wintle and Huntley 1979), and the first optically stimulated luminescence ages in 2003 (Jakobsson *et al.* 2003; Stokes *et al.* 2003). However, there are relatively few published luminescence chronologies for deep-sea sediments, since accurate determination of both the environmental dose rate and equivalent dose are often problematic. Nonetheless, more conventional methods for establishing chronologies for deep-sea sediments such as orbital tuning (Lisiecki and Raymo 2005), identification of known-age event horizons (Collins *et al.* 2012; Matthews *et al.* 2015) and/or radiocarbon dating are not universally applicable, meaning that OSL dating of these deposits could be of considerable scientific value. This section outlines the main considerations relating specifically to luminescence dating of deep-sea sediments, though it should be noted that many of the complications associated with dating terrestrial sediments will also apply. The term 'deep-sea' is used here to indicate sediments deposited under an appreciable depth (>~100 m) of seawater, and to distinguish them from the sediments discussed in Section 8.5.

8.6.1 Environmental dose rate

Calculation of the environmental dose rate experienced by deep-sea sediments is both simplified and complicated by the overlying water column. Calculations are simplified since the cosmic-ray dose rate drops to a negligible value (0.006 Gy/ka) by 100 m water depth, and may be omitted without introducing a significant error. Also, deep-sea sediments are water-saturated, meaning that climatic factors do not alter the mean burial water content. In cases where no systematic decrease in water content with depth is observed (Sugisaki *et al.* 2010), present-day measured water contents may be used when calculating the dose rate. However, some cores appear to undergo dewatering as sediments are compacted by the weight of overlying material, in which case it may be necessary to account for this process either by modelling dewatering (Sugisaki *et al.* 2012) or by adopting a large uncertainty term on the water content (Armitage 2015). However, in the majority of cases the environmental dose rate calculation for deep-sea sediments is more complicated than for terrestrial sediments, owing to pronounced disequilibrium in both uranium decay series. This disequilibrium occurs due to the presence of long-lived insoluble isotopes, and to the insolubility of uranium itself in anoxic or suboxic conditions (Henderson and Anderson 2003). These two forms of disequilibrium are independent of one another, and will be discussed separately.

8.6.1.1 Long-lived insoluble isotopes

When considering the effects of disequilibrium on the dose rate, it is only the long-lived isotopes (half-life > 1,000 years) in each decay series that are important: only these isotopes exist for sufficient time for their chemical behaviour to cause a physical separation from their parents or progeny. In the case of the ^{238}U decay series, the long-lived isotopes are ^{238}U, ^{234}U, ^{230}Th and ^{226}Ra, while for the ^{235}U decay series the isotopes of interest are ^{235}U and ^{231}Pa. Of these, only ^{230}Th and ^{231}Pa are insoluble in the oxidising conditions typical of seawater (Henderson and Anderson 2003). Once produced, the insoluble isotopes adhere to particles in the water column and settle to the seafloor over a period of years, leading to the incorporation of 'excess' ^{230}Th and ^{231}Pa into deep-sea sediments. Here the term excess is used to denote the activity of the daughter isotope above that of the parent isotope, i.e. the additional activity associated with the incorporation of unsupported isotopes into the sediment body. Since these excess isotopes (denoted ^{230}Th$_{xs}$ and ^{231}Pa$_{xs}$ hereafter) are not supported by their respective parent isotopes, their activity decreases over time, creating a time-dependent component in the dose rate calculation. Both isotopes have half-lives similar to the timeframe over which luminescence dating is practical, so it is often necessary to measure and account for them so that the accuracy of the age estimate is not affected (Stokes *et al.* 2003; Wintle and Huntley 1979). Failing to account for ^{230}Th$_{xs}$ and ^{231}Pa$_{xs}$ in dose rate calculations will lead to an age overestimate, though the size of this error will depend upon the method used for dose rate determination, and upon the concentration of excess isotopes. The latter is controlled by the depth of water overlying the sample site, particle flux and sedimentation rate, meaning that the effects of excess isotopes upon the dose rate are both spatially and temporally variable. However, data from Ocean Drilling Program (ODP) site 658B illustrate the importance of correcting for excess insoluble isotopes. At this site ^{230}Th$_{xs}$ concentrations over the period 2–18.5 ka average 36.6 ± 9.3 Bq/kg (~0.4 Gy/ka), which would lead to a 22 ± 2% age overestimate if secular equilibrium in the uranium decay series was assumed (Adkins *et al.* 2006; Armitage 2015).

Measurement of ^{230}Th$_{xs}$ is more straightforward than ^{231}Pa$_{xs}$ since the former is more abundant. Originally ^{230}Th$_{xs}$ was measured using thick-source alpha counting (Huntley and Wintle 1981; Wintle and Huntley 1979), and this technique is still a convenient method for determining whether appreciable ^{230}Th$_{xs}$ is present (Berger 2006). More precise alternatives such as alpha spectrometry (Stokes *et al.* 2003) and mass spectrometry (Armitage 2015) have been preferred in recent studies. Having determined ^{230}Th$_{xs}$ activity, it is probably sufficient to calculate ^{231}Pa$_{xs}$ by assuming it is incorporated into the sediment at the seawater production activity ratio of ^{231}Pa/^{230}Th = 0.093 (Armitage 2015; Henderson and Anderson 2003). However, where excess isotopes make a large contribution to the total dose rate, direct measurement of ^{231}Pa$_{xs}$ may be advisable, since the ^{231}Pa/^{230}Th activity ratio in modern deep-sea sediments can vary from 0.03–0.30 (Henderson and Anderson 2003).

Because the conventional age-equation assumes secular equilibrium in the uranium and thorium decay series, it must be modified to account for the time-dependent dose rate changes which occur as excess activity in both the uranium decay series decrease. This calculation is described in detail by Stokes *et al.* (2003). Briefly, an iterative model is used whereby the cumulative dose deposited in the sample by (a) decay series in equilibrium and (b) ^{230}Th$_{xs}$, ^{231}Pa$_{xs}$ and their daughters, is calculated for a number of time periods. The age of the sample is determined by calculating the time at which the sum of these two doses is equal to the measured equivalent dose. There are a number of important points to consider when applying this model. First, it assumes that the ^{232}Th decay series is in equilibrium, since the parent is the only long-lived isotope present. Second, considerable amounts of uranium may be incorporated into deep-sea sediments within detrital mineral grains, and this uranium will be in equilibrium with all its daughters. In extreme cases, where deep-sea sediments receive a large terrestrial input, this may obviate the need to account for ^{230}Th$_{xs}$ and ^{231}Pa$_{xs}$ (Jakobsson *et al.* 2003). Third, both ^{230}Th and ^{231}Pa are the last long-lived isotopes in their decay series, meaning they attain equilibrium with their decay products geologically instantaneously. This considerably simplifies the calculation of their contribution to the total dose rate. Lastly, application of the Stokes *et al.* (2003) calculation method requires dose rates due to parts of the two uranium decay series (e.g. for ^{230}Th–^{206}Pb to calculate the cumulative dose due to ^{230}Th$_{xs}$), rather than for the entire decay series as is usual. Stokes *et al.* (2003) provide these values, based on dose rate conversion factors which are now outdated (Adamiec and Aitken 1998). The updated values presented by Armitage and Pinder (2017) based on new dose rate conversion factors (Guérin *et al.* 2011) should be used instead.

8.6.1.2 Authigenic uranium

Although anoxic and suboxic sediments are known to be the largest sink of oceanic uranium (Henderson and Anderson 2003), the implication of this for OSL dating of deep-sea material has only recently been identified. In material from ODP site 658B, Armitage (2015) observed a 27 ± 5% age underestimate due to the incorporation of seawater uranium into anoxic/suboxic sediments. Under the oxic conditions typical of most oceanic waters, uranium exists in its soluble hexavalent state, whereas under anoxic and suboxic conditions it is reduced to its insoluble hexavalent state. Since anoxic and suboxic conditions frequently occur at and immediately below the sediment-water interface – due to the decomposition of organic material – deep-sea sediments incorporate uranium isotopes from seawater. This 'authigenic' uranium (U_{auth} hereafter) is incorporated without decay products. In the case of the ^{238}U decay series, authigenic ^{238}U and ^{234}U (denoted $^{238}U_{auth}$ and $^{234}U_{auth}$ hereafter) will be incorporated into the sediment; but since the intervening isotopes are short lived, the ^{238}U-^{234}U portion instantly attains secular equilibrium. The immediate decay product of ^{234}U is ^{230}Th, which accumulates in the sediment at a rate governed by the half-life of ^{230}Th. As an added complication, U_{auth} will be incorporated into deep-sea sediment at the seawater activity ratio of $^{234}U/^{238}U = 1.14$ (Robinson *et al.* 2004), though fortunately the half-life of ^{234}U is sufficiently long (245 ka) for this ratio to be regarded as constant for the purpose of calculating dose rates. It should be noted that the 1.14 seawater activity ratio of $^{234}U/^{238}U$ only applies to U_{auth}, and not to detrital uranium. However, partitioning total measured uranium between authigenic and detrital sources is not straightforward. Because ^{232}Th is entirely detrital, it is possible to use it as a proxy for detrital uranium input. Assuming that the $^{238}U/^{232}Th$ activity ratio of crustal rocks and pelagic marine sediments of 0.8 ± 0.2 (Anderson *et al.* 1989) applies to deep-sea sediments, the $^{238}U_{auth}$ content of the sample may be determined from measured quantities of ^{238}U and ^{232}Th in the sample (Armitage 2015). This approach appears to yield satisfactory results (Armitage 2015; Armitage and Pinder 2017), although because the large uncertainty on the $^{238}U/^{232}Th$ activity ratio is propagated into the OSL age, it would be advantageous if a more precise method for determining U_{auth} content could be devised.

The presence of U_{auth} in a deep-sea sediment necessitates two modifications to the dose rate calculation described by Stokes *et al.* (2003). Firstly, total ^{238}U content must be partitioned between the detrital and authigenic components. Detrital ^{238}U is in equilibrium with its decay products, and the dose rate due to this component may be calculated using the standard conversion factors (Guérin *et al.* 2011). $^{238}U_{auth}$ is in equilibrium with the next two decay products, and at the seawater $^{234}U/^{238}U$ activity ratio of 1.14 with $^{234}U_{auth}$. For practical purposes, the dose rate due to the decay series $^{238}U–^{234}U$ is not time dependent, and may be calculated using the conversion factors of Armitage and Pinder (2017). Secondly, the dose rate due to the ingrowth of ^{230}Th and its decay products from $^{234}U_{auth}$ must be calculated. This dose rate varies over time, and must be calculated iteratively (Armitage and Pinder 2017). It should be noted at this point that no published OSL dating study presents direct measurements of U_{auth} content. The presence of U_{auth} in deep-sea sediments is geochemically plausible, and in the two studies where the correction described above has been applied, it yields better agreement between OSL and independent ages (Armitage 2015; Armitage and Pinder 2017). Nonetheless, this evidence is circumstantial, and the importance of U_{auth} in OSL dating is far less well established than that of $^{230}Th_{xs}$ and $^{231}Pa_{xs}$.

8.6.2 Equivalent dose

Specific considerations for measurement of equivalent dose from deep-sea sediment relate to the drilling equipment used (sample size), core analysis, core curation, and subaqueous sediment movement. Although deep-sea sediments may be recovered by dredge or

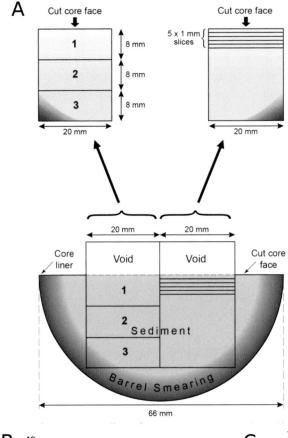

Figure 8.11 Sampling strategy for deep-sea ODP core 658B (modified from Armitage and Pinder 2017). A) The split core was sampled by pushing opaque tubes into the split core face. Segments 1 and 3 were discarded due to bleaching during storage and barrel smearing, respectively. B) Equivalent dose for two samples, plotted against the depth below the split core face. Sediment taken from >1 mm depth was inferred to have been shielded from light during core sampling and storage. C) Equivalent dose for two samples as a function of grain size. Coarser grains are less likely to be reworked on the sea floor, and so they provide a smaller and more accurate Equivalent dose.

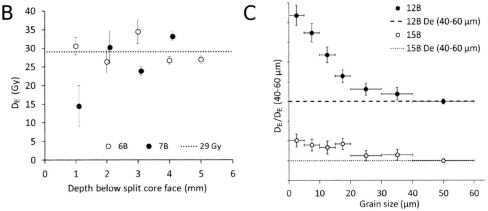

remotely operated vehicle, the majority of such materials available for research purposes are recovered as part of scientific drilling projects. These cores tend to have a relatively small diameter, such as the ODP's Advanced Piston Corer (62 mm internal diameter), and are often only available to the geochronologist after the core has been halved for inspection. Furthermore, the process of inserting a core barrel into soft sediments tends to drag younger sediments to deeper levels at the edges of the core in a phenomenon known as 'barrel-smear'. Consequently, the volume of sample available to the geochronologist is generally small e.g. Armitage and Pinder (2017) calculated that a 20 mm diameter sampling tube inserted into a split ODP core might yield ~2.5 cm³ of datable sample once light-exposed and barrel-smeared material had been removed. Depending on the sediment source, a high proportion of this material will consist of biogenic carbonate, opal and organic matter, meaning that in many cases the principal impediment to measuring an equivalent dose may be the absence of datable mineral grains.

Core analysis and curation procedures may alter the luminescence properties of deep-sea sediments. Several routine core scanning procedures involve irradiation. For example, ODP and its successor Integrated Ocean Drilling Program (IODP) routinely assess downcore variations in bulk density using gamma-ray attenuation measurements (Blum 1997), while split cores are frequently analysed using X-ray core scanning procedures. Couapel and Bowles (2006) demonstrated that radiation exposure during gamma-ray attenuation measurements had no measurable impact on the equivalent dose for a test sample, and grossly erroneous ages have not been observed for cores which have undergone this procedure (e.g. Armitage 2015). X-ray scanning is known to deliver a measurable radiation dose to the surface of the core (Davids et al. 2010a). Core curation procedures vary between institutions, but unless a core was collected for the purpose of OSL dating (e.g. Jakobsson et al. 2003) it is unlikely that subsequent core handling was carried out with regard to preserving light-sensitive signals. However, provided that a core is stored in cool and moist conditions which prevent desiccation or shrinkage, only the cut face of a core half should be exposed to laboratory light. In the case of ODP 658B, a core which had been stored for 16 years at the point of sampling, Armitage and Pinder (2017) were unable to detect bleaching below 1–2 mm from the cut face (Fig. 8.11). However, since light transmission through sediment is largely controlled by grain size (Ollerhead 2001) it would be sensible to test this for each core analysed.

In deep-sea sediments recovered from open ocean settings, it is likely that the dominant (possibly only) minerogenic component will be derived from far-travelled atmospheric dust, probably in the fine silt size fraction. However, where coarser fractions are present in sufficient abundance for their use to be viable, their use may be advantageous. This is because coarser grain sizes are easier to test for incomplete bleaching prior to deposition (Berger 2011; Olley et al. 2004) and because fine silt is prone to mobilisation and redeposition at depth without light exposure (Armitage 2015; Berger 2009). A number of authors (e.g. Stokes et al. 2003) have assumed that because the minerogenic fraction of deep-ocean sediments is usually aeolian in origin, it is unlikely to be incompletely bleached. However, Olley et al. (2004) applied single-grain OSL techniques to material from core Fr10/95-GC17, retrieved 60 km from the coast of Australia, and found evidence for incomplete bleaching in two of the seven samples measured. Although this core was located relatively close to land, it is clear that incomplete bleaching may occur in deep-sea settings, and coarser size fractions are required to detect this using single-grain techniques. If only fine-grained material is present, it has been proposed that incomplete bleaching may be detected in deep-sea sediments by dating both quartz and feldspar fractions (Yang et al. 2015). Because feldspar bleaches more slowly than quartz, agreement between the two dosimeters should indicate complete bleaching of both signals prior to deposition (see Section 8.5).

Once sediment has reached the sea floor, it is also prone to remobilisation either downslope in turbidity currents or along slope in contour currents. Since both processes occur underwater, the luminescence signal is not bleached during transportation, and grains are redeposited with a palaeodose which relates to the previous period of burial. Owing to the slower settling rate of finer grains, these grains are retained in suspension for longer and travel further than coarser material remobilised by the same current. Consequently, it would be expected that fine silt-sized grains are more likely to yield age overestimates than coarser grains (Berger 2006, 2009). Seafloor reworking of sediments has been used to explain gross age overestimates from sediment core tops from the Alaskan margin (Berger 2009) and a large but temporally variable offset between ages for paired fine silt (4–11 μm) and coarse silt (40–63 μm) fractions from a core from the north African margin (Armitage 2015). Despite these considerations, it is likely that in the majority of deep-sea settings, the geochronologist will be required to work with whatever size fraction is present, and in open ocean settings this is likely to be fine silt.

8.7 PLEISTOCENE BARRIERS

Raised beaches and fossilised barrier systems are identifiable in many regions, and play crucial role in the understanding of sea-level change and glacial cycles over the mid–late Pleistocene. As with other coastal sediments, they can prove extremely difficult to date with any method, and so the growth of luminescence methods over recent decades has seen repeated applications to barrier systems. Depending on the site, questions may concern eustatic sea-level change, glacio-isostacy, the formation process and sediment sources, and tectonic uplift rates (Murray-Wallace and Woodroffe 2014). Pleistocene shorelines may also have archaeological significance, with luminescence dates helping to answer questions in human evolution. A set of extensive raised beaches in Sussex, southern England, relate to interglacial sea-level highstands of the mid-Pleistocene, with the oldest beach containing a wealth of Palaeolithic artefacts and remains of *Homo heidelbergensis* (Bates *et al.* 2010). Fossil footprints in coastal sediment near East London, South Africa, have been dated with quartz OSL to the last interglacial (Jacobs and Roberts 2009).

In warm climates, carbonate-rich barriers preserve evidence of former sea-level highstands. Warm and shallow coastal waters promote the growth of marine invertebrates, with their shells breaking down into sand-sized particles which are pushed onshore (Bourman et al 2016). Dunes formed through the mixture of calcium carbonate and siliclastic sand become solidified over time, as the carbonate dissolves and re-precipitates through contact with meteoric water. The cementation of the barriers enhances their preservation potential, in some cases preventing the barrier from being eroded during subsequent marine transgressions. Carbonate also has low radionuclide content, meaning that the carbonate-rich barriers provide very low environmental dose rates to mineral grains – of the order of 0.50 Gy ka^{-1} (e.g. Huntley *et al.* 1994). This is highly advantageous for OSL dating, as it increases the upper age limit of any luminescence dating protocol, while the preservation of the barrier means that uncertainty over the sample depth over time (and thus the cosmic dose rate) is limited.

Luminescence dating of quartz has been crucial for dating the formation of carbonate-rich barriers in relation to global sea-level change. In southeast South Australia, a long sequence of coastal barriers has been stranded on the slowly uplifting Coorong Coastal plain. With the barriers each 1–3 km deep and formed during interglacial highstands, the uplift ensures that the barriers are physically separate. TL and early OSL methods, along with Amino Acid dating, have allowed many of the barriers to be correlated with the marine isotope record, forming a crucial piece of evidence on Pleistocene sea-level change from a far-field location (Banerjee *et al.* 2003; Huntley and Prescott 2001; Murray-Wallace *et al.* 2002). In Western Australia (Brooke *et al.* 2014) and South Africa (Bateman *et al.* 2011), Quaternary uplift is almost absent, and individual carbonate barriers may form over multiple interglacial high-stands. The identification of compound barriers may indeed depend on OSL dating, if the stratigraphical evidence is absent.

Provided the natural dose rates are less than ~1.5 Gy ka^{-1}, the last interglacial should be within the age range of quartz OSL dating, and comfortably within range of IR and post-IR methods. Nevertheless, there has been a long-standing concern that OSL does not perform as accurately as expected for samples of 100 ka or older. A tendency for quartz SAR ages to underestimate was noted in a review by Murray and Olley (2002), and observed by others (e.g. Stokes *et al.* 2003). An MIS 5e site in northern Russia was selected by Murray *et al.* (2007) to test quartz OSL methodology. This site had a reliable water-content estimate due to the presence of permafrost during most of the burial period. Using HRGS for dose rate estimation and multi-grain quartz OSL for D_E, the 16 ages from the site underestimated the known age by ~14%.

Some possible sources of D_E underestimates can be easily prevented: first, by ensuring that the net OSL signal is reflecting the quartz fast component, so the thermally less stable medium or slow components do not contribute to the equivalent dose estimate. Second, by ensuring that the dose response curve grows to a sufficiently high value so that the natural signal can be projected onto it, through use of a D_0 threshold as an acceptance criterion. While these checks are fairly common, it is evident that OSL ages of likely interglacial sediment are still over-scattered. A compilation of published data by Lamothe (2016) shows an abundance of quartz OSL ages in the range of 100–130 ka, suggesting a slight underestimate of MIS 5e is the norm – and is evident in carbonate-rich barriers in South Africa and Australia with otherwise good OSL properties (Bateman *et al.* 2011; Brooke *et al.* 2014; Carr *et al.* 2010). OSL dates of raised shorelines can prove very contentious where they conflict with other geological evidence of sea-level change (see Lamothe (2016) and references therein), so it is responsibility of the geochronologist to ensure known sources of error have been considered, and that steps have been taken to check accuracy of all measurements.

Evaluating D_E becomes problematic as the natural dose approaches saturation. For an aliquot or grain, slight errors in sensitivity correction of regenerative dose points can lead to significant error in D_E. Some grains or aliquots may be appear 'oversaturated' – when their natural OSL signal cannot be projected onto the dose response curve. Rejection of such aliquots, however, introduces a bias towards younger ages. A further bias is introduced when the CAM is used to estimate the burial dose. The CAM tends to underestimate the mean burial dose, most severely when the D_E distribution is highly overdispersed (OD), and dose estimates are derived from the non-linear portion of the dose-response curve (Guérin *et al.* 2017). The use of single-grain OSL dating for old, well-bleached material is then unlikely to offer any advantages over small aliquots.

Estimation of dose rates for interglacial sites can be problematic, with uncertainties surrounding water content and burial depth requiring special consideration. Dramatic changes in moisture are possible over glacial cycles, and the present depth of the sample may have been affected by recent erosion or sedimentation. In low-dose rate sediments, the estimated dose rate is likely to be very sensitive to the depth and water content assumptions. The likely timing of carbonate sedimentation should also be considered (Section 8.2.2). These issues concern the unknown conditions of the past, which require reasoned assumptions to be made (with an assessment of their uncertainty). However, there are also some fundamental problems of measuring the present-day dose rate. The presence of calcium carbonate can affect the dose rate estimates derived from some measurement techniques: radiation detectors that rely on the emission of gamma photons (gamma spectrometry) or beta particles (beta counting) are sensitive to the composition of the sample, due to the differences in self-attenuation. Elements with higher atomic numbers are more effective at stopping ionising radiation, meaning that more gamma photons or beta particles are absorbed in the material before reaching the detector. The effect is minor for quartz-rich samples, if the detector is calibrated using standards that have a similar matrix material. However, samples rich in other materials, such as calcium carbonate or iron oxide, will need special consideration.

A good strategy for dose rate estimation is to collect data using different methods, and assess the quality of the datasets according to the strengths and weaknesses of each method. For example, thick-source alpha counting is a precise method of estimating the *combined* beta and gamma dose rate from U and Th, but far less precise for assessing U and Th concentrations individually. This strategy can be assisted by a quality control procedure for each of type of instrument used. As an example, Carr *et al.* (2007) used several types of measurement for dating the Wilderness barrier in South Africa. They employed thick-source alpha counting, ICP-MS, and field gamma spectrometry, providing three independent estimates of the U and Th concentrations. By comparing the three measurements, they could identify an error in their ICP-MS measurements of U. It should be cautioned, however, that inferences on the state of U-series equilibrium should be made using data from a single sprectroscopic method (e.g. HRGS, alpha spectroscopy), and not from differences observed between unrelated methods.

The discussion above focuses on systematic errors in the equivalent dose and dose rate. It should also be recognised that random, sample-to-sample uncertainty in age can be significant for older samples, and larger than usually stated in the uncertainty terms. The reproducibility of quartz OSL has been tested at the MIS 5e site of Gammelmark, Denmark (Fig. 8.12; Buylaert *et al.* 2011; Murray and Funder 2003). Twenty samples from a 10 m section of marine sand produced a mean quartz OSL age of 114 ± 4 ka (random uncertainty only), slightly underestimating the expected age of 125–133 ka. However, the standard deviation in ages was 13 ka, and so the random uncertainty for a single sample at the site is ~11%. Random errors can be reduced by dating multiple OSL samples from the same stratigraphic unit.

Buylaert *et al.* (2011) also tested a feldspar IRSL protocol at the Gammelmark site. Feldspar IRSL dose–response curves saturate at a much higher dose than the quartz OSL, and could permit the routine dating of mid-Pleistocene sediment. The fading-corrected IRSL ages at Gammelmark are consistent with the quartz ages, despite the relatively severe correction that must be applied to samples >50 ka, and similar success was reported by Balescu *et al.* (2015) for an MIS 7 shoreline in Tunisia. However, the development of post-IR protocol now allows feldspar-based age estimates to be obtained with little or no need for fading correction (Thomsen *et al.* 2008), and tests of the method against independent age control have proven successful (Buylaert *et al.* 2012; Kars *et al.* 2012). An initial

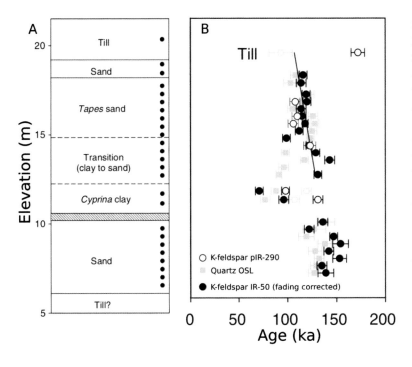

Figure 8.12
Dating results
from the coastal
MIS 5e site in
Gammelmark,
Denmark
(modified from
Buylaert *et al.*
2011, 2012;
Murray and
Funder, 2003).
A) The main
sedimentary units
– shallow marine
sands and clays
bracketed by till.
B) OSL ages using
quartz, and IR
and post-IR ages
using K-feldspar.

application to interglacial coastal sediments beyond MIS 5 has also been very encouraging (Thiel *et al.* 2012).

The cautions expressed above should not detract from the utility of OSL in dating interglacial barrier or beach deposits. Standard SAR OSL with fast-component-dominated quartz should be used with confidence to date MIS 5 sediments, provided the natural dose rate is relatively low. Due consideration should be given to the dose rate side of the age equation, which is equally important to get right as the equivalent dose. Importantly, the development of a post-IR protocol for feldspars provides a means to date sediments up to 500 ka and perhaps further. The method is supported by theoretical models, and has been well tested using samples with independent age control; its potential for dating mid-Pleistocene coastal sediment is clear.

8.8 SUMMARY AND FUTURE PROSPECTS

Luminescence dating has been applied with particular success to coastal sediments, and is now a valuable tool for coastal geoscientists. Clastic coastal sediment is both well suited for OSL methods, and lacking in datable material suitable for other radiometric methods. Beach and dune sediment is likely to have received enough sunlight before burial to reset the quartz OSL signal, and so dating of very recent sediment, e.g. 10–300 a, is fairly routine. Holocene material is comfortably within the applicable age range, and OSL dating is now commonly applied to beach-barrier systems, and back-barrier overwash records of storms and tsunamis. Where environmental dose rates are relatively low, OSL dating has helped to compare raised beach and barrier deposits with records of global sea-level change.

Development in instrumentation, measurement protocols, and analytical methods is helping to increase the range of sediments to which OSL dating can be applied. Encouraging results have emerged from studies of deltaic and shallow marine sediments, where the sometimes reduced quality of bleaching is manageable through the use of statistical or signal processing methods. In this setting, high-resolution sampling can add detail to a chronology, which may otherwise depend on the sporadic occurrence of radiocarbon-datable material. The OSL dating of deep-sea sediment cores remains an enticing prospect, given the paucity of absolute methods beyond the radiocarbon age range; dose rate estimation remains a tricky, but not insurmountable problem.

The development of the post-IR protocol now permits the dating of feldspar with little or no need for fading correction. The far-higher dose saturation level of feldspar, compared to quartz, permits the dating of sediment to ~500 ka, and perhaps further; this places a number of mid-Pleistocene sea-level highstands newly within the upper dating limit.

REFERENCES

Adamiec, G., Aitken, M.J. 1998. Dose rate conversion factors: update. *Ancient TL* 16, 37–50.

Adkins, J., deMenocal, P., Eshel, G. 2006. The 'African humid period' and the record of marine upwelling from excess 230Th in Ocean Drilling Program Hole 658C. *Paleoceanography* 21, PA4203.

Aitken, M.J., 1985. *Thermoluminescence Dating*. Academic Press.

Andersen, T.J., Pejrup, M., Nielsen, A.A. 2006. Long-term and high-resolution measurements of bed level changes in a temperate, microtidal coastal lagoon. *Marine Geology* 226, 115–125.

Anderson, R.F., LeHuray, A.P., Fleisher, M.Q., Murray, J.W. 1989. Uranium deposition in saanich inlet sediments, vancouver island. *Geochimica et Cosmochimica Acta* 53, 2205–2213.

Armitage, S.J. 2015. Optically stimulated luminescence dating of Ocean Drilling Program core 658B: Complications arising from authigenic uranium uptake and lateral sediment movement. *Quaternary Geochronology* 30, 270–274.

Armitage, S.J., Pinder, R.C. 2017. Testing the applicability of optically stimulated luminescence dating to Ocean Drilling Program cores. *Quaternary Geochronology* 39, 124–130.

Balescu, S., Huot, S., Mejri, H., Barré, M., Forget Brisson, L., Lamothe, M., Oueslati, A. 2015. Luminescence dating of Middle Pleistocene (MIS 7) marine shoreline deposits along the eastern coast of Tunisia: A comparison of K-feldspar and Na-feldspar IRSL ages. *Quaternary Geochronology, LED14 Proceedings* 30, 288–293.

Ballarini, M., Wallinga, J., Murray, A.S., van Heteren, S., Oost, A.P., Bos, A.J.J., Van Eijk, C.W.E. 2003. Optical dating of young coastal dunes on a decadal time scale. *Quaternary Science Reviews* 22, 1011–1017.

Ballarini, M., Wallinga, J., Duller, G.A.T., Brouwer, J.C., Bos, A.J.J., Van Eijk, C.W.E. 2005. Optimizing detection filters for single-grain optical dating of quartz. *Radiation Measurements* 40, 5–12.

Ballarini, M., Wallinga, J., Wintle, A.G., Bos, A.J.J. 2007. A modified SAR protocol for optical dating of individual grains from young quartz samples. *Radiation Measurements* 42, 360–369.

Banerjee, D., Murray, A.S., Foster, I.D.L. 2001. Scilly Isles, UK: optical dating of a possible tsunami deposit from the 1755 Lisbon earthquake. *Quaternary Science Reviews* 20, 715–718.

Banerjee, D., Hildebrand, A.N., Murray-Wallace, C.V., Bourman, R.P., Brooke, B.P., Blair, M. 2003. New quartz SAR-OSL ages from the stranded beach dune sequence in south-east South Australia. *Quaternary Science Reviews* 22, 1019–1025.

Bateman, M.D., Carr, A.S., Dunajko, A.C., Holmes, P.J., Roberts, D.L., McLaren, S.J., Bryant, R.G., Marker, M.E., Murray-Wallace, C.V. 2011. The evolution of coastal barrier systems: A case study of the Middle–Late Pleistocene Wilderness barriers, South Africa. *Quaternary Science Reviews* 30, 63–81.

Bates, M.R., Briant, R.M., Rhodes, E.J., Schwenninger, J.-L., Whittaker, J.E. 2010. A new chronological framework for Middle and Upper Pleistocene landscape evolution in the Sussex/Hampshire Coastal Corridor, UK. *Proceedings of the Geologists' Association* 121, 369–392.

Berger, G.W. 2006. Trans-arctic-ocean tests of fine-silt luminescence sediment dating provide a basis for an additional geochronometer for this region. *Quaternary Science Reviews* 25, 2529–2551.

Berger, G.W. 2009. Zeroing tests of luminescence sediment dating in the Arctic Ocean: Review and new results from Alaska-margin core tops and central-ocean dirty sea ice. *Global and Planetary Change* 68, 48-57.

Berger, G.W. 2011. Surmounting luminescence age overestimation in Alaska-margin Arctic Ocean sediments by use of 'micro-hole' quartz dating. *Quaternary Science Reviews* 30, 1750-1769.

Bishop, P., Sanderson, D., Hansom, J.I.M., Chaimanee, N. 2005. Age-dating of tsunami deposits: lessons from the 26 December 2004 tsunami in Thailand. *The Geographical Journal* 171, 379–384.

Blum, P. 1997. Physical properties handbook: a guide to the shipboard measurement of physical properties of deep-sea cores. ODP Tech. Note 26.

Bourman, R.P., Murray-Wallace, C.V., Harvey, N. 2016. *Coastal Landscapes of South Australia.* University of Adelaide Press.

Brill, D., Klasen, N., Brückner, H., Jankaew, K., Scheffers, A., Kelletat, D., Scheffers, S. 2012. OSL dating of tsunami deposits from Phra Thong Island, Thailand. *Quaternary Geochronology* 10, 224–229.

Brill, D., Jankaew, K., Brückner, H. 2015. Holocene evolution of Phra Thong's beach-ridge plain (Thailand) – chronology, processes and driving factors. *Geomorphology* 245, 117–134.

Brill, D., May, S.M., Shah-Hosseini, M., Rufer, D., Schmidt, C., Engel, M. 2017. Luminescence dating of cyclone-induced washover fans at Point Lefroy (NW Australia). *Quaternary Geochronology* 41, 134–150.

Bristow, C.S., Pucillo, K. 2006. Quantifying rates of coastal progradation from sediment volume using GPR and OSL: the Holocene fill of Guichen Bay, south-east South Australia. *Sedimentology* 53, 769–788.

Brooke, B.P., Olley, J.M., Pietsch, T., Playford, P.E., Haines, P.W., Murray-Wallace, C.V., Woodroffe, C.D. 2014. Chronology of Quaternary coastal aeolianite deposition and the drowned shorelines of southwestern Western Australia – a reappraisal. *Quaternary Science Reviews* 93, 106–124.

Buylaert, J.P., Murray, A.S., Vandenberghe, D., Vriend, M., De Corte, F., Van den haute, P. 2008. Optical dating of Chinese loess using sand-sized quartz: Establishing a time frame for Late Pleistocene climate changes in the western part of the Chinese Loess Plateau. *Quaternary Geochronology* 3, 99–113.

Buylaert, J.-P., Huot, S., Murray, A.S., Van Den Haute, P. 2011. Infrared stimulated luminescence dating of an Eemian (MIS 5e) site in Denmark using K-feldspar. Boreas 40, 46–56.

Buylaert, J.-P., Jain, M., Murray, A.S., Thomsen, K.J., Thiel, C., Sohbati, R. 2012. A robust feldspar luminescence dating method for Middle and Late Pleistocene sediments. *Boreas* 41, 435–451.

Buynevich, I.V., FitzGerald, D.M., Goble, R.J. 2007. A 1500 yr record of North Atlantic storm activity based on optically dated relict beach scarps. *Geology* 35, 543–546.

Carr, A.S., Bateman, M.D., Holmes, P.J. 2007. Developing a 150ka luminescence chronology for the barrier dunes of the southern Cape, South Africa. *Quaternary Geochronology*, LED 2005 2, 110–116.

Carr, A.S., Bateman, M.D., Roberts, D.L., Murray-Wallace, C.V., Jacobs, Z., Holmes, P.J. 2010. The last interglacial sea-level high stand on the southern Cape coastline of South Africa. *Quaternary Research* 73, 351–363.

Chamberlain, E.L., Wallinga, J., Reimann, T., Goodbred, S.L., Steckler, M.S., Shen, Z., Sincavage, R. 2017. Luminescence dating of delta sediments: Novel approaches explored for the Ganges-Brahmaputra-Meghna Delta. *Quaternary Geochronology* 41, 97–111.

Chaumillon, E., Bertin, X., Fortunato, A.B., Bajo, M., Schneider, J.-L., Dezileau, L., Walsh, J.P., Michelot, A., Chauveau, E., Créach, A., Hénaff, A., Sauzeau, T., Waeles, B., Gervais, B., Jan, G., Baumann, J., Breilh, J.-F., Pedreros, R. 2017. Storm-induced marine flooding: Lessons from a multidisciplinary approach. *Earth-Science Reviews* 165, 151–184.

Clemmensen, L.B., Glad, A.C., Kroon, A. 2016. Storm flood impacts along the shores of micro-tidal inland seas: A morphological and sedimentological study of the Vesterlyng beach, the Belt Sea, Denmark. *Geomorphology* 253, 251–261.

Collins, L.G., Hounslow, M.W., Allen, C.S., Hodgson, D.A., Pike, J., Karloukovski, V.V. 2012. Palaeomagnetic and biostratigraphic dating of marine sediments from the Scotia Sea, Antarctica: First identification of the Laschamp excursion in the Southern Ocean. *Quaternary Geochronology* 7, 67–75.

Couapel, M.J., Bowles, C.J. 2006. Impact of gamma densitometry on the luminescence signal of quartz grains. *Geo-Marine Letters* 26, 1–5.

Cunha, P.P., Buylaert, J.-P., Murray, A.S., Andrade, C., Freitas, M.C., Fatela, F., Munhá, J.M., Martins, A.A., Sugisaki, S. 2010. Optical dating of clastic deposits generated by an extreme marine coastal flood: The 1755 tsunami deposits in the Algarve (Portugal). *Quaternary Geochronology* 5, 329–335.

Cunningham, A.C. 2016. External beta dose rates to mineral grains in shell-rich sediment. *Ancient TL* 34, 1–5.

Cunningham, A.C., Wallinga, J. 2009. Optically stimulated luminescence dating of young quartz using the fast component. Radiation Measurements, *Proceedings of the 12th International Conference on Luminescence and Electron Spin Resonance Dating (LED 2008)* 44, 423–428.

Cunningham, A.C., Wallinga, J. 2010. Selection of integration time intervals for quartz OSL decay curves. *Quaternary Geochronology* 5, 657–666.

Cunningham, A.C., Bakker, M.A., van Heteren, S., van der Valk, B., van der Spek, A.J., Schaart, D.R., Wallinga, J. 2011. Extracting storm-surge data from coastal dunes for improved assessment of flood risk. *Geology* 39, 1063–1066.

Cunningham. A.C., Murray, A.S., Armitage, S.J., Autzen, M. 2018. High-precision natural dose rate estimates through beta counting. *Radiation Measurements*, in press.

Davids, F., Duller, G.A., Roberts, H.M. 2010a. Testing the use of feldspars for optical dating of hurricane overwash deposits. Quaternary Geochronology 5, 125–130.

Davids, F., Roberts, H.M., Duller, G.A. 2010b. Is X-ray core scanning non-destructive? Assessing the implications for optically stimulated luminescence (OSL) dating of sediments. *Journal of Quaternary Science* 25, 348–353.

Donnelly, J.P., Roll, S., Wengren, M., Butler, J., Lederer, R., Webb, T. 2001. Sedimentary evidence of intense hurricane strikes from New Jersey. *Geology* 29, 615–618.

Fruergaard, M., Kroon, A. 2016. Morphological response of a barrier island system on a catastrophic event: the AD 1634 North Sea storm. *Earth Surface Processes and Landforms* 41, 420-426.

Fruergaard, M., Andersen, T.J., Johannessen, P.N., Nielsen, L.H., Pejrup, M. 2013. Major coastal impact induced by a 1000-year storm event. *Sci Rep* 3, 1051.

Fruergaard, M., Andersen, T.J., Nielsen, L.H., Johannessen, P.N., Aagaard, T., Pejrup, M. 2015a. High-resolution reconstruction of a coastal barrier system: Impact of Holocene sea-level change. *Sedimentology* 62, 928–969.

Fruergaard, M., Pejrup, M., Murray, A.S., Andersen, T.J. 2015b. On luminescence bleaching of tidal channel sediments. *Geografisk Tidsskrift-Danish Journal of Geography* 115, 57–65.

Galbraith, R.F., Roberts, R.G., Laslett, G.M., Yoshida, H., Olley, J.M. 1999. Optical dating of single and multiple grains of quartz from Jinmium rock shelter, northern Australia: Part I, experimental design and statistical models. *Archaeometry* 41, 339–364.

Gao, L., Long, H., Shen, J., Yu, G., Liao, M., Yin, Y. 2017. Optical dating of Holocene tidal deposits from the southwestern coast of the South Yellow Sea using different grain-size quartz fractions. *Journal of Asian Earth Sciences* 135, 155–165.

Giosan, L., Donnelly, J.P., Constantinescu, S., Filip, F., Ovejanu, I., Vespremeanu-Stroe, A., Vespremeanu, E., Duller, G.A. 2006. Young Danube delta documents stable Black Sea level since the middle Holocene: Morphodynamic, paleogeographic, and archaeological implications. *Geology* 34, 757–760.

Guérin, G., Mercier, N., Adamiec, G. 2011. Dose rate conversion factors: update. *Ancient TL* 29, 5–8.

Guérin, G., Christophe, C., Philippe, A., Murray, A.S., Thomsen, K.J., Tribolo, C., Urbanova, P., Jain, M., Guibert, P., Mercier, N., Kreutzer, S., Lahaye, C. 2017. Absorbed dose, equivalent dose, measured dose rates, and implications for OSL age estimates: Introducing the Average Dose Model. *Quaternary Geochronology* 41, 163–173.

Hall, A.M., Hansom, J.D., Williams, D.M., Jarvis, J. 2006. Distribution, geomorphology and lithofacies of cliff-top storm deposits: Examples from the high-energy coasts of Scotland and Ireland. *Marine Geology* 232, 131–155.

Henderson, G.M., Anderson, R.F. 2003. The U-series Toolbox for Paleoceanography. *Reviews in Mineralogy and Geochemistry* 52, 493–531.

Hesp, P.A. 1984. Foredune formation in southeast Australia. *Coastal Geomorphology in Australia*. Academic Press, Sydney 69–97.

Huntley, D.J., Clague, J.J. 1996. Optical dating of tsunami-laid sands. *Quaternary Research* 46, 127–140.

Huntley, D.J., Godfrey-Smith, D.I., Thewalt, M.L. 1985. Optical dating of sediments. *Nature* 313, 105–107.

Huntley, D.J., Prescott, J.R. 2001. Improved methodology and new thermoluminescence ages for the dune sequence in south-east South Australia. *Quaternary Science Reviews* 20, 687–699.

Huntley, D.J., Wintle, A.G. 1981. The use of alpha scintillation counting for measuring Th-230 and Pa-231 contents of ocean sediments. *Canadian Journal of Earth Sciences* 18, 419–432.

Huntley, D.J., Hutton, J.T., Prescott, J.R. 1993. The stranded beach-dune sequence of south-east South Australia: A test of thermoluminescence dating, 0–800 ka. *Quaternary Science Reviews* 12, 1–20.

Huntley, D.J., Hutton, J.T., Prescott, J.R. 1994. Further thermoluminescence dates from the dune sequence in the southeast of South Australia. *Quaternary Science Reviews* 13, 201–207.

Isla, F.I., Bujalesky, G.G. 2000. Cannibalisation of Holocene gravel beach-ridge plains, northern Tierra del Fuego, Argentina. *Marine Geology* 170, 105–122.

Jacobs, Z., Roberts, D.L. 2009. Last Interglacial Age for aeolian and marine deposits and the Nahoon fossil human footprints, Southeast Coast of South Africa. *Quaternary Geochronology* 4, 160–169.

Jakobsson, M., Backman, J., Murray, A., Løvlie, R. 2003. Optically Stimulated Luminescence dating supports central Arctic Ocean cm-scale sedimentation rates. *Geochemistry, Geophysics, Geosystems* 4.

Kars, R.H., Busschers, F.S., Wallinga, J. 2012. Validating post IR-IRSL dating on K-feldspars through comparison with quartz OSL ages. *Quaternary Geochronology* 12, 74–86.

Kars, R.H., Reimann, T., Ankjærgaard, C., Wallinga, J. 2014. Bleaching of the post-IR IRSL signal: new insights for feldspar luminescence dating. *Boreas* 43, 780–791.

Kim, J.C., Cheong, D., Shin, S., Park, Y.-H., Hong, S.S. 2015. OSL chronology and accumulation rate of the Nakdong deltaic sediments, southeastern Korean Peninsula. *Quaternary Geochronology* 30, 245–250.

Lamothe, M., 2016. Luminescence dating of interglacial coastal depositional systems: Recent developments and future avenues of research. *Quaternary Science Reviews* 146, 1–27.

Lisiecki, L.E., Raymo, M.E. 2005. A Pliocene–Pleistocene stack of 57 globally distributed benthic $\delta 18O$ records. *Paleoceanography* 20, PA1003.

López, G.I., Rink, W.J. 2007. Characteristics of the burial environment related to quartz SAR-OSL dating at St. Vincent Island, NW Florida, USA. *Quaternary Geochronology* 2, 65–70.

Madsen, A.T., Murray, A.S. 2009. Optically stimulated luminescence dating of young sediments: A review. *Geomorphology* 109, 3–16.

Madsen, A.T., Murray, A.S., Andersen, T.J., Pejrup, M., Breuning-Madsen, H. 2005. Optically stimulated luminescence dating of young estuarine sediments: A comparison with ^{210}Pb and ^{137}Cs dating. *Marine Geology* 214, 251–268.

Madsen, A.T., Murray, A.S., Andersen, T.J., Pejrup, M. 2007a. Temporal changes of accretion rates on an estuarine salt marsh during the late Holocene – reflection of local sea level changes? The Wadden Sea, Denmark. *Marine Geology* 242, 221–233.

Madsen, A.T., Murray, A.S., Andersen, T.J., Pejrup, M. 2007b. Optical dating of young tidal sediments in the Danish Wadden Sea. *Quaternary Geochronology* 2, 89–94.

Madsen, A.T., Duller, G.A.T., Donnelly, J.P., Roberts, H.M., Wintle, A.G. 2009. A chronology of hurricane landfalls at Little Sippewissett Marsh, Massachusetts, USA, using optical dating. *Geomorphology* 109, 36–45.

Martin, L., Mercier, N., Incerti, S., Lefrais, Y., Pecheyran, C., Guérin, G., Jarry, M., Bruxelles, L., Bon, F., Pallier, C. 2015. Dosimetric study of sediments at the beta dose rate scale: Characterization and modelization with the DosiVox software. *Radiation Measurements* 81, 134–141.

Maselli, V., Trincardi, F. 2013. Man made deltas. *Sci Rep* 3.

Matthews, I.P., Trincardi, F., Lowe, J.J., Bourne, A.J., MacLeod, A., Abbott, P.M., Andersen, N., Asioli, A., Blockley, S.P.E., Lane, C.S., Oh, Y.A., Satow, C.S., Staff, R.A., Wulf, S. 2015. Developing a robust tephrochronological framework for Late Quaternary marine records in the Southern Adriatic Sea: new data from core station SA03-11. *Quaternary Science Reviews, Synchronising Environmental and Archaeological Records using Volcanic Ash Isochrons* 118, 84–104.

Mauz, B., Hoffman, D. 2014. What to do when carbonate replaced water: Carb, the model for estimating the dose rate of carbonate-rich samples. *Ancient TL* 32, 24–32.

Mauz, B., Baeteman, C., Bungenstock, F., Plater, A.J. 2010. Optical dating of tidal sediments: Potentials and limits inferred from the North Sea coast. *Quaternary Geochronology* 5, 667–678.

Mauz, B., Vacchi, M., Green, A., Hoffmann, G., Cooper, A. 2015. Beachrock: A tool for reconstructing relative sea level in the far-field. *Marine Geology* 362, 1–16.

May, S.M., Brill, D., Leopold, M., Callow, J.N., Engel, M., Scheffers, A., Opitz, S., Norpoth, M., Brückner, H. 2017. Chronostratigraphy and geomorphology of washover fans in the Exmouth Gulf (NW Australia): A record of tropical cyclone activity during the late Holocene. *Quaternary Science Reviews* 169, 65–84.

McKeever, S.W.S., Bøtter-Jensen, L., Agersnap Larsen, N., Mejdahl, V., Poolton, N.R.J. 1996. Optically stimulated luminescence sensitivity changes in quartz due to repeated use in single aliquot readout: experiments and computer simulations. *Radiation Protection Dosimetry* 65, 49–54.

Murari, M.K., Achyuthan, H., Singhvi, A.K. 2007. Luminescence studies on the sediments laid down by the December 2004 tsunami event: Prospects for the dating of palaeo tsunamis and for the estimation of sediment fluxes. *Current Science* 92, 367–371.

Murray, A., Buylaert, J.-P., Henriksen, M., Svendsen, J.-I., Mangerud, J. 2008. Testing the reliability of quartz OSL ages beyond the Eemian. *Radiation Measurements* 43, 776–780.

Murray, A., Buylaert, J.-P., Thiel, C. 2015. A luminescence dating intercomparison based on a Danish beach-ridge sand. *Radiation Measurements* 81, 32–38.

Murray, A.S., Funder, S. 2003. Optically stimulated luminescence dating of a Danish Eemian coastal marine deposit: A test of accuracy. *Quaternary Science Reviews* 22, 1177–1183.

Murray, A.S., Olley, J.M. 2002. Precision and accuracy in the optically stimulated luminescence dating of sedimentary quartz: A status review. *Geochronometria* 21, 1–16.

Murray, A.S., Svendsen, J.I., Mangerud, J., Astakhov, V.I. 2007. Testing the accuracy of quartz OSL dating using a known-age Eemian site on the river Sula, northern Russia. *Quaternary Geochronology*, LED 2005 2, 102–109.

Murray, A.S., Thomsen, K.J., Masuda, N., Buylaert, J.-P., Jain, M. 2012. Identifying well-bleached quartz using the different bleaching rates of quartz and feldspar luminescence signals. *Radiation Measurements* 47, 688–695.

Murray-Wallace, C.V., Woodroffe, C.D. 2014. *Quaternary Sea-Level Changes: A Global Perspective.* Cambridge University Press.

Murray-Wallace, C.V., Banerjee, D., Bourman, R.P., Olley, J.M., Brooke, B.P. 2002. Optically stimulated luminescence dating of Holocene relict foredunes, Guichen Bay, South Australia. *Quaternary Science Reviews* 21, 1077–1086.

Nathan, R.P., Mauz, B. 2008. On the dose rate estimate of carbonate-rich sediments for trapped charge dating. *Radiation Measurements* 43, 14–25.

Nentwig, V., Tsukamoto, S., Frechen, M., Bahlburg, H. 2015. Reconstructing the tsunami record in Tirúa, Central Chile beyond the historical record with quartz-based SAR-OSL. *Quaternary Geochronology* 30, 299–305.

Nichol, S.L., Lian, O.B., Carter, C.H. 2003. Sheet-gravel evidence for a late Holocene tsunami run-up on beach dunes, Great Barrier Island, New Zealand. *Sedimentary Geology* 155, 129–145.

Nielsen, A., Murray, A.S., Pejrup, M., Elberling, B. 2006. Optically stimulated luminescence dating of a Holocene beach ridge plain in Northern Jutland, Denmark. *Quaternary Geochronology* 1, 305–312.

Ollerhead, J. 2001. Light transmittance through dry, sieved sand: Some test results. *Ancient TL* 19, 13–17.

Ollerhead, J., Huntley, D.J., Nelson, A.R., Kelsey, H.M. 2001. Optical dating of tsunami-laid sand from an Oregon coastal lake. *Quaternary Science Reviews* 20, 1915–1926.

Olley, J.M., Murray, A., Roberts, R.G. 1996. The effects of disequilibria in the uranium and thorium decay chains on burial dose rates in fluvial sediments. *Quaternary Science Reviews* 15, 751–760.

Olley, J.M., De Deckker, P., Roberts, R.G., Fifield, L.K., Yoshida, H., Hancock, G. 2004. Optical dating of deep-sea sediments using single grains of quartz: A comparison with radiocarbon. *Sedimentary Geology* 169, 175–189.

Orford, J.D., Murdy, J.M., Wintle, A.G. 2003. Prograded Holocene beach ridges with superimposed dunes in north-east Ireland: Mechanisms and timescales of fine and coarse beach sediment decoupling and deposition. *Marine Geology* 194, 47–64.

Pietsch, T.J., Olley, J.M., Nanson, G.C. 2008. Fluvial transport as a natural luminescence sensitiser of quartz. *Quaternary Geochronology* 3, 365–376.

Plater, A.J., Stupples, P., Roberts, H.M. 2009. Evidence of episodic coastal change during the Late Holocene: The Dungeness barrier complex, SE England. *Geomorphology* 104, 47–58.

Prendergast, A.L., Cupper, M.L., Jankaew, K., Sawai, Y. 2012. Indian Ocean tsunami recurrence from optical dating of tsunami sand sheets in Thailand. *Marine Geology* 295, 20–27.

Preusser, F., Muru, M., Rosentau, A. 2014. Comparing different post-IR IRSL approaches for the dating of Holocene coastal foredunes from Ruhnu Island, Estonia. *Geochronometria* 41, 342–351.

Reimann, T., Tsukamoto, S. 2012. Dating the recent past (<500 years) by post-IR IRSL feldspar – Examples from the North Sea and Baltic Sea coast. *Quaternary Geochronology, 13th International Conference on Luminescence and Electron Spin Resonance Dating – LED 2011 Dedicated to J. Prescott and G. Berger* 10, 180–187.

Reimann, T., Tsukamoto, S., Harff, J., Osadczuk, K., Frechen, M. 2011 Reconstruction of Holocene coastal foredune progradation using luminescence dating – An example from the Świna barrier (southern Baltic Sea, NW Poland). *Geomorphology* 132, 1–16.

Rink, W.J. 1999. Quartz luminescence as a light-sensitive indicator of sediment transport in coastal processes. *Journal of Coastal Research* 148–154.

Rink, W.J., Pieper, K.D. 2001. Quartz thermoluminescence in a storm deposit and a welded beach ridge. *Quaternary Science Reviews* 20, 815–820.

Roberts, H.M., Plater, A.J. 2007. Reconstruction of Holocene foreland progradation using optically stimulated luminescence (OSL) dating: an example from Dungeness, UK. *The Holocene* 17, 495–505.

Robinson, L.F., Belshaw, N.S., Henderson, G.M. 2004. U and Th concentrations and isotope ratios in modern carbonates and waters from the Bahamas. *Geochimica et Cosmochimica Acta* 68, 1777–1789.

Rodriguez, A.B., Meyer, C.T. 2006. Sea-level variation during the Holocene deduced from the morphologic and stratigraphic evolution of Morgan Peninsula, Alabama, USA. *Journal of Sedimentary Research* 76, 257–269.

Sanderson, D.C., Murphy, S. 2010. Using simple portable OSL measurements and laboratory characterisation to help understand complex and heterogeneous sediment sequences for luminescence dating. *Quaternary Geochronology* 5, 299–305.

Sanderson, D.C.W., Bishop, P., Stark, M., Alexander, S., Penny, D. 2007. Luminescence dating of canal sediments from Angkor Borei, Mekong Delta, Southern Cambodia. *Quaternary Geochronology*, LED 2005 2, 322–329.

Sawai, Y., Namegaya, Y., Okamura, Y., Satake, K., Shishikura, M. 2012. Challenges of anticipating the 2011 Tohoku earthquake and tsunami using coastal geology. *Geophysical Research Letters* 39.

Sawakuchi, A.O., Blair, M.W., DeWitt, R., Faleiros, F.M., Hyppolito, T., Guedes, C.C.F. 2011. Thermal history versus sedimentary history: OSL sensitivity of quartz grains extracted from rocks and sediments. *Quaternary Geochronology* 6, 261–272.

Simms, A.R., DeWitt, R., Kouremenos, P., Drewry, A.M. 2011. A new approach to reconstructing sea levels in Antarctica using optically stimulated luminescence of cobble surfaces. *Quaternary Geochronology* 6, 50–60.

Sommerville, A.A., Hansom, J.D., Sanderson, D.C.W., Housley, R.A. 2003. Optically stimulated luminescence dating of large storm events in Northern Scotland. *Quaternary Science Reviews* 22, 1085–1092.

Spiske, M., Piepenbreier, J., Benavente, C., Kunz, A., Bahlburg, H., Steffahn, J. 2013. Historical tsunami deposits in Peru: Sedimentology, inverse modeling and optically stimulated luminescence dating. *Quaternary international* 305, 31–44.

Stokes, S., Ingram, S., Aitken, M.J., Sirocko, F., Anderson, R., Leuschner, D. 2003. Alternative chronologies for Late Quaternary (Last Interglacial–Holocene) deep sea sediments via optical dating of silt-sized quartz. *Quaternary Science Reviews* 22, 925–941.

Sugisaki, S., Buylaert, J.-P., Murray, A., Tada, R., Zheng, H., Ke, W., Saito, K., Chao, L., Li, S., Irino, T. 2015. OSL dating of fine-grained quartz from Holocene Yangtze delta sediments. *Quaternary Geochronology* 30, 226–232.

Sugisaki, S., Buylaert, J.-P., Murray, A., Tsukamoto, S., Nogi, Y., Miura, H., Sakai, S., Iijima, K., Sakamoto, T. 2010. High resolution OSL dating back to MIS 5e in the central Sea of Okhotsk. *Quaternary Geochronology* 5, 293–298.

Sugisaki, S., Buylaert, J.P., Murray, A.S., Harada, N., Kimoto, K., Okazaki, Y., Sakamoto, T., Iijima, K., Tsukamoto, S., Miura, H., Nogi, Y. 2012. High resolution optically stimulated luminescence dating of a sediment core from the southwestern Sea of Okhotsk. *Geochem. Geophys. Geosyst.* 13.

Tamura, T. 2012. Beach ridges and prograded beach deposits as palaeoenvironment records. *Earth-Science Reviews* 114, 279–297.

Tamura, T., Horaguchi, K., Saito, Y., Nguyen, V.L., Tateishi, M., Ta, T.K.O., Nanayama, F., Watanabe, K. 2010. Monsoon-influenced variations in morphology and sediment of a mesotidal beach on the Mekong River delta coast. *Geomorphology* 116, 11–23.

Tamura, T., Cunningham, A.C., Oliver, T.S.N. 2018a. Two-dimensional chronostratigraphic modelling of OSL ages from recent beach-ridge deposits, SE Australia. *Quaternary Geochronology*, in press.

Tamura, T., Nicholas, W.A., Oliver, T.S.N., Brooke, B.P. 2018b. Coarse-sand beach ridges at Cowley Beach, north-eastern Australia: Their formative processes and potential as records of tropical cyclone history. *Sedimentology* 65, 721–744.

Tamura, T., Saito, Y., Bateman, M.D., Nguyen, V.L., Ta, T.O., Matsumoto, D. 2012a. Luminescence dating of beach ridges for characterizing multi-decadal to centennial deltaic shoreline changes during Late Holocene, Mekong River delta. *Marine Geology* 326, 140–153.

Tamura, T., Saito, Y., Nguyen, V.L., Ta, T.O., Bateman, M.D., Matsumoto, D., Yamashita, S. 2012b. Origin and evolution of interdistributary delta plains: Insights from Mekong River delta. *Geology* 40, 303–306.

Tamura, T., Sawai, Y., Ito, K. 2015. OSL dating of the AD 869 Jogan tsunami deposit, northeastern Japan. *Quaternary Geochronology* 30, 294–298.

Taylor, M., Stone, G.W. 1996. Beach-ridges: a review. *Journal of Coastal Research* 12, 612–621.

Thiel, C., Buylaert, J.-P., Murray, A.S., Elmejdoub, N., Jedoui, Y. 2012. A comparison of TT-OSL and post-IR IRSL dating of coastal deposits on Cap Bon peninsula, north-eastern Tunisia. *Quaternary Geochronology, 13th International Conference on Luminescence and Electron Spin Resonance Dating – LED 2011 Dedicated to J. Prescott and G. Berger* 10, 209–217.

Thomsen, K.J., Murray, A.S., Jain, M., Bøtter-Jensen, L. 2008. Laboratory fading rates of various luminescence signals from feldspar-rich sediment extracts. *Radiation Measurements* 43, 1474–1486.

Truelsen, J.L., Wallinga, J. 2003. Zeroing of the OSL signal as a function of grain size: investigating bleaching and thermal transfer for a young fluvial sample. *Geochronometria* 22, 1–8.

van Heteren, S., Huntley, D.J., van de Plassche, O., Lubberts, R.K. 2000. Optical dating of dune sand for the study of sea-level change. *Geology* 28, 411–414.

Vespremeanu-Stroe, A., Zăinescu, F., Preoteasa, L., Tătui, F., Rotaru, S., Morhange, C., Stoica, M., Hanganu, J., Timar-Gabor, A., Cârdan, I., Piotrowska, N. 2017. Holocene evolution of the Danube delta: An integral reconstruction and a revised chronology. *Marine Geology* 388, 38–61.

Wintle, A.G., Huntley, D.J. 1979. Thermoluminescence dating of a deep-sea sediment core. *Nature* 279, 710–712.

Yang, L., Long, H., Yi, L., Li, P., Wang, Y., Gao, L., Shen, J. 2015. Luminescence dating of marine sediments from the Sea of Japan using quartz OSL and polymineral pIRIR signals of fine grains. *Quaternary Geochronology* 30, 257–263.

Zander, A., Degering, D., Preusser, F., Kasper, H.U., Brückner, H. 2007. Optically stimulated luminescence dating of sublittoral and intertidal sediments from Dubai, UAE: Radioactive disequilibria in the uranium decay series. *Quaternary Geochronology* 2, 123–128.

9 APPLICATIONS OF LUMINESCENCE DATING TO ACTIVE TECTONIC CONTEXTS

EDWARD J. RHODES[1] AND RICHARD T. WALKER[2]

[1]Geography Department, University of Sheffield, Winter St., Sheffield S10 2TN. Email: ed.rhodes@sheffield.ac.uk

[2] Department of Earth Sciences, University of Oxford, South Parks Road, Oxford, OX1 3AN

ABSTRACT: Tectonic forces generated within the Earth cause earthquakes, and these are sometimes associated with surface rupture. Repeated sequences of these events lead to significant landscape modification, including mountain building. Our understanding of these events and the processes involved comes from direct measurement and observations of the present day, combined with geological data from past events. Luminescence dating of sediments associated with tectonic events can contribute significantly to constraining their timing. Sediments may pre-exist an earthquake, and become deformed or offset by the fault slip and ground shaking. In this case they provide a maximum age estimate for an event. Alternatively, samples may be formed in direct response to ground movements; for example, small colluvial wedges of sediment built up against a newly formed fault scarp. Sediments may also be deposited over deformation structures, or within depressions produced during the earthquake; in both these cases, sediment age provides a minimum age estimate for the earthquake. Measurements of offset features such as terrace risers that have developed over several earthquake events can provide estimates of fault slip rates. There are particular issues associated with the application of luminescence dating of these contexts, but recent developments have produced results with important implications for the study of fault mechanics and seismic hazard.

KEYWORDS: surface rupture, palaeoseismology, slip rate, fault, terrace, riser, pIR-IRSL

9.1 INTRODUCTION

The primary focus of this chapter is to introduce and discuss the ways in which luminescence dating, primarily using single-grain pIR-IRSL dating of K-feldspar, or quartz OSL (see Chapter 1 for explanation of terms and introduction to the techniques), can be applied to determine fault slip rates, or the ages of ancient earthquake events, termed palaeoseismology. There are other applications to contexts that relate directly to tectonics, for example the dating of uplifted marine terraces or tsunami deposits (see

293

Chapter 8), and the determination of rock exhumation rates using low temperature thermochronometry (see Chapter 11), but these do not form the primary focus of this chapter. It is the particular issues that surround the application of luminescence dating to situations where the surface topography and/or superficial sediments were modified as a result of ground movements that take place during an earthquake that form the basis of the chapter.

Luminescence dating has some obvious advantages in comparison to other dating techniques for application to tectonic contexts. The clearest of these is the potential to date sediments directly, without the requirement to find rare constituents such as organic material for ^{14}C dating. There are also fewer limitations regarding post-event site evolution than associated with cosmogenic nuclide techniques including surface erosion or subsequent deposition. However, for reasons described below, these methods have been less widely applied in active tectonic settings than might be expected. Many tectonic contexts are characterised by sediment that is not well-suited for luminescence, both in terms of the sediment sources and the depositional environments encountered. Combined with a relatively small number of suitable geomorphic locations that preserve unambiguous information about fault slip or past earthquakes, these negative factors have tended to mitigate against the application of luminescence methods in active tectonic contexts. Recent developments in single-grain pIR-IRSL of potassium feldspar (Brown *et al.* 2015; Reimann *et al.* 2012; Rhodes 2015; Smedley and Duller 2013; Smedley *et al.* 2015; Trauerstein *et al.* 2014) have led to greater applicability, particularly in overcoming some of the limitations encountered with quartz OSL.

There are basically two main problems that must be overcome in tectonic contexts; these issues are not restricted to these locations, but are often greater in magnitude than in other settings. These are:

1. problems relating to the *luminescence characteristics* of mineral grains within available sediments
2. the preponderance of high energy environments associated with changing topography, typically combined with short transport distances, *limiting opportunities for signal zeroing* by daylight exposure.

Between them, these two considerations have led to these contexts being unattractive targets for luminescence specialists. In some cases, researchers have attempted to date sediments in active tectonic settings using standard approaches; in particular, using a conventional multiple-grain quartz SAR (single-aliquot regenerative dose; see Chapter 1) protocol without achieving satisfactory results. This apparent lack of success previously tended to relegate luminescence dating to a technique of last resort. In contrast, the apparent recent success of single-grain K-feldspar p-IR-IRSL dating is generating renewed interest in luminescence as a tool in active tectonic applications.

Luminescence sediment dating can contribute to our understanding of tectonic landscape development, fault movement, and the timing of ancient earthquakes in several different ways (Fattahi 2009). The following are four specific areas in which luminescence is of particular significance for these studies:

1. **Timescale** – there are particular needs to date outside the usual ^{14}C age range; that is, for age estimates younger than 250 years and older than 40,000 years. In areas without detailed historical records, such as the western USA and Alaska, seismic hazard estimation can benefit from age estimates for events within the last 250 years. For slow-moving faults characterised by fewer earthquakes, longer time range dating is useful.

2. **Independent age estimates** – luminescence does not depend on similar constraints or inputs to ^{14}C, cosmogenic nuclides and U-series, providing the opportunity to assess age control for tectonic processes with an independent chronometer, often useful when a high degree of reliability is required (for example in seismic hazard assessment).

3. **Material availability** – the most important benefit of luminescence dating is the wider range of geomorphic and sedimentary tectonic situations that can be dated in comparison to other dating approaches, and that finding material suitable for sampling typically requires much smaller-scale excavations, often simply a hand-dug pit, in surfaces that do not have the same strict requirement of being unchanged since deposition as is the case for cosmogenic nuclide dating. These aspects broaden the scope of possible applications besides reducing the time and cost involved.

4. **Additional information** – there is the possibility of developing increased understanding of the veracity of each age estimate and of the sedimentary environment relevant to interpretation of the site directly from the measured luminescence signals; in particular, using the measured degree of signal zeroing as an indicator of prior grain history.

As with all luminescence studies, several issues are regularly encountered (see Chapter 1 for details of these). Additionally, grains may be moved within sediment bodies after deposition, for example by bioturbation, including movement within animal burrows or root holes. Finally, there remains the possibility in all luminescence studies that the particular material measured from a site does not perfectly fulfil all the requirements for the dating protocol applied. Although some tests may be available to assess this (e.g. the dose recovery test; Wintle and Murray 2006), it is always possible that an error in the age estimate or associated uncertainty has been unwittingly introduced. Different means to assess the performance of luminescence methods have been devised, but perhaps the most important is comparison with fully independent age control (e.g. Rhodes 2015).

In principle, the potential and limitations of luminescence dating applied in active tectonic contexts are similar to those encountered in other environments. However, the particular limitation of low signal sensitivity for quartz OSL mentioned above, and discussed in further detail below, is important. Dating depositional events on timescales of 10–200,000 years, often with 1 σ uncertainties of ± 5–15%, is possible in typical environments. It is worth noting that for many previous tectonic studies, large uncertainties were not a significant limitation, as the problems addressed often lacked any alternative chronological control. This is changing, as more detailed questions are investigated, and a greater reliance on dating precision is required. The optimal target material is primarily sand-sized quartz or feldspar grains, although very fine-grained silts may be dated in certain circumstances too (Rizza *et al.* 2011).

Many different sedimentary environments have been dated successfully with OSL or IRSL, and this can provide some flexibility in sample selection during research planning or sample collection. Typical sediments used in active tectonic applications are sandy fluvial deposits. They have a widespread distribution and can provide discrete time markers in the form of fluvial terraces and channel features. They also contain sand grains of a size suitable for luminescence single-grain determination. Other sediments commonly encountered include colluvial or slope deposits, and more rarely aeolian (wind-blown) deposits. Single-grain OSL or IRSL approaches can overcome limitations of incomplete signal zeroing at deposition (but note that this approach cannot be used for silt-sized grains, and is also not useful when most grains are insensitive). Where multiple samples are dated from the same feature, or from a series of related contexts, Bayesian statistical techniques may be used to reduce the magnitude of uncertainties on each age estimate considerably (Rhodes *et al.* 2003; Zinke *et al.* 2017), and this has implications for the optimisation of sampling strategies (see Chapters 2 and 3).

9.2 APPLICATION TO ACTIVE TECTONIC CONTEXTS

There are many ways in which the deposition of a sedimentary unit can be related to tectonic events. The two most common applications that will be discussed in detail in this chapter are (a) palaeoseismology, the determination of the timing of specific earthquake events, and (b) the determination of fault slip rates, using offset geomorphic features such as fluvial terrace risers. However, under certain circumstances these two approaches may be combined, so that both earthquake timing and some indication of event magnitude or local slip displacement are determined. Several common applications are illustrated by Figure 9.1, which shows a typical strike slip fault slip rate site (Fig. 9.1a), an idealised palaeoseismic site (Fig. 9.1b), and a site where both palaeoseismic information and details of fault slip may be determined (Fig. 9.1c) for a blind (hidden) thrust fault, with locations useful for dating samples indicated with numbered circular symbols.

In Figure 9.1a, the contemporary surface and active channel are indicated by the letters 'SC'. S1 and S2 are abandoned fluvial terrace surfaces corresponding to terrace units T1 and T2 (note the US system for naming terraces is used here). Since deposition of unit T1, and the erosion of terrace riser R1–2, the fault has slipped by displacement d1, indicated with a red arrow. This terrace may be dated using OSL or IRSL by collecting samples within the fluvial sediments, illustrated by samples 3 and 4. Note that this older terrace surface has been subject to erosion, soil formation, and the deposition of post-abandonment sediments such as overbank deposits, locally derived fluvial and colluvial sediments, plus additional windblown components. This cover is illustrated by the brown line in the exposed face of T1, and may be dated as indicated by sample 5. Terrace T2 (with surface S2) is younger, and was deposited after a phase of fluvial incision; it has been offset by a smaller displacement d2 (red arrow). The age of T2 terrace sediments may be dated as illustrated by samples 1 and 2. The age of terrace riser R2-C (offset d2) is younger than samples 1 and 2 and the age of terrace riser R1–2 (offset d1) must be younger than samples 3 and 4, but older than 1 and 2; see Section 9.2.2 below for discussion of this key point.

In Figure 9.1b, the thin black lines represent small faults along which sediment units are offset, and the numbered circles indicate notional OSL or IRSL samples, representing a simplified palaeoseismic site where deposits have been disturbed by the effects of two

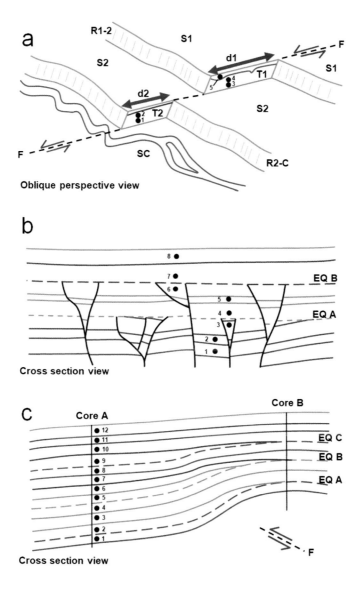

Figure 9.1 a) Typical simple slip rate site on a left-lateral strike slip fault (F–F) sketched in oblique perspective view from above. T1, T2: Fluvial terraces; S1,S2: Terrace surfaces; SC Contemporary surface of active channel. R1–2, R2-C: Terrace risers formed by erosion between T1 and T2, and between T2 and contemporary surface; d1, d2: Displacements at offset edges of T1 and T2 respectively caused by fault slip. Note T1 (older) displays a greater slip displacement as it has been subject to more earthquake events; numbered circles indicate notional OSL or IRSL samples. b) Simplified cross-section view of palaeoseismic site displaying sediments disturbed by two earthquakes (EQA and B). Thin black lines represent small faults with sediment offsets, numbered circles indicate notional OSL or IRSL samples. c) Differing stratigraphic records preserved across a fault bend fold or hanging wall anticline, shown in cross-section. A thrust fault dies out at depth (F), overlying soft sediments are deformed by folding. Sedimentation continues into accommodation space on the downslope side of the fold until a relatively planar surface morphology is achieved. Dashed lines: stratigraphic positions of three earthquakes, EQA, EQB and EQC; numbered circles indicate notional OSL or IRSL samples. See text for discussion.

earthquakes (EQA and B), shown in cross-section. The youngest disturbed sediment is the horizon containing dating sample 6, while the layers with samples 7 and 8 are undisturbed; the age of EQB is older than samples 7 and 8, but younger than samples 1 to 6. Similarly, EQA is older than 4 to 8, but younger than 1 to 3.

In Figure 9.1c differing stratigraphic records are preserved across a fault bend fold or hanging wall anticline, shown in cross-section. A thrust dies out at depth (F), with deformed sediments above. Following an earthquake associated with tightening of the fold and uplift of the hanging wall ground surface (in this case to the right of the figure), an increased rate of sedimentation occurs on the downslope side of the fold (left side) until a relatively planar surface morphology is achieved; in the contemporary situation this is illustrated by the layers containing OSL or IRSL dating sample 10 from Core A. However, correlation of sediment units between Cores A and B demonstrate the presence of some material preserved only in Core A; layers with dating samples 9, 5, 6, and 2 are of this nature. These layers correspond to the filling of new accommodation space generated by movement during an earthquake on the thrust fault at depth; their thickness can be related to the magnitude of slip in that event. The age of each event (EQs A, B and C) can be determined from the age estimates of samples taken above and below the base of these infill sediments preserved only in Core A; this is illustrated by the dashed sediment boundaries. For example, for the age estimation of event EQ C, this must be older than the ages of samples 9–12, but younger than samples 1–8.

9.2.1 Palaeoseismology

For palaeoseismological applications, the development of a sedimentary succession close to the fault, capable of preserving evidence of disturbance by seismic shaking or direct truncation by fault slip, is required (e.g. Fig. 9.1b; Fig. 9.2). This situation is typically found in small lake basins or swampy areas located across or immediately adjacent to the fault. In many cases, fault slip was responsible for creating the depression that has subsequently developed into a lake; in strike slip fault contexts, this may represent a pull-apart basin or a valley dammed by a shutter ridge. This is exemplified by the site of El Paso Peaks, California, USA, studied by Dawson *et al.* (2003), which represents a small ephemeral playa lake trapped between a shutter ridge and a developing alluvial fan (Fig. 9.2). The site is located on the left-lateral strike slip central Garlock fault, and provides an excellent mid to late Holocene record of earthquakes, firmly dated by a superb series of radiocarbon ages on charcoal preserved within the fine sand and silty sediments. This site was re-opened by the one of the present authors (EJR) with colleagues as a location to develop and assess new luminescence dating approaches (Lawson *et al.* 2012; Roder *et al.* 2012). Ages in agreement with the ^{14}C age control were determined using samples from the upper parts of these sediments based on K-feldspar multiple-grain isothermal TL (Roder *et al.* 2012) and single-grain pIR-IRSL at 225°C (Rhodes 2015). In contrast to the feldspar IRSL results, quartz from this site was characterised by low sensitivity, and provided age underestimates.

Perhaps the biggest single issue, beyond considerations for luminescence dating, is the correct identification of suites of disturbance that are truncated at a single horizon, as required to identify an earthquake event. Sometimes poorly preserved sedimentary features, or structures such as those created by burrowing or root activity, render this task highly challenging. Additional complications arise when there is a low or highly

Figure 9.2 Close up photograph of disturbed ephemeral lake sediments from the palaeoseismic site of El Paso Peaks, California, USA, previously dated using ^{14}C by Dawson *et al.* (2003), and used to develop K-feldspar luminescence dating approaches (Roder *et al.* 2012). A restricted zone of heavily disturbed sediment may be observed in the centre of the photograph. Shovel approximately 1m in length. Photography taken by Edward Rhodes.

episodic sedimentation rate; as the undisturbed layer above the inferred earthquake event provides a post-event limit for the age; that is, the earthquake must be older than the age of this sediment (e.g. layer with sample 7 for EQ B in Fig. 9.1b), then a significant delay in deposition of this next layer after the earthquake can lead to increased uncertainty in age constraint. It is also possible that one or more additional earthquakes might occur before sufficient sediment has accumulated; in this case, the apparent palaeoseismic record may be incomplete. If the selected site is close to more than one fault, it is possible that the sediment may record the effects of ground shaking from an earthquake along a different fault from the one intended, leading to more events than are representative for the target fault. The degree of resolution that can be achieved depends on the sedimentation rate, the precision of each date, and how many age estimates are used to constrain each event. In these contexts, Bayesian statistical analysis can be very powerful, and contribute significantly to how well constrained the earthquake ages are.

Sedimentary structures associated with earthquakes include cracking, faulting, rolling, water escape forms including flame structures and load casts, ball and pillow structures, and folding. Loss of sedimentary lamination can occur, with elongate grains sometimes rotated to a sub-vertical position. As other processes can also be responsible for some of these structures (e.g. glaciation, periglaciation, rapid deposition of overlying material etc.) care must be taken in their interpretation. A good introduction to the identification of these structures and how they may be interpreted is provided by Rudersdorf *et al.* (2015).

A common situation is to have disturbed sediments, possibly including direct evidence of faulting in the form of offset or truncated sedimentary units, overlain by sediments with no apparent disturbance. Dating the undisturbed sediments as close as is practical to their lower boundary provides a minimum age for earthquake shaking. Dating the upper disturbed units will help provide a maximum age for this earthquake event; where these are closely spaced in time, the event has been constrained effectively. If the ages are relatively widely spaced, this site can provide no further constraint for the age of the earthquake, which must have occurred between these age estimates (remembering to take into account their uncertainties). Resolution may be improved by dating additional samples above and below the event; for example, measuring samples 4, 5 and 8, as well as samples 6 and 7, in the dating of EQ B in Figure 9.1b.

9.2.2 Fault slip rate studies

As faults slip during earthquakes, or as they experience slow slip or creep events, they often produce changes to the ground surface. Given that seismogenic depths at which earthquakes are triggered in events that affect the surface are typically 10 to 15 km, and the scaling relationships that relate earthquake magnitude (representing energy release) to fault slip area (Scharer *et al.* 2014; Wells and Coppersmith 1994), surface rupture is usually only observed for earthquake events greater than around magnitudes 5 to 6. However, many large magnitude deep earthquakes such as those on the lower parts of subduction megathrusts may cause no surface rupture. Faults typically fall into one of three categories, namely normal faults, associated with crustal extension and thinning, thrust or reverse faults associated with crustal shortening and thickening, and strike slip faults representing the boundaries between crustal blocks that move horizontally in relation to each other. These categories correspond to vertical or horizontal attitudes for the three principal stress orientations within the crust. However, combinations of these categories can also be found; oblique slip may be observed on an individual fault plane, for example the Papatea Fault involved in the 2016 Kaikoura earthquake in New Zealand, on which left lateral strike slip motion and thrusting occurred (Fig. 9.3). Alternatively, different slip directions may be partitioned onto different faults (e.g. in Owens Valley, California, USA) or different sections of the same fault with varying attitudes may display different styles of slip (e.g. Death Valley – Fish Lake fault system). These more complex combinations of slip are often located in areas subject to widespread transtensional or transpressive tectonic regimes. It should be noted that many different fault types can be found within relatively close proximity (for example in Southern California or the Marlborough district, New Zealand), and that crustal deformation can also be expressed in other ways, such as fold development or distributed deformation. This means that even precise determination of slip on an individual fault may have a complex relationship to the wider tectonic regime, and great care in developing an understanding of the complexities of each area is required.

Vertical movements are produced by normal or reverse (thrust) faulting (Fig. 9.3; Fig. 9.4), and by oblique strike slip movements. This may produce a clear fault scarp (Fig. 9.3), and the relatively unambiguous offset of horizontal or sub-horizontal surfaces such as fluvial or marine terraces (Fig. 9.4b). In this case, the terrace sediments (and their geomorphic surface) *predate* the slip event, and so dating these can provide a *maximum* age for the earthquake event. In turn, combining this age estimate for sediment deposition with a determination of the total fault slip (derived from the vertical offset of the terrace and the angle of the fault) provides a *minimum* slip rate estimate (e.g. Dolan *et al.* 2016).

Figure 9.3 Photograph of recent surface rupture of a fluvial terrace surface and underlying alluvial sediments by the Papatea Fault, Marlborough, New Zealand, during the 14th November 2016 M_w 7.8 Kaikoura earthquake, photographed in late February 2017. The terrace surface has been elevated beyond this strand of the fault by around 1.5 m at this point by thrusting, producing a clear fault scarp in a previously flat surface, and has also undergone several metres of left-lateral strike slip movement not visible in this photograph. Note the coarse sediment texture comprising cobbles and boulders set in a pebble and sand matrix. With an absence of discrete sand lenses, this unit would be difficult to sample for OSL or IRSL dating using a conventional approach of inserting opaque tubes horizontally, but see later discussion for ways to overcome this problem. Photograph taken by Edward Rhodes

Even in the relatively simple example described above, potential complexities abound. Was the offset derived in one or more earthquakes? Note, this is not necessarily important in deriving a fault slip rate estimate, but when earthquake events are rare, as is the case for many slow slip rate faults, or an event occurred in the recent geological past, the relationship to the earthquake cycle may be important. This point is discussed in greater depth below. Perhaps the most important question in this context is whether the observed sub-horizontal surface represents the same thing on either side of the fault scarp. For example, the side that was lowered by the fault movement may have experienced renewed sediment deposition, such as by more frequent overbank deposition events, and the upper side may have been subject to increased erosion. Note that 'surfaces' themselves cannot be dated directly by luminescence sediment dating techniques (at least this is the case at present, although surface-exposure dating programmes for cobbles, boulders and exposed rock surfaces by OSL and IRSL dating are being developed; see Chapter 11). Geomorphic surfaces may represent simply the upper boundary of a sedimentary unit, or may be produced by subsequent planar erosion; their 'age' must post-date sediment deposition. Note also that if the upper and lower surfaces (on either side of the fault scarp) have suffered differential deposition and/or erosion (for example by post-event overbank flood deposition), the estimate of fault throw (the amount moved) may be in error.

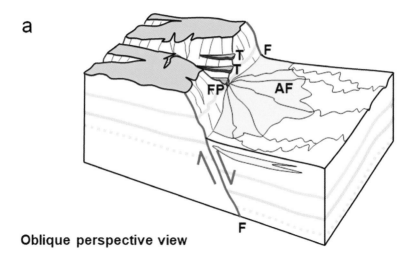

a

Oblique perspective view

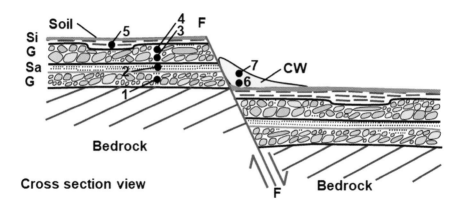

b

Cross section view

Figure 9.4 Geomorphic and sedimentary structures associated with a large normal fault. a) Broader-scale oblique perspective view, showing geological structures (lithological horizons, fault marked 'F') in section, and in surface expression. AF: alluvial fan, FP: flood plain, or contemporary surface of the river responsible for deposition of the alluvial fan, T: two abandoned fluvial terrace treads preserved by incision. Note the buried alluvial fan sediments visible in the near cross-section. A fault slip rate may be derived by dating the abandoned terrace sediments, though see text for discussion of the assumptions implicit in this. The sketch illustrates how each time the fault moves, the upper side (foot wall) to the left is subject to increased erosion and valley incision, while on the hanging wall side to the right, more accommodation space is made where sediments may be deposited and preserved. b) Sketch cross-section illustrating typical sediments associated with a fluvial terrace offset by recent normal fault activity (F). G: gravel, Sa: sand, Si: silt, CW: colluvial wedge comprising reworked components of the fluvial sequence plus material transported in from other locations; black dots represent possible OSL or IRSL sample locations. While the samples within the sand and gravel (1 to 4) provide firm maximum age estimates for fault movement, the two samples in the colluvial wedge (6 and 7) provide a minimum estimate. Sample 5 is ambiguous, as it sits in a silty channel fill on the surface of the terrace, and may represent modification of the terrace surface after faulting.

Where vertical fault movement occurs, sediments may be produced as a direct result of the movement; for example, the colluvial wedge developed against the fault surface in Figure 9.4b. These can include scarp collapse sediments, or other deposits that represent material released or moved by the seismic shaking. They can also represent sediment that is deposited in depressions adjacent to the fault plane, or otherwise within the fault zone. Whilst in principle these sediments may provide a close temporal relationship to the earthquake, they may include very poorly bleached sediment (e.g. if the earthquake occurs at night, or if they collapsed over a very short distance), or may be ambiguous in their attribution to immediate deposition. These sediments are assumed to post-date fault movement, so any luminescence residual signal caused by incomplete zeroing that is not correctly attributed, or a misinterpretation of pre-existing sediments that are interpreted as post-event deposition will lead to incorrect conclusions. In most cases, a trench excavated through the fault zone and the related deposits may reveal much detail about the relationships of different units and structures. An example of this is provided by Middleton *et al.* (2016).

In many cases, offset fluvial terraces are chosen to provide fault slip rate estimates, in particular for strike slip faults where offset is entirely or dominantly horizontal. However, there are significant issues and debates about the procedures adopted that relate both to chronological considerations and also to geomorphic issue; in particular, whether the fluvial system in question is likely to have incised into the upper terrace shortly after its deposition, or alternatively whether it trimmed the terrace riser shortly before deposition of the lower terrace sediments is a question of longstanding and continual dispute between different workers (e.g. see Cowgill 2007). In reality, the uncertainties associated with this type of issue are rarely included formally within the quoted slip rate error estimates. These debates are discussed in greater detail below.

In most slip-rate study situations, the number of offset markers such as terrace risers is typically limited by the preservation of geomorphic features in the landscape. That is, in many locations, the gradual on-going denudation of the landsurface tends eventually to remove fluvial terraces by a combination of direct erosion; for example, by the incision of tributary streams, or by the lateral erosion associated with the main channel. These erosive processes may include rotational slides and slumps, downslope colluviation, as well as direct erosive undercutting by fluvial action. It should be noted that unconsolidated fluvial terrace deposits typically composed of gravel, sand and silt, with a soil developed on the surface, are often significantly more susceptible to erosion than the bedrock valleys within which they are deposited. As a result of these processes, even in situations where fluvial channels conveniently cross faults at a significant angle, providing the potential opportunity to preserve markers of past slip, there are rarely more than a couple of terraces sufficiently well preserved to allow this.

In relatively rare cases, where the earthquake event frequency is not significantly in excess of the landform development rate such that offset geomorphic records exist with different slip displacement corresponding to individual earthquakes, a slip-per-event record may be approached (Cowie *et al.* 2017; Dolan *et al.* 2016). This latter type of record is particularly valuable, as it offers the opportunity to assess models of fault behaviour, providing insights not available by other means.

9.3 LUMINESCENCE DATING

As explained in Chapter 1, the development of the SAR and the first routine application of single-grain OSL methods led to a significant increase in the applicability of luminescence sediment dating, initially using quartz OSL, and subsequently for feldspar. The primary improvements offered by the SAR protocol are increased precision, and the ability to detect the presence of incomplete zeroing depending on aliquot size and single-grain sensitivity distribution characteristics (Rhodes 2007, 2011).

A second development that tends to render luminescence 'competitive' with other chronological methods in active tectonic settings is the application of Bayesian statistical methods to reduce the size of measurement uncertainties (see Chapter 3). This approach is useful when samples with known stratigraphic relationships have age uncertainty distributions that overlap with each other. Simply by measuring multiple samples closely spaced in terms of depositional age, overall age uncertainties for specific horizons or events such as fluvial incision episodes can be reduced significantly. Note, however, that the 'cost' of this Bayesian approach is a significantly increased number of samples used for the age determination of each geomorphic or sedimentary target element. The rate of uncertainty improvement tends to follow the square root of the number of samples. That is, the measurement of four times as many samples may lead roughly to a halving in event age uncertainties.

The final luminescence dating development that is important is the single-grain K-feldspar pIR-IRSL$_{225}$ approach that has become relatively widely used in active tectonic contexts since around 2012. The significant feature of this approach is that it is often

Table 9.1 Typical parameters used in the single-aliquot regenerative dose (SAR) measurement of single-grain pIR-IRSL. The signal used for age estimation came from the initial 0.5 s of step 4 (IRSL at 225 °C), subtracting background (last 0.5 s of that measurement), and then corrected for sensitivity change using the background-subtracted initial 0.5 s pIR-IRSL from step 8.

SAR STEP	TYPICAL MEASUREMENT PARAMETERS
1 Beta dose irradiation	0 (Nat), 20, 6.4, 64, 200, 640, 0, 20 Gy in turn
2 Preheat	60s whilst held at 250 °C
3 IRSL 1	2.5 s stimulation with IR laser at 90% power whilst at 50 °C
4 IRSL 2 (pIR-IRSL)	2.5 s stimulation with IR laser at 90% power whilst at 225 °C Signal integrated over 0–0.5 s of stimulations time, Background integrated over 2.0–2.5 s.
5 Beta test dose	8 Gy
6 Preheat	60 s whilst held at 250 °C
7 IRSL 1 Sensitivity measurement	2.5 s stimulation with IR laser at 90% power whilst at 50 °C
8 IRSL 2 Sensitivity measurement	2.5 s stimulation with IR laser at 90% power whilst at 225 °C signal integrated over 0–0.5 s of stimulations time, background integrated over 2.0–2.5 s.
9 Hot bleach – then return to 1	40 s IRSL (using LED diodes) whilst at 290 °C

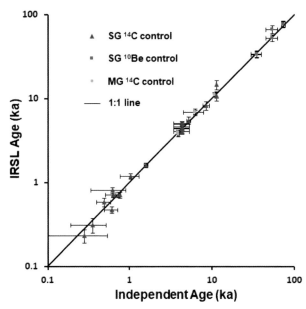

Figure 9.5 Independent age comparison for 35 pIR-IRSL at 225 °C samples from different locations and contexts spanning the last 80,000 years, plotted on logarithmic age axes. The dataset includes three multiple-grain ages from Tibet (circles), plus single-grain results from many sites in California, USA, with ^{14}C control (triangles), besides three samples from Baja California Sud, Mexico, and two from Mongolia with ^{10}Be depth profile control (squares). Figure updated from Rhodes (2015). Note the high degree of agreement between pIR-IRSL age estimate and independent chronological control using this approach.

applicable in contexts where quartz OSL is unsuitable because of poor luminescence characteristics. The technique is a direct translation for single grains of the conventional aliquot pIR-IRSL method of Buylaert *et al.* (2009). Both approaches incorporate a 60 s preheat at 250 °C and two IRSL measurements, one at 50 °C and a subsequent one at 225 °C (Table 9.1).

At some sites, Rhodes (2015) discovered that restricting the potassium feldspar fraction to only those grains with the highest potassium content by using a more selective heavy liquid density separation at 2.565 g cm^{-3} provided a significantly higher yield in terms of the proportion of grains giving a signal, and also the total light sum collected. However, subsequent research shows that at some sites, this approach (termed 'Super-K' by Rhodes 2015) can restrict the number of grains that provide useful IRSL signals. This single-grain pIR-IRSL approach is not different in any substantive way from that used by other workers for different types of application (e.g. Nian *et al.* 2012, Reimann *et al.* 2012), but was tested by comparison with radiocarbon and ^{10}Be depth profiles at a significant number of slip rate, palaeoseismic and palaeoenvironmental sites (Fig. 9.5).

9.3.1 Key considerations

9.3.1.1 Poor luminescence characteristics

Both quartz and feldspar can suffer from poor luminescence characteristics. For quartz, the primary limitation encountered in sedimentary contexts where many grains have been eroded directly out of bedrock, or have been through a low number of shallow sedimentary cycles over the past few millions of years, is low OSL sensitivity (Fitzsimmons *et al.* 2010; Pietsch *et al.* 2009; Preusser *et al.* 2009), and this can be a particular problem in active tectonic contexts. This condition is not fully understood, but causes two problems:

1. *poor counting statistics* associated with each measurement, resulting high uncertainty values for the natural OSL and the growth with added dose (e.g. Porat *et al.* 2009).

2. a significantly increased risk of *OSL signal contamination* from mineral inclusions of different composition such as feldspar within quartz grains (e.g. Nissen *et al.* 2009).

Standard contamination tests for quartz OSL can be applied, including presence of an IRSL signal or OSL signal depletion by IR exposure. A good assessment of the magnitude and impact of potential contamination can also be made using the test developed by Lawson *et al.* (2015), developed specifically for tectonic contexts, which combines OSL signal depletion by IR exposure with an assessment of thermal quenching and thermal assistance. Porat *et al.* (2009) measured a series of age estimates for fault-adjacent colluvial wedges associated with a small fresh-looking normal fault scarp along the Dead Sea transform, Elat, Israel. These were dated with a SAR approach using quartz single grains and also with regular aliquots (5 mg, ~1000–2000 grains). Approximately 5–10% of the single grains provided a useful OSL signal, but very few of these were well-bleached, severely limiting the number of grains contributing to the final quartz age estimates for deposition. The single-grain OSL ages were not strictly in stratigraphic order, and were associated with large age uncertainties, but were in broad agreement with independent fault-slip age estimation based on erosional scarp retreat modelling, demonstrating both the potential and utility of luminescence in this context, but also the limitations imposed by low signal sensitivity. Multiple-grain small (150–200 grains) and regular aliquots of K-feldspar provided age overestimates using IRSL measured at 50 °C, thought to be caused by strong incomplete signal zeroing (Porat *et al.* 2009).

Poor characteristics for feldspar include low sensitivity, high or complex fading behaviour, and a condition observed in single-grain measurements referred to by Rhodes (2015) as 'declining base' behaviour. In addition to these, we should note that the pIR-IRSL signal in feldspar is reduced by light much less rapidly than the fast component of quartz OSL (Lawson *et al.* 2012; Smedley *et al.* 2015). This means that samples from any sedimentary environment are at risk of incomplete pIR-IRSL signal zeroing, and caution should be used when making conventional multiple-grain feldspar measurements. However, in contrast to quartz, feldspar grains extracted from bedrock, or from sediments formed dominantly of grains that were recently eroded from bedrock, often display intense IRSL and pIR-IRSL signals (e.g. Brown *et al.* 2015). As K-feldspar single-grain pIR-IRSL is now a significant feature of dating in tectonic contexts, these potential limitations need to be borne in mind.

9.3.1.2 Sample optimisation

A point often particularly relevant to the dating of active tectonic contexts is that the luminescence characteristics of the material available at the location of interest was already established by geological and environmental factors at the time of deposition, but these characteristics are usually unknown to the dating specialist and project team at the time that choices about sampling and preparation must be made. In some instances, it is sensible to increase the number and range of samples collected, or to revisit a site and collect new samples in the light of laboratory measurements, or to prepare subsamples for multiple approaches (e.g. quartz grains for OSL, K-feldspar for single-grain pIR-IRSL). The key point is that given the wide variation in luminescence characteristics observed for both quartz (Preusser *et al.* 2009) and the feldspar family minerals (Krbetschek *et al.* 1997), there is a degree of unpredictability about the quality of results that it may be possible to achieve at a given site. To some degree, these problems can be reduced by adopting a sensible sampling strategy, ample research planning, and careful selection of the dating approach in the laboratory based on initial determinations. This is particularly the case in luminescence dating of tectonic contexts, as many different issues posing serious technical challenges are often encountered, but may be overcome by judicious means.

When choosing detailed sample locations, the characteristic depositional energy of the deposits selected for collection is often considered. However, more important than this is the implication of the age of each unit for the timing of the earthquake or fault slip event in question. While, in broad terms, lower energy units are characterised by finer grained sediments, better sorting and clearer lamination may be associated with a higher degree of luminescence signal zeroing. This is of little benefit if the unit dated is ambiguous with respect to the target event. This warning is provided specifically because it is not uncommon for samples to be taken in the finer-grained material above terrace gravels, as a way to determine the terrace age, but not within or below the terrace gravels themselves. This in part stems from an issue of scale and different uses of the term 'terrace'. The term terrace can be used to refer both to the geomorphic unit (an approximately flat surface) as well as the sediments (in this case fluvial deposits). Where a fluvial terrace is offset by fault movement, and its eroded edge, the terrace riser, is used as a piercing point to assess the magnitude of fault slip; it is the offset of the larger scale gravel unit that is of primary importance, rather than simply its surface. The surface of a gravel terrace continues to evolve after abandonment by the river following incision. This can include deposition by overbank flood events, local subsidiary fluvial flow, and both aeolian and colluvial deposition. The surface can also become eroded by incision of secondary channels or by other means (such as glacial erosion). Weathering will also eventually lead to gradual surface lowering. The surface can also be modified by soil formation, which includes in-mixing of grains from the surface and bioturbation by tree-throw and animal burrows can be significant. Sampling only the finer grained cover sediments above the gravel is represented by collecting only sample 5 in Figures 9.1a and 9.4b. This approach is not recommended; these cover sediments can often be significantly lower in age than the underlying gravels (see Zinke *et al.* 2017) for several clear examples of this), and therefore have the potential to provide misleading fault slip-rate estimates.

9.3.1.3 Incomplete bleaching

A surprising finding of the application of single-grain K-feldspar pIR-IRSL$_{225}$ dating is how well zeroed many of the samples are, even when the depositional environment represents a high energy event likely to involve rapid deposition in turbid conditions following relatively short transport distances. Zinke *et al.* (2017) show data for each of 34 samples collected in high energy deposits and lower energy cover sediments; the terrace gravel and sand deposits vary between only 8% and 92% of the grains well bleached (consistent with the minimum apparent age value), and cover sediments (silt) vary from 15 and 89%, but can be up to 10,000 years younger than the underlying gravel deposits. It is likely that the grains used to determine the age of a sediment were in many cases well zeroed with respect to their pIR-IRSL$_{225}$ signal before incorporation into the flow that deposited the unit subsequently sampled for dating. Grain histories are envisaged that include gradual reworking over a number of separate transport events within the active channel, and periods spent on the surface of bars within and beside the channel (both above and beneath the water surface), as responsible for this relatively high degree of signal zeroing in many samples. This observation (of relatively good pIR-IRSL$_{225}$ signal zeroing) can provide some increased confidence when applying conventional multiple-grain OSL dating of quartz OSL, characterised by a much more rapid reduction by light, to high energy fluvial systems, including those from active tectonic sites as well as other contexts.

9.4 GEOMORPHIC CONTEXTS SUITABLE FOR LUMINESCENCE APPLICATIONS

In palaeoseismology, OSL or IRSL may complement other sediment age estimation approaches, in particular ^{14}C dating; however, there are many locations where organic material is rare or absent, and where luminescence dating may be very valuable if it can be shown to be reliable in that context. These locations include desert environments, for example Mojave Desert, California, Iran, parts of central Asia including locations in Kazakstan, Turkmenistan, Mongolia, Tibet and China. However, organic material can also be rare or poorly preserved in high mountain environments.

The typical approach used in palaeoseismology is to excavate a trench into sediments located close to a fault that recorded violent ground shaking by crack propagation and offset strata. As mentioned above, palaeoseismologists often select small ephemeral lakes or bogs as sites to trench when these are present, as these may provide a semi-continuous record of sediment deposition over some extended time period; longer periods have the possibility of recording more earthquake events, and more continuous deposition reduces the chance that earthquake events were missed from the record. In locations without access to lakes or pools, trenching through scarps may reveal soil or sediment layers or lenses that were displaced by fault movement. Each earthquake that occurs may cause damage (cracking, offsets)

Figure 9.6 Photograph of a palaeoseismic trench at Bila Voda, Czech Republic. OSL/IRSL samples have been collected in metal tubes with black plastic caps approximately 5 cm in diameter, and are left in place so that their location can be recorded accurately. Notice the equipment case containing a portable NaI gamma spectrometer being used to record the environmental gamma dose rate within the sediments. A truncated sand lens is visible in the central part of the photograph. Photograph taken by Edward Rhodes.

to existing strata, but of course cannot affect future sediments that are not yet deposited. Groups of features (cracks, small faults) extending to a particular horizon are identified and considered to represent a single earthquake event. Where samples are collected from the first horizon unaffected by a group of damage features, these provide an age estimate that pre-dates the earthquake event, whilst the uppermost or youngest sediments that are affected by that group of damage features represent material deposited before the event.

Fault slip rates may be determined where a feature crosses a fault that has slipped, and has been offset as a result. The slip rate is determined by OSL or IRSL dating of sediments that relate directly to the offset structures (e.g. Fattahi *et al.* 2006, 2007). Examples of potential structures include (a) offset fluvial or marine terraces that include a constructional phase comprising sediments, and (b) offset thalwegs, sometimes with offset channel fill sediments within them. Depending on the type of fault, and on the context, suitable sites and sediments may be located from high resolution topographic data, such as DEMs constructed using LiDAR or photogrammetry data. Alternatively, these may be located in large trenches, or a series of interlocking trenches (Ferrater *et al.* 2016). One of the most common situations is where an eroded edge of a fluvial terrace, often called a terrace riser, has been offset; these situations are relatively common on strike slip faults, especially in places with a significant difference in topography on one side of the fault, forming a potential energy difference that

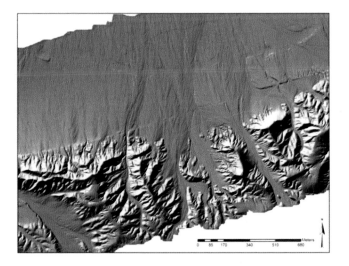

Figure 9.7 High resolution digital elevation model (DEM) based on LiDAR data for the central Garlock fault, California, USA. The fault runs from around 250° to 70° azimuth through the centre of this view. This area is called Christmas Canyon West, and was the focus of the slip rate study presented by Dolan *et al.* (2016), based on single-grain K-feldspar pIR-IRSL$_{225}$ dating of offset alluvial fan sediments. Several left lateral offsets can be seen in the central part of this view. The digital topographic data values were used to estimate slip displacement using a least squares fitting approach of offset terrace features. The nature of one feature was studied in detail using an excavated trench through the geomorphic expression exposing the relationship of the surface to the sediments, demonstrating that the small terrace riser did indeed truncate the fluvial units. IRSL samples were collected in steel tubes hammered horizontally into sandy sediments in hand-dug pits. Data from OpenTopography. Data processing and image prepared by E. Wolf.

leads to stream channels crossing the fault at a relatively high angle. This is illustrated in Figure 9.1a. Examples include the Garlock fault, California (Figure 9.7; Dolan *et al.* 2016), and the Awatere Fault, New Zealand (Zinke *et al.* 2015, 2017).

There are many potential complexities and subtleties to achieving meaningful fault slip rates. One of these is when terrace risers are used as offset markers, either the age of the upper terrace sediments (older) or the lower terrace sediments (younger) may be closer in age to the timing of the erosional incision event that cut the riser. This depends on the nature of the river or stream, on climate fluctuations that may have driven geomorphic change, but also on the detail of the context; for example, whether the main fluvial channel was close to this location and actively eroding the terrace edge prior to abandonment of the lower terrace. In some cases, additional useful information may be available from high resolution DEMs (e.g. LiDAR; see Fig. 9.7) or from subsurface observations made by excavation. Ground penetrating radar (GPR) determinations can be useful to help distinguish buried sediment units or features such as channels in this context, as well as in other active tectonic settings. In the absence of additional information, it is not justified simply to refer to a slip rate based on a single dated terrace without taking into consideration the implicit uncertainties of the relationship between the depositional age and the timing of the incision event.

Another important consideration, discussed by Cowgill (2007), is how different parts of offset terrace risers preserve the record of displacement. When a stream channel crosses a fault that experiences strike slip movement, for example, on the downstream side of the fault, one bank is pushed into the active channel while the other moves away from the position previously occupied by the channel (Fig. 9.8). This part of the bank, and associated terrace riser, is at significantly increased risk of erosion, indicated in Figure 9.8

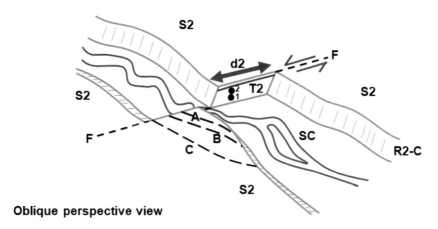

Oblique perspective view

Figure 9.8 Sketch showing the vulnerability to erosion of one side of a fluvial terrace immediately after a significant strike slip displacement. The part of the terrace T2, with surface S2, nearest to the viewer in this oblique perspective view has been pushed into the channel 'SC', almost blocking the stream course. The black dashed lines A, B and C show possible sequential positions for the terrace edge as it is eroded back. Note that the upstream part of S2 that now protrudes into the channel including OSL sample positions 1 and 2 is also at increased risk of erosion, threatening the validity of the estimate of displacement d2.

by the dashed lines marked A, B and C. Observations following the Kaikoura earthquake suggest that this process of erosion can be very rapid, taking place on timescales of weeks to months, though clearly this depends on the power of the stream as well as the frequency of flow within it, besides the composition of the bank material. This situation will be further modified where the stream crossing the fault sits within a landscape with an average surface slope that is not at 90 degrees to the channel. In this case, there is an increased erosion risk to the protruding bank on the larger scale downslope side; where this is also the downstream side, that protrusion may well be removed entirely, though the upstream side protrusion may preserve the full slip displacement, as the stream has an increased tendency to erode the opposite bank. Where this situation is reversed, neither side may represent the full slip displacement, and slip reconstruction can be difficult. In summary, the geomorphic response to the changed topography will itself modify that topography, rendering all reconstructions potentially limited in their accuracy.

Where fault movement involves vertical displacement, as is the case for normal faulting and thrust or reverse faults, or in the case of oblique strike slip faulting, the power of all surface processes is increased, including the erosional power of streams and the rate at which material is transported by slides, slumps and diffusion processes. This tends to exaggerate the magnitude of difficulties experienced in reconstructing total slip displacement.

In summary, very careful consideration is required to find locations that preserve useful evidence of fault displacement. In addition, these need to be constructed from sediments suitable for OSL or IRSL dating, and a clear relationship between sediment deposition and the morphological feature used to determine slip displacement needs to be established. The particular advantage of luminescence is its ability to date a wide range of different sediment types using almost ubiquitous materials, but this does not alter the requirement to find locations that are secure in their geomorphic, sedimentary and tectonic relationships.

9.5 SAMPLING AND ANALYSIS CONSIDERATIONS

9.5.1 Sample collection

Sampling in active tectonics contexts is generally not performed at natural exposures, as these tend to be rare, and are sometimes ambiguous regarding the relationship of exposed sediment to the target geomorphic or sedimentary units. Typically, depending on the scale and nature of the project, pits are dug by hand (Fig. 9.9) or by mechanical excavator, or larger scale trenches are excavated, providing access to improved visibility of stratigraphic relationships. The nature of contacts between different sedimentary units can represent an aspect of paramount importance in these studies, as can the detail of the features caused by seismic disturbance. As the target material is usually fine to medium sand grains, sand layers or lenses are usually selected where available. Note, however, that other modal sediment grades often contain sufficient sand-sized grains within them (e.g. silt, loam, coarse sand, gravel) for successful dating, and that sampling deposits with different energies can have benefits when it comes to interpreting results. Similarly, sampling the full succession of sediments at an exposure usually contributes to reducing age uncertainties, besides making interpretation of the site and luminescence results easier.

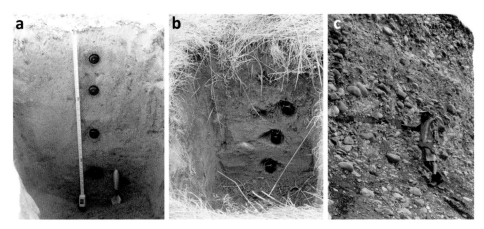

Figure 9.9 Photographs of IRSL sample collection for fault slip rate studies. a) Four samples collected in sandy gravel in vertical succession in a hand-dug pit approximately 80 cm deep at Christmas Canyon West, on the Garlock fault, California, USA (Dolan *et al.* 2016). b) Three samples in sandy silts with cobbles at Branch River, Wairau Fault, New Zealand. c) Sand lens within coarse gravels, Branch River, Wairau Fault, New Zealand. Dr Russ Van Dissen for scale. Photographs taken by Edward Rhodes.

One feature of the recently developed single-grain pIR-IRSL$_{225}$ approach is its ability to provide apparently reliable age estimates for high energy fluvial contexts (Fig. 9.9). It is usually the gravel of a fluvial terrace that defines the depositional unit that is subsequently eroded to form a terrace riser, rather than overlying, lower energy deposits, so it is important to sample this unit directly. However, these can be very hard to sample for OSL or IRSL, as metal or plastic tubes cannot easily be driven into gravel without bending or breaking, and suitable sand lenses within gravel deposits can be relatively rare, depending on the conditions prevailing during deposition. In these cases, two specific solutions have been developed, termed, for want of available alternative names, 'scrivelling' and 'scrumbling'. The former consists of collecting samples under dark conditions, either at night, or by covering a pit with a large wooden board. In theory, an opaque tarpaulin can also be used, but is perhaps harder to manage with respect to wind, rain and light leaks from the side. A safe-light (typically a red LED cycle rear light unit, or a specially made filtered amber or red LED torch) is used to locate the correct layer to sample, and to ensure full removal of light-exposed surfaces (which may be marked with spray paint), before carefully placing material into a light-proof bag from the correct location with a small trowel. Unsuitable material such as rocks may be discarded, although allowance for their dose rate contributions must be made.

The second approach, 'scrumbling', can be used in certain sandy gravel deposits, and comprises the driving in of a metal tube into the gravel, parallel to, but about 10 cm from, the external surface of the gravel within a pit or trench. To do this, a small subsidiary slot or trench is dug into a section, and then a tube is hammered into the side of that slot, parallel to the inner wall of the main trench. As the tube is driven in, larger stones that it encounters are pushed away laterally, into the main trench. In some cases, the operator collecting the sample can control the movement of material in this trench wall with their foot, whilst holding the tube in one hand, and the hammer in the other. This avoids collapse of the wall, which might lead to light exposure of the forward-moving tube end. In this manner, with some serendipity, a sample tube may be filled with the finer fraction of a gravel deposit without it being exposed to light.

9.5.2 Analysis and interpretation

Analysis of results requires some degree of care and experience at each level. For the purposes of this discussion, issues arising in feldspar single-grain pIR-IRSL data will be considered, though similar parallel considerations exist for each different approach taken. Single-grain methods are adopted specifically to overcome problems encountered for samples that display multiple dose values. These multiple dose values may be caused by incomplete zeroing, leading to some (or potentially all) grains having a residual IRSL or OSL signal, or the in-mixing of grains from a younger deposit or the surface, or by other causes such as differential anomalous fading. As different mechanisms can have taken place, very few samples are truly unambiguous in their apparent age, perhaps only those rare samples that display a single dose value within measurement uncertainties and the expected over-dispersion.

Rather than go into many possible scenarios and possible interpretations, what follows is an attempt to define a recommended set of main principles and approach. In many samples, based on experience of something like 400 results from tectonic related contexts, the minimum shared dose estimate represents a close estimate for the depositional age. However, some samples do contain grains with lower dose values, either as a result of post-depositional in-mixing or poor grain characteristics (e.g. high fading rate), and the (probably very rare) possibility of sample contamination at very low levels within the laboratory cannot be ruled out. Where dose values cluster into discrete populations, the lowest of these is likely to represent depositional age, but this is not necessarily so; grains may be deposited rapidly under conditions with little light (e.g. at night), and can incorporate grains dominantly from a single input such as a collapsing river bank which may contain mostly well-bleached grains. In this case, a small number of lower dose values might represent unit age, rather than being intrusive grains. This scenario illustrates the type of potential ambiguity that may arise.

This type of ambiguity may be overcome in two main ways. First, more grains may be measured, and an assessment of the degree to which lower dose estimates are duplicated by new measurements considered. Although shared lower dose values do not necessarily indicate depositional age, the probability may increase as some possible causes of low values are less likely (such as differential fading or laboratory contamination). The second way is to compare the apparent age results from the ambiguous sample with the results of samples from closely related stratigraphic positions; for example, from immediately above and below in the same section. It is unusual to encounter different samples with similar issues, and in this way, it is usually the case that ambiguities such as that described above can be resolved. In practice, where such ambiguities arise, both methods (measuring more grains and additional samples) may be used to reduce ambiguity.

Experience suggests use of a uniform value of over-dispersion of 15% appears to provide a reliable age estimate for many samples, in comparison to independent age control (Brown et al. 2015; Rhodes 2015). Pleistocene samples with [10]Be depth profile control

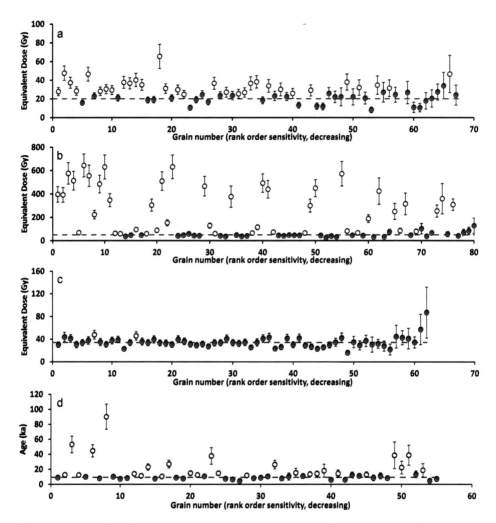

Figure 9.10 Examples of single-grain pIR-IRSL$_{225}$ equivalent dose and age values, plotted in order of decreasing grain sensitivity. Grains consistent with a shared minimum value based on measurement uncertainties and an over-dispersion of 15% added in quadrature shown in red, combined value red dashed line. a) Typical partially bleached sediment result; sample from the San Andreas fault, Carrizo Plain, California, USA. b) Example of dramatic incomplete zeroing; fine-grained sediment from glacial moraine, Pyrenees Mountains, Spain. c) An almost perfectly zeroed sample from a fault slip rate study, Changma, China. d) Single-grain D_E values are divided by the mean dose rate to provide apparent age values; 53% of measured grains are well zeroed, providing a combined age estimate of 8,460 ± 610 years; see Zinke *et al.* (2017) for other results, and Rhodes (2015) for methods employed.

require fading correction using the pIR-IRSL$_{225}$ approach, whereas many Holocene samples do not. Clarification of this point awaits further research, but the authors' advice when using this approach is to date a subset of known-age samples from the project site, within the same catchment, or the same lithology, and assess age performance using these. Also, the use of 'infinite age' cobbles or pebbles of the dominant bedrock lithology can provide an independent means to assess steady state fading behaviour for any signal (Brown *et al.*

2015; Kars *et al.* 2008), and it is usual to include an assessment of fading for every grain within these protocols. Rhodes (2015) lists five ways to make fading assessment, though it should be stressed that the possibility of apparent fading in the laboratory, while little or no fading occurred in nature, needs to be considered.

9.6 CHECKLIST FOR PLANNING CAMPAIGN

When planning a project within an active tectonic context, the following guidelines may be useful. Consider the number of target 'events' that are likely to be encountered and, for each, plan several luminescence samples. This information is often not clear before detailed digital elevation models are constructed or trenches dug for each location, so it is useful to plan sufficient time and sample collection equipment to cope with better preserved sites with extensive potential for dating. Not every sample collected needs to be dated, but where ambiguities arise, immediate access to additional samples may be very valuable. Where possible, take a dating specialist to help collect samples. It is recommended to make *in situ* gamma measurements whenever possible. Ensure to record sample longitude, latitude, elevation and burial depth, and record the stratigraphy and relative sample locations in notes as well as photographically.

The poorest areas for both quartz and feldspar characteristics tend to be volcanic areas. This may be caused by the relatively rapid cooling of both quartz and feldspar in volcanic bedrock leading to a more disordered crystal lattice (Daniels 2016). In these contexts, and in carbonate provinces, reliance on far-travelled aeolian grains of quartz or feldspar may be necessary. These grains are likely to be at the finest end of what may be dated with conventional 'coarse grained' (with respect to beta dose rate) luminescence approaches at around 75–100 μm, and special considerations for single-grain dating may be needed. Where target grains are expected to be relatively rare in the sediment, collection of larger samples, or duplicates for key samples, may help.

In the laboratory, running a small subset of samples as a 'pilot' helps determine whether quartz grains have sufficient sensitivity for OSL, whether Super-K or conventional K-feldspar separation will provide better pIR-IRSL signals, what yields can be expected for different grain sizes and mineral fractions, and the degree of signal zeroing that may be typical. However, experience is that the yields and zeroing can vary widely between apparently similar samples from the same site. Where quartz OSL can be undertaken with single grains, this is the preferred approach.

Wherever possible, do collect samples for radiocarbon dating, and make use of every opportunity to cross-check the different dating techniques including cosmogenic isotopes and U-series dating of carbonate crusts. For larger projects, collect samples from the modern channel to assess signal zeroing in that environment, and consider collecting bedrock pebbles for feldspar fading determinations.

9.7 ADDITIONAL APPLICATIONS RELEVANT TO TECTONICS

There are additional applications of luminescence that have less direct relevance to the understanding of processes in tectonics or continental dynamics. These include the study of many different geomorphic features or sedimentary contexts affected in some way by tectonic processes such as raised beaches (Coutard *et al.* 2006; Ree *et al.* 2003) or fluvial terraces abandoned as a result of incision driven by uplift (Bates *et al.* 2010; Lewis *et al.* 2017). The dating issues encountered in studies of sediments in these contexts do not usually go beyond the problems and limitations faced in other luminescence sediment-dating applications, discussed elsewhere in this book. However, in some cases the particular issues identified above, namely poor mineral characteristics and sub-optimal conditions for signal zeroing, are also found, and the discussion within this chapter may be relevant. In many cases, the interpretation of the geomorphic and tectonic context depends on a simple model, as is the case for the specific applications to fault slip rate and palaeoseismology that we discuss in detail below. It should be borne in mind that, however robust the sediment dating undertaken, the interpretation depends fundamentally on the veracity of these models and, in some cases, these receive rather brief presentation in publications, particularly with regard to the assumptions that underlie the model.

A second group of applications represents attempts to date fault surfaces directly. Most of these focus on the expectation that frictional heating or increased stress during fault movement will reset the luminescence signal (Banerjee *et al.* 1999; Ding and Lai 1997; Mukul *et al.* 2007; Singhvi *et al.* 1994). The partial success of a similar approach attempted using ESR signals in quartz (e.g. Fukuchi 1996; Fukuchi and Imai 1998) associated with some promising results illustrates the potential for parallel luminescence studies. However, the relatively high trap depth for quartz ESR centres was equated by Lee and Schwartz (1994) to a required overburden thickness of around 70 m at the time of resetting. Clearly, this approach could not be applied to samples collected at, or close to, the present surface that had been affected by recent events over the last few thousand years, as rates of denudation are rarely sufficiently high. The expectation of lower thermal stability associated with quartz TL or OSL signals, or feldspar TL, IRSL, or pIR-IRSL signals, leaves this as an area for possible future research. Exploration of the signals in feldspar were undertaken at the SAFOD (San Andreas Fault Observatory at Depth) in California, USA (Spencer *et al.* 2012), on grains collected from the fault plane at *c.* 2600 m depth, and surprisingly, given the ambient temperature of 112 °C, both TL and IRSL provide D_E values in the correct range to correspond to the last major earthquake in 1906.

Three basic mechanisms may be operating in these cases, (1) direct frictional heating of grains within or close to the fault plane, (2) indirect heating, where fluids are heated and may migrate some distance from the fault, and (3) stress-induced detrapping, either on the fault plane or at some distance from it. In different studies, authors have explored these mechanisms, but there appears to be some variation in findings, and little consensus has emerged as to the best approach, or the environmental requirements for such a technique to be successful. In fact, the difficulty of finding exposed sections of fault plane in consolidated material likely to have generated frictional heating makes this a technical development with potential, but of somewhat specialist application.

A third group of luminescence–tectonic studies are those applying low temperature thermochronometry using TL or OSL for quartz or feldspar, or feldspar IRSL. These techniques measure the time since a sample was cooled through a temperature range that allows the trapping of charge to exceed the thermal losses. Brown *et al.* (2017) summarise early studies for feldspar, including an early attempt by Prokein and Wagner (1994) to assess TL thermochronometry systematics. An application using quartz OSL was also made in New Zealand (Herman *et al.* 2010), although the mineral origin of the measured OSL signals were later questioned (Guralnik *et al.* 2015a). Guralnik *et al.* (2015b) quantified the systematics of Na-feldspar IRSL signals providing a theoretical and mathematical treatment (Guralnik *et al.* 2013) complementing that developed by Li and Li (2012). Recent application of a feldspar MET (multiple elevated temperature) IRSL approach to determine more complex cooling histories has been made by King *et al.* (2016), while Brown and Rhodes (2017) explore the systematics of feldspar TL, subsequently applied by Brown *et al.* (2017).

9.8 SUMMARY AND CONCLUSIONS

In summary, many developments have contributed to a suite of techniques that may be applied in active tectonic environments. Where quartz properties are suitable, and samples display intense rapidly decaying OSL signals characteristic of the fast component (Smith and Rhodes 1994), this is the approach of choice, using single-grain measurements where incomplete zeroing may be a problem. In locations with poor quartz sensitivity, single-grain K-feldspar pIR-IRSL measured at 225 °C can often work well; the signal sensitivity to light is too low to use conventional multiple-grain aliquots in most active tectonic contexts. This approach still has aspects that require further research, such as the extent of fading of natural IRSL signals, but problems can often be overcome by comparison to independent age control for a subset of samples within a project, and completed projects including independent age control suggest that Holocene and some late Pleistocene results rarely require fading correction for this signal. More intensive campaigns are helping to answer fundamental questions about fault behaviour and earthquake generation (e.g. Dolan *et al.* 2016; Zinke *et al.* 2017); for example, suggesting that faults undergo significant variations in slip rate over several earthquake cycles.

REFERENCES

Banerjee, D., Singhvi, A.K., Pande, K., Gogte, V.D. and Chandra, B.P. 1999. Towards a direct dating of fault gouges using luminescence dating techniques e methodological aspects. *Current Science* 77, 256–268.

Bates, M.R., Briant, R.M., Rhodes, E.J., Schwenninger, J.-L. and Whittaker, J.E. 2010. A new chronological framework for Middle and Upper Pleistocene landscape evolution in the Sussex/Hampshire Coastal Corridor, UK. *Proceedings of the Geologists' Association* 121, 369–392.

Brown, N.D. and Rhodes, E.J. 2017. Thermoluminescence measurements of trap depth in alkali feldspars extracted from bedrock samples. *Radiation Measurements* 96, 53–61.

Brown, N.D., Rhodes, E.J., Antinao, J.L., McDonald, E.V. 2015. Single-grain post-IR IRSL signals of K-feldspars from alluvial fan deposits in Baja California Sur, Mexico. *Quaternary International* 362, 132–138.

Brown, N.D., Rhodes, E.J., Harrison, T.M. 2017. Using thermoluminescence signals from K-feldspars for low-temperature thermochronology. *Quaternary Geochronology* 41, 31–41.

Buylaert, J.P., Murray, A.S., Thomsen, K.J., Jain, M. 2009. Testing the potential of an elevated temperature IRSL signal from K-feldspar. *Radiation Measurements* 44, 560–565.

Constraints from luminescence dating. *Quaternary International* 199, 15–24.

Coutard, S., Lautridou, J.-P., Rhodes, E.J. and Clet, M. 2006. Tectonic, eustatic and climatic significance of raised beaches of Cotentin (Val de Saire, Normandy, France). *Quaternary Science Reviews* 25, 595–611.

Cowie, P.A., Phillips, R.J., Roberts, G.P., McCaffrey, K., Zijerveld, L.J.J., Gregory, L.C., Faure Walker, J., Wedmore, L.N.J., Dunai, T.J., Binnie, S.A., Freeman, S.P.H.T., Wilcken, K., Shanks, R.P., Huismans, R.S., Papanikolaou, I., Michetti, A.M., Wilkinson, M. 2017. Orogen-scale uplift in the central Italian Apennines drives episodic behaviour of earthquake faults. *Scientific Reports* 7:44858.

Cowgill, E. 2007. Impact of riser reconstructions on estimation of secular variation in rates of strike–slip faulting: Revisiting the Cherchen River site along the Altyn Tagh Fault, NW China. *Earth and Planetary Science Letters*, v. 254, pp. 239–255.

Daniels, J.T.M. 2016. Mineralogic controls on the infrared stimulated luminescence of feldspars: An exploratory study of the effects of Al, Si order and composition on the behavior of a modified post-IR IRSL signal. Unpublished MS thesis, UCLA.

Dawson, T.E., McGill S.F. and Rockwell, T.K. 2003. Irregular recurrence of paleoearthquakes along the central Garlock fault near El Paso Peaks, California: *Jour. Geophys. Res.* 108, 2356–2385.

Ding, Y.Z., Lai, K.W. 1997. Neotectonic fault activity in Hong Kong: evidence from seismic events and thermoluminescence dating of fault gouge. *Journal of the Geological Society*, London 154, 1001–1007.

Dolan, J.F., McAuliffe, L.J., Rhodes, E.J., McGill, S.F. and Zinke, R. 2016. Extreme multi-millennial slip rate variation on the Garlock fault, California: Strain super-cycles, potentially time-variable fault strength, and implications for system-level earthquake occurrence. *Earth and Planetary Science Letters* 446, 123–136.

Fattahi M. 2009. Dating past earthquakes and related sediments by thermoluminescence methods: A review. *Quaternary International* 199, 104–146.

Fattahi M., Walker, R., Hollingsworth, J., Bahroudi, A., Nazari, H., Talebian, M., Armitage, S. and Stokes, S. 2006. Holocene slip-rate on the Sabzevar thrust fault, NE Iran, determined using optically stimulated luminescence (OSL), *Earth and Planetary Science Letters*, 245, 673–684.

Fattahi, M., Walker, R. T., Khatib, M. M., Dolati, A. and Bahroudi, A 2007. Slip-rate estimate and past earthquakes on the Doruneh fault, eastern Iran, *Geophysical Journal International*, v. 168, p. 691–709.

Ferrater, M., Ortuño, M., Masana, E., Pallàs, R., Baize, S., García-Meléndez, E., Martínez-Díaz, J.J., Echeverria, A., Rockwell, T.K., Sharp, W.D., Medialdea, A. and Rhodes, E.J. 2016. Refining seismic parameters in low seismicity areas by 3D trenching: The Alhama de Murcia fault, SE Iberia. *Tectonophysics* 680, 122–128.

Fitzsimmons, K.E., Rhodes, E.J. and Barrows, T.T. 2010. OSL dating of southeast Australian quartz: A preliminary assessment of luminescence characteristics and behaviour. *Quaternary Geochronology* 5, 91–95.

Fukuchi, T. 1996. Direct ESR dating of fault gouge using clay minerals and the assessment of fault activity. *Engineering Geology* 43, 201–211.

Fukuchi, T. and Imai, N.,1998 Resetting experiment of E' centers by natural faulting – the case of the Nojima Earthquake fault in Japan. *Quaternary Geochronology* 17,1063–1068.

Guralnik, B., Ankjærgaard, C., Jain, M., Murray, A.S., Müller, A., Wälle, M., Lowick, S.E., Preusser, F., Rhodes, E.J., Wu, T.-S., Mathew, G., Herman, F. 2015a. OSL thermochronometry using bedrock quartz: a note of caution, *Quaternary Geochronology* 25, 37–48.

Guralnik, B., Jain, M., Herman, F., Ankjærgaard, C., Murray, A.S., Valla, P.G., Preusser, F., King, G.E., Chen, R., Lowick, S.E., Kook, M. and Rhodes, E.J. 2015b. OSL-thermochronometry of feldspar from the KTB borehole, Germany. *Earth and Planetary Science Letters* 423, 232–243.

Guralnik, B., Jain, M., Herman, F., Paris, R.B., Harrison, T.M., Murray, A.S., Valla, P.G. and Rhodes, E.J. 2013. Effective closure temperature in leaky or saturating thermochronometers. *Earth and Planetary Science Letters* 384, 209–218.

Herman, F., Rhodes, E.J., Braun, J. and Heinihger, J. 2010. Uniform erosion rates and relief amplitude during glacial cycles in the Southern Alps of New Zealand, as revealed from OSL-thermochronology. *Earth and Planetary Science Letters* 297, 183–189.

Kars, R.H., Wallinga, J., Cohen, K.M. 2008. A new approach towards anomalous fading correction for feldspar IRSL dating – tests on samples in field saturation. *Radiation Measurements* 43, 786 – 790.

King, G.E., Herman, F. and Guralnik, B. 2016. Northward migration of the eastern Himalayan syntaxis revealed by OSL thermochronometry. *Science* 353, 800–804.

Krbetschek, M.R., Götze, J., Dietrich, A. and Trautmann, T. 1997. Spectral information from minerals relevant for luminescence dating. *Radiation Measurements* 27, 695–748.

Lawson, M.J., Roder, B.J., Stang, D.M. and Rhodes, E.J. 2012. Characteristics of quartz and feldspar from southern California, USA. *Radiation Measurements* 47, 830–836.

Lawson, M.J., Daniels, J.T.M., Rhodes, E.J. 2015. Assessing Optically Stimulated Luminescence (OSL) signal contamination within small aliquots and single grain measurements utilizing the composition test. *Quaternary International* 362, 34-41.

Lee, H.K. and Schwartz, H.P. 1994. Criteria for complete zeroing of ESR signals during faulting of the San Gabriel fault zone, southern California. *Tectonophysics* 235, 317–337.

Lewis, C.J., Sancho, C., McDonald, E.V., Peña-Monné, J.L., Pueyo, E.L., Rhodes, E.J., Calle, M., Soto, R. 2017. Post-tectonic landscape evolution in NE Iberia using a staircase of terraces: combined effects of uplift and climate. *Geomorphology* 292, 85–103.

Li, B. and Sheng-Hua Li, S.-H. 2012. Determining the cooling age using luminescence-thermochronology. *Tectonophysics* 580, 242–248.

Middleton, T.A., Walker, R.T., Rood, D.H., Rhodes, E.J., Parsons, B., Lei, Q., Zhou, Y., Elliott, J.R., Ren, Z. 2016. The tectonics of the western Ordos Plateau, Ningxia, China: Slip rates on the Luoshan and East Helanshan Faults. *Tectonics Tectonics*, 35, 2754–2777.

Mukul, M., Jaiswal, M., Singhvi, A.K. 2007. Timing of recent out-of-sequence active deformation in the frontal Himalayan wedge: insights from the Darjiling sub-Himalaya, India. *Geology* 35, 999–1002.

Nian, X, Bailey, R.M. and Zhou, L. 2012. Investigations of the post-IR IRSL protocol applied to single K-feldspar grains from fluvial sediment samples. *Radiation Measurements* 47, 703–709.

Nissen E., Walker, R. T., Bayasgalan, A., Carter, A., Fattahi, M., Molor, E., Schnabel, C., West, A.J., Xu, S. 2009. The late Quaternary slip-rate of the Har-Us-Nuur fault (Mongolian Altai) from cosmogenic ^{10}Be and luminescence dating. *Earth and Planetary Science Letters* 286, 467–478.

Pietsch, T., Olley, J., and Nanson, G. 2008. Fluvial transport as a natural luminescence sensitiser of quartz, *Quaternary Geochronology* 3, 365–376.

Porat, N., Duller, G. A. T., Amit, R., Zilberman, E. and Enzel, Y. 2009. Recent faulting in the southern Arava, Dead Sea Transform: Evidence from single grain luminescence dating. *Quaternary International* 199: 34–44.doi:10.1016/j.quaint.2007.08.039

Preusser, F., Chithambo, M.L., Götte, T., Martini, M., Ramseyer, K., Sendezera, E.J., Susino, G.J. and Wintle, A.G. 2009. Quartz as a natural luminescence dosimeter. *Earth-Science Reviews* 97, 184–214.

Prokein, J. and Wagner, G.A. 1994. Analysis of thermoluminescent glow peaks in quartz derived from the KTB-drill hole. *Radiation Measurements* 23, 85–94.

Porat, M., Levi, T. and Weinberger, R. 2007. Possible resetting of quartz OSL signals during earthquakes: Evidence from late Pleistocene injection dikes, Dead Sea basin, Israel, *Quaternary Geochronology* 2, 272– 277.

Porat, N., Duller, G.A.T., Amit, R., Zilberman, E. and Enzel, Y. 2009. Recent faulting in the southern Arava, Dead Sea Transform: Evidence from single grain luminescence dating. *Quaternary International* 199, 34–44.

Ree, J.H., Lee, Y.J., Rhodes, E.J., Park, Y., Kwon, S. T., Chwae, U., Jeon, J. S. and Lee, B. 2003. Quaternary reactivation of Tertiary faults in southeastern Korean peninsula: Age constraint by optically stimulated luminescence dating. *Island Arc* 12, 1–12.

Reimann, T., Thomsen, K.J., Jain, M., Murray, A.S. and Frechen, M. 2012. Single-grain dating of young sediment using the pIRIR signal from feldspar. *Quaternary Geochronology* 11, 28–41.

Rhodes, E.J. 2007. Quartz single grain OSL sensitivity distributions: Implications for multiple grain single aliquot dating. *Geochronometria* 26, 19–29.

Rhodes, E.J. 2011. Optically stimulated luminescence dating of sediments over the past 200,000 years. *Annual Review of Earth and Planetary Sciences* 39, 461–488.

Rhodes, E.J. 2015. Dating sediments using potassium feldspar single-grain IRSL: Initial methodological considerations, *Quaternary International* 362, 14–22.

Rhodes, E.J., Bronk-Ramsey, C., Outram, Z., Batt, C., Willis, L., Dockrill, S. and Bond, J. 2003. Bayesian methods applied to the interpretation of multiple OSL dates: high precision sediment age estimates from Old Scatness Broch excavations, Shetland Isles. *Quaternary Science Reviews* 22, 1231–1244.

Rizza, M., Ritz, J.F., Braucher, R., Vassallo, R., Prentice, C., Mahan, S.A., McGill, S., Chauvet, A., Marco, S., Todbileg, M., Demberel, S. and Bourlès, D. 2011. Slip rate and slip magnitudes of past earthquakes along the Bogd left-lateral strike-slip fault (Mongolia), *Geophys. J. Int.*186, 897–927.

Roder, B.J., Lawson, M.J., Rhodes, E.J., Dolan, J.F., McAuliffe, L. and McGill, S.F. 2012. Assessing the potential of luminescence dating for fault slip rate studies on the Garlock fault, Mojave Desert, California, USA. *Quaternary Geochronology* 10, 285–290.

Rudersdorf, A., Hartmann, K., Yu, K., Stauch, G., Reicherter, K. 2015. Seismites as indicators for Holocene seismicity in the northeastern Ejina Basin, Inner Mongolia. In Landgraf, A., Kuebler,

S., Hintersberger, E. and Stein, S. (eds), *Seismicity, Fault Rupture and Earthquake Hazards in Slowly Deforming Regions*. Geological Society, London, Special Publications, 432.

Scharer, K., Weldon, R., Streig, A., T. Fumal, T. 2014. Paleoearthquakes at Frazier Mountain, California delimit extent and frequency of past San Andreas Fault ruptures along 1857 trace, *Geophys. Res. Lett.* 41, 4527–4534.

Singarayer, J.S. and Bailey, R.M. 2004. Component-resolved bleaching spectra of quartz optically stimulated luminescence: Preliminary results and implications for dating. *Radiation Measurements* 38, 111 – 118.

Singhvi, A.K., Banerjee, D., Pande, K., Gogte, V., Valdiya, K.S. 1994. Luminescence studies on neotectonic events in south-central Kumaun Himalaya: A feasibility study. *Quaternary Science Reviews (Quaternary Geochronology)* 13, 595–600.

Smedley, R.K. and Duller, G.A.T. 2013. Optimising the reproducibility of measurements of the post-IR IRSL signal from single-grains of feldspar for dating. *Ancient TL* 31, 49–58.

Smedley, R.K., Duller, G.A.T., Roberts, H.M. 2015. Bleaching of the post-IR IRSL signal from individual grains of K-feldspar: Implications for single-grain dating. *Radiation Measurements* 79, 33–42.

Smith, B.W, Rhodes, E.J. 1994. Charge movements in quartz and their relevance to optical dating. *Radiation Measurements* 23, 329-333.

Spencer, J.Q.G., Hadizadeh, J., Jean-Pierre Gratier, J.-P. and Doan, M.-L. 2012. Dating deep? Luminescence studies of fault gouge from the San Andreas Fault zone 2.6 km beneath Earth's surface. *Quaternary Geochronology* 10, 280–284.

Trauerstein, M., Lowick, S.E., Preusser, F., Schlnegger, F. 2014. Small aliquot and single grain IRSL and post-IR IRSL dating of fluvial and alluvial sediments from the Pativilca valley, Peru. *Quaternary Geochronology* 22, 163–174.

Wells, D.L., and Coppersmith, K.J. 1994. New empirical relationships among magnitude, rupture length, rupture width, rupture area, and surface displacement. *Bull. Seismol. Soc. Am.*, 94, 974–1002.

Wintle, A.G., Murray, A.S. 2006. A review of quartz optically stimulated luminescence characteristics and their relevance in single-aliquot regeneration dating protocols. *Radiation Measurements* 41, 369–391.

Zinke, R., Dolan, F.D., Rhodes, E.J., Van Dissen, R., McGuire, C.P. 2017. Highly variable latest Pleistocene-Holocene incremental slip rates on the Awatere Fault at Saxton River, South Island, New Zealand, revealed by Lidar mapping and luminescence dating. *Geophysical Research Letters* 44, 41–61.

Zinke, R., Dolan, J.F., Van Dissen, R., Grenader, J.R., Rhodes, E.J., McGuire, C.P., Langridge, R.M., Nicol, A. and Hatem, A.E. 2015. Evolution and progressive geomorphic manifestation of surface faulting: A comparison of the Wairau and Awatere faults, South Island, New Zealand. *Geology* 43, 1019–1022.

10 APPLICATIONS IN ARCHAEOLOGICAL CONTEXTS

IAN K. BAILIFF

Department of Archaeology, Durham University, South Road., Durham DH1 3LE. Email: ian.bailiff@dur.ac.uk

ABSTRACT: As a chronometric tool, luminescence has a potentially important role to play on archaeological sites where organic samples are lacking or where sites are beyond the effective range of the more widely used radiocarbon dating. The variety and complexity of archaeological sites and artefacts, combined with the requirements of the dating technique, mean that close collaboration between field and laboratory researchers is essential. The selection of the most appropriate laboratory techniques to date a particular archaeological material and/or process depends on various technical factors, including the type of mineral(s) present, how and when the chronometer mechanism was reset (by heating or exposure to daylight), and the anticipated age of the depositional contexts from which the dating samples were extracted. Successful application of the method to archaeological materials relies on an understanding of both the experimental issues and the nature of the fieldwork undertaken to collect the primary archaeological data, with the integrity of the dating samples forming an essential component in the construction of a reliable chronological framework.

KEYWORDS: archaeological remains, Palaeolithic, coastal sites, agricultural features, structures and buildings, pottery

10.1 INTRODUCTION

Although the origins of luminescence dating lie in applications to archaeology, its scope, as reflected in other chapters, has broadened considerably during the intervening 50 years, largely accelerated by the introduction of OSL techniques. As a result, archaeological dating forms a relatively small part of the now global engagement in use of the method for dating various types of depositional processes in Quaternary research. While the output of luminescence dates in archaeological research has been dwarfed by the routine access to radiocarbon dating, the method nonetheless fulfils an important role as a means of dating archaeological materials where suitable organic samples are lacking and, in particular, beyond the range of radiocarbon within the domain of Palaeolithic archaeology. As the luminescence method is complex and costly, applications of luminescence are often the result of research collaboration between fieldwork investigator and laboratory to

321

ensure a mutual understanding of site and laboratory issues. These enable adjustment of archaeological fieldwork to obtain optimal samples (see also Chapter 2).

This chapter is necessarily selective and highlights recent applications of OSL and TL to archaeological research where the impact of luminescence dating results is gaining more traction amongst archaeologists. The examples discussed are drawn across the wide chronometric range of application of the method, from the Middle Palaeolithic (broadly *c*. 300,000 to 40, 000 years ago) occupation of cave and open-air sites to the comparatively recent buildings of the medieval period. These case studies are then used to provide additional notes on sampling and technical issues related specifically to the application of luminescence to archaeological contexts.

10.2 PALAEOLITHIC CAVE, ROCKSHELTER AND OPEN-AIR SITES

By dating materials from sites with evidence of Palaeolithic occupation, luminescence has made significant contributions to our understanding of the chronology of hominin evolution and migration. The luminescence techniques applied are rarely routine to execute and require detailed and specialised work. Sites that contain cultural evidence of lithic tool fabrication and use by hominins over many millennia where the sequence of deposits contains changes in technology are of particular importance to evolutionary studies, and such sites are of singular value where the changes can be placed on a chronological timescale. The trapped charge methods (luminescence and electron spin resonance, ESR) have a potentially important role to play on sites older than the later phases of the Middle Palaeolithic (300–30 ka), most of which are beyond the range of radiocarbon (~50 ka). The typological characteristics of lithic assemblages that survive on many Palaeolithic sites play an important role in forming the basis of chronological frameworks established in the study of later hominin evolution. However, due to inter-site variability in their lithic typologies, reliable chronometric dating of sites, for example with evidence of Neanderthal and early anatomically modern human (AMH) occupation in the Southern Levant, is proving to be of critical importance (Porat *et al.* 2018). Where there is evidence of the use of fire (e.g. hearths), it is likely that burnt flint fragments, whether accidentally or deliberated heated, will be found nearby, and this may also include pebbles and cobbles. Heated materials are often preferred as chronological markers because of their direct association with anthropogenic activity. Unheated sediment deposits that form the burial medium for cultural remains provide markers for reconstruction of the site formation processes. Whereas the latter can be dated using OSL techniques, heated flint or chert are dated using TL techniques as the OSL of flint generally has unfavourable characteristics. However, Schmidt and Kreutzer (2013) recently found that some of the flints they tested were potentially suitable for dating using OSL measurement procedures, provided experimental screening procedures were applied. The dating of crystalline cobbles (quartz and granite) optically reset before burial using OSL techniques has also been demonstrated (see Chapter 11.), which extends the range of potentially suitable materials from early sites. Unheated calcite, in the form of speleothems, is often abundant in karstic caves and was the subject of early investigations using TL, but it was found to be too complex to apply routine procedures and has been largely superseded by uranium-series dating (Pike 2017; Pike *et al.* 2017).

Palaeolithic sites often present challenging issues when applying luminescence, the nature of which depend on the type of dating material tested and the technique applied.

A laboratory will base its choice of technique(s) to determine the equivalent dose (D_E) on the type of mineral(s) present, whether it was heated (i.e. above *c.* 400 °C), and the anticipated age of the depositional context from which the dating samples were extracted. Although most of the published luminescence dates have been produced for materials from Palaeolithic contexts in cave and rockshelter sites, the number of studies of surviving open-air sites is increasing. The following examples illustrate a range of issues that have arisen in testing material from cave, rock shelter and open-air sites which are also common to many other archaeological applications where there is a strong reliance on the interpretation of stratigraphy and site formation processes when applying luminescence techniques. The recent reassessment of the chronostratigraphy at the Liang Bua site, Flores Indonesia (Sutikna *et al.* 2016), for example, illustrates the potentially complex nature of cave sites that should not be underestimated.

10.2.1 Applications

10.2.1.1 Jebel Irhoud

Located in western Morocco, the cave site of Jebel Irhoud is well known for its Middle Stone Age (MSA) lithic assemblages characterised by Levallois technology and lacking the presence of any Acheulean or Aterian elements (Hublin *et al.* 2017). Recent TL dating results by Richter *et al.* (2017), obtained from burnt flint found in association with *H. sapiens* remains, indicate that the fossils represent the earliest known presence of *H. sapiens*. In this work, a TL age of 315 ± 34 ka was obtained from Layer 7, which contained the highest density of archaeological remains (including the most recently discovered hominin fossils). This was supported by an age of 302 ± 32 ka for burnt flint from the overlying deposit (Layer 6). Although materials providing independent dating markers such as tephra or speleothems were absent, the TL ages are supported by a revised combined uranium series/electron spin resonance age of 286 ± 32 ka for a human tooth fragment (Irhoud 3; Smith *et al.* 2007) that had been recovered from deposits reported to be equivalent to layer 7. OSL dating of the sediments in Layers 6 and 7 was not attempted as the antiquity of the site was expected to be beyond the limit of the conventional SAR procedure. Whilst this assumes that the heat treatment of the flint occurred at the same time as the *H. sapiens* occupation, indirect dating of the fossils has important implications for our understanding of *H. sapiens* evolution as it places them as an early pre-modern phase of the species (Stringer and Galway-Witham 2017).

Historical excavations have long suffered from sampling of curated materials many years after their original excavation. This has been the case with many important Palaeolithic sites that were excavated during the early 20th century, before the development of scientific dating methods. Such an approach often leads to compromised chronometric data. Contemporary fieldwork often includes better planning of dating sample collection and therefore results in more robust chronologies. In the case of Jebel Irhoud, re-excavation of the site allowed for the deployment of 47 dosimeters in advance of the excavation to establish a detailed picture of background dose rate variability (Richter *et al.* 2017). The use of 3D recording techniques during the excavation also allowed accurate reconstruction of the positions of artefacts later selected for luminescence testing within the stratigraphy. This aimed to reduce the overall uncertainty in TL ages by providing a precise record of the spatial relationship between the dated lithics and locations at which the gamma dose rate had been either measured directly or

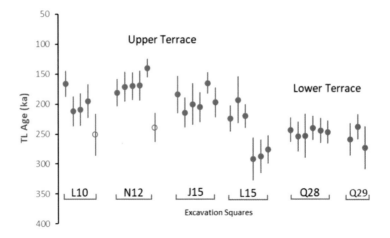

Figure 10.1 TL ages obtained for burnt flint samples from the Upper Terrace and Lower Terrace deposits of Mislaya Cave, near Haifa, which contained late Lower Palaeolithic (LP) and early Middle Palaeolithic (MP) lithic assemblages respectively. The recovery of the two assemblages from the same site is important because they show a distinctive change in lithic technology, from Acheuleo-Yabrudian (LP) to Mousterian (MP) industries, identified by the TL chronology to have occurred *c.* 250 ka ago. The two TL ages in the Upper Terrace deposits indicated by open symbols were discounted as being overestimates of the age, as discussed in the main text. Data reproduced from Valladas *et al.* (2013).

calculated on the basis of an analysis of the radionuclide content. The variability in individual TL ages obtained for layers 6 (relative standard deviation, RSD, 9%; n = 4) and 7 (RSD, 6%; n = 6), respectively, was found to be comparable to that observed in the dosimeter measurements (RSD 8–14%). The variability, attributed to dose rate heterogeneity (see Section 10.2.3), is commonly encountered on karstic features containing limestone and further improvements in overall precision in TL ages from the site would require an even finer resolution approach to the reconstruction of the gamma dose rate.

10.2.1.2 Misliya Cave

This cave site, located in the vicinity of Tabun cave on Mount Carmel, near Haifa, contains a rich sequence of early Middle Palaeolithic (MP) and Lower Palaeolithic (LP, Acheulo-Yabrudian) cultural remains, including very large assemblages of flint artefacts and many hearths that indicate intensive use of the cave over a prolonged period. A programme of dating burnt flints by Valladas *et al.* (2013), recovered from the layers containing early MP and LP artefacts, produced 32 TL ages (Fig. 10.1). From an analysis of the ages it was concluded that a rapid transition from the late LP (244 ± 27 ka) to early MP (212 ± 27 ka) had occurred and that the use of the cave during the early MP had persisted for at least 75 ka. The uncertainties in the TL ages (with errors of 12–14% of the age) were comparable to those obtained for the Jebel Irhoud burnt flints and, although a similar experimental approach was employed in determining the equivalent dose (D_E), the dose rate assessment in this study additionally included an isochron analysis of the LP samples. Isochron analysis, devised to provide an estimate of age independent of the external dose rate, can be applied to obtain an average age for a group of flint specimens where there is a sufficient spread of internal dose rates between samples, but where the external dose rate within the sampled volume is uniform. An isochron age of 244 ± 30 ka obtained for the LP samples (Fig. 10.1, Q28 and Q29) showed good consistency with the individual ages calculated using the individual external dose rates, reinforcing confidence in the assessment of the external dose rate for the section of the site sampled. However, it should also be noted that some of the TL ages for the early MP samples were discounted as being overestimates of the age (Fig. 10.1, open symbols) based on their inconsistency with the chronostratigraphy, flint typology and sediment micromorphology. While these particular results remain enigmatic, they signal the importance of obtaining a sufficient number of age determinations to both detect and accommodate the occurrence of outliers of this type. Apart from technical investigations to examine more deeply the robustness of D_E and dose rate assessments, the dating of the burial of the artefacts by applying the recently developed extended range OSL techniques (see Chapter 1) to sediments may be appropriate in future work, as discussed below in the case of Combe Brune 2.

10.2.1.3 Diepkloof Rockshelter

Amongst the relatively few instances of direct comparisons of age results obtained by independent laboratories that provide an internal test of robustness, the studies undertaken at the Diepkloof Rockshelter (DR), located in the Western Cape of South Africa, provide an interesting example, where three laboratories undertook independent studies of samples from the site, some of which were taken from equivalent stratigraphic contexts. Diepkloof Rockshelter is regarded as a key site in understanding the MSA in the region, containing a preserved sequence of cultural deposits that includes the onset and evolution of both the Still Bay (SB) and Howiesons Poort (HP) typologies, spanning the period *c.* 40–100 ka according to the current interpretation of the stratigraphy (Parkington *et al.* 2013). The luminescence dates obtained for sediments (Jacobs *et al.* 2008; Wollongong laboratory) and burnt quartzite (Tribolo *et al.* 2009; Bordeaux laboratory) had indicated some differences in the chronologies produced by each laboratory. The results of further investigation were subsequently reported by Tribolo *et al.* (2013), adding their single-grain OSL dates for sediment samples, and Jacobs and Roberts (2015) who undertook a detailed reassessment of their earlier data. Contributions to the debate were also made by Guérin *et al.* (2013; Bordeaux lab.) and subsequently further single-grain OSL age estimates for sediment samples obtained in 1995 (related to a 1973 excavation pit) were published by Feathers (2015; Seattle laboratory).

The issues raised in these papers are instructive by revealing, not only significant disagreement in ages produced for some of the stratigraphically coeval deposits (Fig. 10.2), but the inner workings of the method. Whilst OSL ages for some of the directly comparable sample locations are in agreement, the OSL chronology obtained by the Bordeaux laboratory for the lower levels of the sequence indicates generally earlier ages than those obtained by the Wollongong lab (Fig. 10.2, DRS 11, 13–16). The difference appears to be greatest in the case of DRS 13 and 14, and this proved to be pivotal in the proposal of an earlier chronology for the SB (Parkington *et al.* 2013). In their reassessment, Jacobs and Roberts (2015) point to differences in the measurement of the beta dose rate by each laboratory as the potential cause of the main differences in the OSL ages obtained for sediment using single-grain (SG) analysis. In this particular section of the rockshelter sequence, the beta particle emission from ^{40}K makes a dominant contribution to the total dose rate.

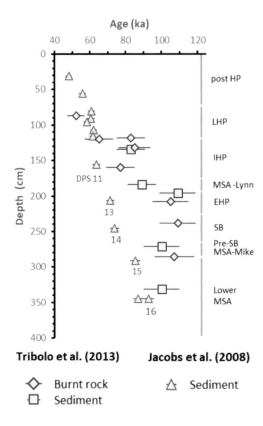

Figure 10.2 Luminescence chronostratigraphy for Diepkloof Rockshelter, Western Cape, South Africa with comparative data produced by the Bordeaux and Wollongong laboratories. The mean luminescence age estimates, plotted as function of depth (cm), were obtained for burnt rock (shaded diamonds) and sediment (shaded squares) by the Bordeaux laboratory (Tribolo *et al.* 2013) and for sediment (shaded triangles; those samples discussed in the main text with DRS sample numbers indicated) obtained by the Wollongong laboratory (Jacobs *et al.* 2008). In the case of the upper six age estimates by Wollongong, the depths of the top six samples were adjusted (by Tribolo *et al.*) to take the cave topography into account. The main cultural assignments are indicated on the RHS where, HP, Howiesons Poort; MSA, Middle Stone Age; SB, Stillbay, and where Jack and Mike represent the names assigned to the stratigraphic units during the original test pit excavations. (Modified from Tribolo *et al.* 2013).

Each laboratory had employed an indirect approach to calculating the beta dose rate, by determining concentrations of the radionuclides in the sediment (Bordeaux, high resolution gamma spectrometry; Wollongong, beta gas counting) and applying published conversion factors to obtain the infinite medium beta dose rate. The measurement of radionuclide concentrations with homogenised samples necessarily produces a volume-averaged measure of the beta dose rate. As these laboratories have well-established calibration procedures for their instrumentation, there remains an as-yet unresolved difference, and in the case of sample DRS 13 (DRS-OSL6) the difference between the estimates of beta dose rate obtained by the two laboratories is beyond the limits of measurement uncertainty. The three sediment samples obtained by the Seattle laboratory, although stratigraphically not directly correlated with the samples obtained from the later excavations, were interpreted to be from strata containing artefacts of the LHP/Post-HP (UW247; 73±5 ka), LHP/IHP (UW325; 63±4 ka) and IHP (UW260; 80±6 ka) MSA lithic technology cultural periods. When compared with the Bordeaux and Wollongong ages shown in Figure 10.2, these additional ages can be seen to be closer to the earlier ages produced by Bordeaux in this particular section of the stratigraphy. If the differences are eventually resolved, it is likely that that factors related to both the determination of the equivalent dose and dose rate will be involved. However, one of the key aspects revealed by work at Diepkloof is a potential methodological weakness when current methods for dose rate assessment are applied to single-grain analysis of sediment sample where the distribution of radionuclides is likely to be heterogeneous (see Section 10.2.3).

10.2.1.4 Open-air sites

Where cultural remains have been preserved in fluvial or colluvial contexts (Chauhan *et al.* 2017), open-air sites may offer greater flexibility in terms of sampling. The sands and silts provide sediment suitable for luminescence measurements and also provide a more uniform burial medium that simplifies dose rate assessment. Recent studies of MP open air sites that have successfully produced OSL chronologies (e.g. Arnold *et al.* 2012; 2015; Been *et al.* 2017; Bueno *et al.* 2013; Duller *et al.* 2015; Frouin *et al.* 2014; Zaidner *et al.* 2018) report a mixture of deposits, some containing grains that had been uniformly reset and others containing partially reset grains. This is dependent on the particular geomorphological and sedimentological processes involved in site formation. In these circumstances the availability of single-grain measurement techniques is important, but saturation effects usually limit the utility of conventional OSL procedures for sites of MP age. Where this arises, the 'extended range' techniques can be applied using TT-OSL and pIR-IRSL procedures to quartz and feldspar extracts respectively (see Chapter 1). The luminescence signal measured using these procedures saturates at much higher levels of dose although, in the case of TT-OSL, the signal detected is much weaker (Fig. 10.3). The TT-OSL technique, although at a developmental stage, is being applied to materials from archaeological sites to explore its potential. Where feasible, both TT-OSL and pIR-IRSL techniques are being applied to test for 'internal consistency'. To make these evaluations more robust, sites with independent dating controls are preferable (Arnold *et al.* 2015).

Excavation of a transitional Stone Age open-air site at Kalambo Falls in northern Zambia recovered an important archive of Stone Age artefacts, assessed to span the late Acheulean to early MSA transition in south-central Africa (*c.* 400–500 ka). The OSL chronology that Duller *et al.* (2015), produced by dating the deposition of fluvial deposits, extended over 500 ka and identified four punctuated phases of sediment deposition (~500–300 ka; ~300–50 ka; ~50–30 ka; 1.5–0.49 ka). A single-grain technique was applied to quartz extracted from sediment samples taken from the younger deposits and the TT-OSL technique was applied to date the older depositional events, but using multiple grain aliquots to obtain sufficiently strong OSL signals. This technique had also been applied to

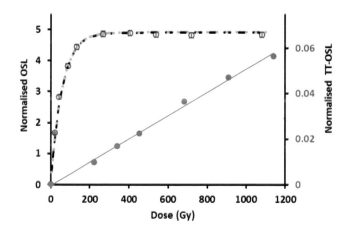

Figure 10.3 Dose response characteristics measured with quartz grains extracted from fluvial deposits on the open-air site at Kalambo Falls, Zambia. Single-grain measurements of the conventional OSL (open circles) exhibit the onset and approach to saturation of the signal between *c.* 100 and 250 Gy. In contrast, the dose response of the TT-OSL signal (filled circles) is linear to beyond 1000 Gy. (Modified from Duller *et al.* 2015).

fluvial terrace deposits from the MP open-air site of Hotel California in the region of the Sierra de Atapuerca, Burgos, north-central Spain (Arnold *et al.* 2012).

Open-air sites within doline features in karstic landscapes have also provided suitable repositories of Middle and Lower Palaeolithic artefacts. An assemblage of Levallois lithic artefacts was recovered from the site of Combe Brune 2, located in the Bergerac region of south-western France, which contained a continuous sequence with Levallois lithic assemblages buried in loessic colluvium deposits formed by two merging dolines (Frouin *et al.* 2014). The sediment infill and good preservation of the lithic artefacts indicated highly localised deposition following discard. A combination of techniques was applied to date burnt flint (TL) present within the deposits and to coarse grains extracted from sediment (TT-OSL, quartz; pIR-IR, feldspar). The closely grouped TL dates for four burnt flints (183 ± 20; 185 ± 23; 187 ± 21; 195 ± 16 ka) were stratigraphically consistent with the OSL dates obtained for sediment deposits in the underlying (234 ± 25 ka; pIR-IRSL) and overlying (161 ± 18 ka, pIR-IRSL; 184 ± 19 ka, TT-OSL) units. The four additional OSL ages for deposits higher in the sequence were also in stratigraphic order. Also, the preservation of Lower Palaeolithic remains were recovered within a more deeply developed doline at the site of Nahal Hesi in the north-western Negev, reported by Zaidner et al. (2018). An OSL age of 430 ± 35 ka was obtained for the deposition of a layer containing Lower Palaeolithic limestone artefacts of Acheulian typology using the extended range techniques. However, the colluvial processes within a doline feature may not always be as benign. The OSL dating of sediments from the upland site of West Cliffe (Kent, UK) located on a doline feature and containing typologically Lower Palaeolithic artefacts and debitage (>300 ka), placed the deposition of the artefacts in their excavated context to between *c.* 140 and 80 ka ago (Bailiff *et al.* 2013). It was concluded that the cultural deposits were probably displaced from their primary context by colluvial processes associated with the development of the solution feature. It is suggested that the latter caused displacement of the artefacts from their location of original deposition, and that this may have occurred during the last interglacial when the solution feature was active.

10.2.2 Sampling and on-site measurement issues

The above cases studies serve to illustrate a number of important issues which require consideration when applying luminescence in archaeological contexts. These include:

1. **Separation of artefacts from burial medium.** One of the potential difficulties in sampling artefacts during archaeological excavations is associated with the basic process of the recovery archaeological data, which normally requires the physical unfolding of the site formation processes (i.e. by excavation). This invariably leads to the disassociation of artefacts and their immediate burial medium which is of primary interest in the assessment of the background dose rate for luminescence dating. With the exception of static features such as hearths, access to artefacts must often wait until a section has been excavated and recorded. In these circumstances, at least half of the burial medium is likely to have been removed once potentially suitable dating samples have been identified. This is in contrast to sampling sediment where access to the layers of

interest usually can be obtained in an unexcavated section of a trench and where an examination of the stratigraphy can be more readily examined.

2. The composition of the burial medium is also relevant to the assessment of the beta dose rate, particularly where single-grain techniques are to be applied. Whereas sediment samples are usually obtained by insertion of an opaque tube (Chapter 2), where the sampled deposits are uniform, the process of sample preparation leads to disaggregation and loss of information concerning the environments of individual grains. The extraction of samples where the sediment structure is preserved (e.g. excised sediment blocks) is common practice in micromorphological analysis (Goldberg and Berna 2010). It is being increasingly recognised that, for complex burial environments, similar approaches are required in OSL sampling to provide the opportunity to conduct more detailed examination and analysis in the laboratory. For example, this may include thin section analysis of the sediment microstructure and dosimetry measurements, such as a detailed spatially resolved assessment of the radioactive environment from those grains yielding D_E values in SG analyses. Measurement of the beta dose rate in this way (e.g. Jankowski and Jacobs 2018; Lebrun *et al.* 2018), once fully developed, is likely to require a transformational change in technique equivalent to that made from single-aliquot to single-grain analysis in OSL.

3. **Insufficient material for dating.** When sampling burnt flint, laboratories commonly apply a 'coarse grain' preparation procedure to extract particles in the size range 100–200 mm and typically specimens of at least 5–10 g weight are required to yield a sufficient quantity of material within the interior once the outer 2–3 mm layer of the flint sample has been removed. This sample size requirement may limit the number of samples that are available for testing.

4. **On-site dose rate reconstruction.** When sampling artefacts or sediments, the volume of the burial medium within *c.* 30 cm of the sample is of primary interest in the assessment of the gamma dose rate. Where sediment samples are obtained by tube extraction, measurement of the gamma dose rate at the sample location is usually achieved by performing dose rate measurements using dosimeters or a portable gamma ray spectrometer (see Chapter 2). However, this is rarely the case with artefacts and the laboratory usually has the task of reconstructing the gamma dose rate at the sample location(s) based on measurements with dosimeters or placement of the spectrometer probe at an equivalent stratigraphic location within the site. For both artefact and sediment samples, samples (e.g. *c.* 50 g) are usually taken from each of the major units and subsequently analysed in the laboratory for the concentrations of the lithogenic radionuclides, including an examination for the presence of radioactive disequilibrium using high resolution gamma spectrometry. Where the burial medium contains one or more layers of uniform deposits, and providing sedimentological examination indicates

the absence of significant taphonomic change, such assessments based on a 'representative' environment is expected to provide a reasonable estimate of the gamma dose rate. However, in these circumstances there is an attendant risk that the measurements may not be representative of the burial environments of the samples tested. As illustrated in the Jebel Irhoud and Diepkloof studies, cave and rockshelter contexts require particularly careful attention to detail where the stratigraphy is complex (e.g. Mercier *et al.* 1995), particularly where the deposits contain a heterogeneous mixture of materials of significantly differing concentrations of lithogenic radionuclides. Large spatial variations in gamma dose rate can be expected to arise when deposits contain materials of contrasting radionuclide content, such as carbonate rock fragments typically containing very low radionuclide content, are mixed with clay-bearing alluvial sediment, for example. This issue has been addressed in previous studies by using many passive dosimeters inserted at multiple location points to measure the spatial variation in gamma and cosmic dose rate (e.g. Mercier *et al.* 2007). The deployment of dosimeter measurements in advance of excavation aims to improve the reliability of gamma dose estimation (e.g. as at Jebel Irhoud). Similarly, data collected from measurements performed with a portable gamma spectrometer at the exposed face of a section prior to excavation (Guérin and Mercier 2012) can be used to calculate dose rates within the interior of a section that is expected to yield dating samples.

10.2.3 Technical issues

As can be seen from the above case studies, a wide range of different luminescence approaches are available to be applied to archaeological contexts. Each has different characteristics making them more or less suitable according to site configuration, artefacts/material to be dated and the potential antiquity of the site. What is also clear is that there are various technical issues to be considered. These include:

1. **Evaluating whether burnt flint samples were reset by sufficient heating.** The TL glow curve (detected in the blue/UV region) typically contains a broad peak in the temperature range 300–400 °C and the TL signal recorded in this range forms the basis of the D_E evaluation. The 'plateau' test is one of the critical tests that is applied to determine whether heating of the flint prior to burial was sufficient to eliminate an inherited trapped charge population. In the laboratory such resetting can be achieved with prepared flint samples by annealing at 360 °C for 90 minutes in air. For aliquots passing the plateau test, a multiple aliquot additive dose procedure (Fig. 10.4; Richter and Krbetschek 2015; Mercier *et al.* 1992) is commonly applied to determine the equivalent dose, D_E, in conjunction with a 'slide' procedure (Prescott *et al.* 1993), where a dose response curve generated with thermally reset portions of sample is produced. Samples where the equivalent dose (D_E) exceeds twice the characteristic dose (D_0)

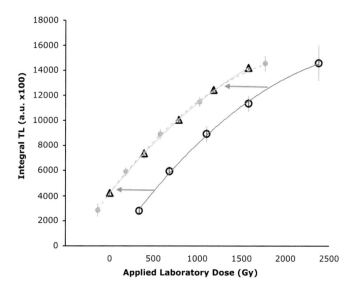

Figure 10.4 Dose response curves measured on a burnt flint sample (LUM- 08/07) from the cave site of Jebel Irhoud, Morocco, for which a TL date of 328 ± 28 ka was determined. The curves are not linear but best fitted by a saturating exponential function. In the approach used to determine the palaeodose for burnt flint, the data are obtained by applying to a series of sample portions in two stages: (a) an additive dose procedure (open triangles) and (b) a regenerative dose procedure (open circles). The filled data points and fitted curve (broken line) represent the regenerative dose response shifted along the dose axis using the slide procedure; essentially the palaeodose corresponds to the translation along the dose axis required to match the regenerative and additive DRCs. (modified from Richter *et al.* 2017).

are deemed to have approached a saturation level (Chapter 1) and can provide only an estimate of minimum age. The dose response of aliquots is measured using a calibrated laboratory beta radiation source, but the response to alpha radiation must also be established as the 'internal' dose rate includes a component arising from alpha radiation emitted by lithogenic radionuclides within the lithic fabric.

2. **Heterogeneity of background dose rate**. The term is used to refer to significant departures from uniformity in the distribution of lithogenic radionuclides within materials, which is often correlated with the type and form of the material within the burial medium (See 10.2.2). Heterogeneity of this type potentially affects the dispersion in D_E values and in the dose rate, the extent of which depends on the technique applied to determine D_E and requires assessment on an appropriate spatial scale (defining a volume of interest). The latter varies according to the range (penetrating power) in matter of the three basic types of radiation (α, β and ϒ) emitted by the lithogenic radionuclides (<50 mm, 2–3 mm and ~30–50 cm respectively; Chapter 2). Consequently, the volume of interest for the gamma dose rate is up to 50 cm from the sample, potentially

extending across several depositional layers, whereas the volume(s) of interest in case of the beta and alpha dose rates is highly localised. Sparing the technical detail, how does this affect the dating of burnt flint and sediment?

In the case of lithic samples, the total dose rate comprises an 'internal' component arising from alpha and beta radiation emitted by radioactive sources distributed within the lithic and an 'external' component comprising beta and gamma radiation emitted by sources within the surrounding medium and in addition, cosmic radiation. By physically removing the outer layer of flint during sample preparation, the external beta dose rate can be eliminated from the assessment. Although the specific (radio)activity of flint and chert is relatively low, it is subject to natural variation and the sources within samples can contribute a relatively high proportion of the total dose rate (typically 20–30%), and in the case of the Jebel Irhoud samples, for example, it ranged between 20 and 50%. In principle, a proportionately higher internal component can be expected to reduce the effect of large uncertainties in the external component that may exist where a gamma dose rate has been reconstructed. However, full advantage is only gained if the radioactive sources are uniformly distributed within the lithic sample. Additionally, those samples that contain a heterogeneous distribution of sources, arising from the clustering of uranium for example, are usually avoided (Schmidt and Kreutzer 2013; Tribolo *et al.* 2013) as they are expected to give rise to increased dispersion in the measured values of D_E.

For sediment samples, the primary source of uncertainty in the dose rate is likely to be associated with the beta dose rate. The effect of heterogeneity on the beta dose rate differs according to the technique applied to determine D_E. Where D_E is evaluated using single aliquots containing multiple brightly emitting coarse grains, the use of an average beta dose rate to evaluate the age is appropriate, providing the analysis is performed on material of composition similar to that from which the coarse grains were extracted for OSL measurements. Where a single-grain technique is applied, the immediate environment within 2–3 mm of each grain yielding a D_E value determines the beta dose rate to that grain, but by disaggregating the sediment during sample preparation that information is lost. The beta dose rate that is typically derived by laboratories, as in the case of Diepkloof for example, is an average value for a volume of several cm^3, at best, and consequently may not be representative on the scale of an individual grain. However, where beta dose rate heterogeneity is to be examined in the laboratory on a sub-mm scale, samples collected in the field must remain structurally intact.

10.3 AGRICULTURE

The archaeological record provides abundant evidence of the potential human vulnerability to adverse climatic change and its effect on the viability of maintaining long-term settlement in marginal habitats, which is dependent on maintaining an agricultural system for the production of foods. Several aspects of agricultural activity, including the construction and use of terraces in upland areas, irrigation features at the edge of mountainous catchment regions and sustaining cultivation within the coastal zone, have been the subject of recent OSL studies, examples of which are included in the following sections. OSL techniques have been shown to lend themselves well to dating deposits formed naturally by colluvial, fluvial (Chapter 7) and coastal dune formation processes (Chapter 8), and in cases where they were modified by anthropogenic activity (e.g. Lang and Wagner 1996; Liritzis *et al.* 2013; Chapter 6), which is of specific relevance to archaeological applications.

10.3.1 Applications

10.3.1.1 Agricultural terraces

Terraces within dryland farming zones are of central importance to sustaining many upland settlements by creating areas of near flat ground on hillsides that reduce soil erosion and assist the retention of moisture in the soils. Reliable dating of their construction and subsequent use is often difficult to obtain using conventional archaeological techniques. The possibility of applying OSL to date the deposition of terrace soils is consequently of particular interest in the study of settlement and a number of studies (Avni *et al.* 2006; Beckers *et al.* 2013; Davidovich *et al.* 2012; Meister *et al.* 2017) conducted in the southern Levant appear to have produced promising results. The selection of locations for sampling is an important issue, with the excavation of test pits or sections revealed by road cuts being essential to allow a reliable geomorphological assessment and confirmation of the nature of deposits associated with agricultural activity overlying any previously established soils. The degree of resetting of grains, prior to and following construction of the terrace, should be dependent on a number of factors, including the nature of the terrace tread fill and subsequent mixing processes where cultivation processes have been applied to the terrace soils. In particular, luminescent grains within the deeper tread fill of run-off terraces, transported by fluvial processes, are likely to have different exposure histories from equivalent deposits within terrace filled with quarried soils.

Davidovich *et al.* (2012), examined a terraced slope on Ramat Rahel, near Jerusalem, by investigating nine terrace walls and excavating 11 test pits within an area of *c.* 2 hectares. OSL samples were obtained at various depths in the test pits located adjacent to the terrace retaining walls (risers); the excavation had revealed four types of riser construction (Fig. 10.5, I–IV; see caption). Analysis of the tread fill indicated that the soil had been obtained from a local source of slope sediment with a relatively fine composition arising from its aeolian origin. This study applied a single-aliquot measurement procedure to relatively fine coarse grain quartz (75–125 μm). The distribution of D_E values from these measurements indicated partial resetting of grains (OD ranging widely from 5 to 60%). However, there was little evidence of either downward percolation of quartz grains from much younger reworking of the soils or mixing with the *in situ* basal geogenic soil lying in pockets above the bedrock. The OSL burial ages obtained for each pit ranged from *c.* 1600 to 400 years. The ages, when grouped by riser type (Fig. 10.5), are interesting in that, with one exception, while Type I can be associated with Late Byzantine/Early Islamic settlement, examples of Type II occur in all three of the main settlement periods. The double leaf construction built on bedrock (Type III) is relatively recent (Ottoman), whereas a similarly robust form of riser construction with soil footings is assigned to the same early settlement phase as Type I. Overall, the findings of this study, discussed in detail in Davidovich *et al.* (2012), illustrates the complex landscape history that is likely to underlie an investigation of cultivation history and a potentially large number of OSL determinations may be required to achieve a satisfactory outcome.

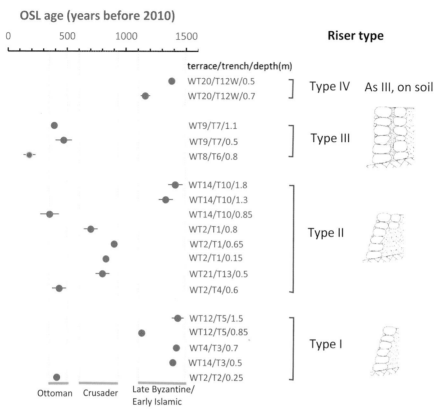

Figure 10.5 OSL ages for sediment samples from excavated contexts within the terraced landscape investigated near the site of Ramat Rahel, Israel. The OSL ages are grouped by terrace riser type (I–IV), where the labels adjacent to each OSL age plotted indicate the terrace and trench references and the sample depth below the ground surface (m). The shaded areas indicate the cultural periods assigned on the basis of archaeological evidence of the settlement in the area. (modified from Davidovich *et al.* 2012).

The study by Avni *et al.* (2006) provides an example of a broader landscape approach within the same geographic region applied to run-off agricultural terraces formed by check dams constructed within a wadi basin that trapped run-off water within the sandy loessic soils. The terraces were in use during the Roman and the Early Islamic periods and then abandoned. The aeolian origin of the luminescent minerals within most of the soils underpinned an assumption of predominantly fully reset grains and the central dose model was applied in the analysis of D_E values obtained with single aliquots. However, the finding of a 'residual' OSL age estimate of *c.* 380 years for surface deposits within a modern cultivated field system flagged the need to sample contemporary soils. Although samples of the latter were not available in the study of run-off terraces by Beckers *et al.* (2013), stratigraphically consistent OSL ages were obtained for samples of the basal and upper fill, and also deposits underlying the risers. This indicated construction and use of the terrace during the 1st millennium AD that is consistent with archaeological evidence of settlement in the area (Petra, Jordan).

To provide a means of enabling more productive sampling strategies for terrace contexts, Kinnaird *et al.* (2017) developed a combined approach of OSL profiling using a portable reader and laboratory-based dating measurements (Sanderson and Murphy 2010; Chapter 12). This approach aimed to provide on-site evaluation of the bulk luminescence characteristics of sediment samples taken from an excavated section, producing a sufficient number of determinations to create section profiles of the luminescence signal intensities from quartz and feldspars within the deposits. This information was used to detect marked changes in depositional processes within the terrace and to identify optimal locations of samples for more detailed dating measurements in the laboratory. Further use of the profile results was made to obtain interpolated OSL ages (using a reduced set of measurements to obtain D_E estimates) and information regarding the depositional processes between the chronological markers obtained with the latter. Applying this approach to late medieval terraces in Catalonia, Spain, Kinnaird *et al.* (2017) produced a *terminus post quem* (TPQ) for the construction of each of three terraces, ranging from the 13th to 17th centuries AD.

10.3.1.2 Irrigation features

The application of luminescence to the dating of canal irrigation features associated with past human settlement has been shown to be successful in several studies (Berger *et al.* 2004, 2009; Huckleberry and Rittenour 2014; Huckleberry *et al.* 2012). Although upcast sediment dumped either side of a canal during construction and subsequent cleaning episodes is potentially suitable, these deposits have been found to contain a high proportion of grains that were incompletely reset before final burial (Rittenour 2008) and it is the strata preserved within channel fills that are usually sampled. Some success has been obtained dating the deposition of sediment associated with the construction of other types of irrigation features, including wells (Khasswneh *et al.* 2016) and shaft and gallery irrigation systems, such as qanat networks (Bailiff *et al.* 2018). Although these applications are relatively new and have yet to be tested more widely, luminescence offers a potentially important means of dating irrigation features, the age of which have proved to be particularly difficult to establish using conventional archaeological techniques. In the case of qanats, the characteristic ventilation shafts are surrounded by a small mound formed from sediment upcast that is drawn from the construction deposits and sediment collected during subsequent episodes of cleaning the subterranean gallery (Manuel *et al.* 2018). Excavation of the mound allows the depositional

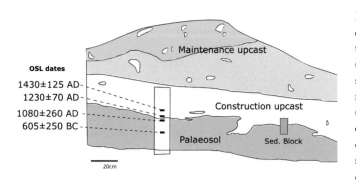

OSL dates

1430±125 AD
1230±70 AD
1080±260 AD
605±250 BC

Maintenance upcast

Construction upcast

Palaeosol   Sed. Block

20cm

Figure 10.6 A section of an excavated qanat mound showing the positions of OSL samples (within white rectangle) and a sediment block extracted for micromorphological analysis (Sed. Block). The vertical line on the LHS of the section corresponds to the ventilation shaft wall (modified from Bailiff *et al.* 2018).

sequence to be examined and sampled for OSL and micromorphological analysis (Fig. 10.6). The mode of deposition of sediment and its effect on the resetting process, the underlying geology, sediment grain size, mineralogy and bioturbation are physical factors that potentially influence the success of the OSL testing of a ventilation shaft mound and these may differ between mounds in a network. A critical issue in sampling a mound for OSL is the identification of the contact between the ancient ground surface and the initial deposit of construction upcast (see Fig. 10.6). OSL samples taken from both the palaeosol underlying this contact and the major units of overlying upcast, comprising construction and maintenance deposits, enable a chronostratigraphy to be developed, supported by detailed micromorphological analysis of the sediment structure.

10.3.1.3 Coastal zone

As discussed in Chapter 8, the development of dune chronologies using OSL techniques has contributed substantially to the study of climate change in coastal regions of NW Europe by applying sand mobilisation in coastal regions as a proxy record (e.g. Madsen and Murray 2009). The processes of aeolian transport and dune formation, linked to storm events within the North Atlantic region during the Holocene is of particular relevance to the study of the settlement of prehistoric and later communities in coastal zones. Various OSL studies undertaken on sites in the Western Isles, Orkney, Shetland and on the NE English coast have identified periods of significant aeolian sand mobilisation of sufficient severity to cause abandonment of settlements during periods of climate deterioration. This approach has been further developed and applied to the recent study of landscape evolution on Herm, one of the Channel Islands that lies off the coast of Guernsey (Bailiff *et al.* 2014). This small island contains a relatively high concentration of important megalithic monuments that were constructed during the Neolithic period, some of which have been completely buried by dune formation. OSL, in conjunction with geomorphological analysis, was applied to date key stages in the development of the landscape, including the formation of stabilised dunes, the burial of land surfaces associated with prehistoric settlement and major horizons in the prehistoric palaeosol sequence. Preserved evidence of cultivation was found in several of the excavation trenches, with ard marks in the form of shallow linear furrows and Neolithic pot and flint fragments scattered in upper buried soil horizons that indicate the practice of intensive manuring to improve the fertility of the soils. Three phases of significant aeolian activity during the prehistoric period led to dune formation (OSL age determinations of *c.* 4 ka, 3 ka and 2.3 ka), the first marking the start of a long-term trend of aggradation of soils that persisted for the following two millennia.

A critical stage was reached where the cumulative aeolian activity affected the capability of the soils to support cultivation, and this appears to have led to the abandonment of the northern part of the island. Evidence of these changes was captured in one of the excavation trenches and the OSL dates identify a prolonged period of aeolian activity during the first millennium BC (Fig. 10.7) that delayed the resumption of cultivation until the Roman period. OSL dates obtained for basal dune sands sampled in seven locations confirmed that the onset of accumulation of sand that now blankets the north part of the island commenced during the early 13th century and persisted until the late 16th century. The loss of at least a millennium in the sedimentary record evident in one of the trenches (Trench E; comparing the OSL dates for the two uppermost samples) reflects an instability of landscape, which is likely to have occurred during the early medieval period.

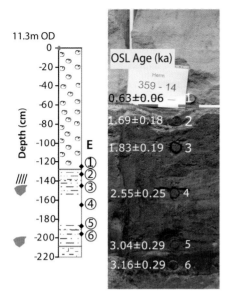

Figure 10.7 A sequence of OSL ages (ka) shown with their sand-rich coastal contexts in Trench E, on the island of Herm, Guernsey. Six OSL sample cores (E.1–6) were positioned to sample four stages of land surface development: a) E.1, the formation of the overlying dune sand, b) the deposition of the upper palaeosol (E.2), c) the stabilisation and duration of wind-blown sand (E.3, E.4 and E.5), and d) the later phase of development of the lower palaeosol (E.6). The upper palaeosol contained structural features attributed to ard-like plough marks. On the basis of the OSL dates for samples E.3–5, the rate of aggradation of the wind-blown sand was about 4 cm per century (modified from Bailiff *et al.* 2014).

10.3.2 Sampling and on-site measurement issues – terraces

Where the sampling of contexts is performed immediately behind terrace retaining walls, careful examination of the stratigraphy is necessary to avoid areas disturbed by repairs to the wall that may have also resulted in the incorporation of younger material into earlier contexts during the process of backfilling. Gibson (2015) argues that the application of a landscape approach is essential to identify optimum locations for the deployment of OSL, including prior topographic analysis and fieldwalking and the recovery of diagnostic artefacts for phasing of land use for cultivation within the terrace system. These recommendations are in keeping with modern geoarchaeological fieldwork practice (French 2015). In addition, an assessment of the depositional processes within sampled contexts using micromorphological analysis of sediment thin sections, as discussed above, is appropriate to develop a robust approach to the assessment of the particular site-formation processes.

10.3.3 Technical issues – terraces

In semi-arid regions with high levels of insolation, the application of a single-aliquot SAR technique appears to produce satisfactory age determinations where the grains were fully reset. However, where less favourable conditions for the resetting process arise (Kouki 2006), the D_E estimates are expected to exhibit significant overdispersion (OD) and may produce overestimates of the burial age. In these circumstances the use of a single-grain approach, although requiring more advanced instrumentation and generally greater experimental time, is advisable.

An interesting extension of the analysis of D_E values obtained with samples from terrace contexts has been proposed in cases where the use of a central dose model is not appropriate. To extract information from those D_E values not assigned to the minimum dose group and that would normally be discarded as being obtained from partially reset grains, Porat *et al.* (2017) used these data to derive age estimates for 'relict' grains deposited by previous episodes of agricultural activity, structural evidence of which remained extant within the terrace deposits. It should be noted that their measurements were performed with very small aliquots (1 and 2 mm diameter) and not with a single-grain reader. The testing of quartz grains extracted from the soils within the contemporary surface soil and confirmation of a modern date (±40 years), point to the conditions where full resetting of grains within the cultivation zone is likely to be essential for this type of analysis to yield reliable results. From a number of ancient terrace systems in the Jordan Highlands they found that most of the worked terrace deposits were dated to the last 600 years, but that the D_E components extracted from the application of a finite mixtures model (FMM) identified earlier phases of resetting associated with agricultural activity dating back to the late Hellenistic period, which is consistent with the archaeological evidence of settlement in the region.

10.4 STRUCTURES AND BUILDINGS

Ceramic brick, more generally referred to as a CBM, has been used extensively in the construction of buildings and structures during the last 2 millennia and even earlier in certain regions of the world. Although the luminescence techniques employed to date brick (and tile) are similar to those applied to pottery, their exploitation to date brick buildings remained largely dormant for over 20 years. Examples of application can now be found for individual buildings and structures in Europe constructed with brick (e.g. the Czech Republic, Denmark, England, Finland, France, Germany, Italy and Poland), in Asia (e.g. Cambodia, India, Sri Lanka, Thailand and Uzbekistan) and South America (Brazil). In NW Europe dating the construction of high status secular and ecclesiastical buildings of the last millennium can be achieved often to within couple of decades based on documentary records and stylistic features. However, dating the construction of lower status vernacular buildings to better than 50–100 years is frequently difficult to achieve and there is the potential for luminescence to make an impact on the study of such buildings. There are also many types of brick structure in other geographic regions where stylistic dating is problematic because of the absence of a diagnostic typology.

10.4.1 Applications

In terms of precision, the performance obtained with brick can be expected to vary according to region and building, and even at the level of individual bricks. Both accuracy and precision are influenced by the characteristics of luminescent minerals extracted for measurement and the nature of the brick fabric. Published studies of testing medieval and Roman bricks, using quartz coarse and also fine grains, indicate that where a fabric is uniform and manufactured using a sand temper, there are good prospects of obtaining an uncertainty (±1σ) for individual ages within the commonly applied yardstick of ±5–10% of the luminescence age. When tested against a group of late medieval English buildings dated on the basis of documentary and stylistic evidence, encouraging agreement was obtained (Bailiff 2007), and for six samples taken from dating 'control' buildings constructed within the range *c.* AD 1400–1720 the mean difference between the central values of luminescence and assigned ages was 5±10 years (σ, $n = 6$). Other studies that have addressed research questions extending beyond a single building provide an indication of the likely utility of the method when applied to new areas of buildings research. These include the testing of a longstanding assumption that, following the collapse of the Roman Empire in the early 5th century AD, brick making in NW Europe ceased and was not resumed until at least six centuries later in France, and even later in England. This was partly supported by evidence of the widespread quarrying and reuse of abundant quantities of high quality Roman brick. Luminescence dates obtained for bricks from a selection of ecclesiastical buildings in NW France and southern England, albeit a small number, indicate that brickmaking had been resumed during the Early Medieval period (Fig. 10.8) in France, but not in England until much later, during the 11th century AD (Blain *et al.* 2007, 2014). It had also been generally assumed that buildings were constructed with new bricks during the Late Medieval period when brick production expanded significantly across Europe. Yet in one English building examined, the practice of reusing brick appeared to have continued (Bailiff *et al.* 2010) and this is of particular relevance to the study of vernacular buildings when assessing changes to, and loss of, buildings within the built landscape. It also provides a further note of caution, signalling the need for a careful structural examination of walls for the presence of reused brick when selecting samples for testing to establish the date of construction of a building.

The detection of the reuse of bricks that emerged from the study of late medieval English buildings identified a potentially interesting tool for the study of the construction practice of masons and the development of a brick commerce. While there is much archaeological

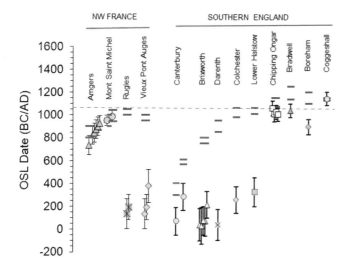

Figure 10.8 A compilation of OSL dates for bricks sampled from a range of medieval ecclesiastical buildings located in NW France and southern England. The architectural age range assigned by stylistic and documentary evidence, where available, is indicated by red bars. As discussed in the main text, the middle group of dates reflects the reuse of Roman brick in the fabric (Rugles – Lower Halstow). The data were extracted from Blain *et al.* (2010) and Bailiff *et al.* (2014). The horizontal broken line indicates the Norman Conquest. Below: a portable drill coupled to a water-lubricated core drill in use to sample a ceramic quoin.

evidence for the reuse of Roman brick within Medieval European buildings that usually can be readily identified by brick typology and fabric, elsewhere its practice may be more difficult to detect. The reuse of brick is likely to have been geographically widespread and their occurrence may present a confounding factor (e.g. Stark *et al.* 2006), given that the method does not (economically) lend itself to testing a large number of bricks. A more direct means of dating the emplacement of brick, rather than its manufacture, can be achieved by testing minerals (a) within the brick surface exposed to daylight and subsequently covered by mortar (Viellevigne *et al.* 2007), and (b) within the mortar binding medium (e.g. Feathers *et al.* 2008; Goedicke 2011; Solongo *et al.* 2014; Stella *et al.* 2013). Where sand is contained in the mortar mix, the application of techniques for dating partially optically reset grains is appropriate, with quartz likely to be the better candidate. The use of a single-grain technique provides a means of identifying a sub-population of grains that were more thoroughly reset before mixing and use of the mortar (Urbanova and Guibert 2017). It is worth noting that lime mortars may also contain thermally reset minerals if they were incorporated with crushed limestone during the preparation of quicklime. This approach, which is at a developmental stage, is clearly not confined to brick and opens up the much wider aspect of dating the emplacement of lithic masonry in built structures, in addition to application of techniques that have been developed for dating the insertion of worked lithic surfaces in structures (discussed in Chapter 11).

10.4.2 Sampling and on-site measurement issues

In addition to the general sampling guidance discussed in Chapter 2, there are further aspects to take into account when sampling a building. The development of a sound sampling strategy benefits from close cooperation with a building specialist who has phased the construction history of the building. This should include careful assessment of the brick fabric for repairs and/or evidence of the reuse of brick (Bailiff *et al.* 2010) to avoid producing results that are not related to the construction history of the building. Brick fabrics that include large rock fragments may have a strongly heterogeneous distribution of radionuclide sources, giving rise to a complex spatial variation of the beta dose rate and fabrics of this type should be approached with caution (Guibert *et al.* 2009).

A reduction in the overall uncertainty in the luminescence age is expected to be obtained by the selection of samples that have maintained a low average moisture content (<5%), and hence are likely to be above ground level in the case of standing walls located in a temperate climate. While the majority of dating applications are performed with bricks sampled from such locations, the method is equally capable of producing dates for excavated brick, for example from foundations. For samples taken from such contexts, an adjustment to the dose rate made for moisture content increases the overall uncertainty in the age (e.g. St James Church, Toruń; Chruścińska *et al.* 2014). However, in some cases brick associated with the foundations in drier conditions may be accessible where cellars were installed.

The sampling of buildings identified to be of historic importance is likely to be restricted to areas where cosmetic damage is minimised, unless the building is undergoing extensive renovation. The use of a diamond-tipped core drill (38 mm diameter) is usually sufficient to provide a relatively surgical means of sampling brick in the walls of standing buildings and core drill bits suitable for dry and wet cutting are available (Fig. 10.8). The cavity can be backfilled with lime mortar, finishing using either mortar coloured with brick dust or a cap comprising a surface section cut from the core. An alternative procedure to coring is the extraction of whole bricks by removing the surrounding mortar. This allows a section to be cut from the rear of the brick and subsequent replacement achieved without damaging the front face. However, it is a considerably lengthier extraction process and usually requires a specialist mason to perform the work. The extraction of solid cores extending to the inner part of the brick (e.g. ~8–10 cm from the surface) reduces the uncertainty in the dose rate from sources external to the wall (e.g. the ground surface, or an internal stone flag floor), the environment of which may have changed since original construction.

The insertion of a dosimeter capsule (Chapter 2) enables direct measurement of the combined gamma and cosmic dose rate at, or close to, the sampled location, where γ radiation is emitted by radionuclides both in the wall and in the immediate environment of the sampled location. The capsule is usually placed in a hole drilled into a mortar layer to a depth that is sufficient to place it within the rear half of the brick (e.g. 10 cm for NW European late medieval bricks). The period of measurement is normally at least several months, although shorter periods are feasible. *In situ* measurements (~1–2 h at each location) can also be made using a portable gamma ray spectrometer, with the detector either held against the wall or inserted into a cavity if the detector probe is sufficiently small. The conversion of the recorded spectra to dose rate within the wall requires the use of a detector that has been calibrated for measurements in a wall geometry. While this type of measurement generally gives rise to a higher uncertainty in dose rate compared with that obtained using a dosimeter, deposition of the latter may not be practicable for a short-term study. Alternatively, samples of brick and other materials within the local environment are analysed for their radionuclide content, and the dose rate calculated using published conversion factors and a geometry factor applied to account for the structural form of the sampled wall.

Different bricks within the same wall have been found to yield minerals with markedly different characteristics, some yielding coarse grain fractions with luminescence signals that are too dim to measure and lacking any bright grains (Bailiff 2007). This may occur because, in addition to the dependence of the luminescence sensitivity of minerals on geological source and transport history (Pietsch *et al.* 2008), thermal treatment also has a strong effect and its severity during firing can be expected to vary according to location in the brick kiln. Although yet to be systematically investigated with brick, the reduction is likely to be related to prolonged high temperature treatment (Bøtter-Jensen *et al.* 1995) and also stacking according to the kiln technology used (Blain *et al.* 2014). In the absence of advanced testing, it is consequently advisable to obtain samples from more than one brick in a section of interest.

10.4.3 Technical issues

As for pottery, well-fired clay bricks can be expected to contain fully reset luminescent grains of the mineral selected for equivalent dose determination, and both TL and OSL measurement techniques are potentially suitable. The choice of technique is usually governed by the signal intensity and dose response characteristics, and the preferred choice in current applications is likely to be an OSL SAR procedure applied to quartz coarse grains, or a fine-grain technique (4–11 μm) where the fabric texture is very fine. The latter requires an assessment of the fading characteristics (Chapter 1) of feldspar grains within the mix of luminescent minerals unless they are selectively removed using chemical treatments (Stella *et al.* 2013). The fine-grain technique provides a technical advantage over coarse quartz grains since the proportion of the equivalent dose due to radioactive sources within the sampled brick core is maximised, reducing the effect of changes in the environment external to the brick that may affect the gamma dose rate.

The application of plaster to the outer surface of rubble walls is known to have been practised in the construction of early medieval buildings, but often it does not survive. In the case of an external wall, such plaster would have provided additional shielding from radiation emitted by sources in the ground. Fortunately, lime-based plaster and mortar usually contain very low concentrations of radionuclides and the effect on the dose rate due to differences in shielding, although slight, should be taken into account.

10.5 POTTERY

As discussed in Chapter 1, the dating of pottery was initially a primary focus of the formative developmental years of luminescence during the 1960s (Aitken 1985; Wintle 2008), and application to other types of ceramic artefact, such as curated works of art (Fleming 1979) and ceramic building materials (CBM), soon followed. Yet adoption of the method by the archaeological community for dating pottery has been generally limited, despite the common occurrence of pottery on many sites from the Neolithic onwards (Orton *et al.* 1993). Although more recent research has revealed the earliest known pottery to be of Late Pleistocene origin, and well within the chronological range of the method (Kuzmin *et al.* 2001), the destructive nature of the experimental procedure steered excavators to seek radiocarbon ages of carbon residues or associated organic deposits for these rare artefacts. For less exotic applications, the specialist advice needed for sampling has made its deployment less convenient compared with radiocarbon testing. Moreover, the typical level of uncertainty obtained (±5–10% of the age, 1σ) for individual dates was generally considered insufficient to improve chronologies developed on the basis of pottery typologies that had been tested indirectly by radiocarbon dating of organic samples in coeval sealed deposits. However, the dating of many sites relies on pottery typology, particularly fabric composition. There is a potential role for luminescence where diagnostic pottery is lacking, or suitable organic samples are unavailable, or where temporal resolution is limited as a result of calibration issues (e.g. the flat spot at *c.* 2500 BP and the multiple ranges obtained during the last 500 years). The types of application can be broadly grouped into:

1. Establishing the date of the deposition of a layer or feature from which the pottery was recovered
2. Testing chronologies developed for pottery typologies
3. Examining methodological issues, such as the feasibility of dating previously untested types of pottery that lack diagnostic features.

The potential risk in (a) is that the pottery tested is residual and consequently not associated with the depositional event of interest. Frequently found scattered both spatially and temporally within the stratigraphy of a site, sherds in the fills of large features such as ditches, pits and middens may be re-deposited. If tested, samples of this type can be expected to produce misleading results where the processes of manufacture and deposition occurred at significantly different times. The sampling of sealed contexts where the use of the pottery before deposition was short-lived is likely to yield more reliable results. In regions where luminescence has not been undertaken previously, a preliminary evaluation of the suitability of the characteristics of the luminescent minerals within the pottery fabrics would normally be performed as this can be expected to vary according to the geological source of the minerals and the firing conditions during manufacture. A further type of application that has emerged is the dating of pottery recovered from surface sites where the deposits form palimpsests, an approach that practitioners had been previously advised to avoid (Aitken 1985).

10.5.1 Applications

Detailed dating studies by Barnett (2000) and (Feathers 2009) provide good examples of validation exercises applied to test the veracity of the method with pottery. Barnett, in comparing TL dates for British Iron Age diagnostic pottery (c. 800 BC–AD 100) with independent archaeological dating evidence, obtained an average difference of ~100 years between the luminescence dates and the mid-point of the assigned age range in each case. Feathers (2000, 2003) tested pottery recovered from short-lived Navajo sites of c. AD 1700 in eastern North America and found them to compare well with independent dating evidence provided by dendrochronology.

An aspect where luminescence has demonstrated a constructive role in pottery research is in testing chronologies for the onset and persistence of particular fabric types where the forms lack diagnostic features. The luminescence dates obtained in several studies on pottery from sites in Britain (Barnett 2000; Cramp et al. 2006) and North America (Feathers 2009) have challenged aspects of previously established fabric chronologies, in particular for shell-tempered pottery. Although an attempt to cross-check the luminescence results for shell-tempered pottery from Mississippi by applying AMS radiocarbon dating to carbonate temper (Peacock and Feathers 2009) encountered difficulties in establishing an accurate value for the reservoir offset, the outcome underlined the potential importance of the availability of luminescence testing for this type of pottery. Notwithstanding this issue, the availability of independent chronometric techniques when dating pottery and other cultural artefacts is of central importance (Bonsall et al. 2002). The extent of application of the method to prehistoric pottery from other regions is gradually being expanded, and recent examples include prehistoric pottery from Romania (Benea et al. 2007), Poland (Czopek et al. 2013) and Syria (Sanjurjo-Sánchez and Montero 2012).

Of potentially wider geographical interest is the application of the method (Dunnell and Feathers 1994; Sampson et al. 1997) to sites where pottery surface scatters form a key component of the surviving archaeological evidence and where the testing of organic matter by radiocarbon is highly problematic because of post-depositional disturbance. Where pottery is sampled from contexts that are very shallow or within erosional surfaces, the burial history of the sample following discard is unknown and consequently there a relatively high margin of uncertainty associated with the dose rate due to sources of radiation external to the pottery (i.e. the gamma and cosmic components). However, as deployed in authenticity testing of ceramic works of art (Fleming 1979), the proportion of the total dose rate arising from external sources typically can be reduced to c. 20% or less by applying a fine-grain technique (Chapter 1). Although there is the attendant issue of anomalous fading associated with feldspar minerals that are usually present in prepared fine-grain samples, a combined OSL/IRSL procedure can be applied to reduce the extent of this effect.

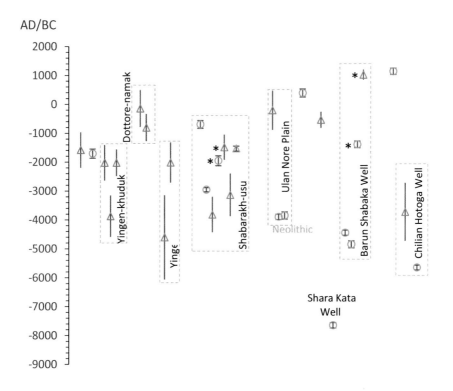

Figure 10.9 OSL (open triangles) and calibrated radiocarbon dates (open circles) obtained for Gobi Desert pottery sherds, excluding radiocarbon samples with carbon yields below 0.1%. All dates are expressed on a calendrical timescale (BC/AD) with an indicated 2σ range of uncertainty. The asterisk indicates where both OSL and radiocarbon dates were obtained from the same sherd. Radiocarbon samples comprised organic and carbonaceous residues extracted from the ceramic fabric. The results have been grouped according to site, as indicated and discussed in the main text. Modified from data presented in Janz *et al.* (2015).

In a recent study (Janz *et al.* 2015), collections of surface pottery and ostrich eggshell finds from sites in the Gobi Desert region of Mongolia and China were dated using luminescence (pottery) and AMS radiocarbon (organic material in pottery and eggshell), made available from curated collections of surface scatters obtained by scientific expeditions during the 1920s and 1930s. The radiocarbon analysis of the ostrich shell revealed that while the carbon yield was sufficient, the shell from the sites tested was not a reliable material for dating the occupation because the wide-ranging dates obtained indicated that the Neolithic inhabitants of the region had recovered and reused ancient shell. A formative regional chronology for the occupation sites was assembled with the radiocarbon and luminescence dating results for the pottery finds (Fig. 10.9). Taking a broader interpretation of the results shown in the figure, pottery appears to have been produced much earlier than the accepted date for the inception of the Neolithic in the region (4000–4200 BC), with pottery from the Shara Kata well representing the earliest known example of pottery in Mongolia and the Gobi Desert. The dates obtained for ceramic assemblages from some sites (e.g. Shabarakh-usu and Yingen Khuduk) point to occupation episodes that persist over very long periods, extending to millennia. A direct comparison of the two methods applied to the same (ceramic) samples, restricted to three sherds in this study, yielded good agreement with one (Fig. 10.9, indicated by asterisk). However, for the other two sherds, the significantly earlier radiocarbon AMS ages (Fig. 10.9, Barun Shabaka Well and Chilian Hotoga Well, radiocarbon age off scale) were judged unreliable due to a very low carbon content (*c.* 0.1%). If pottery with higher levels of carbon can be identified, a programme of further development of this combined approach and direct comparison would provide a valuable opportunity to test the luminescence methodology for dating surface finds. Moreover, in regions where pottery lacks organic content sufficient to produce reliable radiocarbon ages, this study indicates that luminescence is a viable alternative, albeit with generally lower resolution.

10.6 SUMMARY

The range of dating applications discussed in this chapter reflects a progressive expansion in the scope for applying luminescence to archaeological research. As the technical aspects of research progress, a better understanding of the performance of the method where complex burial environments are encountered can be expected, which is a particular issue in the case of caves sites. As will be evident elsewhere in the literature, execution of the method is rarely 'routine' and this has generally limited the number of luminescence dates in circulation when compared with radiocarbon. However, the further development of the application of the method has and will continue to be driven by addressing challenging research questions and these should be generated equally by aspects related to site formation and anthropogenic processes as well as by technical methodological issues.

REFERENCES

Aitken, M.J. 1985. *Thermoluminescence Dating.* Academic Press, Oxford.

Arnold, L.J., Demuro, M., Navazo, M., Benito-Calvo, A., Pérez-González, A. 2012. OSL dating of the Middle Palaeolithic Hotel California site, Sierra de Atapuerca, north-central Spain. *Boreas* 42, 285–305.

Arnold, L.J., Demuro, M., Parés, J.M., Pérez-González, A., Arsuaga, J.L., Bermúdez de Castro, J.M., Carbonell, E. 2015. Evaluating the suitability of extended-range luminescence dating techniques over early and Middle Pleistocene timescales: Published datasets and case studies from Atapuerca, Spain. *Quaternary International* 389, 167–190.

Avni, Y., Porat, N., Plakht, J., Avni, G. 2006. Geomorphic changes leading to natural desertification versus anthropogenic land conservation in an arid environment, the Negev Highlands, Israel. *Geomorphology* 82, 177–209.

Bailiff, I.K. 2007. Methodological developments in the luminescence dating of brick from English late-medieval and post-medieval buildings. *Archaeometry* 49, 827–851.

Bailiff, I.K., Blain, S., Graves, C.P., Gurling, T., Semple, S. 2010. Uses and recycling of brick in medieval English buildings: Insights from the application of luminescence dating and new avenues for further research. *The Archaeological Journal* 167, 165–196.

Bailiff, I.K. Lewis, S., Drinkall, H., White, M. 2013. Luminescence dating of sediments from a Palaeolithic site associated with a solution feature on the North Downs of Kent, UK. *Quaternary Geochronology* 18, 135–148.

Bailiff, I.K., French, C.A., Scarre, C.J. 2014. Application of luminescence dating and geomorphological analysis to the study of landscape evolution, settlement and climate change on the Channel Island of Herm. *Journal of Archaeological Science* 41, 890–903.

Bailiff, I.K., Jankowski, N., Gerrard, C.M., Gutiérrez, A., Snape, L.M., Wilkinson, K.N. 2018 Luminescence dating of qanat technology: Prospects for further development. *Water History* 10, 73–84.

Barnett, S.M. 2000. Luminescence dating pottery from later prehistoric Britain. *Archaeometry* 42, 431–457.

Beckers, B., Schütt, B., Tsukamoto, S., Frechen, M. 2013. Age determination of Petra's engineered landscape: Optically stimulated luminescence (OSL) and radiocarbon ages of runoff terrace systems in the Eastern Highlands of Jordan. *Journal of Archaeological Science* 40, 333–348.

Been, E., Hovers, E., Ekshtain, R., Malinsky-Buller, A., Agha, N., Barash, A., Bar-Yosef, D., Benazzi, S., Hublin, J-J., Levin, L., Greenbaum, N., Mitki, N., Oxilia, G., Porat, N., Roskin, J., Soudack, M., Yeshurun, R., Shahack-Gross, R., Nir, N., Barzilai, O. 2017. The first Neanderthal remains from an open-air Middle Palaeolithic site in the Levant OPEN. *Scientific Reports* 7.

Benea, V., Vandenberghe, D., Timar, A., Van den Haute, P., Cosma, C, Gligor, M., Florescu, C. 2007. Luminescence dating of Neolithic ceramics from Lumea Noua, Romania. *Geochronometria* 28, 9–16.

Berger, G.W., Henderson, T.K., Banerjee, D., Nials, F.L. 2004. Photonic dating of prehistoric irrigation canals at Phoenix, Arizona, U.S.A. *Geoarchaeology* 19, 1–19.

Berger, G.W., Post, S., Wenker, C. 2009. Single and multiple-grain quartz-luminescence dating of irrigation-channel features in Santa Fe, New Mexico. *Geoarchaeology* 24, 383–401.

Blain, S., Guibert, P., Bouvier, A., Vieillevigne, E., Bechtel, F., Sapin, C., Baylé, M. 2007. TL dating applied to building archaeology: The case of the medieval church Notre-Dame-Sous-Terre (Mont-Saint-Michel, France). *Radiation Measurements* 42, 1483–1491.

Blain, S., Bailiff, I.K., Guibert P., Bouvier A., Baylé, M. 2010. An intercomparison study of luminescence dating protocols and techniques applied to medieval brick samples from Normandy (France). *Quaternary Geochronology* 5, 311-316.

Blain, S., Lanos, P., Bailiff, I., Guibert, P., Sapin, C. 2014. Early medieval brickmaking: a cross-Channel perspective based on recent luminescence and archaeomagnetic dating results. In Ratilainen, T., Bernotas, R., Herrmann, C. (eds) Fresh Approaches to the Brick production and Use in the Middle Ages. (Eds) Archaeopress, Oxford.

Bøtter-Jensen, L., Agersnap Larsen, N., Mejdahl, V., Poolton, N.R.J, Morris, M.F., McKeever S.W.S. 1995. Luminescence sensitivity changes in quartz as a result of annealing. *Radiation Measurements* 24, 535–541.

Bonsall, C., Cook, G., Manson, J. L., Sanderson, D. 2002. Direct dating of Neolithic pottery: progress and prospects. *Documenta Praehistorica* 29, 47–59.

Bueno, L., Feathers, J., Plasis, P.D. 2013. The formation process of a paleoindian open-air site in Central Brazil: integrating lithic analysis, radiocarbon and luminescence dating, *Journal of Archaeological Science* 40, 190–203.

Chauhan, P., Bridgland, D., Moncel, M.-H., Antoine, P., Bahain, J.-J., Briant, R., Cunha, P.P., Locht, J.-L., Martins, A., Schreve, D., Shaw, A., Voinchet, P., Westaway, R., White, M., White, T. 2017. Fluvial deposits as an archive of early human activity: progress during the 20 years of the Fluvial Archives Group. *Quaternary Science Reviews* 166, 114–149.

Chruścińska, A., Cicha, A., Kijek, K., Palczewski, P., Przegiętka, K., Sulkowska-Tuszyńska, K. 2014. Luminescence dating of bricks from the Gothic Saint James church in Toruń. *Geochronology* 41, 352–360.

Cramp, R. J., Bettess, G., Bettess, F. 2006. *Wearmouth and Jarrow Monastic Sites Vol II*. English Heritage, Swindon.

Czopek, S., Kusiak, J., Trybała-Zawiślak, K. 2013. Thermoluminescent dating of the Late Bronze and Early Iron Age pottery on sites in Kłyżów and Jarosław (SE Poland). *Geochronometria* 40, 113–125.

Davidovich, U., Porat, N., Gadot, Y., Avni, Y., Lipschits, O. 2012. Archaeological investigations and OSL dating of terraces at Ramat Rahel, Israel. *Journal of Field Archaeology* 37, 192–208.

Duller, G.A.T., Tooth, L., Barham, L. Tsukamoto. S. 2015. New investigations at Kalambo Falls, Zambia: Luminescence chronology, site formation, and archaeological significance. *Journal of Human Evolution* 85, 111–125.

Dunnell, R.C., Feathers, J.K. 1994. Thermoluminescence dating of surficial archaeological material.

In Beck, C. (ed), *Dating in Exposed and Surface Contexts*, University of New Mexico Press, Albuquerque, 115–137.

Feathers, J.K. 2000. Date List 7: Luminescence dates for prehistoric and protohistoric pottery from the American southwest. *Ancient TL* 18, 51–61.

Feathers, J.K. 2003. Use of luminescence dating in archaeology. *Measurement Science and Technology* 14, 1493–1509.

Feathers, J.K., Johnson, J., Kembel, S.R. 2008. Luminescence dating of monumental stone architecture at Chavín De Huántar, Perú. *Journal of Archaeological Method Theory* 15, 266–296.

Feathers, J.K. 2009. Problems of ceramic chronology in the Southeast: Does shell-tempered pottery appear earlier than we think? *American Antiquity* 74, 113–142.

Feathers, J.K. 2015. Luminescence dating at Diepkloof Rock Shelter – new dates from single-grain quartz. *Journal of Archaeological Sci*ence 63, 164–174.

Fleming, S.J. 1979. *Thermoluminescence Techniques in Archaeology*. Clarendon Press, Oxford.

French, C. 2015. *A Handbook of Geoarchaeological Approaches for Investigating Landscapes and Settlement Sites*. Studying Scientific Archaeoelogy 1, Oxbow Books, Oxford, UK.

Frouin, M., Lahaye, C., Hernandez, M., Mercier, N., Guibert, P., Brenet, M., Folgado-Lopez, M., Bertran, P. 2014. Chronology of the middle palaeolithic open-air site of Combe Brune 2 (Dordogne, France): A multi luminescence dating approach. *Journal of Archaeological Science* 52, 524–534.

Gibson, S. 2015. The archaeology of agricultural terraces in the Mediterranean zone of the southern Levant and the use of the optically stimulated luminescence dating method. In Lucke, B., Bäumler, R., Schmidt, M. (eds) *Soils and Sediments as Archives of Environmental Change. Geoarchaeology and Landscape Change in the Subtropics and Tropics*. Erlanger Geographische Arbeiten Band 42, 295–314.

Goedicke, C. 2011. Dating mortar by optically stimulated luminescence: A feasibility study. *Geochronometria* 38, 42–49.

Goldberg, P., Berna, F. 2010. Micromorphology and context. *Quaternary International* 214, 56–62.

Guérin, G., Mercier, N. 2012. Field gamma spectrometry, Monte Carlo simulations and potential of non-invasive measurements. *Geochronometria* 39, 40–47.

Guérin, G., Murray, A.S., Jain, M., Thomsen, K.J., Mercier, N. 2013. How confident are we in the chronology of the transition between Howieson's Poort and Still Bay? *Journal of Human Evolution* 64, 314–317.

Guibert P., Bailiff I. K, Blain S., Gueli A. M., Martini, M., Sibilia, E., Stella, G., Troja, S. 2009. Luminescence dating of architectural ceramics from an early medieval abbey: The St-Philbert intercomparison (Loire Atlantique, France). *Radiation Measurements* 44, 488–493.

Hublin, J.J., Ben-Ncer, A., Bailey, S.E., Freidline, S.E., Neubauer, S., Skinner, M.M., Bergmann, I., Le Cabec, A., Benazzi, S., Harvati, K., Gunz, P. 2017. New fossils from Jebel Irhoud, Morocco and the pan-African origin of *Homo sapiens*. *Nature* 546, 289–292.

Huckleberry, G., Hayshida, F., Johnson, J. 2012. New insights into the evolution of an intervalley prehistoric irrigation canal system, north coastal Peru. *Geoarchaeology* 27, 492–520.

Huckleberry, G., Rittenour, T. 2014. Combining radiocarbon and single-grain optically stimulated luminescence methods to accurately date pre-ceramic irrigation canals, Tuscon, Arizona. *Journal of Archaeological Science* 41, 156–170.

Jacobs, Z., Roberts, R.G., Galbraith, R.F., Deacon, H.J., Grün, R., Mackay, A., Mitchell, P., Vogelsang, R., Wadley, L. 2008. Ages for the Middle Stone Age of Southern Africa: Implications for Human Behaviour and Dispersal. *Science* 322, 733–735.

Jacobs, Z., Roberts, R.G. 2015. An improved single grain OSL chronology for the sedimentary deposits from Diepkloof Rockshelter, Western Cape, South Africa. *Journal of Archaeological Science* 63, 175–192.

Jankowski, N.R., Jacobs, Z. 2018. Beta dose variability and its spatial contextualisation in samples used for optical dating: An empirical approach to examining beta microdosimetry. *Quaternary Geochronology* 44, 23–37.

Janz L., Feathers J.K., Burr, G.S. 2015. Dating surface assemblages using pottery and eggshell: Assessing radiocarbon and luminescence techniques in Northeast Asia. *Journal of Archaeological Science* 57, 119–129.

Khasswneh, S. al., Murray, A. S., Gebel, H. G., Bonatz, D. 2016. First application of OSL dating to a chalcolithic well structure in Qulban Bani Murra, Jordan. *Mediterranean Archaeology and Archaeometry* 16, 127–134.

Kinnaird, T., Bolos, J., Turner, A. 2017. Optically-stimulated luminescence profiling and dating of historic agricultural terraces in Catalonia (Spain). *Journal of Archaeological Science* 78, 66–78.

Kouki, P. 2006. Environmental change and human history in the Jabal Harun area, Jordan. Unpublished Licentiate thesis, University of Helsinki

Kuzmin, Y.V., Hall, S., Tite, M.S., Bailey, R., O'Malley, J.M., Medvedev, V.E. 2001. Radiocarbon and thermoluminescence dating of the pottery from the early Neolithic site of Gasya (Russian Far East): initial results. *Quaternary Science Reviews* 20, 945–8.

Lang, A., Wagner, G.A. 1996. Infrared stimulated luminescence dating of archaeosediments. *Archaeometry* 38, 129–141.

Lebrun B., Martin, L., Tribolo, C., Mercier, N. 2018. Direct and indirect study of beta dose rate heterogeneities: a case study of West African archaeological sediments. *Radiation Measurements* (submitted).

Liritzis, I., Singhvi, A.K., Feathers, J. K., Wagner, G.A., Kadereit, A., Zacharias, N., Li, S-H. 2013. *Luminescence Dating in Archaeology, Anthropology, and Geoarchaeology. An Overview.* SpringerBriefs in Earth System Sciences, 97.

Madsen, A.T., Murray, A.S. 2009. Optically stimulated luminescence dating of young sediments: A review. *Geomorphology* 109: 3–16.

Manuel, M., Lightfoot, D., Fattahi, M. 2018. The sustainability of ancient water control techniques in Iran: an overview. *Water History* 10, 13–20.

Meister J., Krause J., Müller-Neuhof B., Portillo M., Reimann T., Schütt, B. 2017. Desert agricultural systems at EBA Jawa (Jordan): Integrating archaeological and paleoenvironmental records. *Quaternary International* 434, 33–50.

Mercier, N., Valladas, H., Valladas, G., Reyss, J.-L., Jelinek, A., Meignen, L., Joron, J.-L. 1995. TL dates of burnt flints from Jelinek's excavations at Tabun and their implications. *Journal of Archaeological Science* 22, 495–509.

Mercier, N., Valladas, H., Valladas, G. 1992. Observations on palaeodose determination with burnt flints. *Ancient TL* 10, 28–32.

Mercier, N., Valladas, H., Froget, L., Joron, J.-L., Reyss, J.-L., Weiner, S., Goldberg, P., Meignen, L., Bar-Yosef, O., Kuhn, S.L., Stiner, M.C., Tillier, A.-M., Arensburg, B., Vandermeersch, B. 2007. Hayonim Cave: A TL-based chronology for this Levantine Mousterian sequence. *Journal of Archaeological Science* 34, 1064–1077.

Orton, C., Tyers, P., Vince, A. 1993. *Pottery in Archaeology. Cambridge Manuals in Archaeology.* Cambridge University Press, Cambridge.

Parkington, J.E., Rigaud, J.-Ph., Poggenpoel, C., Porraz, G., Texier, P.-J. 2013. Introduction to the project and excavation of Diepkloof Rock Shelter (Western Cape, South Africa): A view on the Middle Stone Age. *Journal of Archaeological Sciences* 40, 3369–3375.

Peacock, E, Feathers, J.K.2009. Accelerator mass spectrometry radiocarbon dating of temper in Shell-tempered ceramics: Test cases from Mississippi, southeastern United States. Amer. Antiquity 74, 351–369.

Pietsch, T.J., Olley, J.M., Nanson, G.C. 2008. Fluvial transport as a natural luminescence sensitiser of quartz. *Quaternary Geochronology* 3, 365–376.

Pike, A.W.G. 2017. Uranium–thorium dating of cave art. In David, B. and McNiven, I.J. (eds) *The Oxford Handbook of the Archaeology and Anthropology of Rock Art*. Oxford University Press, Oxford.

Pike, A.W.G., Hoffmann, D.L., Pettitt, P.B., García-Diez, M., Zilhão, J. 2017. Dating Palaeolithic cave art: Why U–Th is the way to go. *Quaternary International* 432, 41–49.

Prescott, J.R., D.J. Huntley, D.J., Hutton, J.T. 1993. Estimation of equivalent dose in thermoluminescence dating – the Australian slide method. *Ancient TL* 11, 1–5.

Porat, N., Davidovich, U., Avni, Y., Avni, G., Gadot, Y. 2017. Using OSL measurements to decipher soil history in archaeological terraces, Judean Highlands, Israel. *Land Degradation and Development* 29, 643–650.

Porat, N., Jain, M., Ronen, A., Horwitz, L.K. 2018. A contribution to late Middle Paleolithic chronology of the Levant: New luminescence ages for the Atlit Railway Bridge site, Coastal Plain, Israel. *Quaternary International* 464, 32–42.

Richter, D., Krbetschek, M. 2015. Luminescence dating of the Lower Palaeolithic occupation at

Schöningen. *Journal of Human Evolution* 89, 46–56.

Richter, D., Grün, R., Joannes-Boyau, R., Steele, T.E., Amani, F., Rué, M., Fernandes, P., Raynal, J., Geraads, D., Ben-Ncer, A., Hublin, J., McPherron, S.P. 2017. The age of the hominin fossils from Jebel Irhoud, Morocco, and the origins of the Middle Stone Age. *Nature* 546, 293–296.

Rittenour, T.M. 2008. Luminescence dating of fluvial deposits: applications to geomorphic, palaeoseismic and archaeological Research. *Boreas* 37, 613–635.

Sampson, C.G., Bailiff, I., Barnett, S. 1997. Thermoluminescence dates from Later Stone Age pottery on surface sites in the Upper Karoo. *South African Archaeological Bulletin* 52, 38–42.

Sanderson, D.C.W., Murphy, S. 2010. Using simple portable OSL measurements and laboratory characterization to help understand complex and heterogeneous sediment sequences for luminescence dating. *Quaternary Geochronology*. 5, 299–305.

Sanjurjo-Sánchez, J., Montero Fenollós, J.L. 2012. Chronology during the Bronze Age in the archaeological site Tell Qubr Abu Al-'Atiq, Syria. *Journal of Archaeological Science* 39: 163–174.

Schmidt, C., Kreutzer, S. 2013. Optically stimulated luminescence of amorphous/microcrystalline $SiO2$ (silex): Basic investigations and potential in archaeological dosimetry. *Quaternary Geochronology* 15, 1–10.

Smith, T.M., Tafforeau, P., Reid, D.J., Grün, R., Eggins, S., Boutakiout, M., Hublin, J-J. 2007. Earliest vidence of modern human life history in north African early Homo sapiens. *Proceedings of the National Academy of Sciences USA* 104, 6128–6133.

Solongo, S., Ochir, A., Tengis, S., Fitzsimmons, K., Hublin, J.-J. 2014. Luminescence dating of mortar and terracotta from a Royal Tomb at Ulaankhermiin Shoroon Bumbagar, Mongolia. *STAR* 2, 235–242.

Stark, M.T., Sanderson, D., Bingham, R.G. 2006. Monumentality in the Mekong delta: Luminescence dating and implications. *Indo-Pacific Prehistory Association Bulletin* 26, 110–20.

Stella, G., Fontana, D., Gueli, A.M., Troja, S.O. 2013. Historical mortars dating from OSL signals of fine grain fraction enriched in quartz. *Geochronometria* 40, 153–164.

Stringer, C., Galway-Witham, J. 2017. On the origin of our species. *Nature* 546, 212–214.

Sutikna T., Tocheri M.W., Morwood M.J., Saptomo E.W., Jatmiko, Awe R.D., Wasisto, S., Westaway K.E., Aubert, M., Li, B., Zhao, J.X., Storey, M., Alloway, B.V., Morley, M.W., Meijer, H.J., van den Bergh, G.D., Grün, R., Dosseto, A., Brumm, A., Jungers, W.L., Roberts, R.G. 2016. Revised stratigraphy and chronology for Homo floresiensis at Liang Bua in Indonesia. *Nature* 532, 366–369.

Tribolo, C., Mercier, N., Valladas, G., Joron, J.L., Guibert, P., Lefrais, Y., Selo, M., Texier, P.-J., Rigaud, J.-P., Porraz, G., Poggenpoel, C., Parkington, J., Texier, J.-P., Lenoble, A. 2009. Thermoluminescence dating of a Stillbay-Howiesons poort sequence at Diepkloof rock shelter (Western Cape, South Africa). *Journal of Archaeological Science* 36, 730–739.

Tribolo, C., Mercier, N., Douville, E., Joron, J.-L., Reyss, J.-L., Rufer, D., Cantin, N., Lefrais, Y., Miller, C.E., Parkington, J., Porraz, G., Rigaud, J.-P., Texier, P.-J. 2013. OSL and TL dating of the middle stone age sequence of Diepkloof rock shelter (Western Cape, South Africa): A clarification. *Journal of Archaeological Science* 40, 3401–3411.

Urbanova, P., Guibert, P., 2017. Methodological study on single grain OSL dating of mortars: Comparison of five reference archaeological sites. *Geochronometria* 44, 77–97.

Valladas, H., Mercier, N., Hershkovitz, I., Zaidner, Y., Tsatskin, A., Yeshurun, R., Vialettes, L., Joron, J-L, Reyss, J.-L., Weinstein-Evron, M. 2013. Dating the Lower to Middle Paleolithic transition in the Levant: a view from Misliya Cave, Mount Carmel, Israel. *Journal of Human Evolution*. 65, 585–593.

Vieillevigne, E., Guibert, P., Bechtel, F. 2007. Luminescence chronology of the medieval citadel of Termez, Uzbekistan: TL dating of brick masonries. *Journal of Archaeological Science* 34, 1402–1416.

Wintle, A. G. 2008. Fifty years of luminescence dating. *Archaeometry* 50, 276–312.

Zaidner, Y., Porat, N., Zilberman, E., Herzlinger, G., Almogi-Labin, A., Roskin, J. 2018. Geo-chronological context of the open-air Acheulian site at Nahal Hesi, northwestern Negev, Israel. *Quaternary International* 464, 18–31.

11 ROCK SURFACE BURIAL AND EXPOSURE DATING

GEORGINA E. KING[1], PIERRE G. VALLA[2] AND BENJAMIN LEHMANN[1]

[1] Institute of Earth Surface Dynamics, University of Lausanne, Switzerland
[2] Institute of Earth Sciences (ISTerre), Université Grenoble-Alpes CNRS, France

ABSTRACT: In addition to the versatile ways in which luminescence dating can be applied to sediments, it can also be used for dating rock surfaces. This can either be done using a similar approach to sediment burial dating, i.e. constraining the last exposure to daylight for grains extracted from a buried rock surface, or it can be used to date the exposure history of a surface to daylight. In this chapter we discuss the practicalities of rock surface and rock surface-exposure dating using luminescence techniques as well as the latest model(s) for describing signal bleaching of rock surfaces, a range of different applications, and the challenges that need to be overcome before these methods can become as routine as luminescence dating of sediments.

KEYWORDS: rock surface, bleaching model, heterogeneous dosimetry

11.1 INTRODUCTION

Within the Earth and Archaeological Sciences there are many examples of rock surfaces with unknown absolute ages. In archaeology there are numerous buildings, pavements and artefacts of unknown age, the dating of which would contribute to our understanding of human–environment interactions and the emergence of societies and culture. Dating glacially-polished bedrock, moraine boulders and glacial erratics gives insight into ice-dynamics, whereas measuring the age of rock-fall boulder debris can indicate rates of mass wastage and the frequency of hazardous events. Furthermore, in some environments, unconsolidated sedimentary deposits are absent or not representative of the depositional setting and/or processes of deposition. The ability to directly date the surfaces of larger clasts (pebbles/cobbles) or polished bedrock increases the applicability of luminescence dating, allowing new research questions to be addressed.

11.1.1 Key principles

In a rock surface continuously exposed to daylight, bleaching of the luminescence signal is expected not only to occur for minerals at the rock surface, but also to progressively propagate deeper into the surface with time (e.g. Habermann *et al.* 2000; Polikreti *et al.* 2002; Vafiadou *et al.* 2007). When this rock surface is buried, for example following a rock fall (Chapot *et al.* 2012) or within a channel (Sohbati *et al.* 2012a), or when it simply becomes positioned facedown within a relict geomorphological feature protected from daylight (Simms *et al.* 2011; Sohbati *et al.* 2011), a luminescence signal can begin to accumulate. Dating of this rock surface can yield a burial age in the same way as for luminescence dating of sediments, with the only contrast being that the rock surface must be prepared either as rock slices of ~1 mm thickness (e.g. Sohbati *et al.* 2011), or following crushing and preparation of a specific mineral and grain-size fraction (e.g. Simms *et al.* 2011). In contrast, in rock surface–exposure dating, the luminescence bleaching profile is measured with depth through/into the rock sample, and a model which describes the rate of bleaching with time is used to determine the duration of daylight exposure and thus the surface-exposure age (Sohbati *et al.* 2011; Lehmann *et al.* 2018).

11.1.2 Method development

The potential of luminescence dating of rock surfaces was initially recognised for archaeological applications (e.g. Liritzis 1994; Richards 1994). Whilst the dating of burnt lithics such as flint, which can become buried after thermal resetting in a fire, has been widely investigated (e.g. Aitken 1985; Gösku *et al.* 1974), efforts to use optically reset signals from rock surfaces are relatively recent and were initially predominantly linked to the investigation of archaeological monuments (e.g. megalithic structures). Richards (1994) used optically stimulated luminescence (OSL) to date quartzite pebble hand axes, whereas the potential of dating calcitic (limestone, marble) megalithic structures using thermoluminescence (TL) was recognised by Liritzis (1994; Liritzis *et al.* 1996; Polikreti *et al.* 2003). This method works on the principle that a rock surface can be bleached optically, like sediment, and then accumulates a luminescence signal again after subsequent burial. Burial could be by sediment, or alternatively by another rock such as within a building.

An important consideration, applicable in any luminescence dating study, is that the luminescence signal is effectively zeroed before/during the event investigated. Where a megalithic structure is under investigation, sufficient daylight exposure must have occurred prior to burial; similarly, where a transport event that led to the deposition of a glacial/fluvial cobble is to be measured, daylight exposure must have been sufficient to bleach the luminescence signal of the cobble surface. Thermoluminescence dating can be limited by relatively high uncertainties related to multiple-aliquot measurement protocols and bleaching rates which are less rapid than for optically stimulated luminescence signals (Tribolo *et al.* 2003). Habermann *et al.* (2000), Greilich *et al.* (2005) and Greilich and Wagner (2006) confirmed that bedrock infrared stimulated luminesence (IRSL) and OSL signals can be fully reset by daylight and so are appropriate for dating of rock surfaces. These observations have paved the way for later applications (see Section 11.7 for details), such as to lithic artefacts (Morgenstein *et al.* 2003), soil floors and overlaying pebbles (Vafiadou *et al.* 2007), cobble surfaces (e.g. Simms *et al.* 2011; Sohbati *et al.* 2011; Jenkins *et al.* 2018; Rades *et al.* 2018) and rock-fall deposits (Chapot *et al.* 2012).

Richards (1994) first showed that the optically stimulated signal of quartz changes with depth through quartzite pebble hand-axes (Fig. 11.1A). Using a 'micro-stratigraphic' approach, whereby multiple layers 250 μm thick were removed from the pebbles in a series of 30-minute hydrofluoric acid (HF) etches, Richards (1994) was able to determine that the rate of light transmittance was reduced to 1–8% after 0.25–0.6mm depth. Polikreti *et al.*

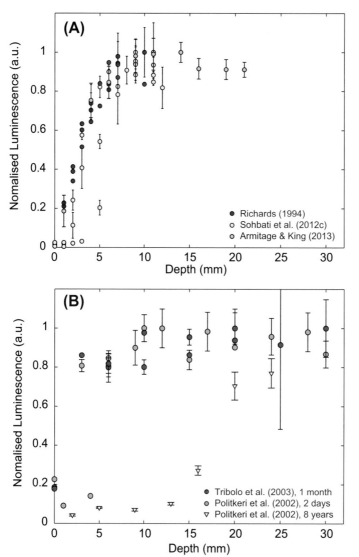

Figure 11.1 A) Quartz OSL signal change with depth for grains extracted from quartzite (Richards 1994, data reported in Roberts 1997) and sandstone (Armitage and King 2013; Sohbati *et al.* 2012c). Note that the samples have been exposed for different durations and are from different geological and/or archaeological settings. B) TL signal reduction with depth for rock slices of marble (Polikreti *et al.* 2002) and quartzite (Tribolo *et al.* 2003) exposed to daylight for different durations.

(2002; 2003) investigated how TL signals changed with depth following daylight exposure through exposing blocks of marble to daylight for different durations of up to 70 days. They observed that the luminescence signal can be reset to increasing depths, following increasing bleaching durations (Fig. 11.1B). Polikreti *et al.* (2002; 2003) suggested that these bleaching profiles could be used for the authentication of marble sculptures, with rock surfaces of sufficient antiquity experiencing deeper maximum bleaching depths (Fig. 11.1B). Following these earlier studies, Sohbati *et al.* (2011) proposed that OSL signals from quartz and feldspar could be used for rock surface-exposure dating. In the remainder of this chapter, the models of luminescence signal bleaching through bedrock are presented, as well as the various methodological considerations necessary for bedrock surface and surface-exposure dating. A number of detailed case studies are given as well as an outlook for this developing technique.

11.2 MODEL OF LUMINESCENCE SIGNAL BLEACHING

A rock surface that has not been exposed to daylight has a luminescence signal reflecting an equilibrium between electron trapping, due to ambient radiation (cosmic rays, high-energy solar particle flux and radioactive decay in the rock matrix) and electron detrapping due to anomalous fading and/or thermal signal losses. This condition is often termed field saturation or field steady-state and is the starting point (i.e. the initial condition) prior to any signal loss through exposure to daylight. If this rock surface is then continuously exposed to daylight, luminescence signal bleaching is expected to propagate deeper into the surface with time (Polikreti *et al.* 2002; Sohbati *et al.* 2011). This can be seen schematically for an example from a glacial setting in Figure 11.2. Polikreti *et al.* (2002) were the first to develop a model for luminescence signal bleaching with depth following daylight exposure. More recently, Sohbati *et al.* (2011) have presented a model for the depletion of optically stimulated luminescence signals with depth. Although this model has been recently challenged (Laskaris and Liritzis, 2011), here we focus on the model of Sohbati *et al.* (2011) as it remains the most widely used approach in recent applications.

The intensity of a luminescence signal can be considered to reflect the number of trapped electrons, *n*. For a rock surface exposed to daylight, the trapped electron concentration (and thus the luminescence signal intensity) is governed by competing processes of electron detrapping due to daylight exposure, and of electron trapping in response to ambient

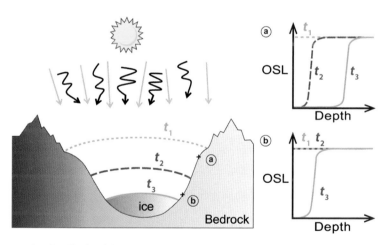

Figure 11.2 Sketch of bedrock luminescence signal evolution following exposure to daylight caused by glacial retreat, modified from Lehmann *et al.* (2018). Prior to bedrock surface exposure when the glacier is at its maximum extent (time t_1), the luminescence signals of both surfaces (a) and (b) are in field-saturation. Following progressive glacial retreat (time t_2), surface (a) becomes exposed and the luminescence signal begins to bleach at depth. In contrast, surface (b) remains covered and its luminescence signal remains unchanged. Following further glacial retreat (time t_3), surface (a) remains exposed and its luminescence signal is more deeply bleached whereas surface (b) is only now exposed to daylight, with its luminescence signal bleached to a shallow depth. Straight arrows (grey) represent cosmic rays and the flux of high-energy solar particles, which together with radioactive decay in the rock-matrix build the latent luminescence signal. Black arrows represent low energy electromagnetic radiation from the sun, which bleach the latent luminescence signal following bedrock exposure.

radiation. Thus the concentration of trapped electrons $n(x,t)$ at a given depth x(mm) and time t(s) can be described by the following differential equation (Sohbati et al. 2012b):

$$\frac{\partial n(x,t)}{\partial t} = -E(x)n(x,t) + F(x)[N(x) - n(x,t)] \tag{1}$$

where $N(x)$ is the maximum possible number of trapped electrons at depth x, and $E(x)$ (s^{-1}) and $F(x)$ (s^{-1}) are the rates of charge detrapping and trapping respectively. The rate of charge detrapping is given by:

$$E(x) = \overline{\sigma\varphi_0}\ e^{-\mu x} \tag{2}$$

where $\sigma(\lambda)$ is the luminescence photoionisation cross-section (cm^2) and φ_0 is the photon flux (cm^{-2} s^{-1}) as a function of wavelength at the surface of the rock ($x = 0$). The product of these two terms, $\overline{\sigma\varphi_0}$ is thus the effective decay rate (s^{-1}) of luminescence at the rock surface, following exposure to a particular spectrum of light. The final parameter of equation (2), μ, is the light attenuation coefficient of the rock (mm^{-1}).

In addition to electron detrapping, equation (1) contains the term $F(x)$ to describe electron trapping, which whilst of lesser importance for young terrestrial surfaces (i.e. <10 ka) may be significant for extra-terrestrial applications where radiation dose rates, particularly from cosmic rays, are greater (Sohbati et al. 2012b). The rate of charge trapping due to ionising radiation is given by:

$$F(x) = \frac{\dot{D}(x)}{D_0} \tag{3}$$

where $\dot{D}$ is the environmental dose rate (Gy s^{-1}) and D_0 is the characteristic dose of saturation (Gy) for the luminescence signal being investigated.

It is assumed that the number of trapped electrons at a given depth $n(x)$ is proportional to the measured luminescence signal $L(x, t)$, and thus assuming $F(x) \approx 0$, the luminescence can be described following Sohbati et al. (2012c) as:

$$L(x, t) = L_0\ e^{\overline{\sigma\varphi_0}te^{-\mu x}} \tag{4}$$

where L_0 is the field-saturation luminescence signal (i.e. equilibrium level for a trapped electron population of N), assumed to have been constant at all depths prior to bleaching (i.e. at $t = 0$). For the case where $F(x) > 0$, this equation becomes (Sohbati et al. 2012b):

$$L(x, t) = \frac{\overline{\sigma\varphi_0}\ e^{-\mu x}e^{-t(\overline{\sigma\varphi_0}\ e^{-\mu x} + \frac{\dot{D}}{D_0})} + \frac{\dot{D}}{D_0}}{\overline{\sigma\varphi_0}\ e^{-\mu x} + \frac{\dot{D}}{D_0})} \tag{5}$$

Or where a surface has been completely bleached, i.e. $L_0(x) = 0$, for example following a meteorite impact (Sohbati et al. 2012b):

$$L(x, t) = \frac{\frac{\dot{D}(x)}{D_0}\left\{1 - e^{-t[\overline{\sigma\varphi_0}\ e^{-\mu x} + \frac{\dot{D}(x)}{D_0}]}\right\}}{\overline{\sigma\varphi_0}\ e^{-\mu x} + \frac{\dot{D}(x)}{D_0}} \tag{6}$$

Applications of luminescence surface-exposure dating are likely to be more common in terrestrial settings for rock-surface exposures of <10 ka; therefore, we will consider equation (4) whilst discussing the model. However, for older rock surfaces, an equilibrium between electron detrapping by daylight exposure and electron retrapping due to environmental radiation may occur (Sohbati *et al.* 2012c). Equation (4) contains three unknown parameters: $\overline{\sigma\varphi_0}$, the luminescence signal decay rate at the rock surface; μ the rock attenuation coefficient; and t, the exposure time. A range of values have been reported for $\overline{\sigma\varphi_0}$ and μ for different minerals and rock samples; however, here for illustration we use the values from Sohbati *et al.* (2012c) for quartz minerals to explore the luminescence signal behaviour described in equation (4).

Equation (4) describes how bleaching of a luminescence signal propagates with time and depth into a rock surface. Figure 11.3A shows how luminescence signals may bleach to different depths for $\overline{\sigma\varphi_0}$ of 6.8×10^{-9} s^{-1} and μ of 1.01 mm^{-1} for different time periods. As the exposure duration increases, the depth to which the luminescence signals are reset within the sample increases. In order to calculate an exposure age, both $\overline{\sigma\varphi_0}$ and μ must be constrained, which remains the largest barrier to the routine application of luminescence surface-exposure dating (Sohbati *et al.* 2011; 2015). Before considering the ways through which this challenge can be addressed, we first consider the variables that control these two parameters.

11.2.1 Controls on luminescence signal bleaching rate

The luminescence signal bleaching rate $\overline{\sigma\varphi_0}$ is controlled by the photon flux at the sample location (i.e. $\varphi(\lambda,0)$) which is affected by variability in the daylight spectrum. The effect of varying $\overline{\sigma\varphi_0}$ on the measured luminescence bleaching profile is shown in Figure 11.3B. Spooner (1994) investigated the optical bleaching of feldspar minerals in response to different wavelengths of light. Using these data, Sohbati *et al.* (2011) attempted to determine $\overline{\sigma\varphi_0}$ from first principles, by calculating the annual wavelength dependent flux of incident photons at the Earth's surface for the latitude and longitude of the Danish cobble that they were investigating. Unfortunately, the value of $\overline{\sigma\varphi_0}$ they determined of $\sim 1.2 \times 10^{-4}$ s^{-1} equated to an exposure time of ~ 30 minutes, considerably shorter than anticipated. They attributed this difference to the fact that the bleaching of feldspar is known not to be exponential for small residual signals (Kars *et al.* 2014), and thus their calculated value likely underestimated the time required for luminescence signal bleaching to occur.

The attenuation coefficient of a rock sample (μ) is dependent upon its lithology. The effect of varying μ on the measured luminescence bleaching profile is shown in Figure 11.3C. Rock-types with large amounts of dark minerals (i.e. melanocratic lithologies such as basalts, mica-rich metasediments etc.) will have higher μ values than those comprising mainly translucent minerals (i.e. leucocratic lithologies such as sandstone or quartz-rich lithologies). Average grain size and density of grain packing are also expected to affect the light attenuation. However, in their successful validations of OSL surface-exposure dating, Sohbati *et al.* (2012c) and more recently, Lehmann *et al.* (2018) have shown that it is possible to assume that the attenuation factor (μ) is similar for rock samples of the same lithology.

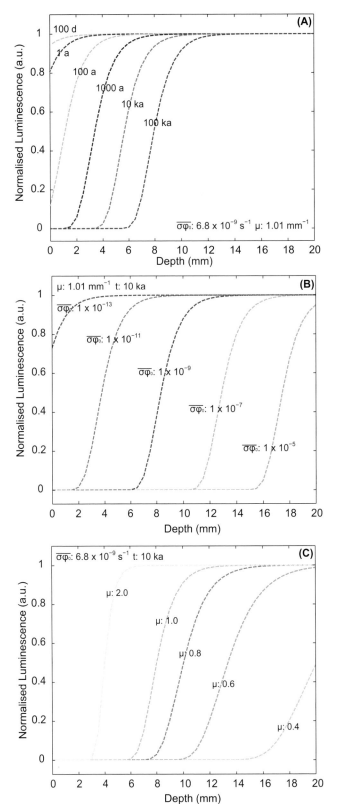

Figure 11.3 Modelled luminescence bleaching profiles using equation (4) and parameters for quartz OSL following Sohbati *et al.* (2012c). A) Changing luminescence signal bleaching depth with increasing exposure time from 100 days to 100 ka for fixed μ and $\overline{\sigma\varphi}_0$ parameters. B) Assuming an exposure time of 10 ka and keeping μ = 1.01 mm⁻¹ (after Sohbati *et al.* 2012c), changing $\overline{\sigma\varphi}_0$ causes the depth of luminescence signal bleaching to shift, with smaller values indicative of lower bleaching rates resulting in a shallower bleaching profile. C) Assuming an exposure time 10 ka and keeping $\overline{\sigma\varphi}_0$ as 6.89×10^{-9} s⁻¹ (after Sohbati *et al.* 2012c) changing μ values result in different depths and gradients of signal bleaching, with larger values indicating higher levels of attenuation and shallowing signal bleaching.

11.3 ROCK SURFACE-EXPOSURE DATING

Whilst only a limited number of studies have been published to date, a large degree of variability in μ and $\overline{\sigma\varphi_0}$ has been recorded (e.g. Sohbati *et al.* 2012c; Lehmann *et al.* 2018). Site-specific calibration may be the only possibility for constraining $\overline{\sigma\varphi_0}$, whilst it is possible to determine μ in the laboratory, although recent studies have shown this to be challenging (see Gliganic *et al.* in press; Meyer *et al.* in press). Constraining these parameters is the major limitation affecting the widespread uptake of luminescence surface-exposure dating. Site-specific calibration samples can comprise bedrock exposed for a known duration, e.g. in a road-cut (Sohbati *et al.* 2012c) or quarry (Polikreti *et al.* 2002), or surfaces of known-age constrained through historical records (Lehmann *et al.* 2018). Alternatively, Polikreti *et al.* (2002) determined the TL signal bleaching parameters of marble through exposing bedrock samples to natural daylight for known durations of between 2 and 70 days in an experiment (Fig. 11.1B). A known exposure-age allows t to be fixed in equation (4), $\overline{\sigma\varphi_0}$ and μ can then be derived through fitting the data and solving the equation for only these two unknown parameters. However, it should also be noted that for a rock with large heterogeneous crystals, μ may vary spatially across the rock surface and with depth (Meyer *et al.* 2018). Alternatively, μ can be estimated using numerical modelling. Sohbati *et al.*

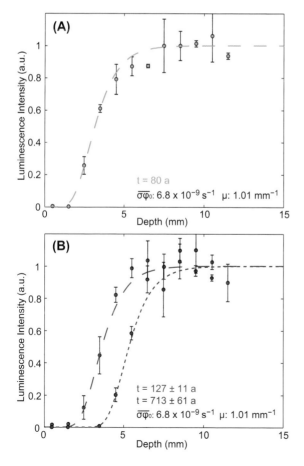

Figure 11.4 A) Quartz OSL signal increase with depth for a sandstone sample of known age (HS-OSL-29) taken from Sohbati *et al.* (2012c). Fitting these data with equation (4) and known exposure time, t, allows parameters μ and $\overline{\sigma\varphi_0}$ to be estimated.

B) Using these parameters, it is possible to fit the quartz OSL data of two samples of the same lithology and from the same location to derive unknown exposure time, t. The raw data of sample HS-OSL-25 ($t = 713 \pm 61$ yr) have been adjusted to remove a burial dose (Chapot *et al.* 2012; Sohbati *et al.* 2012c).

(2015) treated μ as a shared parameter when fitting luminescence bleaching profiles of the same core, but for signals which have different bleaching rates. Once these parameters are constrained, equation (4) can be solved again with t as the only unknown parameter, through fitting the luminescence data for the sample of unknown age; this is illustrated in Figure 11.4 for a specific case study detailed below.

Sohbati et al. (2012c) used luminescence surface-exposure dating to determine the age of Barrier Canyon Style rock art from Canyonlands National Park, Utah, USA. Through fitting the measured luminescence signals of a known-age sample of the Navajo Sandstone and determining $\overline{\sigma\varphi_0}$ and μ (Fig. 11.4A), Sohbati et al. (2012c) were able to input these values into equation (4), and to fit the luminescence data of samples of unknown age to determine the exposure time (Fig. 11.4B). Subsequent work by Lehmann et al. (2018) also employed a similar approach. Working on glacially-polished bedrock, Lehmann et al. (2018) took a transect of known-exposure-age samples from near to the Mer de Glace (Mont Blanc massif, France). They found that a minimum of 4 different calibration samples were required in order to accurately constrain the parameters for a further sample of the same lithology. Further details of these two case studies are given in Section 11.4.

11.3.1 Identifying complex daylight exposure histories and incomplete bleaching

In common with luminescence sediment dating, rock surface dating is reliant on the assumption that a rock surface has been fully bleached prior to burial. This can be evaluated in a number of different ways including using laboratory experiments and through the measurement of modern analogue samples (e.g. Simms et al. 2011), as well as through contrasting luminescence ages of overlaying sediments and rock surfaces (e.g. Chapot et al. 2012; Sohbati et al. 2012c). The consistency between ages of different rock surfaces sampled from the same depositional setting can also be evaluated (e.g. Simms et al. 2011). A complementary approach is to model the luminescence bleaching profile of a rock sample, and to extrapolate back to the initial level of signal resetting (Freiesleben et al. 2015; Sohbati et al. 2015).

The variation of luminescence signals with depth through a cobble or a clast may record evidence of multiple bleaching events (Freiesleben et al. 2015; Sohbati et al. 2012a). This is because following initial signal resetting, if the clast or surface is buried again, a new luminescence signal will accumulate, which can then be reset by a second period of daylight exposure (provided that this is shorter than the previous bleaching event to preserve information about the older episode). Such multiple exposure histories result in a 'stepped' succession of luminescence level plateaus (Fig. 11.5). Using quartz OSL, Sohbati et al. (2012a) determined two bleaching events for a cobble from the Tapada do Montinho archaeological site (east-central Portugal). Measurements using a single-aliquot regenerative dose protocol revealed that the maximum D_E of 67 Gy at >5 mm depth into the cobble, was not in saturation. This indicated that the cobble had previously been bleached sufficiently to reset the luminescence signal at this depth and a more recent bleaching event affecting only the first 2 mm of the cobble could also be detected. Freiesleben et al. (2015) modified the model of Sohbati et al. (2012b; equation (1) of this chapter) to fit clast and rock surface luminescence profiles generated by multiple daylight exposure and burial events. Their new model works on the premise that the final luminescence signal intensity (L_1) of one event, which can be

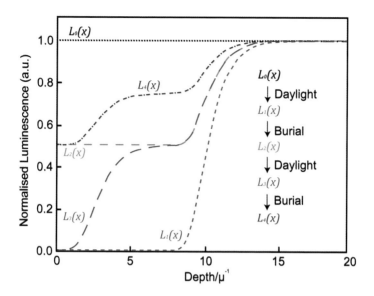

Figure 11.5 Evolution of luminescence signals following multiple burial and exposure cycles, modified after Freiesleben *et al.* (2015). It is assumed that the environmental dose rate is constant with depth, and that no trap filling has occurred during the exposure periods. The depicted burial/exposure history comprises (1) burial of the sample for sufficient duration that all of the electron traps are saturated (L_0), (2) the sample is exposed to daylight for a duration sufficient for complete signal resetting to occur to a depth of 8 mm (L_1), (3) the sample is buried allowing electron trapping and signal accumulation to occur (L_2), (4) the sample is exposed again for a short duration, resulting in complete signal bleaching to a depth of 2 mm (L_3), the sample is buried once more for the same duration as in stage (3) allowing signal accumulation (L_4). The profiles of all the previous exposure/burial periods are recorded in profile (L_4).

thought of as its final condition ($n_{f,1}$), becomes the initial condition of the following burial or exposure event ($n_{i,2}$); thus the event series in Figure 11.5 can be described as:

$$n_{i,1} \xrightarrow{\text{Exposure } (t_{e1})} n_{f,1} = n_{i,2} \xrightarrow{\text{Burial } (t_{b1})} n_{f,2} = n_{i,3} \xrightarrow{\text{Exposure } (t_{e2})} n_{f,3} = n_{i,4} \xrightarrow{\text{Burial } (t_{b2})} n_{f,4} \quad (7)$$

and following Sohbati *et al.* (2015) as:

$$L(x) = \left(\left(\left(e^{-t_{e1}\overline{\sigma\varphi_0}\, e^{-\mu x}} - 1 \right) e^{-\frac{\dot{D}(x)}{D_0}t_{b1}} + 1 \right) e^{-t_{e2}\overline{\sigma\varphi_0}\, e^{-\mu x}} - 1 \right) e^{-\frac{\dot{D}(x)}{D_0}t_{b2}} + 1 \quad (8)$$

where $t_{e1(2)}$ and $t_{b1(2)}$ are the exposure period and burial period 1 (2), respectively.

Freiesleben *et al.* (2015) successfully used this model to fit the $IRSL_{50}$ and post-IR $IRSL_{290}$ signals of feldspars in rock slices from a granite cobble excavated from an archaeological site near Aarhus (Denmark). Sohbati *et al.* (2015) were also able to use this approach to model whether the $IRSL_{50}$ and post-IR $IRSL_{225}$ signals of their cobble samples from the Negev desert, Israel had been fully reset. Thus, the ability to measure luminescence depth profiles through bedrock surfaces may offer an advantage over conventional sediment luminescence dating approaches for determining whether samples have been fully reset. However, sediment dating also provides insights into deposit bleaching histories (see Chapter 1) through, for example, the degree of overdispersion (OD) of equivalent dose values (cf. Duller 2008), or the ratio of IRSL and post-IR IRSL signals (Buylaert *et al.* 2013).

11.3.2 Environmental dose rate determination

A major challenge for rock surface and surface-exposure dating is to quantitatively constrain the environmental dose rate, which determines the rate of luminescence signal accumulation. In many sedimentary applications of luminescence dating, it is possible to make an infinite matrix assumption (Aitken 1985; Durcan *et al.* 2015; Chapter 1). This is because the chemical composition of surrounding sediments is the same as the sample under investigation; in the case of cobble or rock surface(-exposure) dating this is almost never the case (although see Simms *et al.* 2011). Instead the principle of superposition must be applied, which uses the geometry of the cobble/rock surface and the surrounding material to scale the relative dose contributions. For example, Sohbati *et al.* (2012a) followed the equations of Appendix H of Aitken (1985) in order to scale infinite matrix beta and gamma dose rates for a range of geometries to obtain the dose rate of their investigated cobbles (Fig. 11.6). If we assume that a buried rock surface has a given thickness h, and infinite lateral extent, the beta-dose contribution to the environmental dose rate can be approximated following Freiesleben *et al.* (2015) as:

$$\dot{D}(x)_\beta^{Cobble} = \dot{D}_{Rock,\beta}^{inf}\left[1 - 0.5\left(e^{-bx} + e^{-b(h-x)}\right)\right] + \dot{D}_{Sed,\beta}^{inf}\, 0.5(e^{-bx} + e^{-b(h-x)}) \quad (9)$$

where b is the beta dose grain-size attenuation factor (e.g. Guérin *et al.* 2012) and $\dot{D}_{Rock,\beta}^{inf}$ and $\dot{D}_{Sed,\beta}^{inf}$ are the water content corrected infinite matrix beta-dose rates for the rock and sediment respectively (see Durcan *et al.* 2015 for a detailed description of environmental dose-rate calculations). The same approach can be used for the gamma, alpha and cosmic dose-rate contributions and the final dose rate with depth is given from the sum:

$$\dot{D}(x)^{Cobble} = \dot{D}(x)_\alpha^{Cobble} + \dot{D}(x)_\beta^{Cobble} + \dot{D}(x)_\gamma^{Cobble} + \dot{D}_{Cosmic}^{Cobble} \quad (10)$$

where the cosmic dose rate is assumed not to vary significantly over the mm-scale depths typically investigated in rock surface dating. Changes in U, Th and K content and grain size may result in significant changes in the environmental dose rate with depth and must be corrected for.

Grain-size variability between minerals that contribute to measured luminescence signals is a major source of uncertainty in luminescence dating applications using rock samples (Simkins *et al.* 2016; Sohbati *et al.* 2013). Increasing grain sizes result in greater external radiation dose attenuation and thus for grains without a significant internal dose rate (see below), a reduced overall dose. In order to estimate the environmental dose rate effectively, the grain size of the minerals that contribute to the measured luminescence signal

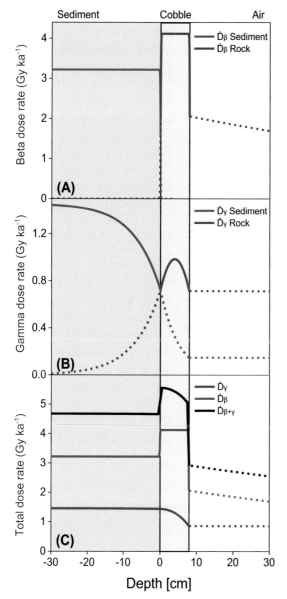

Figure 11.6 Schematic showing how the environmental dose rate ($\dot{D}$) changes between sediment, a cobble and air, modified from Sohbati *et al.* (2015). A) Beta, B) gamma and C) total (i.e. beta + gamma) environmental dose rates. The dose rate in each of the three mediums is shown as a solid line, whereas the dose rate due to the adjacent medium is shown as a dashed line.

must be known. Simms *et al.* (2011) took care to crush their quartzite rock slices gently, so that the constituent grains were not broken, and thus the obtained 90–250 μm fraction could be used to calculate the dose rate. An alternative approach was used by King *et al.* (2016a) when determining the dose rate of grains extracted from bedrock samples for investigation using OSL-thermochronometry. Instead, they used software that could determine the grain-size distribution from high-resolution thin-section images (Buscombe 2013), and calculated their environmental dose rates using maximum and minimum grain-size values.

Feldspar minerals are commonly used in rock surface(-exposure) dating, because of their greater luminescence sensitivity relative to quartz grains extracted from bedrock, and also because of the difficulty of isolating quartz grains that are not contaminated by other minerals (e.g. Guralnik *et al.* 2015). However, the internal dose rate of feldspars varies considerably, dependent upon their chemical composition (e.g. Smedley and Pearce 2016; Smedley *et al.* 2012). For example, the internal K-content ranged from 0.1 to 15% measured using electron probe micro-analysis for feldspars with density <2.58 g cm^{-3} extracted from the same rock sample (NB126; King *et al.* 2016b). Furthermore, for grains with a significant internal dose contribution, such as K-feldspars, increasing grain size results in an increasing dose rate, which may significantly affect age calculations from rock surfaces (Greilich and Wagner 2006; Sohbati *et al.* 2011). Sohbati *et al.* (2013) suggested that because Na-feldspar grains have much lower internal K contents and thus avoid this inter-grain variability, they could comprise a more suitable target mineral for rock surface dating. However, Sohbati *et al.* (2013) found that the IRSL emission detected in the blue part of the emission spectrum may derive from K-feldspar inclusions within Na-feldspar extracts, and thus instead suggested that the yellow emission could be targeted to isolate the Na-feldspar luminescence signal.

11.3.3 Sample preparation and measurement protocols

Sample preparation and luminescence measurement protocols for rock surface dating are still under development with procedures less established compared to sediment dating. Rock surface sample preparation is also dependent on the lithology studied: whereas sandstones can be gently abraded/rubbed (e.g. Sohbati *et al.* 2012c; Liritzis *et al.* 2013; Liritzis and Vafiadou 2015), granitic/metamorphic lithologies require sample coring and are generally sliced to produce 1-mm thick rock slices (e.g. Vafiadou *et al.* 2007; Simms *et al.* 2011; Sohbati *et al.* 2011; 2012a). Following rock slicing for granitic or metamorphic lithologies, subsequent sample preparation then differs between studies. Rock slices can either be crushed to extract specific target minerals following classical chemical/physical procedures (i.e. quartz or K-feldspar, Simms *et al.* 2011; 2012; Sohbati *et al.* 2011; 2012a) or whole rock slices can be measured directly without any further treatment (e.g. Freiesleben *et al.* 2015; Sohbati *et al.* 2015; Lehmann *et al.* 2018). Sohbati *et al.* (2011) showed better IRSL D_E reproducibility between intact rock slices compared to K-feldspar extracts from bedrock, which could not be related to crushing or to partial bleaching. Conversely, Sobhati *et al.* (2012a) found good agreement between equivalent doses and luminescence characteristics of rock slices and quartz grains extracted from the same quartzite lithology. These different outcomes may be explained by feldspar microdosimetry (e.g. Smedley and Pearce 2016). This is because whereas the luminescence signal from a rock slice may be averaged across many grains, such differences may become visible during measurement of a purified separate. However, comparison between these two experimental approaches requires further investigation and rock-slice crushing could result in the loss of information regarding the original grain-size distribution (Section 11.5).

Rock surface dating measurement protocols have been developed based on existing procedures established for sediment dating and vary depending on sample preparation and the scientific question that is to be addressed. TL MAAD (multi-aliquot additive dose) protocols have been applied to both marble (Polikreti *et al.* 2003) and sandstone (Liritzis and Vafiadou 2015) lithologies. OSL SAR (single-aliquot regenerative) measurement protocols have been applied to quartz mineral extracts from crystalline pebbles (Simms *et al.* 2011;

2012; Sohbati *et al.* 2012a; Simkins *et al.* 2013), and to quartzite pebbles using solid rock slices (Vafiadou *et al.* 2007; Sohbati *et al.* 2012a). However, bedrock quartz often shows poor luminescence sensitivity and measurement reliability (e.g. Guralnik *et al.* 2015), and thus equivalent dose scatter (e.g. Simms *et al.* 2011). Recent investigations by Simkins *et al.* (2016) suggest that this scatter in D_E values may originate from heterogeneity in the environmental dose rate due to changing water content and/or grain-size effects, as well as post-crystallisation transport histories (Sawakuchi *et al.* 2011). IRSL SAR measurement protocols have also been successfully applied to date rock slices, by using either one IR stimulation at 50°C (Sohbati *et al.* 2011; Lehmann *et al.* 2018) or different post-IR IRSL protocols at 225°C (Sohbati *et al.* 2015) or 290°C (Freiesleben *et al.* 2015; Liu *et al.* 2016). Such post-IR IRSL protocols can provide two datasets of information for rock surface (-exposure) dating, as the different temperature IR signals have been shown to have different bleaching rates (e.g. Sohbati *et al.* 2015). Furthermore, IRSL stimulation of rock slices may allow the investigation of specific feldspars using either blue (K-feldspar) or yellow (Na-feldspar) emissions (Sohbati *et al.* 2013). However, IRSL protocols on rock slices may be associated with high residual doses (Vafiadu *et al.* 2007), and post-IR IRSL protocols may be problematic due to high levels of recuperation after IR stimulation, which appear to be dependent on the temperature of the first IR stimulation (Liu *et al.* 2016).

11.4 APPLICATIONS

11.4.1. Archaeological case studies

Rock surface dating has been widely used to study archaeological artefacts, especially megalithic structures, with first attempts applying TL dating. Liritzis (1994) investigated the optical bleaching properties of TL in limestone experimentally (275°C TL peak), and used these properties to date megalithic structures in Peloponnese (Greece), yielding ages of around 3 ka, which are in agreement with independent archaeological ages. Theocaris *et al.* (1997) were also able to successfully use this approach to date two Hellenic pyramids, with the resultant ages showing them to be prehistoric. Polikreti *et al.* (2002; 2003) investigated marble objects for authentication purposes, and experimentally determined TL bleaching properties and bleaching depth following daylight exposure (Fig. 11.7A). They proposed using the 290°C TL peak because of favourable luminescence characteristics for dating marble artefacts, which were commonly used for construction throughout antiquity. Using the MAAD protocol, they provided a burial age for a temple artefact (Macedonia, Greece) of 2.6±0.4 ka, in relatively close agreement with the archaeological age. Building on these results, more recent archaeological studies have applied OSL dating.

Vafiadu *et al.* (2007) investigated SAR OSL protocols on whole rock slices from granitic, ultramafic and quartz metamorphic rocks sampled from archaeological sites (Greece, Denmark and Sweden). Their results showed good OSL signal bleaching properties and successful tests regarding the application of the SAR protocol on whole rock slices, with obtained burial ages in agreement with independent archaeological estimates. The SAR OSL protocol has also been applied in combination with the SAAD (single-aliquot additive dose) OSL protocol, to provide new dates for megalithic structures from Egypt and Saudi Arabia (Liritzis *et al.* 2013). Using quartz grains extracted from sandstones and granitic rocks, Lirtizis *et al.* (2013) provided burial ages between 3–4 ka for these structures and suggested that

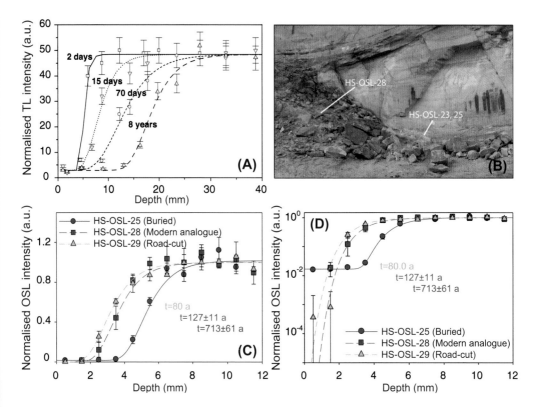

Figure 11.7 A) Experimental quantification of TL signal bleaching within a marble surface for various exposure times (Pentelic marble); modified from Polikreti *et al.* (2003). B) Picture of the Great Gallery rock art panel (southeastern Utah, USA) showing rock art, the rockfall and sample locations of the buried rock art (HS-OSL-25, ~40-cm thick), the underlying sediments (HS-OSL-23) and the modern analogue sample (HS-OSL-28); modified from Chapot *et al.* (2012). C) and D) OSL data from different samples (average of at least three aliquots for each depth, error bars represent one standard error) and fitted bleaching model to constrain the bleaching parameters (road cut) and quantify the exposure time (modern analogue and buried sample). See locations in (B) and note the logarithmic scale in (D) from which the OSL signal build up in sample HS-OSL-25 can be assessed and related to post-rockfall burial; modified from Sohbati *et al.* (2012c).

cross-checking following combined OSL protocols is recommended for applying rock surface dating in archaeological investigations. This approach has recently been used by Liritzis and Vafiadou (2015), who obtained ages between 3 and 8 ka for several archaeological sites located in Egypt. They extracted grains from various lithologies (granite, limestone, sandstone, dacite etc.) and used different OSL (SAR and SAAD) and TL (MAAD) protocols. Their strategy allowed them to compare the bleaching properties of different lithologies and stimulation types (e.g. the OSL signal is bleached more rapidly in sandstone than in granite, whereas the opposite trend can be observed for the TL signal). They also confirmed the potential of dating old archaeological monuments composed of carved granites, sandstones or limestones, using a combined TL and OSL approach, although the latter may present several advantages (except for limestone) regarding dating accuracy and measurement efficiency.

Another original application of rock surface dating, coupled with surface-exposure dating, is the investigation of historical rock art from the Great Gallery rock art panel in the Canyonlands National Park (southeastern Utah, USA; Fig. 11.7B). The age of the paintings has been highly debated, with hypotheses spanning the entire Holocene epoch (Pederson *et al.* 2014). Some of the paintings were damaged by a rockfall, burying them under sediment (HS-OSL-25 and -23 on Fig. 11.7B respectively). This configuration allowed investigation of both (1) the exposure time of the painted rock surface, and (2) the burial time of the rock and sediments using OSL dating (Chapot *et al.* 2012; Sohbati *et al.* 2012c). Rock samples (sandstones) were gently abraded to extract quartz grains from different depths below the painted surface (HS-OSL-25, Fig. 11.7B) and below the exposed surface of a modern analogue rock fall (HS-OSL-28, Fig. 11.7B). The OSL signals were then measured and showed bleaching within the first 2–5 mm below the exposed surface (Fig. 7C). Using a road cut of known age (HS-OSL-25, Fig. 11.7C) to constrain the bleaching rate for this specific site and lithology, Sohbati *et al.* (2012c) were able to quantify the exposure time of both the modern analogue (~130 years) and the rock art (~700 years, Fig. 11.7C). By further investigating the dose within the first ~2 mm below the surface, they were also able to quantify a finite OSL signal that had accumulated during burial subsequent to the rockfall (Fig. 11.7D, note the logarithmic-scale of the *y*-axis) which was not observed for the road cut nor the modern analogue (Fig. 11.7D). A burial age of ~900 yr was obtained for this sample, in agreement with the OSL age of the underlying sediment (Chapot *et al.* 2012). This original approach provided a precise time range (i.e. between 900 and 1600 years) for the origin of the Great Gallery rock art panel, coinciding with the development of the local Fremont culture (Pederson *et al.* 2014).

The application of OSL rock surface dating to buried cobbles/pebbles, in association with classical OSL dating of surrounding sediments, has also been successfully investigated in various archaeological contexts. Sohbati *et al.* (2012a) analyzed quartzite cobbles deposited by alluvial processes within an archaeological pavement (Tapada do Montinho, Portugal). Based on OSL rock surface dating they identified different potential resetting events (i.e. daylight exposure of the cobbles) of as old as 40–45 ka and between 20 and 14 ka. Sohbati *et al.* (2012a) thus proposed a complex evolutionary history for this pavement, with surficial erosion due to anthropogenic activity. They further investigated this complex history with potential multiple exposure/burial events for the pavement cobbles, their results (1) explained the younger than expected OSL ages obtained for the overlying sediments, and (2) provided a chronology for the evolution of this archaeological site.

Freiesleben *et al.* (2015) and Sohbati *et al.* (2015) have developed and successfully applied a mathematical model to quantify multiple exposure/burial events from the OSL rock surface dating of a single cobble (see Section 11.4). Freiesleben *et al.* (2015) investigated IRSL signals ($IRSL_{50}$ and $pIR-IRSL_{290}$) from a cross-section of a granitic cobble excavated from an archaeological site (Aarhus, Denmark). The measured IRSL profiles through the cobble (~70 mm in total) allowed them to identify a first exposure event of ~0.5 ka (cobble usage) before burial during 1.3–1.7 ka and recent excavation. Moreover, they also showed that different cobble surfaces (i.e. bottom *vs.* top surfaces) might yield complementary information about the cobble's full exposure/burial history (Freiesleben *et al.* 2015).

Sohbati *et al.* (2015) investigated a prehistoric cult site in the Negev desert (Israel) using a pavement cobble (Fig. 11.8B) and underlying sediments for OSL dating. They

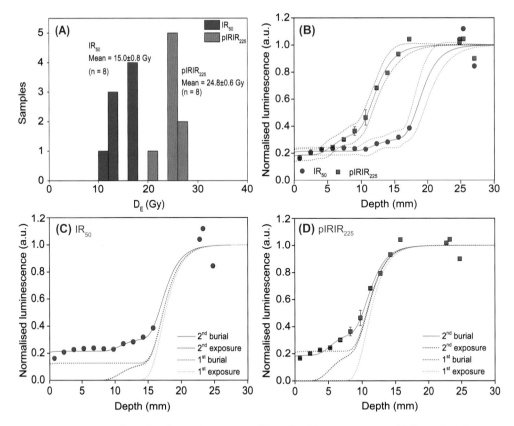

Figure 11.8 OSL rock surface dating (exposure and burial) of desert pavement. A) Equivalent doses using two IRSL signals from surface rock slices extracted from the buried cobble in the pavement. B) Measured IRSL signals with depth into the buried surface of the cobble, showing different bleaching characteristics between the two IRSL signals. The envelopes represent the model fits shown in C and D. C) and D) Model fits comprising multiple exposure/burial events which are able to reproduce the measured IRSL signals with depth; modified from Sohbati *et al.* (2015).

used two complementary IRSL signals (Fig. 11.8) to first derive equivalent doses from surface rock slices (Fig. 11.8A) and thus to determine the latest burial event at ~4 ka, consistent with OSL ages from underlying sediments. However, these well-constrained burial ages differed by 3–4 ka from the expected construction age of this archaeological site, indicating a potentially complex exposure history of the studied pavement. Using the full IRSL profiles from the cobble (Fig. 11.8B), Sohbati *et al.* (2015) showed that it had experienced at least two exposure events (with the earlier event having a longer duration than the more recent one) separated by an intermediate burial event which is of similar duration to the latest burial event (4–5 ka). This complex history is in agreement with the expected construction age of the site (~7–8 ka) and indicates later human intervention with the pavement (~4 ka ago). This highlights the potential of OSL rock surface dating and luminescence-depth profiles to provide tight temporal constraints on the bleaching history of rock surfaces that may not be available from conventional OSL dating on the underlying unconsolidated sediments.

11.4.2 Palaeoenvironmental applications

One novel and promising application of OSL rock surface dating has been proposed by Simms *et al.* (2011) for relative sea-level reconstructions from raised beaches in Antarctica (Fig. 11.9). They proposed the application of OSL dating on quartz mineral extracts from beach cobbles. Simms *et al.* (2011) focused on small granitic cobbles (maximum 1 dm³, Fig. 11.9A) to ensure that the cobbles had been rotated in the intertidal zone before beach fossilisation and thus that the OSL signal has been fully bleached before beach abandonment. Simms *et al.* (2011) obtained OSL ages for beach fossilisation over the last ~2 ka that were in good agreement with independent ¹⁴C dating (Fig. 11.9B), confirming the potential of the method for investigating beach dynamics, especially in high-latitude environments where organic matter is scarce and thus ¹⁴C dating may not be applicable.

Based on these promising results, Simms *et al.* (2012) quantified the timing of the most recent Neoglacial ice advance during the Little Ice Age (~300–500 yr ago) and subsequent glacio-isostatic adjustment of raised beaches in Antarctica (South Shetland Islands), revealing an increase in surface uplift rates (from ~2 to 12 mm yr⁻¹) following the ice retreat. Simkins *et al.* (2013) also investigated raised beaches in Antarctica (Marguerite Bay, western Antarctic Peninsula), showing that raised beaches more than 21 m above sea level might be of pre-LGM age and have been reworked during late-Pleistocene glacial advances. As a result, they proposed that the relative sea-level fall in the area has been only half that previously quantified for the Holocene period. Moreover, their study indicates potential complexities for luminescence dating of modern and recently-raised beaches due to post-depositional reworking of cobbles by storm waves. The impact of storm events requires further investigation for studies focused on Holocene sea-level reconstruction which use fossil beach ridges as well as the modern beach surface to quantitatively evaluate the bleaching variability in cobbles and their associated burial ages.

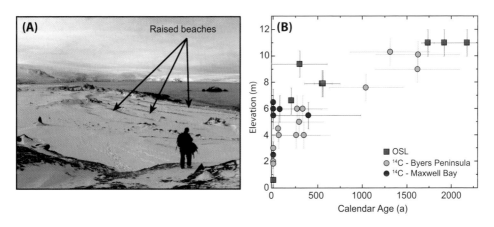

Figure 11.9 A) Picture of raised beaches (South Shetland Islands, Antarctica). B) Relative sea-level reconstruction from raised marine features in the South Shetland Islands, showing good agreement between OSL rock surface dating and independent ¹⁴C dating. Modified from Simms *et al.* (2011).

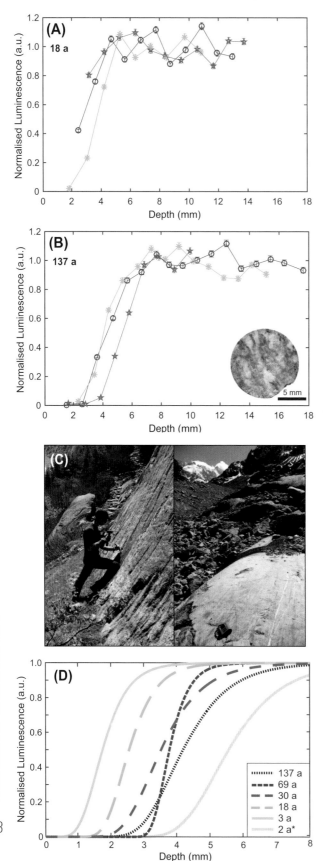

Figure 11.10 Palaeo-glacier reconstruction of the Mer de Glace (Mont-Blanc massif, France) since the Little Ice Age (LIA). A) and B) IRSL signal (measured at 50°C) with depth for two samples with exposure durations of 18 yr (A) and 137 yr (B). Exposure times are reconstructed from historical archives (Vincent *et al.* 2014). Each coloured data point represents an individual rock slice (example shown as inset in B), with three cores for each sample. C) Pictures of sampling sites for LIA (left) and modern (right) glacially polished bedrock surfaces. D) Best-fitting models for IRSL data for all samples studied (raw data are shown for two samples in panels A–B; the reader is referred to Lehmann *et al.* 2018 for the complete dataset), showing a propagating bleaching front within the first 1–4 mm with increasing exposure time (3–137 yr). Note that the granitic sample (*) with 2-yr exposure shows a completely different bleaching profile, highlighting the influence of lithology for OSL rock surface-exposure dating. Modified from Lehmann *et al.* (2018).

Finally, OSL rock surface (-exposure) dating has also been recently investigated in glacial and proglacial environments, with interesting potential for dating palaeo-glacier fluctuations and associated sediment deposits. These applications are especially promising as conventional OSL sediment dating can be challenging in such environments (Fuchs and Owen 2008; Chapter 6). Lehmann *et al.* (2018) worked on glacially-polished bedrock surfaces from the Mer de Glace (Mont-Blanc massif, France; Fig. 11.10) to investigate whether post-Little Ice Age (LIA) glacier retreat can be constrained from OSL rock surface-exposure dating. They collected glacially-polished bedrock surfaces with morphological evidence for glacier erosion (i.e. glacial striations; Fig. 11.10C) and measured the IRSL signal at 50°C on individual rock slices (see example in Fig. 11.10B). Their results showed bleaching of the IRSL signal at different depths, dependent on the exposure time (Fig. 11.10A–B). Using the bleaching model proposed by Sohbati *et al.* (2012c), they were able to reproduce the observed IRSL data and bleaching depths for exposed glacial bedrock surfaces from 3 to 137 years (Fig. 11.10D). Their results also confirmed the strong dependence of IRSL signal bleaching on rock lithology, with significant differences in the bleaching profiles between gneissic and granitic rock surfaces with similar exposure times (Fig. 11.10D). This study highlights the potential of using OSL rock surface-exposure dating to reconstruct recent (i.e. post-LIA) glacier fluctuations, especially because the technique is sensitive over the first 1 cm of the bedrock (Fig. 11.10), which would be eroded even during a short (decades long) glacial re-advance. Further investigation is needed to evaluate the possibility of extending this approach over longer timescales (i.e. Late Pleistocene to Holocene) for which surface weathering and thus erosion might be problematic (c.f. Sohbati *et al.* 2018).

OSL rock surface-exposure dating applied to glacially-polished bedrock can also be combined with another application of OSL rock surface dating focused on glacial/proglacial deposits, opening new directions for palaeoglacier reconstructions. Rades *et al.* (2018) sampled morainic boulders in the Malta valley (Austrian Alps) and measured $IRSL_{50}$ profiles, revealing that the IRSL signals had been fully bleached before deposition and thus demonstrating the potential of cobble surface dating for constraining the timing of moraine deposition. Jenkins *et al.* (2018) followed the same approach by collecting fluvio-glacial cobbles within a late-glacial sandur (Isle of Man and Scotland). They measured both $IRSL_{50}$ and post-IR $IRSL_{225}$ signal depth profiles, showing that some of the collected pebbles have experienced bleaching of the IRSL signals into the cobble subsurfaces to ~4–10 mm depth, allowing the dose acquired post-burial to be quantified and thus to evaluate the depositional age of those cobbles. These two preliminary studies have revealed the potential of OSL rock surface dating in glacial and periglacial environments, where conventional sediment dating can be difficult due to challenges of partial bleaching within unconsolidated sediments (Chapter 6). However, further investigations are required regarding dose rate determination in such settings, where sediment grain size as well as water content is highly variable (see Section 11.3.2).

11.5 CONCLUSIONS AND OUTLOOK

Luminescence rock surface and rock surface-exposure dating are techniques that are in their early stages of development. They offer the potential to constrain surface processes and archaeological events that were previously challenging to date, and it is anticipated that both methods will be widely developed and applied over the coming years. Sample preparation protocols must become more firmly established including identifying the causes of variability between luminescence measurements from rock slices and grains extracted from bedrock. Once this has been achieved the more significant challenges can be addressed. For luminescence rock surface dating, this comprises developing strategies for robust environmental dose rate determination, where complex geometries, varied clast sizes, varied water contents and microdosimetry influence the precision of determined ages. For luminescence rock surface-exposure dating, a means of robustly determining the light attenuation and luminescence signal bleaching rates for samples of different lithologies and locations is essential for this method to be widely adopted. Furthermore, the effect of erosion on the determination of luminescence surface-exposure ages must also be quantified. Nonetheless, both rock surface and rock surface-exposure dating are incredibly promising, and represent exciting new frontiers in luminescence dating research.

REFERENCES

Aitken, M.J. 1985. *Thermoluminescence Dating*. Academic Press.

Armitage, S.J. and King, G.E. 2013. Optically stimulated luminescence dating of hearths from the Fazzan Basin, Libya: A tool for determining the timing and pattern of Holocene occupation of the Sahara. *Quaternary Geochronology* 15, 88–97.

Buscombe, D. 2013. Transferable wavelet method for grain-size distribution from images of sediment surfaces and thin sections, and other natural granular patterns. *Sedimentology* 60, 1709–1732.

Buylaert, J.P. Murray, A.S. Gebhardt, C. Sohbati, R. Ohlendorf, C. Thiel, C. and Zolitschka, B. 2013. Luminescence dating of the PASADO 5022-1D core using IRSL signals from feldspar. *Quaternary Science Reviews* 71, 70–80.

Chapot, M.S. Sohbati, R. Murray, A.S. Pederson, J.L. Rittenour, T.M. 2012. Constraining the age of rock art by dating a rockfall event using sediment and rock-surface luminescence dating techniques. *Quaternary Geochronology* 13, 18–25.

Duller, G.A. 2008. Single-grain optical dating of Quaternary sediments: Why aliquot size matters in luminescence dating. *Boreas* 37, 589–612.

Durcan, J.A. King, G.E. and Duller, G.A. 2015. DRAC: Dose rate and age calculator for trapped charge dating. *Quaternary Geochronology* 28, 54–61.

Freiesleben, T. Sohbati, R. Murray, A. Jain, M. Al Khasawneh, S. Hvidt, S. and Jakobsen, B. 2015. Mathematical model quantifies multiple daylight exposure and burial events for rock surfaces using luminescence dating. *Radiation Measurements* 81, 16–22.

Fuchs, M. Owen, L.A. 2008. Luminescence dating of glacial and associated sediments: review, recommendations and future directions. *Boreas* 37, 636–659.

Gliganic, L.A., Meyer M.C., Sohbati, R., Jain, M., Barrett, S. (in press). OSL surface exposure dating of a lithic quarry in Tibet: Laboratory validation and application. *Quaternary Geochronology*.

Gösku, H.Y. Fremlin, J.H. Irwin, H.T. and Fryxell, R. 1974. Age determination of burned flint by a thermoluminescent method. *Science* 183, 651–654.

Greilich, S. Wagner, G. A. 2006. Development of a spatially resolved dating technique using HR-OSL. *Radiation Measurements* 41, 738–743.

Greilich, S. Glasmacher, U.A. Wagner, G.A. 2005. Optical dating of granitic stone surfaces. *Archaeometry* 47, 645–665.

Guérin, G. Mercier, N. Nathan, R. Adamiec, G. and Lefrais, Y. 2012. On the use of the infinite matrix assumption and associated concepts: A critical review. *Radiation Measurements* 47, 778–785.

Guralnik, B. Ankjærgaard, C. Jain, M. Murray, A.S. Müller, A. Wälle, M. Lowick, S.E. Preusser, F. Rhodes, E.J. Wu, T.S. and Mathew, G. 2015. OSL-thermochronometry using bedrock quartz: A note of caution. *Quaternary geochronology* 25, 37–48.

Habermann, J. Schilles, T. Kalchgruber, R. Wagner, G.A. 2000. Steps towards surface dating using luminescence. *Radiation Measurements* 32, 847–851.

Jenkins, G. T. H., Duller, G. A. T., Roberts, H. M., Chiverrell, R. C. and Glasser, N. F. 2018. A new approach for luminescence dating glaciofluvial deposits: High precision optical dating of cobbles. *Quaternary Science Reviews* 192, 263–273.

Kars, R.H. Reimann, T. Ankjærgaard, C. and Wallinga, J. 2014. Bleaching of the post-IR IRSL signal: New insights for feldspar luminescence dating. *Boreas* 43, 780–791.

King, G.E. Herman, F. Lambert, R. Valla, P.G. and Guralnik, B. 2016a. Multi-OSL-thermochronometry of feldspar. *Quaternary Geochronology* 33, 76–87.

King, G.E. Herman, F. and Guralnik, B. 2016b. Northward migration of the eastern Himalayan syntaxis revealed by OSL thermochronometry. *Science* 353, 800–804.

Laskaris, N. Liritzis, I. 2011. A new mathematical approximation of sunlight attenuation in rocks for surface luminescence dating. *Journal of Luminescence* 131, 1874–1884.

Lehmann, B. Valla, P.G. King, G.E. Herman, F. 2018. Reconstruction of glacier vertical fluctuation in the Western Alps using OSL surface exposure dating. *Quaternary Geochronology* 44, 63–74.

Liritzis, I. 1994. A new dating method by thermoluminescence of carved megalithic stone building. *Comptes rendus de l'Académie des sciences. Série 2. Sciences de la terre et des planets* 319, 603–610.

Liritzis, I. Vafiadou, A. 2015. Surface luminescence dating of some Egyptian monuments. *Journal of Cultural Heritage* 16, 134–150.

Liritzis, I. Guibert, P. Foti, F. and Schvoerer, M. 1996. Solar bleaching of thermoluminescence of calcites. *Nuclear Instruments and Methods in Physics Research Section B: Beam Interactions with Materials and Atoms* 117, 260–268.

Liritzis, I. Vafiadou, A. Zacharias, N. Polymeris, G.S. and Bednarik, R.G. 2013. Advances in surface luminescence dating: New data from selected monuments. *Mediterranean Archaeology and Archaeometry* 13, 105–115.

Liu, J. Murray, A. Sohbati, R. and Jain, M. 2016. The effect of test dose and first IR stimulation temperature on post-IR IRSL measurements of rock slices. *Geochronometria* 43, 179–187.

Meyer, M.C., Gliganic, L.A., Jain, M., Sohbati, R., Schmidmair, D. (in press). Lithological controls on light penetration into rock surfaces: Implications for OSL and IRSL surface exposure dating. *Radiation Measurements*.

Morgenstein, M. E. Luo, S. Ku, T. L. and Feathers, J. 2003. Uranium-series and luminescence dating of volcanic lithic artefacts. *Archaeometry* 45, 503–518.

Pederson, J.L. Chapot, M.S. Simms, S.R. Sohbati, R. Rittenour, T.M. Murray, A.S. Cox, G. 2014. Age of Barrier Canyon-style rock art constrained by cross-cutting relations and luminescence dating techniques. *Proceedings of the National Academy of Sciences*, 111(36), 12986–12991.

Polikreti, K. Michael, C.T. and Maniatis, Y. 2002. Authenticating marble sculpture with thermoluminescence. *Ancient TL* 20, 11–18.

Polikreti, K. Michael, C.T. and Maniatis, Y. 2003. Thermoluminescence characteristics of marble and dating of freshly excavated marble objects. *Radiation Measurements* 37, 87–94.

Rades, E.F., Sohbati, T., Lüthgens, C., Jain. M. and Murray, A.S. 2018. First luminescence-depth profiles from boulders from moraine deposits: Insights into glaciation chronology and transport dynamics in Malta valley, Austria. *Radiation Measurements*.

Richards, M.P. 1994. Luminescence dating of quartzite from the Diring Yuriakh site. M.A. thesis, Simon Fraser University, unpublished.

Roberts, R.G. 1997. Luminescence dating in archaeology. *Radiation Measurements* 27, 819–892.

Sawakuchi, A.O. Blair, M.W. DeWitt, R. Faleiros, F.M. Hyppolito, T. Guedes, C.C.F. 2011. Thermal history versus sedimentary history: OSL sensitivity of quartz grains extracted from rocks and sediment. *Quaternary Geochronology* 6, 261–272.

Simkins, L.M. Simms, A.R. DeWitt, R. 2013. Relative sea-level history of Marguerite Bay, Antarctic Peninsula derived from optically stimulated luminescence-dated beach cobbles. *Quaternary Science Reviews* 77, 141–155.

Simkins, L.M. DeWitt, R. Simms, A.R. Briggs, S. and Shapiro, R.S. 2016. Investigation of optically stimulated luminescence behavior of quartz from crystalline rock surfaces: A look forward. *Quaternary Geochronology* 36, 161–173.

Simms, A.R. DeWitt, R. Kouremenos, P. and Drewry, A.M. 2011. A new approach to reconstructing sea levels in Antarctica using optically stimulated luminescence of cobble surfaces. *Quaternary Geochronology* 6, 50–60.

Simms, A.R. Ivins, E.R. DeWitt, R. Kouremenos, P. Simkins, L.M. 2012. Timing of the most recent Neoglacial advance and retreat in the South Shetland Islands, Antarctic Peninsula: Insights from raised beaches and Holocene uplift rates. *Quaternary Science Reviews* 47, 41–55.

Smedley, R.K. Duller, G.A.T. Pearce, N.J.G. and Roberts, H.M. 2012. Determining the K-content of single-grains of feldspar for luminescence dating. *Radiation Measurements* 47, 790–796.

Smedley, R.K. and Pearce, N.J.G. 2016. Internal U, Th and Rb concentrations of alkali-feldspar grains: Implications for luminescence dating. *Quaternary Geochronology* 35, 16–25.

Sohbati, R. Murray, A. Jain, M. Buylaert, J.P. and Thomsen, K. 2011. Investigating the resetting of OSL signals in rock surfaces. *Geochronometria* 38, 249–258.

Sohbati, R. Murray, A.S. Buylaert, J.P. Almeida, N.A. and Cunha, P.P. (2012a). Optically stimulated luminescence (OSL) dating of quartzite cobbles from the Tapada do Montinho archaeological site (east-central Portugal). *Boreas* 41, 452–462.

Sohbati, R. Jain, M. Murray, A.S. (2012b). Surface exposure dating of non-terrestrial bodies using optically stimulated luminescence: A new method. *Icarus* 221, 160–166.

Sohbati, R. Murray, A.S. Chapot, M.S. Jain, M. Pederson, J. 2012c. Optically stimulated luminescence (OSL) as a chronometer for surface exposure dating. *Journal of Geophysical Research: Solid Earth* 117(B9).

Sohbati, R. Murray, A.S. Jain, M. Thomsen, K. Hong, S-C. Yi, K. Choi, J-H. 2013. Na-rich feldspar as a luminescence dosimeter in infrared stimulated luminescence (IRSL) dating. *Radiation Measurements* 51–52, 67–82.

Sohbati, R. Murray, A.S. Porat, N. Jain, M. Avner, U. 2015. Age of a prehistoric 'Rodedian' cult site constrained by sediment and rock surface dating techniques. *Quaternary Geochronology* 30, 90–99.

Sohbati, R., Liu, J., Jain, M., Murray, A., Egholm, D., Paris, R., Guralnik, B. 2018. Centennial- to millennial-scale hard rock erosion rates deduced from luminescence-depth profiles. *Earth and Planetary Science Letters* 493, 218–230.

Spooner, N.A. 1994. The anomalous fading of infrared-stimulated luminescence from feldspars. *Radiation Measurements*, 23, 625–632.

Theocaris, P.S. Liritzis, I. Galloway, R.B. 1997. Dating of two Hellenic pyramids by a novel application of thermoluminescence. *Journal of Archaeological Science*, 24, 399–405.

Tribolo, C. Mercier, N. Valladas, H. 2003. Attempt at using the single-aliquot regenerative-dose procedure for the determination of equivalent doses of Upper Palaeolithic burnt stones. *Quaternary Science Reviews* 22, 1251–1256.

Vafiadou, A. Murray, A.S. Liritzis, I. 2007. Optically stimulated luminescence (OSL) dating investigations of rock and underlying soil from three case studies. *Journal of Archaeological Science* 34, 1659–1669.

Vincent, C. Harter, A. Gilbert, A. Berthier, E. Six, D. 2014. Future fluctuations of Mer de Glace, French Alps, assessed using a parameterized model calibrated with past thickness changes. *Annals of Glaciology* 55, 15–24.

12 FUTURE DEVELOPMENTS IN LUMINESCENCE DATING

JAKOB WALLINGA

Soil Geography and Landscape Group and Netherlands Centre for Luminescence Dating, Wageningen University and Research, PO Box 47, 6700 AA Wageningen, The Netherlands. Email: jakob.wallinga@wur.nl

ABSTRACT: Luminescence dating is a relatively recent technique, and it is rapidly evolving, both in methods and in applications. In this chapter an attempt is made to foresee future developments, building on recent published work and topics discussed amongst specialists. A brief overview is presented of the most recent methodological advances, and expectations for the near future. These developments are expected to allow more detailed geochronological investigations, and further increase the age range attainable with luminescence dating. The emphasis of the chapter is on two exciting developments which are expected to become important in the future. The first is related to the use of luminescence signals to reconstruct pathways of rock and sediment, based on differences in sensitivity to light and heat of different signals. The second encompasses developments towards dating sediments *in situ*, ranging from portable luminescence instrumentation, to spatially resolved measurement of dose rate and palaeodose. These developments may in future allow rapid age-scanning of exposures, boreholes or cores, and offer new opportunities for a wide range of applications, including reconstructing soil reworking rates, sediment transport and provenance, as well as rock uplift rates.

KEYWORDS: *in situ* luminescence dating, spatially resolved luminescence, sediment provenance, sediment transport, bioturbation

12.1 INTRODUCTION

In this chapter, a brief summary will be provided of the main challenges presently faced by the luminescence dating community, discussing the approaches that are explored to overcome these challenges, and reflecting on research questions that could be solved with these improved methods.

The most exciting new developments are expected in three other directions. The first is related to developments in instrumentation, which yield promise with regards to rapid age determination in the field, and/or of sediments in their undisturbed context. Recent developments with regard to field equipment will be reviewed, as well as spatially resolved detection of dose rate and equivalent dose. Building on this, potential applications of these methods will be discussed. The second development is related to new approaches that use luminescence signals to acquire information on the history of grains; where did they come from, how did grains travel prior to deposition, or how did grains move through the sediment after deposition? The rational of such approaches will be outlined, and first applications to determine soil reworking rates, reconstruct sediment transport processes and sediment provenance, as well as rock uplift rates, will all be discussed. A third new development that should be mentioned is surface-exposure dating. This novel method and its applications feature in Chapter 11, and will not be discussed in the present chapter.

12.2 MAIN CHALLENGES

12.2.1 Poor bleaching

A crucial assumption in luminescence dating is that the luminescence signal was reset fully at the time of deposition. If this requirement is not met, it may lead to overestimation of burial age. Two main approaches are taken to avoid such age overestimation:

1. the use of a luminescence signal that is rapidly reset upon light exposure

2. selecting those grains for analysis that were reset best.

Advances are being made in both respects, and these developments will continue in the future.

Since the pioneering work of Godfrey-Smith *et al.* (1988) it is well established that the quartz OSL signal is more rapidly reset than feldspar IRSL signals (see Chapter 6, Fig. 6.2). For this reason, quartz will be the mineral of choice for most studies where poor bleaching may be an issue (e.g. Wallinga 2002). But even the quartz OSL signal in fact consists of multiple signals; Singarayer and Bailey (2003) have shown that the signal is composed of the sum of different signals, each with their own resetting characteristics (Chapter 6, Fig. 6.3). The fast component is desirable for dating, as it is rapidly reset, stable over geological timescales, and the single-aliquot regenerative dose (SAR) protocol was designed for it (Wintle and Murray 2006). Isolation of the fast component is important to avoid contamination of the signal with slower bleaching signals. Although several approaches have been proposed (e.g. Cunningham and Wallinga 2009; Galbraith *et al.* 1999), these are labour intensive and have not been widely adopted. A much simpler approach that provides a reasonable clear fast component signal in most cases is the use of an early background (Cunningham and Wallinga 2010). Here, the net OSL signal used for analysis is obtained by subtracting the signal immediately following the initial signal. It has been shown that this may reduce age overestimation, at least for fluvial deposits (Shen and Mauz 2012), which may be highly relevant especially when dating young deposits. With increased computing power, automated isolation of the fast component may soon become feasible and should become the new standard in luminescence signal analysis.

With respect to the selection of grains that were reset best, major advances were made by the combination of instrumentation to detect luminescence signals from single grains (Duller *et al.* 2000) and advanced statistical methods to analyse equivalent dose distributions (Galbraith *et al.* 1999; Thomsen *et al.* 2016). The youngest generation of

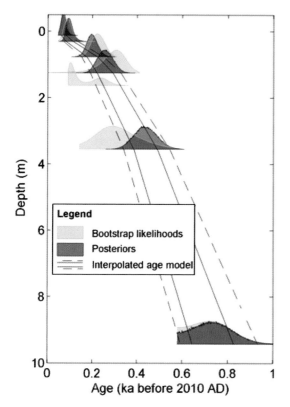

Figure 12.1 Example of the use of the interpretation of a set of poorly bleached fluvial samples using Bayesian approaches. For each of the samples, a probability density function of age is constructed using a bootstrapped version of the minimum age model. Results are combined with stratigraphic information (deposits get younger towards the top) in a Bayesian deposition model (in this case Oxcal) to determine the most likely depositional age. Figure modified from Cunningham and Wallinga (2012).

interpretation models includes Monte-Carlo and Bayesian approaches (e.g. Cunningham and Wallinga 2012; Guerin *et al.* 2017; see also Chapter 3) to include all sources of uncertainty and obtain robust results (Fig. 12.1). The use of Bayesian methods to interpret samples in stratigraphic context (e.g. Cunningham and Wallinga 2012; Nooren *et al.* 2017; Ramsey 2008; Rhodes *et al.* 2003) is, as in the field of radiocarbon dating, rapidly growing, but still not used to full potential (Trachsel and Telford 2017).

12.2.2 Age range

Although the age range covered by luminescence dating is very wide compared to most other geochronological methods, there clearly are limitations. In this section, challenges and recent advances with regard to further stretching the age range of luminescence dating are outlined.

12.2.2.1 Dating very young deposits

When interested in young deposits (<1000 years), several challenges need to be considered. Here a summary is given, loosely based on Madsen and Murray (2009).

1. Natural luminescence signals will be dim, as absorbed doses are expected to be low. This potential problem may be overcome by measuring large aliquots (i.e. the combined luminescence signal of many grains), but at the expense of losing information about grain-to-grain differences.

2. Any remaining OSL signal, upon deposition, may result in a relatively large error. Remnant ages of 100 years may become insignificant when dating pre-Holocene deposits, but will result in grossly inaccurate results when dating deposits formed within the past decade. Solutions may be found in the use of the most light-sensitive signal available , potentially in combination with selecting well-bleached grains.

3. Heating of the grains prior to measurement (preheating) may result in thermal transfer of charge from light-insensitive traps (with large remnant doses) to the light-sensitive traps used for dating. The effect is similar to a remnant dose, discussed above. This issue may be minimised by using low preheat temperatures, making sure that any unstable signal is removed by the preheat.

12.2.2.2 Dating very old deposits

Challenges on the high end of the age range available through luminescence dating, are entirely different. Luminescence signals increase over time, as trapped charge concentrations in the crystal build up. However, after a certain point all sites available for charge trapping are used up, and no further build-up of luminescence signal is possible. This is reflected in graphs that show the luminescence signal as a function of absorbed dose, so-called dose response curves (see Chapter 5, Fig. 5.4). These typically take the shape of a saturating exponential, although more complicated functions may be needed to fit experimental data (e.g. double saturating exponential, exponential plus linear).

As explained in Chapter 1, the equivalent dose is obtained by projecting the natural luminescence signal on the dose response curve reconstructed using laboratory irradiation. However, for the high end of the dose response curve, a small error in the measured luminescence signal, may result in a huge error in equivalent dose. For this reason, Wintle and Murray (2006) have suggested that reliable dating is restricted to the lower part of the dose response curve, using $2*D_0$ as threshold. Recent information suggests that this threshold may be too optimistic (e.g. Timar-Gabor *et al.* 2017; Wintle and Adamiec 2017). Due to this reason, the upper limit for reliable quartz OSL dating is usually around 100 ka, although regionally different limits do apply depending on the combination of quartz OSL properties and local dose rate (e.g. Schokker *et al.* 2005).

There are two potential approaches to extend the age range of luminescence dating. The first is to increase accuracy of the natural OSL measurement and laboratory-constructed dose response curve. If systematic errors are eliminated, it could in principle be possible to use a larger part of the dose response curve. Several investigations are performed, seeking further improvements of the SAR protocol for accurate equivalent dose determination (e.g. Timar-Gabor and Wintle 2013). However, even if successful, only minor expansion of the age range is to be expected, as dating is not possible once full saturation is reached.

The second approach seeks to use luminescence signals which saturate at higher doses. The first obvious candidate is the feldspar IRSL signal, which has been known to saturate less readily than quartz OSL. However, feldspar IRSL signals are also known to be unstable over time due to anomalous fading (Wintle 1973), the escape of electrons from the dosimetric trap due to quantum-mechanical tunnelling (Huntley and Lian 2006). In recent years, novel methods have been developed (Thomsen *et al.* 2008; Li *et al.* 2014) and tested (Buylaert *et al.* 2012; Kars *et al.* 2012) using the (more) stable pIR-IRSL signals. This may extend the dating limit up to about 400 ka (e.g. Joordens *et al.* 2015), again depending on environmental dose rate. A very recent and highly promising development is the use of infrared photoluminescence (IRPL; Kumar *et al.* 2017). Preliminary investigations suggest that the IRPL signal shows negligible anomalous fading. Moreover, the signal can be detected non-destructively, allowing prolonged measurement and thus better signal-to-noise ratios.

Other approaches use quartz luminescence signals, arising from other traps than the fast component quartz OSL signal used for most dating applications. These alternatives include TT-OSL (Duller and Wintle 2012), isothermal TL (Buylaert *et al.* 2006), and violet-stimulated luminescence (VSL; Ankjaergaard *et al.* 2016; Jain 2009). So far, none of these methods has provided a universally applicable approach to extend the upper age limit beyond 400 ka. The research will be continued though, as there is a clear need for reliable dating methods for sediment that cover the entire Quaternary, to reconstruct the evolution of landscapes and humans.

12.3 TOWARDS *IN SITU* DATING

Wouldn't it be great to be able to have direct insight in the age of deposits in the field? Ideally on a grain-to-grain basis, to check for differences in bleaching, or post-depositional mixing? Of course, that is what every Quaternary geologist, archaeologist and/or soil scientists would prefer, rather than the tedious routine of taking samples, and sending them to a luminescence dating lab for costly analyses with long processing time. While developing the necessary equipment is an enormous challenge, some progress has in fact been made, both with regard to field instrumentation and with regard to spatially resolved measurements. In this section, developments in both will be discussed, and an attempt is made to forecast future developments.

12.3.1 Portable instruments

For dating of sediments in the field, both sides of the age equation need to be known. Estimating dose rates in the field is relatively straight forward; field gamma spectrometers have been available for decades, and are widely used in the luminescence community (e.g. Hossain *et al.* 2002). Although results are not as precise as those obtained under controlled laboratory conditions, the instruments provide rapid quantitative information, and have advantages in highly non-uniform settings.

Determining equivalent dose in the field is more challenging. Although compact and transportable luminescence readers were developed (Kook *et al.* 2011; Fig. 12.2); they were in the end not sold. One of the problems with portable readers is the radiation source they require. Health and safety requirements prevent the use of beta-sources, which are used in laboratory instruments. The use of X-ray tubes for irradiation is possible, but its application has shown to be problematic.

An alternative approach was put forward by Sanderson and Murphy (2010) who developed an instrument which could measure infrared stimulated luminescence (IRSL) and blue stimulated luminescence (BSL) in continuous wave or pulsed mode (Fig. 12.2). The instrument does not allow irradiating or heating the sample, but the authors argue

Figure 12.2 Instruments for field luminescence measurements. Left: the Risø prototype, which includes irradiation and heating facilities (Kook *et al.* 2011; photo provided by DTU/Nutech). This instrument is versatile, but is not yet available to the community. Middle and right: two versions of the simpler SUERC luminescence field instrument described by Sanderson and Murphy (2010), which is widely used. The middle picture shows the instrument supplied by SUERC between 2010 and 2015; while the right-hand picture shows the most recent 'one-box design' (Photographs taken by David Sanderson, SUERC).

that the natural intensity of the signals, and the ratio of IRSL to BSL signal may provide sufficient insight to detect major transitions in the stratigraphic record. Indeed, the instrument has been widely used, and in several cases provided such insight (e.g. Bateman *et al.* 2015; Sanderson and Murphy 2010) which may support selection of samples for full analysis (Fig. 12.3). However, even in very uniform settings, results may be more affected by differences in sediment source and/or composition, rather than age (Stone *et al.* 2015).

In more complicated settings, the added advantage of field instruments may be limited. In such cases, more is to be expected from quick and dirty laboratory analysis of samples, or range-finder ages (Durcan *et al.* 2010; Reimann *et al.* 2015; Roberts *et al.* 2009), which does provide a crude estimate of palaeodose. For relatively old samples, laboratory measurement time could also be drastically reduced by using a 'standardised dose curve' approach, as proposed by Roberts and Duller (2004). In this method, the natural OSL signal is measured on each aliquot, and after normalisation projected on a standardised dose response curve. However, differences in growth curve shape between aliquots (Peng *et al.* 2016), regions (Telfer *et al.* 2008), or even grain size (Timar-Gabor *et al.* 2017) may affect accuracy of results.

Although in many cases laboratory measurements for luminescence dating have advantages over field methods, there is a real interest to further possibilities for (even crude) *in situ* estimation of palaeodose. Such measurement would require a radiation source (most likely X-ray), and various stimulation light sources. In combination with similarly crude estimate of dose rate through field gamma spectrometry or potentially hand-held XRF device this could inform sampling strategies, and allow geochronological control in remote areas, where it is not possible to sample or transport material to a lab for full analysis. In the absence of suitable equipment, lab analysis should be preferred in most cases, as interpretation of natural luminescence signals in absence of estimation of dose response is cumbersome.

12.3.2 Spatially resolved luminescence dating

Grain-to-grain variations in dose rate arise from non-uniform distribution of radionuclides, in combination with the limited range of beta-particles in the sediment (2–3 mm).

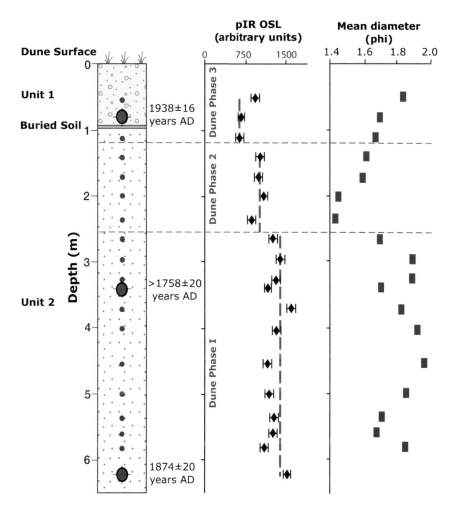

Figure 12.3 Example of application of the field luminescence reader of Sanderson and Murphy (2010). Based on pIR OSL measurements with the portable reader (middle pane) in combination with mean particle size (right-hand pane) different depositional units (Dune Phase 1, 2 and 3) could be identified for a coastal dune. Ages at three depths were obtained using full quartz SAR analysis (shown left). Figure adapted from Bateman *et al.* (2015).

Especially with deposits where a large part of the dose arises from beta radiation, and/ or for sediments with hot-spots in radionuclide concentration, differences in dose rate between grains can be substantial. As a result, palaeodoses of grains will also differ, even for perfectly bleached sediments where no mixing took place.

For many deposits, single-grain palaeodoses may show additional spread, due to heterogeneous bleaching or mixing processes (e.g. bioturbation). Full understanding of single-grain equivalent dose distributions would require insight in both the dose rate distribution and palaeodose distributions of samples in their original context. This, however, poses a great challenge. In this section, the progress made in recent years is outlined, and possibilities for further development are discussed.

12.3.2.1 Imaging dose rate

At the decimetre scale, variations of dose rate can be caused by changes in lithology or sediment source. In such cases, field gamma spectrometry (see sections 1.6.2 and 10.2.2) or calculation of contributions of different layers (Aitken 1985; Wallinga and Bos 2010) may provide good estimates of the gamma dose rate at the sample position. Here I focus on the millimetre scale, which is affected by beta dose heterogeneity. Recently, two methods have been put forward that may allow insight in beta dose heterogeneity of intact samples. Rufer and Preusser (2009) showed how beta-autoradiography can be used to obtain a spatially resolved estimate of total dose rate, for either rock slices or unconsolidated sediment (Fig. 12.4). Although the method was tested on sand samples which were poured onto the imaging plate and spread to form a mono-grain layer, it could also be applied to undisturbed thin slices of sediment in resin. Schmidt *et al.* (2012) combined beta-radiography with alpha-radiography for silex samples, and suggest that highly heterogenous areas in the silex could be avoided for spatially resolved palaeodose estimation. Recently, Romanyukha *et al.* (2017) investigated a more advanced method using a timepix detector; this method allows discrimination of different radiation types and can be applied to resin-impregnated sediment samples. Tests on artificial stratified samples indicated that spatial information on beta dose rate could be obtained (Fig. 12.5).

The studies mentioned above provide insight in dose rate distributions, but also show the challenges that need to be overcome (e.g. signal-to-noise ratio, measurement time, translation to dose rate). Ideally, such imaging of a sediment or rock slice should be combined with spatially resolved estimation of palaeodose, to link dose rate distributions and palaeodose distributions. However, it is important to realise that grains absorb radiation from a sphere, whereas these imaging methods only provide insight in two dimensions. Models have been developed to aid understanding of dose rate distributions in 3D (Martin *et al.* 2015), but application of such models to determine the dose rate at the location of a specific grain would require 3D mapping of radionuclide concentrations. Martin *et al.* (2015) explore such 3D modelling for tooth enamel, but their radionuclide distributions are modelled rather than based on actual measurements.

12.3.2.2 Imaging luminescence and spatially resolved palaeodose estimation

Routinely, luminescence signals are detected with a photo-multiplier tube, which does not allow spatial discrimination. In other words, they do not provide information on what part of the sample produces the luminescence signal that is detected. Conventional single-grain luminescence measurement is based on stimulating grains one-by-one, either by using aliquots consisting of a single grain (Lamothe *et al.* 1994), or by placing grains on a grid in a prepared single-grain disc, and using a steerable laser for stimulation (Duller *et al.* 1999). Downsides of such approaches are that they are not very efficient, grains are no longer in their original context, and no insight is obtained about which parts of the grain luminesce. Imaging techniques hold the promise to overcome these issues, and first attempts to obtain spatially information on luminescence signals were already published in the 1970s (Malik *et al.* 1973). Possibilities for imaging were hugely expanded since the development scientific cameras using charge coupled devices (CCDs), and subsequent technological advances (in particular, the development of electron multiplying (EM) CCDs) that allow detecting low-intensity luminescence emissions typical for natural materials. Recently, such cameras have become available in

Figure 12.4 Images of autoradiography of unconsolidated sand samples, showing dose rate heterogeneity with dark areas indicating high values (modified from Rufer and Preusser, 2009). A glaciofluvial sample (left) shows larger heterogeneity than an aeolian dune sample (right).

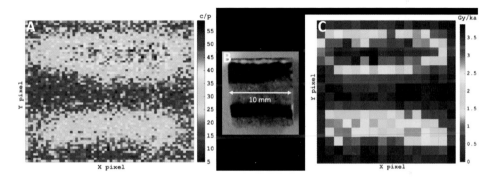

Figure 12.5 Images showing spatially resolved dose rate estimation using the Timepix device. Meaurements on a laboratory prepared sample consisting of layers of radioactive biotite and non-radioactive quartz (B) provided an image of counts per pixel (A), which could be transferred in to an image of dose rate (C). Modified after Romanyukha *et al.* (2017).

automated luminescence instruments (Kook *et al.* 2015; Richter *et al.* 2013), and they are now increasingly used by the dating community.

Although such devices provide high quality images (Fig. 12.6), quantitative determination of equivalent dose is a challenge, due to issues related to image analysis (Greilich *et al.* 2015) and quantifying luminescence signals from individual grains (avoiding 'cross talk' issues; e.g. Gribenski *et al.* 2015). Most feasible approaches seem to be to measure grains with sufficient spacing in between, or using the gridded single-grain discs developed for conventional single-grain dating. Cunningham and Clark-Balzan (2017) show how advanced methods for signal collection and processing may help overcoming some of these issues. It is clear, however, that accurate determination of equivalent dose using spatially resolved measurement of undisturbed sediment (impregnated with resin) and rock slices still awaits considerable methodological advances.

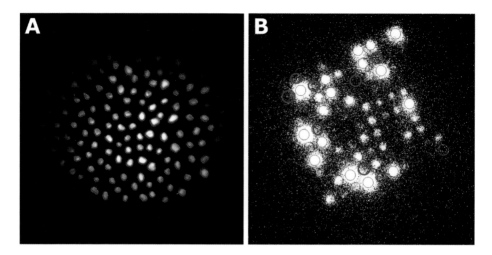

Figure 12.6 EMCCD images of luminescing quartz grains A) placed on a sample disc with spacing between grains (modified from Greilich *et al.* 2015) or B) in a grid (B) (modified from Thomsen *et al.* 2015).

12.3.3 Outlook

The dream of on-site luminescence dating is still alive, and slow but gradual progress is being made to make it possible. However, it is much more complicated than other types of analysis where proximal sensing has become widely applied (e.g. chemical composition, using handheld XRF; subsurface structures through ground-penetrating radar). A major step forward would be to have access to a portable luminescence instrument that supports irradiating samples and thus quantitative estimation of equivalent dose in the field. With respect to spatially resolved luminescence dating, main challenges seem to be in signal processing. Progress is to be expected in that field, as the problems faced are not unique to luminescence dating. In addition, spatially resolved IRPL measurements (see Section 12.2.2.2) could be of great interest, as the non-destructive measurement technique may yield possibilities to overcome signal-to-noise issues and hence improve precision.

12.4 BEYOND DATING; THE USE OF LUMINESCENCE TO UNDERSTAND LANDSCAPE EVOLUTION

Luminescence techniques are now widely used for dating sediments, as outlined in the previous chapters of this book. In recent years, realisation has grown that luminescence signals carry much more information with respect to provenance of sediment, sediment pathways, modes of transportation, rock uplift rates, and even soil mixing processes. In this section, such applications are discussed. The novel application of rock surface-exposure dating (Sohbati *et al.* 2012) should also be mentioned; this exciting new field of research is not further discussed here, as it is presented in Chapter 11.

12.4.1 Sediment provenance and pathway

To understand landscape evolution, one is not only interested in depositional landforms (which may be dated by luminescence), but also wants to know where sediments came from and how they travelled. Obtaining constraints on this is extremely challenging, but may greatly assist conceptual and numerical landscape evolution models.

Luminescence measurements may help with this issue, as it has long been known that luminescence properties vary from site to site, in relation to source material (host rock, sediment). Moreover, during sediment transport, luminescence signals are reset, and the degree of resetting may reveal information on transport mode (e.g. subaerial *vs.* subaqueous). In addition, the remnant age of slow-bleaching signals may yield information on the age of the source material. Finally, luminescence sensitivity may change during cycles of transport and deposition, offering addition information on the history of grains. Below, some inspiring examples from the recent literature on these aspects are discussed.

12.4.1.1 Provenance

Luminescence sensitivity of quartz is highly variable, both with respect to the percentage of grains yielding a useable luminescence signal, and the mean luminescence brightness of grains (e.g. Preusser *et al.* 2009). The luminescence sensitivity of quartz has been shown to vary for different source rocks (e.g. Sawakuchi *et al.* 2011), and to increase with transport distance (e.g. Pietsch *et al.* 2008; Sawakuchi *et al.* 2011; Fig. 12.7). The reasons for the differences and sensitisation are largely unclear, although some mechanisms have been proposed (see e.g. Sawakuchi *et al.* 2011). Feldspar luminescence sensitivity seems to be less variable, does not sensitise (e.g. Reimann *et al.* 2017), and has little value for reconstructing sediment provenance.

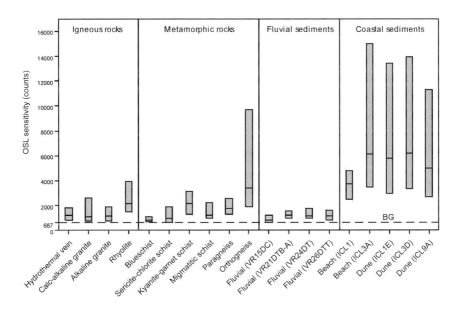

Figure 12.7 Quartz OSL sensitivity for a range of different source rocks and sediments shown as boxplots. OSL sensitivity seems to increase with temperature of rock formation (ordered left to right for both rock types), and with fluvial transport distance (ordered left to right). Modified from Sawakuchi *et al.* (2011).

The use of luminescence sensitivity information for reconstructing provenance and transport distance is still novel, and relatively few applications have been published. With respect to coastal systems, Zular *et al.* (2013) identified differences in provenance for sediment in coastal barriers in southern Brazil, and linked these to climatic shifts during the Late Holocene. With respect to aeolian systems, applications have been limited to loess deposits. Lu and Sun (2011) report TL and OSL sensitivity difference for quartz from different Chinese deserts, which they attribute to regional differences in rock types in the vicinity. In addition, Tertiary aeolian red clay was found to have greater luminescence sensitivity than Quaternary-age loess, indicating a different source of this material. Lu *et al.* (2014) report that quartz OSL sensitivity is greater for palaeosoils than for loess beds. The authors relate this to shifting provenance in different climate regimes during glacial–interglacial cycles. For fluvial systems, do Nascimento *et al.* (2015) use a combination of methods, including quartz OSL sensitivity, to determine sediment sources for the Amazon river. Chamberlain *et al.* (2017) use OSL sensitivity of sand-sized quartz to identify provenance of sediment in the Ganges Brahmaputra delta.

Although clear trends have been reported of luminescence sensitivity with transport distance a better understanding of sensitisation processes and rates would help to use such information to reconstruct sediment sources.

12.4.1.2 Pathway and transport rate

For dating, signals that reset quickly upon exposure to light are most useful, as this maximises the chances of complete resetting at the time of deposition and burial, and thus minimises the chance of age overestimation due to poor bleaching. Yet, there may be added value to comparison with signals that are less rapidly reset. If two signals with different resetting rate provide identical age information, this implies that light exposure was sufficient to reset both signals (Murray *et al.* 2012). Such information thus provides additional evidence with regard to the robustness of age estimates. This approach has been taken by Buylaert *et al.* (2013) who compared palaeodoses derived using $pIR\text{-}IRSL_{290}$ (with the subscript indicating measurement temperature in °C) signals with those of the more rapidly bleaching IRSL signal measured at 50 °C (IR_{50}). Although the latter underestimates the burial dose due to anomalous fading, the ratio between the $pIR\text{-}IRSL_{290}$ and IR_{50} palaeodose allowed the authors to identify samples for which the $pIR\text{-}IRSL_{290}$ signal was not adequately reset. Some of these were related to turbidite deposits, for which poor bleaching was expected.

Slow bleaching luminescence signals can be used to determine depositional environments and sediment transport routes. Forman and Ennis (1991) proposed that remnant feldspar thermoluminescence (TL) signals reflected depositional environment in a proglacial setting, with highest signals found for tills and ice-proximal deposits, and lower signals further away from the ice front. Keizars *et al.* (2008) use a similar approach, and show that residual TL signals decrease with longshore transport along St Joseph Peninsula in Florida (USA). Liu *et al.* (2009, 2014) used feldspar TL signals to study sand transport pathways and depositional environment for an aeolian, fluvial and coastal system in Japan. The feldspar TL signal bleaches very slowly, and is thus only reset for long light exposures. For the young deposits under investigation, the TL signals are dominated by remnant signal, as burial doses are low. The authors find that TL signals are greatest for river tributaries, and high for river sediment. Along the coast,

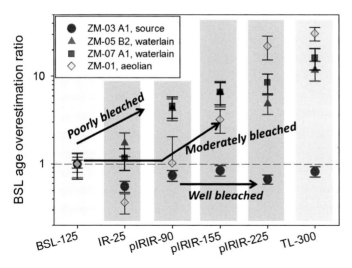

Figure 12.8 Comparison of palaeodoses obtained from signals of different bleachibility may provide information on the degree of bleaching, and hence the depositional environment and transportation history of grains. Signals are plotted from easily bleached on the left, to very difficult to bleach on the right. Modified from Reimann *et al.* (2015).

TL signals were found to decrease from the river mouth, to more distal areas, attributed to light exposure during longshore sediment transport. Aeolian deposits show lowest signals, and can also be identified by a lower high-temperature peak compared to water-lain deposits.

Reimann *et al.* (2015) proposed a method in which a large number of luminescence signals is detected on a multi-mineral sample. By comparing the rapidly bleaching quartz OSL palaeodose, with those obtained from different feldspar IRSL and pIR-IRSL signals, and multi-mineral TL, insight is obtained in the duration of light exposure prior to deposition (Fig. 12.8). The ratio of the palaeodoses relative to the blue stimulated OSL palaeodose of quartz is suggested as measure of light exposure. The method was developed and validated for a coastal site with a large sand nourishment, and ratios were shown to increase with transport distance from this source. The authors propose that the method allows a rapid assessment of approximate age, as well as a degree of bleaching, and may be used to select samples for full analysis. Chamberlain *et al.* (2017) used a similar multi-signal approach to investigate the degree of bleaching for silt samples from the Ganges-Brahmaputra delta. Applications of the technique to fine-grain is especially interesting, as methods based on inter-aliquot scatter to detect poor bleaching cannot be applied due to within-aliquot averaging effects.

McGuire and Rhodes (2015a, 2015b) use a similar approach to investigate sediment transport in a fluvial system. They employ the different bleachability of feldspar IR and pIR-IRSL signals of single aliquots and single grains. Results obtained on river bed material from along the Mojave river show that palaeodoses obtained using each of these signals decrease with distance from the headwaters, but also that the characteristics of these trends are different depending on the bleachability of the signal (Fig. 12.9). A conceptual model is constructed to explain these differences (Fig. 12.10). In subsequent work, Gray *et al.* (2017) developed a mathematical model to derive sediment transport rates from such information.

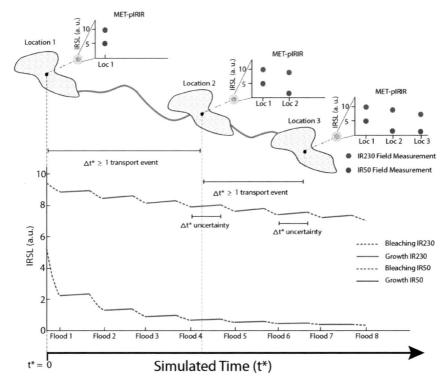

Figure 12.9 For the Mojave River in California, equivalent doses obtained from a range of IR and pIR signals are shown to decrease with downstream distance. Modified from McGuire and Rhodes (2015b).

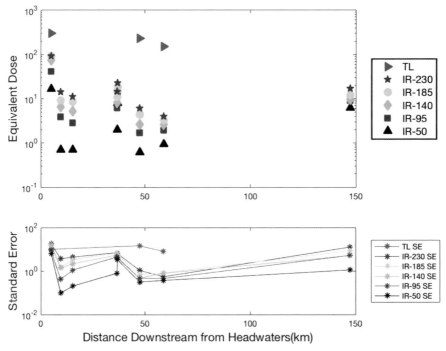

Figure 12.10 Conceptual model of how signals that differ in bleachability evolve during fluvial transport cycles. Modified from McGuire and Rhodes (2015a).

12.4.2 Soil reworking

Sediments close to the surface may be mixed by root growth and decay, animals and humans; such processes are referred to as soil reworking or bioturbation. When interested in the time of deposition and burial of a deposit, such mixing can be a nuisance, as it may cause intrusion of younger grains, causing an age underestimation if not properly analysed (Bateman *et al.* 2003, 2007). Yet, when interested in soil production rates, or bioturbation rates, luminescence signals of these grains may provide valuable information. This field of study was initiated by the pioneering work of Heimsath *et al.* (2002), and a useful introduction in the subject is provided by Wilkinson and Humphreys (2005). The main idea is that the maximum depth at which zero age or young grains are found provides the depth of the mixing zone, while the trend of apparent age with depth reveals information on the reworking rate.

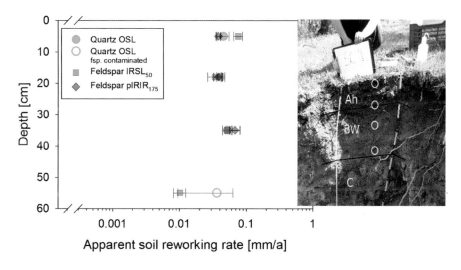

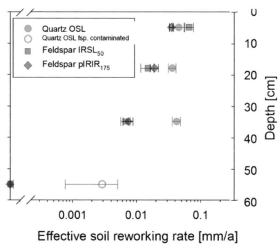

Figure 12.11 Apparent (left) and effective (right) soil reworking rates in a soil profile formed in bedrock (centre). Conventional soil reworking rate analysis is based on dividing the apparent age of the sediments by the sample depth (left). Yet, for soils developed in bedrock, results are affected by grains that never were exposed to daylight and hence are in saturation (feldspar) or have not been sensitised (quartz). When this is taken into account, the effective soil reworking rate can be calculated (right), which shows a marked decrease in mixing rate with depth, as is to be expected for this setting. Modified from Reimann *et al.* (2017).

Quartz single-grain OSL has been the method of choice for luminescence-based studies on bioturbation in tidal flats (Madsen *et al.* 2011), and terrestrial soils (Gliganic *et al.* 2016; Heimsath *et al.* 2002; Johnson *et al.* 2014; Kristensen *et al.* 2015; Stockmann *et al.* 2013). Such methods work well when soils develop in deposits with sensitised quartz grains. However, recent work suggests that for sites where soils develop in weathered bedrock, on-site sensitisation of quartz grains during the bioturbation process may lead to biased results (Reimann *et al.* 2017; Fig. 12.11). In such cases, the use of single-grain feldspar methods has advantages, as feldspar grains do not sensitise during reworking cycles. Another advantage of feldspar-based methods is that they are more widely applicable and less labour intensive than quartz single-grain OSL methods, which are often affected by poor luminescence sensitivity of the vast majority of grains.

12.4.3 Rock uplift

In Section 12.3.1, the use of slowly decaying of luminescence signals to infer past light exposure and thereby reconstruct sediment transport routes and depositional environment was discussed. A similar approach can be taken to reconstruct past exposure to heat, using the thermal- rather than light-resetting of luminescence signals. This property has been exploited to reconstruct rock uplift rates, where tectonic processes cause the rise of bedrock from the deep surface (where temperatures are high) to the surface (with ambient temperatures). Conventionally, methods like fission track dating and Ar/Ar dating are used for this, but the relatively high 'closing temperatures' prevent their use for reconstructing recent uplift rates. Luminescence-based methods have the advantage that different signals of different thermal stability may be used, and that they cover a range of relatively low closure temperatures (see King *et al.* 2016a for review).

Herman *et al.* (2010) were the first to propose and apply luminescence thermochronometry. Based on quartz OSL signals, they inferred that rock uplift rates in the Southern Alps of New Zealand have been constant during the past 100 ka, and are likely tectonic driven. In later work, it was shown that the applied methods were not fully justified, and that quartz OSL thermochronology is hampered by lack of suitable signals in most bedrock types (Guralnik *et al.* 2015a). Improved analytical and numerical solutions were proposed to derive a general expression for the effective closure temperature (Guralnik *et al.* 2013). In addition, methods were expanded to feldspar IRSL signals (Guralnik *et al.* 2015b; King *et al.* 2016c), which were applied to identify northward migration of the eastern Himalayan syntaxis (King *et al.* 2016b).

12.5 CONCLUSIONS

Luminescence methods are still being improved, resulting in better precision and widening the age range to which the method can be applied. Advances are being made towards allowing *in situ* dating at field sites, and spatially resolved dating of grains in their depositional context, but important challenges still need to be overcome. In recent years, the use of luminescence signals for a wide range of applications other than determining the time of deposition has rapidly gained interest. Many of such applications are still in their infancy, and need to be further developed and tested. Nevertheless, recent developments make clear that luminescence methods have a lot to offer outside the field of dating, and provide information on processes at the Earth's surface, such as sediment transport, soil mixing and rock uplift, that cannot be obtained by other methods. Such applications are expected to further gain importance in coming years, and possibly even become more important than luminescence as dating tool.

REFERENCES

Aitken, M.J. 1985. Thermo-luminescence dating – past progress and future trends. *Nucl Tracks Rad Meas* 10, 3–6.

Ankjaergaard, C., Guralnik, B., Buylaert, J.P., Reimann, T., Yi, S.W., Wallinga, J. 2016. Violet stimulated luminescence dating of quartz from Luochuan (Chinese loess plateau): Agreement with independent chronology up to similar to 600 ka. *Quaternary Geochronology* 34, 33–46.

Bateman, M.D., Frederick, C.D., Jaiswal, M.K., Singhvi, A.K. 2003. Investigations into the potential effects of pedoturbation on luminescence dating. *Quaternary Science Reviews* 22, 1169–1176.

Bateman, M.D., Boulter, C.H., Carr, A.S., Frederick, C.D., Peter, D., Wilder, M. 2007. Preserving the palaeoenvironmental record in Drylands: Bioturbation and its significance for luminescence-derived chronologies. *Sedimentary Geology* 195, 5–19.

Bateman, M.D., Stein, S., Ashurst, R.A., Selby, K. 2015. Instant luminescence chronologies? High resolution luminescence profiles using a portable luminescence reader. *Quaternary Geochronology* 30, 141–146.

Buylaert, J.P., Murray, A.S., Huot, S., Vriend, M.G.A., Vandenberghe, D., De Corte, F., Van den haute, P. 2006. A comparison of quartz OSL and isothermal TL measurements on Chinese loess. *Radiation Protection Dosimetry* 119, 474–478.

Buylaert, J.P., Jain, M., Murray, A.S., Thomsen, K.J., Thiel, C., Sohbati, R. 2012. A robust feldspar luminescence dating method for Middle and Late Pleistocene sediments. *Boreas* 41, 435–451.

Buylaert, J.P., Murray, A.S., Gebhardt, A.C., Sohbati, R., Ohlendorf, C., Thiel, C., Wastegard, S., Zolitschka, B., Team, P.S. 2013. Luminescence dating of the PASADO core 5022-1D from Laguna Potrok Aike (Argentina) using IRSL signals from feldspar. *Quaternary Science Reviews* 71, 70–80.

Chamberlain, E.L., Wallinga, J., Reimann, T., Goodbred, S.L., Steckler, M.S., Shen, Z.X., Sincavage, R. 2017. Luminescence dating of delta sediments: Novel approaches explored for the Ganges–Brahmaputra–Meghna Delta. *Quaternary Geochronology* 41, 97–111.

Cunningham, A.C. and Clark-Balzan, L. 2017. Overcoming crosstalk in luminescence images of mineral grains. *Radiation Measurements* 106, 498–505

Cunningham, A.C., Wallinga, J. 2009. Optically stimulated luminescence dating of young quartz using the fast component. *Radiation Measurements* 44, 423–428.

Cunningham, A.C., Wallinga, J. 2010. Selection of integration time intervals for quartz OSL decay curves. *Quaternary Geochronology* 5, 657–666.

Cunningham, A.C., Wallinga, J. 2012. Realizing the potential of fluvial archives using robust OSL chronologies. *Quaternary Geochronology* 12, 98–106.

do Nascimento Jr. D.R., Sawakuchi, A.O., Guedes, C.F.C., Giannini, P.C.F., Grohmann, C.H., Ferreira, M.P. 2015. Provenance of sands from the confluence of the Amazon and Madeira rivers based on detrital heavy minerals and luminescence of quartz and feldspar. *Sedimentary Geology* 316, 1–12.

Duller, G.A.T., Botter-Jensen, L., Murray, A.S., Truscott, A.J. 1999. Single grain laser luminescence (SGLL) measurements using a novel automated reader. *Nuclear Instruments and Methods in Physics Research Section B-Beam Interactions with Materials and Atoms* 155, 506–514.

Duller, G.A.T., Botter-Jensen, L., Murray, A.S. 2000. Optical dating of single sand-sized grains of quartz: sources of variability. *Radiation Measurements* 32, 453–457.

Duller, G.A.T., Wintle, A.G. 2012. A review of the thermally transferred optically stimulated luminescence signal from quartz for dating sediments. *Quaternary Geochronology* 7, 6–20.

Durcan, J.A., Roberts, H.M., Duller, G.A.T., Alizai, A.H. 2010. Testing the use of range-finder OSL dating to inform field sampling and laboratory processing strategies. *Quaternary Geochronology* 5, 86–90.

Forman, S.L., Ennis, G. 1991. The effect of light-intensity and spectra on the reduction of thermoluminescence of near-shore sediments from Spitsbergen, Svalbard – Implications for dating Quaternary water-lain sequences. *Geophysical Research Letters* 18, 1727–1730.

Galbraith, R.F., Roberts, R.G., Laslett, G.M., Yoshida, H., Olley, J.M. 1999. Optical dating of single and multiple grains of quartz from jinmium rock shelter, northern Australia, Part 1, Experimental design and statistical models. *Archaeometry* 41, 339–364.

Gliganic, L.A., Cohen, T.J., Slack, M., Feathers, J.K. 2016. Sediment mixing in aeolian sandsheets identified and quantified using single-grain optically stimulated luminescence. *Quaternary Geochronology* 32, 53–66.

Godfrey-Smith, D.I., Huntley, D.J., Chen, W.H. 1988. Optical dating studies of quartz and feldspar sediment extracts. *Quaternary Science Reviews* 7, 373–380.

Gray, H.J., Tucker, G.E., Mahan, S.A., McGuire, C., Rhodes, E.J. 2017. On extracting sediment transport information from measurements of luminescence in river sediment. *Journal of Geophysical Research: Earth Surface* 122, 654–677.

Greilich, S., Gribenski, N., Mittelstrass, D., Dornich, K., Huot, S., Preusser, F. 2015. Single-grain dose-distribution measurements by optically stimulated luminescence using an integrated EMCCD-based system. *Quaternary Geochronology* 29, 70–79.

Gribenski, N., Preusser, F., Greilich, S., Huot, S., Mittestrass, D. 2015. Investigation of cross talk in single grain luminescence measurements using an EMCCD camera. *Radiation Measurements* 81, 163–170.

Guerin, G., Frouin, M., Tuquoi, J., Thomsen, K.J., Goldberg, P., Aldeias, V., Lahaye, C., Mercier, N., Guibert, P., Jain, M., Sandgathe, D., McPherron, S.J.P., Turq, A., Dibble, H.L. 2017. The complementarity of luminescence dating methods illustrated on the Mousterian sequence of the Roc de Marsal: A series of reindeer-dominated, Quina Mousterian layers dated to MIS 3. *Quaternary International* 433, 102–115.

Guralnik, B., Jain, M., Herman, F., Paris, R.B., Harrison, T.M., Murray, A.S., Valla, P.G., Rhodes, E.J. 2013. Effective closure temperature in leaky and/or saturating thermochronometers. *Earth and Planetary Science Letters* 384, 209–218.

Guralnik, B., Ankjaergaard, C., Jain, M., Murray, A.S., Muller, A., Walle, M., Lowick, S.E., Preusser, F., Rhodes, E.J., Wu, T.S., Mathew, G., Herman, F. 2015a. OSL-thermochronometry using bedrock quartz: A note of caution. *Quaternary Geochronology* 25, 37–48.

Guralnik, B., Li, B., Jain, M., Chen, R., Paris, R.B., Murray, A.S., Li, S.H., Pagonis, V., Valla, P.G., Herman, F. 2015b. Radiation-induced growth and isothermal decay of infrared-stimulated luminescence from feldspar. *Radiation Measurements* 81, 224–231.

Heimsath, A.M., Chappell, J., Spooner, N.A., Questiaux, D.G. 2002. Creeping soil. *Geology* 30, 111–114.

Herman, F., Rhodes, E.J., Jean, B.C., Heiniger, L. 2010. Uniform erosion rates and relief amplitude during glacial cycles in the Southern Alps of New Zealand, as revealed from OSL-thermochronology. *Earth and Planetary Science Letters* 297, 183–189.

Hossain, S.M., De Corte, F., Vandenberghe, D., Van den haute, P. 2002. A comparison of methods for the annual radiation dose determination in the luminescence dating of loess sediment. *Nuclear Instruments and Methods in Physics Research Section A: Accelerators Spectrometers Detectors and Associated Equipment* 490, 598–613.

Huntley, D.J., Lian, O.B. 2006. Some observations on tunnelling of trapped electrons in feldspars and their implications for optical dating. *Quaternary Science Reviews* 25, 2503–2512.

Jain, M. 2009. Extending the dose range: Probing deep traps in quartz with 3.06 eV photons. *Radiation Measurements* 44, 445–452.

Johnson, M.O., Mudd, S.M., Pillans, B., Spooner, N.A., Fifield, L.K., Kirkby, M.J., Gloor, M. 2014. Quantifying the rate and depth dependence of bioturbation based on optically-stimulated luminescence (OSL) dates and meteoric Be-10. *Earth Surface Processes and Landforms* 39, 1188–1196.

Joordens, J.C.A., d'Errico, F., Wesselingh, F.P., Munro, S., de Vos, J., Wallinga, J., Ankjaergaard, C., Reimann, T., Wijbrans, J.R., Kuiper, K.F., Mucher, H.J., Coqueugniot, H., Prie, V., Joosten, I., van Os, B., Schulp, A.S., Panuel, M., van der Haas, V., Lustenhouwer, W., Reijmer, J.J.G., Roebroeks, W. 2015. Homo erectus at Trinil on Java used shells for tool production and engraving. *Nature* 518, 228–U182.

Kars, R.H., Busschers, F.S., Wallinga, J. 2012. Validating post IR-IRSL dating on K-feldspars through comparison with quartz OSL ages. *Quaternary Geochronology* 12, 74–86.

Keizars, K.Z., Forrest, B.M., Rink, W.J. 2008. Natural Residual Thermoluminescence as a Method of Analysis of Sand Transport along the Coast of the St. Joseph Peninsula, Florida. *Journal of Coastal Research* 24, 500–507.

King, G.E., Guralnik, B., Valla, P.G., Herman, F. 2016a. Trapped-charge thermochronometry and thermometry: A status review. *Chemical Geology* 446, 3–17.

King, G.E., Herman, F., Guralnik, B. 2016b. Northward migration of the eastern Himalayan syntaxis revealed by OSL thermochronometry. *Science* 353, 800–804.

King, G.E., Herman, F., Lambert, R., Valla, P.G., Guralnik, B. 2016c. Multi-OSL-thermochronometry of feldspar. *Quaternary Geochronology* 33, 76–87.

Kook, M.H., Murray, A.S., Lapp, T., Denby, P.H., Ankjaergaard, C., Thomsen, K., Jain, M., Choi, J.H., Kim, G.H. 2011. A portable luminescence dating instrument. *Nuclear Instruments and Methods in Physics Research Section B-Beam Interactions with Materials and Atoms* 269, 1370–1378.

Kook, M., Lapp, T., Murray, A.S., Thomsen, K.J., Jain, M. 2015. A luminescence imaging system for the routine measurement of single-grain OSL dose distributions. *Radiation Measurements* 81, 171–177.

Kristensen, J.A., Thomsen, K.J., Murray, A.S., Buylaert, J.P., Jain, M., Breuning-Madsen, H. 2015. Quantification of termite bioturbation in a savannah ecosystem: Application of OSL dating. *Quaternary Geochronology* 30, 334–341.

Kumar, A., Srivastava, P., Meena, N.K., 2017. Late Pleistocene aeolian activity in the cold desert of Ladakh: A record from sand ramps. *Quaternary International* 443, 13–28.

Lamothe, M., Balescu, S., Auclair, M. 1994. Natural IRSL intensities and apparent luminescence ages of single feldspar grains extracted from partially bleached sediments. *Radiation Measurements* 23, 555–561.

Li, B., Jacobs, Z., Roberts, R.G., Li, S.H. 2014. Review and assessment of the potential of post-IR IRSL dating methods to circumvent the problem of anomalous fading in feldspar luminescence. *Geochronometria* 41, 178–201.

Liu, H., Kishimoto, S., Takagawa, T., Shirai, M., and Sato, S. 2009. Investigation of the sediment movement along the Tenryu–Enshunada fluvial system based on feldspar thermoluminescence properties. *Journal of Coastal Research* 25, 1096–1105.

Liu, H., Takagawa, T., Sato, S. 2014. Sand Transport and Sedimentary Features Based on Feldspar Thermoluminescence: A Synthesis of the Tenryu–Enshunada Fluvial System, Japan. *Journal of Coastal Research* 30, 120–129

Lü, T., Sun, J. 2011. Luminescence sensitivities of quartz grains from eolian deposits in northern China and their implications for provenance. *Quaternary Research* 76, 181–189.

Lü, T., Sun, J., Li, S-H., Gong, Z., Xue, L. 2014. Vertical variations of luminescence sensitivity of quartz grains from loess/paleosol of Luochuan section in the central Chinese Loess Plateau since the last interglacial. *Quaternary Geochronology* 22, 107–115.

Madsen, A.T., Murray, A.S. 2009. Optically stimulated luminescence dating of young sediments: A review. *Geomorphology* 109, 3–16.

Madsen, A.T., Murray, A.S., Jain, M., Andersen, T.J., Pejrup, M. 2011. A new method for measuring bioturbation rates in sandy tidal flat sediments based on luminescence dating. *Estuarine Coastal and Shelf Science* 92, 464–471.

Malik, S.R., Durran, A., Fremlin, H. 1973. A comparative study of the spatial distribution of uranium and of TL-producing minerals in archaeological materials. *Archaeometry* 15, 249–253.

Martin, L., Mercier, N., Incerti, S., Lefrais, Y., Pecheyran, C., Guerin, G., Jarry, M., Bruxelles, L., Bon, F., Pallier, C. 2015. Dosimetric study of sediments at the beta dose rate scale: Characterization and modelization with the DosiVox software. *Radiation Measurements* 81, 134–141.

McGuire, C., Rhodes, E.J. 2015a. Determining fluvial sediment virtual velocity on the Mojave River using K-feldspar IRSL: Initial assessment. *Quaternary International* 362, 124–131.

McGuire, C., Rhodes, E.J. 2015b. Downstream MET-IRSL single-grain distributions in the Mojave River, southern California: Testing assumptions of a virtual velocity model. *Quaternary Geochronology*, 30, 239–244.

Murray, A.S., Thomsen, K.J., Masuda, N., Buylaert, J.P., Jain, M. 2012. Identifying well-bleached quartz using the different bleaching rates of quartz and feldspar luminescence signals. *Radiation Measurements* 47, 688–695.

Nooren, K., Hoek, W. Z., Winkels, T., Huizinga, A., Van der Plicht, H., Van Dam, R. L., Van Heteren, S., Van Bergen, M. J., Prins, M. A., Reimann, T., Wallinga, J., Cohen, K. M.,

Minderhoud, P., Middelkoop, H. 2017. The Usumacinta–Grijalva beach-ridge plain in southern Mexico: A high-resolution archive of river discharge and precipitation. *Earth Surface Dynamics* 5, 529–556.

Peng, J., Pagonis, V., Li, B. 2016. On the intrinsic accuracy and precision of the standardised growth curve (SGC) and global-SGC (gSGC) methods for equivalent dose determination: A simulation study. *Radiation Measurements* 94, 53–64.

Pietsch, T.J., Olleya, J.M., Nanson, G.C. 2008. Fluvial transport as a natural luminescence sensitiser of quartz. *Quaternary Geochronology* 3, 365–376.

Preusser, F., Chithambo, M.L., Gotte, T., Martini, M., Ramseyer, K., Sendezera, E.J., Susino, G.J., Wintle, A.G. 2009. Quartz as a natural luminescence dosimeter. *Earth-Science Reviews* 97, 184–214.

Ramsey, C.B. 2008. Deposition models for chronological records. *Quaternary Science Reviews* 27, 42–60.

Reimann, T., Notenboom, P.D., De Schipper, M.A., Wallinga, J. 2015. Testing for sufficient signal resetting during sediment transport using a polymineral multiple-signal luminescence approach. *Quaternary Geochronology* 25, 26–36.

Reimann, T., Roman-Sanchez, A., Vanwalleghem, T., Wallinga, J. 2017. Getting a grip on soil reworking: Single-grain feldspar luminescence as a novel tool to quantify soil reworking rates. *Quaternary Geochronology* 42, 1–14.

Rhodes, E.J., Ramsey, C.B., Outram, Z., Batt, C., Willis, L., Dockrill, S., Bond, J. 2003. Bayesian methods applied to the interpretation of multiple OSL dates: High precision sediment ages from Old Scatness Broch excavations, Shetland Isles. *Quaternary Science Reviews* 22, 1231–1244.

Richter, D., Richter, A., Dornich, K. 2013. Lexsyg – A new system for luminescence research. *Geochronometria* 40, 220–228.

Roberts, H.M., Duller, G.A.T. 2004. Standardised growth curves for optical dating of sediment using multiple-grain aliquots. *Radiation Measurements* 38, 241–252.

Roberts, H.M., Durcan, J.A., Duller, G.A.T. 2009. Exploring procedures for the rapid assessment of optically stimulated luminescence range-finder ages. *Radiation Measurements* 44, 582–587.

Romanyukha, A.A., Cunningham, A.C., George, S.P., Guatelli, S., Petasecca, M., Rosenfeld, A.B., Roberts, R.G. 2017. Deriving spatially resolved beta dose rates in sediment using the Timepix pixelated detector. *Radiation Measurements* 106, 483–490.

Rufer, D., Preusser, F. 2009. Potential of autoradiography to detect spatially resolved radiation patterns in the context of trapped charge dating. *Geochronometria* 34, 1–13.

Sanderson, D.C.W., Murphy, S. 2010. Using simple portable OSL measurements and laboratory characterisation to help understand complex and heterogeneous sediment sequences for luminescence dating. *Quaternary Geochronology* 5, 299–305.

Sawakuchi, A.O., Blair, M.W., DeWitt, R., Faleiros, F.M., Hyppolito, T., Guedes, C.C.F. 2011. Thermal history versus sedimentary history: OSL sensitivity of quartz grains extracted from rocks and sediments. *Quaternary Geochronology* 6, 261–272.

Schmidt, C., Pettke, T., Preusser, F., Rufer, D., Kasper, H.U., Hilgers, A. 2012. Quantification and spatial distribution of dose rate relevant elements in silex used for luminescence dating. *Quaternary Geochronology* 12, 65–73.

Schokker, J., Clevering, P., Murray, A.S., Wallinga, J., Westerhoff, W.E. 2005. An OSL dated Middle and Late Quaternary sedimentary record in the Roer Valley Graben (southeastern Netherlands). *Quaternary Science Reviews* 24, 2243–2264.

Shen, Z.X., Mauz, B. 2012. Optical dating of young deltaic deposits on a decadal time scale. *Quaternary Geochronology* 10, 110–116.

Singarayer, J.S., Bailey, R.M. 2003. Further investigations of the quartz optically stimulated luminescence components using linear modulation. *Radiation Measurements* 37, 451–458.

Sohbati, R. Murray, A.S. Chapot, M.S. Jain, M. Pederson, J. 2012. Optically stimulated luminescence (OSL) as a chronometer for surface exposure dating. *Journal of Geophysical Research: Solid Earth* 117 (B9).

Stockmann, U., Minasny, B., Pietsch, T.J., McBratney, A.B. 2013. Quantifying processes of pedogenesis using optically stimulated luminescence. *European Journal of Soil Science* 64, 145–160.

Stone, A.E.C., Bateman, M.D., Thomas, D.S.G. 2015. Rapid age assessment in the Namib Sand Sea using a portable luminescence reader. *Quaternary Geochronology* 30, 134–140.

Telfer, M.W., Bateman, M.D., Carr, A.S., Chase, B.M.2008. Testing the applicability of a standardized growth curve (SGC) for quartz OSL dating: Kalahari dunes, South African coastal dunes and Florida dune cordons. *Quaternary Geochronoly* 3, 137–142.

Thomsen, K.J., Murray, A.S., Jain, M., Botter-Jensen, L. 2008. Laboratory fading rates of various luminescence signals from feldspar-rich sediment extracts. *Radiation Measurements* 43, 1474–1486.

Thomsen, K.J., Kook, M., Murray, A.S., Jain, M., Lapp, T. 2015. Single-grain results from an EMCCD-based imaging system. *Radiation Measurements* 81, 185–191.

Thomsen, K.J., Murray, A.S., Buylaert, J.P., Jain, M., Hansen, J.H., Aubry, T. 2016. Testing single-grain quartz OSL methods using sediment samples with independent age control from the Bordes-Fitte rockshelter (Roches d'Abilly site, Central France). *Quaternary Geochronology* 31, 77–96.

Timar-Gabor, A., Wintle, A.G. 2013. On natural and laboratory generated dose response curves for quartz of different grain sizes from Romanian loess. *Quaternary Geochronology* 18, 34–40.

Timar-Gabor, A., Buylaert, J. P., Guralnik, B., Trandafir-Antohi, O., Constantin, D., Anechitei-Deacu, V., Jain, M., Murray, A.S., Porat, N., Hao, Q., Wintle, A. G. 2017. On the importance of grain size in luminescence dating using quartz. *Radiation Measurements* 106, 464–471.

Trachsel, M., Telford, R.J. 2017. All age–depth models are wrong, but are getting better. *Holocene* 27, 860–869.

Wallinga, J. 2002. Optically stimulated luminescence dating of fluvial deposits: a review. *Boreas* 31, 303–322.

Wallinga, J., Bos, I.J. 2010. Optical dating of fluvio-deltaic clastic lake-fill sediments – A feasibility study in the Holocene Rhine delta (western Netherlands). *Quaternary Geochronology* 5, 602–610.

Wilkinson, M.T., Humphreys, G.S. 2005. Exploring pedogenesis via nuclide-based soil production rates and OSL-based bioturbation rates. *Aust J Soil Res* 43, 767–779.

Wintle, A.G. 1973. Anomalous Fading of Thermoluminescence in Mineral Samples. *Nature* 245, 143–144.

Wintle, A.G., Adamiec, G. 2017. Optically stimulated luminescence signals from quartz: A review. *Radiation Measurements* 98, 10–33.

Wintle, A.G., Murray, A.S. 2006. A review of quartz optically stimulated luminescence characteristics and their relevance in single-aliquot regeneration dating protocols. *Radiation Measurements* 41, 369–391.

Zular, A., Sawakuchi, A.O., Guedes, C.C.F., Mendes, V.R., do Nascimento, D.R., Giannini, P.C.F., Aguiar, V.A.P., DeWitt, R. 2013. Late Holocene intensification of colds fronts in southern Brazil as indicated by dune development and provenance changes in the Sao Francisco do Sul coastal barrier. *Marine Geology* 335, 64–77

INDEX

United States 116
sand wedges 191, 206, 209
secondary infilling 210
sediment mixing 210
sea-level changes 275, 276
shells 264, 273
shipping samples 57
customs 57
signal optimisation 194
single-aliquot regenerative-dose (SAR) technique 16, 17, 112, 159, 160, 168, 175, 176, 180, 197, 304, 363
dose recovery test 18
plateau test 17
quality assurance tests 18
single-grain dating 112
slumps 214
soil creep 224, 227, 237, 241, 242
soil erosion 239
soils 49, 154–158, 160, 164–167, 170, 172, 174–175, 387
solifluction lobes 214, 216
slow components 196–197, 231, 283
spatially resolved luminescence dating 378
standardised dose response/growth curve (SGC) 175, 176, 177, 183
statistical models (see also age models) 201

talus 136

taphonomic processes 67
tectonic contexts 293 et seq.
tectonically active areas 50
terraces 246, 247, 296–297, 303, 310, 312, 337
thermal contraction cracking 211
thermal transfer of charge 118
thermal transfer tests 261
thermally transferred optically stimulated luminescence (TT-OSL) 163, 164
thermoluminescence (TL) 2, 7, 9, 111, 195
till 191, 194

uncertainty (see also age uncertainty) 69,263

violet-stimulated luminescence 164, 377
volcanic areas 50

water content, of sediment 23, 25, 27, 30, 69, 110, 125–126, 142, 166–167, 183, 233, 247, 260, 264, 277

young deposits, dating 261, 375

zircon 121